World Geography
NORTH AMERICA &
THE CARIBBEAN

Regions | Physical Geography | Biogeography and Natural Resources
Human Geography | Economic Geography | Gazetteer

Second Edition

World Geography
NORTH AMERICA & THE CARIBBEAN

Regions | Physical Geography | Biogeography and Natural Resources
Human Geography | Economic Geography | Gazetteer

Second Edition

Volume 5

Editor
Joseph M. Castagno
Educational Reference Publishing, LLC

SALEM PRESS
A Division of EBSCO Information Services, Inc.
Ipswich, Massachusetts
GREY HOUSE PUBLISHING

Cover photo: North America from outer space. Image by 1xpert.

Publisher's Cataloging-In-Publication Data
(Prepared by The Donohue Group, Inc.)

Names: Castagno, Joseph M., editor.
Title: World geography / editor, Joseph M. Castagno, Educational Reference Publishing, LLC.
Description: Second edition. | Ipswich, Massachusetts : Salem Press, a division of EBSCO Information Services, Inc. ; Amenia, NY : Grey House Publishing, [2020] | Interest grade level: High school. | Includes bibliographical references and index. | Summary: A six-volume geographic encyclopedia of the world, continents and countries of each continent. In addition to physical geography, the set also addresses human geography including population distribution, physiography and hydrology, biogeography and natural resources, economic geography, and political geography. | Contents: Volume 1. South & Central America — Volume 2. Asia — Volume 3. Europe — Volume 4. Africa — Volume 5. North America & the Caribbean — Volume 6. Australia, Oceania & the Antarctic.
Identifiers: ISBN 9781642654257 (set) | ISBN 9781642654288 (v. 1) | ISBN 9781642654318 (v. 2) | ISBN 9781642654301 (v. 3) | ISBN 9781642654295 (v. 4) | ISBN 9781642654271 (v. 5) | ISBN 9781642654325 (v. 6)
Subjects: LCSH: Geography—Encyclopedias, Juvenile. | CYAC: Geography—Encyclopedias. | LCGFT: Encyclopedias.
Classification: LCC G133 .W88 2020 | DDC 910/.3—dc23

First Printing
PRINTED IN CANADA

CONTENTS

THE CHALLENGE OF COVID-19

As *World Geography: North America & the Caribbean* goes to press in July 2020, the entire globe is grappling with the worst and most widespread pandemic in more than a century. The cause of the pandemic is a highly contagious viral condition known as Coronavirus Disease 2019, or COVID-19.

The first documented emergence of COVID-19 occurred in December 2019 as an outbreak of pneumonia in Wuhan City, in Hubai Province, China. By January 2020, Chinese health officials had reported tens of thousands of infections and dozens of deaths. That same month, COVID-19 cases were appearing across Asia, Europe, and North America, and spreading rapidly. On March 11, the World Health Organization (WHO) declared COVID-19 viral disease a pandemic.

The rapid spread of COVID-19 viral disease has strong geographical components. The virus emerged in Wuhan, a huge, densely populated city. It spread rapidly among a population living largely indoors during the cold-weather months. But the most significant geographical factor in the spread of the virus may well be the globalization of transportation. Every day, thousands of people fly to destinations near and far. Each traveler carries the potential to unknowingly spread disease.

To curtail COVID-19's spread, countries have closed their borders, air travel has been drastically reduced, and sweeping mitigation policies have been inaugurated. Some densely populated Asian countries, including South Korea, Singapore, and Taiwan, enforced these measures very early, and have, as of July 2020, managed to dampen the effect of the virus and limit the number of confirmed cases and deaths due to COVID-19. Other places, slower to act, such as China and Iran, have been very hard hit.

Many COVID-19 questions remain: Where will the disease strike next? Will the onset of warmer weather reduce the communicability of the virus? Will a vaccine be available soon? Will there be a second wave of infection? When will life be back to normal?

Geographers will continue to play a unique role in answering these questions, applying the tools and techniques of their discipline to achieving the fullest possible epidemiological understanding of the pandemic.

Publisher's Note

North Americans have long thought of the field of geography as little more than the study of the names and locations of places. This notion is not without a basis in fact: Through much of the twentieth century, geography courses forced students to memorize names of states, capitals, rivers, seas, mountains, and countries. Both students and educators eventually rebelled against that approach, geography courses gradually fell out of favor, and the future of geography as a discipline looked doubtful. Happily, however, the field has undergone a remarkable transformation, starting in the 1990s. Geography now has a bright and pivotal significance at all levels of education.

While learning the locations of places remains an important part of geography studies, educators recognize that place-name recognition is merely the beginning of geographic understanding. Geography now places much greater emphasis on understanding the characteristics of, and interconnections among, places. Modern students address such questions as how the weather in Brazil can affect the price of coffee in the United States, why global warming threatens island nations, and how preserving endangered plant and animal species can conflict with the economic development of poor nations.

World Geography, Second Edition, addresses these and many other questions. Designed and written to meet the needs of high school students, while being accessible to both middle school and undergraduate college students, these six volumes take an integrated approach to the study of geography, emphasizing the connections among world regions and peoples. The set's six volumes concentrate on major world regions: South and Central America; Asia; Europe; Africa, North America and the Caribbean; and Australia, Oceania, and the Antarctic. Each volume begins with common overview information related to the geography, maps and mapmaking. The core essays in the volumes begin with an overview section to provide global context and then go on to examine important geographic aspects of the regions in that area of the world: its physical geography; biogeography and natural resources; human geography (including its political geography); and economic geography. These essays range in length from three to ten pages. A gazetteer indicates major political, geographic, and manmade features throughout the region.

A robust appendix found in each volume provides further information:

- The Earth in Space (The Solar System, Earth's Moon, The Sun and the Earth, The Seasons);
- Earth's Interior (Earths Internal Structure, Plate Tectonics, Volcanoes, Geologic Time Scale);
- Earth's Surface (Internal Geological Processes, External Processes, Fluvial and Karst Processes, Glaciation, Desert Landforms, Ocean Margins);
- Earth's Climates (The Atmosphere, Global Climates; Cloud Formation, Storms);
- Earth's Biological Systems (Biomes);
- Natural Resources (Soils, Water);
- Exploration and Transportation (Exploration and Historical Trade Routes, Road Transportation, Railways, Air Transportation);
- Energy and Engineering (Energy Sources, Alternative Energies, Engineering Projects);
- Industry and Trade (Manufacturing, Globalization of Manufacturing and Trade, Modern World Trade Patterns);
- Political Geography (Forms of Government, Political Geography, Geopolitics, National Park Systems);
- Boundaries and Time Zones (International Boundaries, Global Time and Time Zones);
- Global Education (Themes and Standards in Geography Education);
- Global Data (The World Gazetteer of Oceans and Continents, The World's Oceans and Seas, Major Land Areas of the World, Major Islands of the World, Countries of the World (including population and pollution density), Past and Projected World Population Growth, 1950-2050, The World's Largest Countries by Area, The World's Smallest Countries by Area, The World's Largest Countries by Population, The World's Smallest Countries by Population,

The World's Most Densely Populated Countries, The World's Least Densely Populated Countries, The World's Most Populous Cities, Major Lakes of the World, Major Rivers of the World, The Highest Peaks in Each Continent, Major Deserts of the World, Highest Waterfalls of the World).

- A Glossary, General Bibliography, and Index complete the backmatter.

The regional divisions in the set make it possible to study specific countries or parts of the world. Pairing the specific regional information, organized by regions, physical geography, biogeography and natural resources, human geography, economic geography, and a gazetteer, with global information makes it possible for students to see the connections not only between countries and places within the region, but also between the regions and the entire global system, all within a single volume.

To make this set as easy as possible to use, all of its volumes are organized in a similar fashion, with six major divisions—Regions (organized into subregions by volume), Physical Geography, Biogeography and Natural Resources, Human Geography, Economic Geography, and Gazetteer. The number of subregions in each volume varies, depending upon the major world division being examined—North America and the Caribbean, for example, includes the United States, Canada, Mexico, Greenland and the Arctic, and the Caribbean.

Physical geography considers a world region's physiography, hydrology, and climatology. Biogeography and natural resources explores renewable and nonrenewable resources, flora, and fauna. Human geography addresses the people, population distribution, culture regions, urbanization, and political geography of the area. Economic geography considers the region's agriculture, industries, engineering projects, transportation, trade, and communications.

Gazetteers include descriptive entries on hundreds of important places, especially those mentioned in the volume's essays. A typical entry gives the place name and location, indicating the category into which the place falls (mountain, river, city, country, lake, etc).

The entries also include statistics relevant to the categories of place (height of mountains, length of rivers, population of cities and countries).

A feature new to this edition is the discussion questions included throughout the volume. These questions are meant to foster discussion and further research into the topics related to the history, current issues, and future concerns related to physical, human, economic, and political geography.

Both a physical and a social science, geography is unique among social sciences in the demands it makes for visual support. For this reason, *World Geography* contains more than 100 maps, more than 1300 photographs, and scores of other graphical elements. In addition, essays are punctuated with more than 500 textual sidebars and tables, which amplify information in the essays and call attention to especially important or interesting points.

Both English and metric measures are used throughout this set. In most instances, English measures are given first, followed by their metric equivalents in parentheses. It should be noted that in cases of measures that are only estimates, such as the areas of deserts or average heights of mountain ranges, the metric figures are often rounded off to estimates that may not be exact equivalents of the English-measure estimates. In order to enhance clarity, units of measure are not abbreviated in the text, with these exceptions: kilometers are abbreviated as km. and square kilometers as sq. km. This exception has been made because of the frequency with which these measures appear.

Reference works such as this would be impossible without the expertise of a large team of contributing scholars. This project is no exception. Salem Press would like to thank the more than 175 people who wrote the signed essays and contributed entries to the gazetteers. A full list of contributors follows this note. We recognize the efforts of Dr. Ray Sumner, of California's Long Beach City College, for the expertise and insights that she brought to the previous edition of this book, and which have formed the strong foundation for this new edition. We also acknowledge the work of the editor of this current volume, Joseph Castagno, Educational Reference Publishing, LLC.

INTRODUCTION

When Henry Morton Stanley of the *New York Herald* shook the hand of David Livingstone on the shore of Central Africa's Lake Tanganyika in 1871, the moment represented the high point of geography to many people throughout the world. A Scottish missionary and explorer, Livingstone had been out of contact with the outside world for nearly two years, and European and American newspapers had buzzed with speculation about his disappearance. At that time, so little was known about the geography of the interior of Africa that Stanley's finding Livingstone was acclaimed as a brilliant triumph of explorations.

The field of geography in Stanley and Livingstone's time was—and to a large extent still is—synonymous with explorations. Stories of epic journeys, both historic and contemporary, continue to exert a powerful attraction on readers. Mountains, deserts, forest, caves, and glaciers still draw intrepid explorers, while even more armchair travelers are thrilled by accounts and pictures of these exploits and discoveries. We all love to travel—to the beach, into the mountains, to our great national parks, and to foreign countries. In the need and desire to explore our surroundings, we are all geographers.

Numerous geographical societies welcome both professional geographers and the general public into their membership, as they promote a greater knowledge and understanding of the earth. The National Geographic Society, founded in 1888 "for the increase and diffusion of geographical knowledge," has awarded more than 11,000 grants for scientific exploration and research. Each year, the society invests millions of dollars in expeditions and fieldwork related to environmental concerns and global geographic issues. The findings are recorded in the pages of the familiar yellow-bordered *National Geographic* magazine, now produced in 40 local-language editions in many countries around the world, publishing around 6.8 million copies monthly, with some 60 million readers. The magazine, along with the National Geographic International television network, reaches more than 135 million readers and viewers worldwide and has more than 85 million subscribers.

An even older geographical association is Great Britain's Royal Geographical Society, which grew out of the Geographical Society of London, founded in 1830 with the "sole object" of promoting "that most important and entertaining branch of knowledge—geography." Over the century that followed, the Royal Geographical Society focused on exploration of the continents of Africa and Antarctica. In the society's London headquarters adjacent to the Albert Hall, visitors can still view such historic artifacts as David Livingstone's cap and chair, as well as diaries, sketches, and maps covering the great period of the British Empire and beyond. Today the society assists more than five hundred field expeditions every year.

With the aid of satellites and remote-sensing instruments, we can now obtain images and data from almost anywhere on Earth. However, remote and inaccessible places still invite the intrepid to visit and explore them in person. Although the outlines of the continents have now been completed, and their interiors filled in with details of mountains and rivers, cities and political boundaries, remote places still exert a fascination on modern urbanites.

The enchantment of tales about strange sights and courageous journeys has been with us since the ancient voyages of Homer's *Ulysses*, Marco Polo's travels to China, and the nautical expeditions of Christopher Columbus, Ferdinand Magellan, and James Cook. While those great travelers are from the remote past, the age of exploration is far from over—a fact repeatedly demonstrated by the modern Norwegian navigator Thor Heyerdahl. Moreover, new journeys of discovery are still taking place. In 1993, after dragging a sled wearily across the frigid wastes of Antarctica for more than three months, Sir Ranulph Twisleton-Wykeham-Fiennes announced that the age of exploration is not dead. Six years later, in 1999, the long-missing body of British mountain climber George Mallory was found on the slopes of Mount Everest, near whose top he had mysteriously vanished in 1924. That discovery sparked a new wave of admiration and respect for explorers of such courage and endurance.

How many people have been enthralled by the bravery of Antarctic explorer Robert Falcon Scott and the noble sacrifice his injured colleague Lawrence Oates made in 1912, when he gave up his life in order not to slow down the rest of the expedition? There can be no doubt that the thrills and the dangers of exploring find resonance among many modern readers.

The struggle to survive in environments hostile to human beings reminds us of the power of our planet Earth. Significant books on this theme have included Jon Krakauer's *Into Thin Air* (1998), an account of a disastrous expedition climbing Mount Everest, and Sebastian Junger's *The Perfect Storm* (1997), the story of the worst gale of the twentieth century and its effect on a fishing fleet off the East Coast of North America. *Endurance* (1998), the epic of Sir Ernest Shackleton's survival and leadership for two years on the frozen Arctic, attracts the same people who avidly read *Undaunted Courage* (1996) the story of Meriwether Lewis and William Clark's epic exploration of the Louisiana Purchase territories in the early nineteenth century. In 1997 *Seven Years in Tibet* premiered, a popular film about the Austrian Heinrich Harrer, who lived in Tibet in the mid-twentieth century. The more urban people become, the greater their desire for adventurous, remote places, a least vicariously, to raise the human spirit.

There are, of course, also scientific achievements associated with modern exploration. In November 1999, the elevation of Mount Everest, the world tallest peak was raised by 7 feet (2.1 meters) to a new height of 29,035 feet (8,850 meters) above sea level; the previously accepted height had been based on surveys made during the 1950s. This new value was the result of Global Positioning System (GPS) technology enabling a more accurate measurement than had been possible with land-based earthbound surveying equipment. A team of climbers supported by the National Geographic Society and the Boston Museum of Science was equipped with GPS equipment, which enabled a fifty-minute re-

cording of data based on satellite signals. At the same time, the expedition was able to ascertain that Mount Everest is moving northeast, atop the Indo-Australian Plate, at a rate of approximately 2.4 inches (10 centimeters) per year.

In 2000, the International Hydrographic Organization named a "new" ocean, the Southern Ocean, which encompasses all the water surrounding Antarctica up to 60° south latitude. With an area of approximately 7.8 million square miles (20.3 million sq. km.), the Southern Ocean is about twice the size of the entire United States and ranks as the world's fourth largest ocean, after the Pacific, Atlantic, and Indian Oceans, but just ahead of the Arctic Ocean.

Despite the humanistic and scientific advantage of geographic knowledge, to many people today, geography is a subject where one merely memorized longs lists of facts dealing with "where" questions. (Where is Andorra? Where is Prince Edward Island? Where is Kalamazoo?) or "what" questions (What is the highest mountain in South America? What is the capital of Costa Rica?) This approach to the study of geography has been perpetuated by the annual National Geographic Bee, conducted in the United States each year for students in grades four through eight. Participants in the competition display an astonishing recall of facts but do not have the opportunity of showing any real geographic thought. To a geographer, such factual knowledge is simply a foundation for investigating and explaining the much more important questions dealing with "why"—"Why is the Sahara a desert?"

Geographers aim to understand why environments and societies occur where and as they do, and how they change. Geography must be seen as an integrative science; the collection of factual data and evidence, as in exploration, is the empirical foundation for deductive reasoning. This leads to the creation of a range of geographical methods, models, theories, and analytical approaches that serve to unify a very broad area of knowledge—the interaction between natural and human environments. Although geography as an academic discipline became established in nineteenth century Germany, there have always been geographers, in the sense of people curious about their world. Humans have always wanted to know about day and night, the shape of the earth, the nature of climates, differences in plants and animals, as well as what lies beyond the horizon. Today, as we hear about and actually experience the sweeping effects of globalization, we need more than ever to develop our geographic skills. Not only are we connected by economic ties to the countries of the world, but we must also appreciate the consequences of North America's high standard of living.

Political boundaries are artificial human inventions, but the natural world is one biosphere. As concern over global warming escalates, national leaders meet to seek a solution to the emission of greenhouse gases, rising ocean levels, and mass extinctions. Are we connected to our environment? At a time when the rate of species extinction is a hundred times above normal, and the human population is crowding in increasing numbers into huge urban centers, we have, nevertheless, taken time each year in April to celebrate Earth Day since 1970. We need now to realize that every day is Earth Day.

Geography languished in the United States in the 1960s, as social studies was taught with a history emphasis in schools. American students became alarmingly disadvantaged in geographic knowledge, compared with most other countries. Fortunately, members of the profession acted to restore geography to the curriculum. In 1984, the National Geographic Society undertook the challenge of restoring geography in the United States. The society turned to two organizations active in geographic education: The Association of American Geographers, the professional geographers' group with more than 10,000 members, mostly in higher education in the United States; and the National Council for Geographic Education that supports geography teaching at all levels—from kindergarten through university, with members that include U.S. and international teachers, professors, students, businesses, and others who support geography education. The council administers the Geographic Alliances, found in every state of the United Sates, with a national membership of about 120,000 schoolteachers. Together, they produced the "Guidelines in Geographic Education," which introduced the Five Themes of Geography, to

enhance the teaching of geography in schools. Using the themes of Location, Place, Human/Environment Interaction, Movement and Regions, teachers were able to plan and conduct lessons in which students encountered interesting real-world examples of the relevance and importance of geography. Continued research into geographic education led to the inclusion of geography in 1990 as one of the core subjects of the National Education Goals, along with English, mathematics, science, and history.

Another milestone was the publication in 1994 of "Geography for Life," the national Geography Standards. The earlier Five Themes were subsumed under the new Six Essential Elements: The World in Spatial Terms; Places and Regions; Physical Systems; Human Systems; Environment Systems; Environment and Society; and The Uses of Geography. Eighteen geography standards are included, describing what a geographically informed person knows and understands. States, schools, and individual teachers have welcomed the new prominence of geography, and enthusiastically adopted new approaches to introduce the geography standards to new learners. The rapid spread of computer technology, especially in the field of Geographical Information Science, has also meant a new importance for spatial analysis, a traditional area of geographical expertise. No longer is geography seen as an outdated mass of useless or arcane facts; instead, geography is now seen, again, to be an innovative an integrative science, which can contribute to solving complex problems associated with the human-environmental relationship in the twenty-first century.

Geographers may no longer travel across uncharted realms, but there is still much we long to explore, to learn, and seek to understand, even if it is only as "armchair" geographers. This reference work, *World Geography,* will help carry readers on their own journeys of exploration.

Ray Sumner
Long Beach City College

Joseph M. Castagno
Educational Reference Publishing, LLC

CONTRIBUTORS

Emily Alward
Henderson, Nevada Public Library

Earl P. Andresen
University of Texas at Arlington

Debra D. Andrist
St. Thomas University

Charles F. Bahmueller
Center for Civic Education

Timothy J. Bailey
Pittsburg State University

Irina Balakina
Writer/Editor, Educational Reference Publishing, LLC

David Barratt
Nottingham, England

Maryanne Barsotti
Warren, Michigan

Thomas F. Baucom
Jacksonville State University

Michelle Behr
Western New Mexico University

Alvin K. Benson
Brigham Young University

Cynthia Breslin Beres
Glendale, California

Nicholas Birns
New School University

Olwyn Mary Blouet
Virginia State University

Margaret F. Boorstein
C.W. Post College of Long Island University

Fred Buchstein
John Carroll University

Joseph P. Byrne
Belmont University

Laura M. Calkins
Palm Beach Gardens, Florida

Gary A. Campbell
Michigan Technological University

Byron D. Cannon
University of Utah

Steven D. Carey
University of Mobile

Roger V. Carlson
Jet Propulsion Laboratory

Robert S. Carmichael
University of Iowa

Joseph M. Castagno
Principal, Educational Reference Publishing, LLC

Habte Giorgis Churnet
University of Tennessee at Chattanooga

Richard A. Crooker
Kutztown University

William A. Dando
Indiana State University

Larry E. Davis
College of St. Benedict

Ronald W. Davis
Western Michigan University

Cyrus B. Dawsey
Auburn University

Frank Day
Clemson University

M. Casey Diana
University of Illinois at Urbana-Champaign

Stephen B. Dobrow
Farleigh Dickinson University

Steven L. Driever
University of Missouri, Kansas City

Sherry L. Eaton
San Diego City College

Femi Ferreira
Hutchinson Community College

Helen Finken
Iowa City High School

Eric J. Fournier
Samford University

Anne Galantowicz
El Camino College

Hari P. Garbharran
Middle Tennessee State University

Keith Garebian
Ontario, Canada

Laurie A. B. Garo
University of North Carolina, Charlotte

Jay D. Gatrell
Indiana State University

Carol Ann Gillespie
Grove City College

Nancy M. Gordon
Amherst, Massachusetts

Noreen A. Grice
Boston Museum of Science

Johnpeter Horst Grill
Mississippi State University

Charles F. Gritzner
South Dakota State University

C. James Haug
Mississippi State University

Douglas Heffington
Middle Tennessee State University

Thomas E. Hemmerly
Middle Tennessee State University

Jane F. Hill
Bethesda, Maryland

Carl W. Hoagstrom
Ohio Northern University

Catherine A. Hooey
Pittsburg State University

Robert M. Hordon
Rutgers University

Kelly Howard
La Jolla, California

Paul F. Hudson
University of Texas at Austin

Huia Richard Hutton
University of Hawaii/Kapiolani Community College

Raymond Pierre Hylton
Virginia Union University

Solomon A. Isiorho
Indiana University/Purdue University at Fort Wayne

Ronald A. Janke
Valparaiso University

Albert C. Jensen
Central Florida Community College

Jeffry Jensen
Altadena, California

Bruce E. Johansen
University of Nebraska at Omaha

Kenneth A. Johnson
State University of New York, Oneonta

Walter B. Jung
University of Central Oklahoma

James R. Keese
California Polytechnic State University, San Luis Obispo

Leigh Husband Kimmel
Indianapolis, Indiana

Denise Knotwell
Wayne, Nebraska

James Knotwell
Wayne State College

Grove Koger
Boise Idaho Public Library

Alvin S. Konigsberg
State University of New York at New Paltz

Doris Lechner
Principal, Educational Reference Publishing, LLC

Steven Lehman
John Abbott College

Denyse Lemaire
Rowan University

Dale R. Lightfoot
Oklahoma State University

Jose Javier Lopez
Minnesota State University

James D. Lowry, Jr.
East Central University

Jinshuang Ma
Arnold Arboretum of Harvard University Herbaria

Dana P. McDermott
Chicago, Illinois

Thomas R. MacDonald
University of San Francisco

Robert R. McKay
Clarion University of Pennsylvania

Nancy Farm Männikkö
L'Anse, Michigan

Carl Henry Marcoux
University of California, Riverside

Christopher Marshall
Unity College

Rubén A. Mazariegos-Alfaro
University of Texas/Pan American

Christopher D. Merrett
Western Illinois University

John A. Milbauer
Northeastern State University

Randall L. Milstein
Oregon State University

Judith Mimbs
Loftis Middle School

Karen A. Mulcahy
East Carolina University

B. Keith Murphy
Fort Valley State University

M. Mustoe
Omak, Washington

Bryan Ness
Pacific Union College

Kikombo Ilunga Ngoy
Vassar College

Joseph R. Oppong
University of North Texas

Richard L. Orndorff
University of Nevada, Las Vegas

Bimal K. Paul
Kansas State University

Nis Petersen
New Jersey City University

Mark Anthony Phelps
Ozarks Technical Community College

John R. Phillips
Purdue University, Calumet

Alison Philpotts
Shippensburg University

Julio César Pino
Kent State University

Timothy C. Pitts
Morehead State University

Carolyn V. Prorok
Slippery Rock University

P. S. Ramsey
Highland Michigan

Robert M. Rauber
University of Illinois at Urbana-Champaign

Ronald J. Raven
State University of New York at Buffalo

Neil Reid
University of Toledo

Susan Pommering Reynolds
Southern Oregon University

Nathaniel Richmond
Utica College

Edward A. Riedinger
Ohio State University Libraries

Mika Roinila
West Virginia University

Thomas E. Rotnem
Brenau University

Joyce Sakkal-Gastinel
Marseille, France

Helen Salmon
University of Guelph

Elizabeth D. Schafer
Loachapoka, Alabama

Kathleen Valimont Schreiber
Millersville University of Pennsylvania

Ralph C. Scott
Towson University

Guofan Shao
Purdue University

Wendy Shaw
Southern Illinois University, Edwardsville

R. Baird Shuman
University of Illinois, Champaign-Urbana

Sherman E. Silverman
Prince George's Community College

Roger Smith
Portland, Oregon

Robert J. Stewart
California Maritime Academy

Toby R. Stewart
Alamosa, Colorado

Ray Sumner
Long Beach City College

Paul Charles Sutton
University of Denver

Glenn L. Swygart
Tennessee Temple University

Sue Tarjan
Santa Cruz, California

Robert J. Tata
Florida Atlantic University

John M. Theilmann
Converse College

Virginia Thompson
Towson University

Norman J. W. Thrower
University of California, Los Angeles

Paul B. Trescott
Southern Illinois University

Robert D. Ubriaco, Jr.
Illinois Wesleyan University

Mark M. Van Steeter
Western Oregon University

Johan C. Varekamp
Wesleyan University

Anthony J. Vega
Clarion University

William T. Walker
Chestnut Hill College

William D. Walters, Jr.
Illinois State University

Linda Qingling Wang
University of South Carolina, Aiken

Annita Marie Ward
Salem-Teikyo University

Kristopher D. White
University of Connecticut

P. Gary White
Western Carolina University

Thomas A. Wikle
Oklahoma State University

Rowena Wildin
Pasadena, California

Donald Andrew Wiley
Anne Arundel Community College

Kay R. S. Williams
Shippensburg University

Lisa A. Wroble
Redford Township District Library

Bin Zhou
Southern Illinois University, Edwardsville

REGIONS

Overview

The History of Geography

The moment that early humans first looked around their world with inquiring minds was the moment that geography was born. The history of geography is the history of human effort to understand the nature of the world. Through the centuries, people have asked of geography three basic questions: What is Earth like? Where are things located? How can one explain these observations?

Geography in the Ancient World

In the Western world, the Greeks and the Romans were among the first to write about and study geography. Eratosthenes, a Greek scholar who lived in the third century BCE, is often called the "father of geography and is credited with first using the word geography (from the Greek words *ge*, which means "earth," and *graphe*, which means "to describe"). The ancient Greeks had contact with many older civilizations and began to gather together information about the known world. Some, such as Hecataeus, described the multitude of places and peoples with which the Greeks had contact and wrote of the adventures of mythical characters in strange and exotic lands. However, the ancient Greek scholars went beyond just describing the world. They used their knowledge of mathematics to measure and locate. The Greek scholars also used their philosophical nature to theorize about Earth's place in the universe.

One Greek scholar who used mathematics in the study of geography was Anaximander, who lived from 610 to 547 BCE. Anaximander is credited with being the first person to draw a map of the world to scale. He also invented a sundial that could be used to calculate time and direction and to distinguish

the seasons. Eratosthenes is also famous for his mathematical calculations, in particular of the circumference of Earth, using observations of the Sun. Hipparchus, who lived around 140 BCE, used his mathematical skills to solve geographic problems and was the first person to introduce the idea of a latitude and longitude grid system to locate places.

Such early Greek philosophers as Plato and Aristotle were also concerned with geography. They discussed such issues as whether Earth was flat or spherical and if it was the center of the universe, and debated the nature of Earth as the home of humankind.

Whereas the Greeks were great thinkers and introduced many new ideas into geography, the Roman contribution was to compile and gather available knowledge. Although this did not add much that was new to geography, it meant that the knowledge of the ancient world was available as a base to work from and was passed down across the centuries. Geogra-

---

Curiosity: The Root of Geography

The earliest human beings, as they hunted and gathered food and used primitive tools in order to survive, must have had detailed knowledge of the geography of their part of the world. The environment could be a hostile place, and knowledge of the world meant the difference between life and death. Human curiosity took them one step further. As they lived in an ancient world of ice and fire, human beings looked to the horizon for new worlds, crossing continents and spreading out to all areas of the globe. They learned not only to live as a part of their environment, but also to understand it, predict it, and adapt it to their needs.

---

phy in the ancient world is often said to have ended with the great work of Ptolemy (Claudius Ptolemaeus), who lived from 90 to 168 CE. Ptolemy is best known for his eight-volume *Guide to Geography*, which included a gazetteer of places located by latitude and longitude, and his world map.

Geography in China

The study of geography also was important in ancient China. Chinese scholars described their resources, climate, transportation routes, and travels, and were mapping their known world at the same time as were the great Western civilizations. The study of geography in China begins in the Warring States period (fifth century BCE). It expands its scope beyond the Chinese homeland with the growth of the Chinese Empire under the Han dynasty. It enters its golden age with the invention of the compass in the eleventh century CE (Song dynasty) and peaks with fifteenth century CE (Ming dynasty) Chinese exploration of the Pacific under admiral Zheng He during the treasure voyages.

Geography in the Middle Ages

With the collapse of the Roman Empire in the fifth century CE, Europe entered into what is commonly known as the Early Middle Ages. During this time, which lasted until the fifteenth century, the geographic knowledge of the ancient world was either lost or challenged as being counter to Christian teachings. For example, the early Greeks had theorized that Earth was a sphere, but this was rejected during the Middle Ages. Scholars of the Middle Ages believed that the world was a flat disk, with the holy city of Jerusalem at its center.

The knowledge and ideas of the ancient world might have been lost if they had not been preserved by Muslim scholars. In the Islamic countries of North Africa and the Middle East, some of the scholarship of the ancient world was sheltered in libraries and universities. This knowledge was extensively added to as Muslims traveled and traded across the known world, gathering their own information.

Among the most famous Muslim geographers were Ibn Battutah, al-Idrisi, and Ibn Khaldun. Ibn Battutah traveled east to India and China in the fourteenth century. Al-Idrisi, at the command of King Roger II of Sicily, wrote *Roger's Book*, which systematically described the world. Information from *Roger's Book* was engraved on a huge planisphere (disk), crafted in silver; this once was considered a wonder of the world, but it is thought to have been destroyed. Ibn Khaldun (1332-1406) is best known for his written world history, but he also was a pioneer in focusing on the relationship of human beings to their environment.

The Age of European Exploration

Beginning in the fifteenth century, the isolation of Europe came to an end, and Europeans turned their attention to exploration. The two major goals of this sudden surge in exploration were to spread the Christian faith and to obtain needed resources. In 1418 Prince Henry the Navigator established a school for navigators and began to gather the tools and knowledge needed for exploration. He was the first of many Europeans to travel beyond the limits of the known world, mapping, describing, and cataloging all that they saw.

The great wave of European exploration brought new interest in geography, and the monumental works of the Greeks and Romans—so carefully preserved by Muslim scholars—were rediscovered and translated into Latin. The maps produced in the Middle Ages were of little use to the explorers who were traveling to, and beyond, the limits of the known world. Christopher Columbus, for example, relied on Ptolemy's work during his voyages to the Americas, but soon newer, more accurate maps were drawn and, for the first time, globes were made. A particularly famous map, which is still used as a base map, is the Mercator projection. On the world map produced by Gerardus Mercator (born Geert de Kremer) in 1569, compass directions appear as straight lines, which was a great benefit on navigational charts.

When the age of European exploration began, even the best world maps crudely depicted only a few limited areas of the world. Explorers quickly began to gather huge quantities of information, making detailed charts of coastlines, discovering new continents, and eventually filling in the maps of those continents with information about both the natural and human features they encountered. This age of exploration is often said to have ended when Roald Amundsen planted the Norwegian flag at the South Pole in 1911. At that time, the world map became complete, and human beings had mapped and explored every part of the globe. However, the beginning of modern geography is usually associated with the work of two nineteenth century German geographers: Alexander von Humboldt and Carl Ritter.

The Beginning of Modern Geography

The writings of Alexander von Humboldt and Carl Ritter mark a leap into modern geography, because these writers took an important step beyond the work of previous scholars. The explorers of the previous centuries had focused on gathering information, describing the world, and filling in the world map with as much detail as possible. Humboldt and Ritter took a more scientific and systematic approach to geography. They began not only to compile descriptive information, but also to ask why: Humboldt spent his lifetime looking for relationships among such things as climate and topography (landscape), while Ritter was intrigued by the multitude of connections and relationships he observed within human geographic patterns. Both Humboldt and Ritter died in 1859, ending a period when information-gathering had been paramount. They brought geography into a new age in which synthesis, analysis, and theory-building became central.

European Geography

After the work of Humboldt and Ritter, geography became an accepted academic discipline in Europe, particularly in Germany, France, and Great Britain. Each of these countries emphasized different aspects of geographic study. German geographers continued the tradition of the scientific view, using observable data to answer geographic questions. They also introduced the concept that geography could take a chorological view, studying all aspects, physical and human, of a region and of the interrelationships involved.

The chorological view came to dominate French geography. Paul Vidal de la Blache (1845-1918) was

---

NATIONAL GEOGRAPHIC SOCIETY AND GEOGRAPHIC RESEARCH

In 1888 the National Geographic Society was founded to support the "increase and diffusion of geographic knowledge" of the world. In its first 110 years, the society funded more than five thousand expeditions and research projects with more than 6,500 grants. By the 1990s it was the largest such foundation in the world, and the results of its funded projects are found on television programs, video discs, video cassettes, and books, as well as in the *National Geographic* magazine, established in 1888. Its productions are cutting-edge resources for information about archaeology, ethnology, biology, and both cultural and physical geography.

---

the most prominent French geographer. He advocated the study of small, distinct areas, and French geographers set about identifying the many regions of France. They described and analyzed the unique physical and human geographic complex that was to be found in each region. An important concept that emerged from French geography was "possibilism." German geographers had introduced the notion of environmental determinism—that human beings were largely shaped and controlled by their environments. Possibilism rejected the concept of environmental determinism, asserting that the relationship between human beings and the environment works in two directions: The environment creates both limits and opportunities for people, but people can react in different ways to a given environment, so they are not controlled by it.

British geographers, influenced by the French approach, conducted regional surveys. British regional studies were unique in their emphasis on planning and geography as an applied science. From this work came the concept of a functional region—an area that works together as a unit based on interaction and interdependence.

American Geography

Prior to World War II, only a small group of people in the United States called themselves geographers. They were mostly influenced by German ideas, but the nature of geography was hotly debated. Two schools of geographers were philosophical adversaries. The Midwestern School, led by Richard Hartshorne, believed that description of unique regions was the central task of geography.

The Western (or Berkeley) School of geography, led by Carl Sauer, agreed that regional study was important, but believed it was crucial to go beyond description. Sauer and his followers included genesis and process as important elements in any study. To understand a region and to know where it is going, they argued, one must look at its past and how it got to its present state.

In the 1930s, environmental determinism was introduced to U.S. geography but ultimately was rejected. Although geography in both Europe and the United States was essentially an all-male discipline, the United States produced the first famous woman geographer, Ellen Churchill Semple (1863-1932).

World War II illustrated the importance of geographic knowledge, and after the war came to an end in 1945, geographers began to come into their own in the United States. From the end of World War II to the early 1960s, U.S. geographers produced many descriptive regional studies.

In the early 1960s, what is often called the quantitative revolution occurred. The development of computers allowed complex mathematical analysis to be performed on all kinds of geographic data, and geographers began to analyze a wide range of problems using statistics. There was great enthusiasm for this new approach to geography at first, but beginning in the 1970s, many people considered a purely mathematical approach to be somewhat sterile and thought it left out a valuable human element.

In the 1980s and 1990s, many new ways to look at geographic issues and problems were developed, including humanism, behaviorism, Marxism, feminism, realism, structuration, phenomenology, and postmodernism, all of which bring human beings back into focus within geographical studies.

Geography in the Twenty-first Century
Geography increasingly uses technology to analyze global space and answer a wide range of questions related to a host of concerns including issues related to the environment, climate change, population, rising sea levels, and pollution. The Geographic Information System (GIS), in particular, provides a powerful way for people trained in geography to understand geographic issues, solve geographic problems, and display geographic information. Geographers continue to adopt a wide variety of philosophies, approaches, and methods in their quest to answer questions concerning all things spatial.

Wendy Shaw

MAPMAKING IN HISTORY

Cartography is the science or art of making maps. Although workers in many fields have a concern with cartography and its history, it is most often associated with geography.

Maps of Preliterate Peoples
The history of cartography predates the written record, and most cultures show evidence of mapping skills. The earliest surviving maps are those carved in stone or painted on the walls of caves, but modern preliterate peoples still use a variety of materials to express themselves cartographically. For example, the Marshall Islanders use palm fronds, fiber from coconut husks (coir), and shells to make sea charts for their inter-island navigation. The Inuit use animal skins and driftwood, sometimes painted, in mapping. There is a growing interest in the cartography of early and preliterate peoples, but some of their maps do not fit readily into a more traditional concept of cartography.

Mapping in Antiquity

Early literate peoples, such as those of Egypt and Mesopotamia, displayed considerable variety in their maps and charts, as shown by the few maps from these civilizations that still exist. The early Egyptians painted maps on wooden coffin bases to assist the departed in finding their way in the afterlife; they also made practical route maps for their mining operations. It is thought that geometry developed from the Egyptians' riverine surveys. The Babylonians made maps of different scales, using clay tablets with cuneiform characters and stylized symbols, to create city plans, regional maps, and "world" maps. They also divided the circle in the sexigesimal system, an idea they may have obtained from India and that is commonly used in cartography to this day.

The Greeks inherited ideas from both the Egyptians and the Mesopotamians and made signal contributions to cartography themselves. No direct evidence of early Greek maps exists, but indirect evidence in texts provides information about their cosmological ideas, culminating in the concept of a perfectly spherical earth. This they attempted to measure and divide mathematically. The idea of climatic zones was proposed and possibly mapped, and the large known landmasses were divided into first two continents, then three.

Perhaps the greatest accomplishment of the early Greeks was the remarkably accurate measurement of the circumference of Earth by Eratosthenes (276-196 BCE). Serious study of map projections began at about this time. The gnomonic, orthographic, and stereographic projections were invented before the Christian era, but their use was confined to astronomy in this period. With the possible single exception of Aristarchus of Samos, the Greeks believed in a geocentric universe. They made globes (now lost) and regional maps on metal; a few map coins from this era have survived.

Later Greeks carried on these traditions and expanded upon them. Claudius Ptolemy invented two projections for his world maps in the second century CE. These were enormously important in the European Renaissance as they were modified in the light of new overseas discoveries. Ptolemy's work is known mainly through later translations and reconstructions, but he compiled maps from Greek and Phoenician travel accounts and proposed sectional maps of different scales in his *Geographia*. Ptolemy's prime meridian (0 degrees longitude) in the Canary Islands was generally accepted for a millennium and a half after his death.

Roman cartography was greatly influenced by later Greeks such as Ptolemy, but the Romans themselves improved upon route mapping and surveying. Much of the Roman Empire was subdivided by instruments into hundredths, of which there is a cartographic record in the form of marble tablets. In Rome, a small-scale map of the world known to the Romans was made on metal by Marcus Vipsanius Agrippa, the son-in-law of Augustus Caesar, and displayed publicly. This map no longer exists, however.

Cartography in Early East Asia

As these developments were taking place in the West, a rich cartographic tradition developed in Asia, particularly China. The earliest survey of China (Yu Kung) is approximately contemporaneous with the oldest reported mapmaking activity of the Greeks. Later, maps, charts, and plans accompanied Chinese texts on various geographical themes. Early rulers of China had a high regard for cartography—the science of princes. A rectangular grid was introduced by Chang Heng, a contemporary of Ptolemy, and the south-pointing needle was used for mapmaking in China from an early date.

These traditions culminated in Chinese cartographic primacy in several areas: the earliest printed maps (about 1155 CE), early printed atlases, and terrestrial globes (now lost). Chinese cartography greatly influenced that in other parts of Asia, particularly Korea and Japan, which fostered innovations of their own. It was only after the introduction of ideas from the West, in the Renaissance and later, that Asian cartographic advances were superseded.

Islamic Cartography

A link between China and the West was provided by the Arabs, particularly after the establishment of Islam. It was probably the Arabs who brought the magnetized needle to the Mediterranean, where it was developed into the magnetic compass.

Some scholars have argued that the Arabs were better astronomers than cartographers, but the Arabs did make several clear advances in mapmaking. Both fields of study were important in Muslim science, and the astrolabe, invented by the Greeks in antiquity but developed by the Arabs, was used in both their astronomical and terrestrial surveys. They made and used many maps, as indicated by the output of their most famous cartographer, al-Idrisi (who lived about 1100–1165). Some of his work still exists, including a zonal world map and detailed charts of the Mediterranean islands.

At about the same time, the magnetic compass was invented in the coastal cities of Italy, which gave rise to advanced navigational charts, including information on ports. These remarkably accurate charts were used for navigating in the Mediterranean Sea. They were superior to the European maps of the Middle Ages, which often were concerned with religious iconography, pilgrimage, and crusade. The scene was now set for the great overseas discoveries of the Europeans, which were initiated in Portugal and Spain in the fifteenth century.

In the next four centuries, most of the coasts of the world were visited and mapped. The early, projectionless navigational charts were no longer adequate, so new projections were invented to map the enlarged world as revealed by the European overseas explorations. The culmination of this activity was the development of the projection, in 1569, of Gerardus Mercator, which bears his name and is of special value in navigation.

Early Modern Mapmaking

Europeans began mapping their own countries with greater accuracy. New surveying instruments were invented for this purpose, and a great land-mapping activity was undertaken to match the worldwide coastal surveys. For about a century, the Low Countries of Belgium, Luxembourg, and the Netherlands dominated the map and chart trades, producing beautiful hand-colored engraved sheet wall maps and atlases.

France and England established new national observatories, and by the middle of the seventeenth century, the Low Countries had been eclipsed by France in surveying and making maps and charts. The French adopted the method of triangulation of Mercator's teacher, Gemma Frisius. Under four generations of the Cassini family, a topographic survey of France more comprehensive than any previous survey was completed. Rigorous coastal surveys were undertaken, as well as the precise measurement of latitude (parallels).

The invention of the marine chronometer by John Harrison made it possible for ships at sea to determine longitude. This led to the production of charts of all the oceans, with England's Greenwich eventually being adopted as the international prime meridian.

Quantitative, thematic mapping was advanced by astronomer Edmond Halley (1656–1742) who produced a map of the trade winds; the first published magnetic variation chart, using isolines; tidal charts; and the earliest map of an eclipse. The Venetian Vincenzo Coronelli made globes of greater beauty and accuracy than any previous ones. In the German lands, the study of map projections was vigorously pursued. Johann H. Lambert and others invented a number of equal-area projections that were still in use in the twentieth century.

Ideas developed in Europe were transmitted to colonial areas, and to countries such as China and Russia, where they were grafted onto existing cartographic traditions and methods. The oceanographic explorations of the British and the French built on the earlier charting of the Pacific Ocean and its islands by native navigators and the Iberians.

Nineteenth Century Cartography

Cartography was greatly diversified and developed in the nineteenth century. Quantitative, thematic mapping was expanded to include the social as well as the physical sciences. Alexander von Humboldt used isolines to show mean air temperature, a method that later was applied to other phenomena. Contour lines gradually replaced less quantitative methods of representing terrain on topographic maps. Such maps were made of many areas, for example India, which previously had been poorly mapped.

Extraterrestrial (especially lunar) mapping, had begun seriously in the preceding two centuries with the invention of the telescope. It was expanded in the nineteenth century. In the same period, regular national censuses provided a large body of data that could be mapped. Ingenious methods were created to express the distribution of population, diseases, social problems, and other data quantitatively, using uniform symbols.

Geological mapping began in the nineteenth century with the work of William Smith in England, but soon was adopted worldwide and systematized, notably in the United States. The same is true of transportation maps, as the steamship and the railway increased mobility for many people. Faster land travel in an east-west direction, as in the United States, led to the official adoption of Greenwich as the international prime meridian at a conference held in Washington, D.C., in 1884. Time zone maps were soon published and became a feature of the many world atlases then being published for use in schools, offices, and homes.

A remarkable development in cartography in the nineteenth century was the surveying of areas newly occupied by Europeans. This occurred in such places as the South American republics, Australia, and Canada, but was most evident in the United States. The U.S. Public Land Survey covered all areas not previously subdivided for settlement. Property maps arising from surveys were widely available, and in many cases, the information was contained in county and township atlases and maps.

Modern Mapping and Imaging

Cartography was revolutionized in the twentieth century by aerial photography, sonic sounding, satellite imaging, and the computer. Before those developments, however, Albrecht Penck proposed an ambitious undertaking—an International Map of the World (IMW). Cartography historically had been a nationalistic enterprise, but Penck suggested a map of the world in multiple sheets produced cooperatively by all nations at the scale of 1:1,000,000 with uniform symbols. This was started in the first half of the twentieth century but was not completed, and was superseded by the World Aeronautical Chart (WAC) project, at the same scale, during and after World War II.

The WAC project owed its existence to flight information made available following the invention of the airplane. Both photography and balloons were developed before the twentieth century, but the new, heavier-than-air craft permitted overlapping aerial photographs to be taken, which greatly facilitated the mapping process. Aerial photography revolutionized land surveys—maps could be made at less cost, in less time, and with greater accuracy than by previous methods. Similarly, marine surveying was revolutionized by the advent of sonic sounding in the second half of the twentieth century. This enabled mapping of the floor of the oceans, essentially unknown before this time.

Satellite imaging, especially continuous surveillance by Landsat since 1972, allows temporal monitoring of Earth. The computer, through Geographical Information Systems (GIS) and other technologies, has greatly simplified and speeded up the mapping process. During the twentieth century, the most widely available cartographic product was the road map for travel by automobile.

Spatial information is typically accessed through apps on computers and mobile devices; traditional maps are becoming less common. The new media also facilitate animated presentations of geographical and extraterrestrial distributions. Cartographers remain responsive to the opportunities provided by new technologies, materials, and ideas.

Norman J. W. Thrower

MAPMAKING AND NEW TECHNOLOGIES

The field of geography is concerned primarily with the study of the curved surface of Earth. Earth is huge, however, with an equatorial radius of 3,963 miles (6,378 km.). How can one examine anything more than the small patch of earth that can be experienced at one time? Geographers do what scientists do all of the time: create models. The most common model of Earth is a globe—a spherical map that is usually about the size of a basketball.

A globe can show physical features such as rivers, oceans, the continents, and even the ocean floor. Political globes show the division of Earth into countries and states. Globes can even present views of the distant past of Earth, when the continents and oceans were very different than they are today. Globes are excellent for learning about the distributions, shapes, sizes, and relationships of features of Earth. However, there are limits to the use of globes.

How can the distribution of people over the entire world be described at one glance? On a globe, the human eye can see only half of Earth at one time. What if a city planner needs to map every street, building, fire hydrant, and streetlight in a town? To fit this much detail on a globe, the globe might have to be bigger than the town being mapped. Globes like these would be impossible to create and to carry around. Instead of having to hire a fleet of flatbed trucks to haul oversized globes, the curved surface of the globe can be transformed to a flat plane.

The method used to change from a curved globe surface to a flat map surface is called a map projection. There are hundreds of projections, from simple to extremely complex and dating from about two thousand years ago to projections being invented today. One of the oldest is the gnomonic projection. Imagine a clear globe with a light inside. Now imagine holding a piece of paper against the surface of the globe. The coastlines and parallels of latitude and meridians of longitude would show through the globe and be visible on the paper. Computers can do the same thing because there are mathematical formulas for nearly all map projections.

Geometric Models for Map Projections

One way to organize map projections is to imagine what kind of geometric shape might be used to create a map. Like the paper (a plane surface) against the globe described above, other useful geometric shapes include a cone and a cylinder. When the rounded surface of any object, including Earth, is flattened there must be some stretching, or tearing. Map projections help to control the amount and kinds of distortion in maps. There are always a few exceptions that cannot be described in this way, but using geometric shapes helps to classify projections into groups and to organize the hundreds of projections.

Another way to describe a map projection is to consider what it might be good for. Some map projections show all of the continents and oceans at their proper sizes relative to one another. Another type of projection can show correct distances between certain points.

Map Projection Properties

When areas are retained in the proper size relationships to one another, the map is called an equal-area map, and the map projection is called an equal-area projection. Equal-area (also called equivalent or homolographic) maps are used to measure areas or view densities such as a population density.

If true angles are retained, the shapes of islands, continents, and oceans look more correct. Maps made in this way are called conformal maps or conformal map projections. They are used for navigation, topographic mapping, or in other cases when it is important to view features with a good representation of shape. It is impossible for a map to be both equal-area and conformal at the same time. One or the other must be selected based on the needs of the map user or mapmaker.

One special property—distance—can only be true on a few parts of a map at one time. To see how far it is between places hundreds or thousands of miles apart, an equidistant projection should be used. There will be several lines along which distance is true. The azimuthal equidistant projection shows true distances from the center of the map outward. Some map projections do not retain any of these properties but are useful for showing compromise views of the world.

Modern Mapmaking

Modern mapmaking is assisted from beginning to end by digital technologies. In the past, the paper map was both the primary means for communicating information about the world and the database used to store information. Today, the database is a digital database stored in computers, and cartographic visualizations have taken the place of the paper map. Visualizations may still take the form of paper maps, but they also can appear as flashes on computer screens, animations on local television news programs, and even on screens within vehicles to help drivers navigate. Communication of information is one of the primary purposes of making maps. Mapping helps people to explore and analyze the world.

Making maps has become much easier and the capability available to many people. Desktop mapping software and Internet mapping sites can make anyone with a computer an instant cartographer. The maps, or cartographic visualizations, might be quite basic but they are easy to make. The procedures that trained cartographers use to make map products vary in the choice of data, software, and hardware, but several basic design steps should always take place.

First, the purpose and audience for whom the map is being made must be clear. Is this to be a general reference map or a thematic map? What image should be created in the mind of the map reader? Who will use the map? Will it be used to teach young children the shapes of the continents and oceans, or to show scientists the results of advanced research? What form will the cartographic visualization take?

SLIDING ROCKS GET DIGITAL TREATMENT

Dr. Paula Messina studied the trails of rocks that slide across the surface of a flat playa in Death Valley, California. The sliding rocks have been studied in the past, but no one had been able to say for certain how or when the rocks moved. It was unclear whether the rocks were caught in ice floes during the winter, were blown by strong winds coming through the nearby mountains, or were moved by some other method.

Messina gave the mystery a totally digital treatment. She mapped the locations of the rocks and the rock trails using the global positioning system (GPS) and entered her rock trail data into a geographic information system (GIS) for analysis. She was able to determine that ice was not the moving agent by studying the pattern of the trails. She also used digital elevation models (DEM) and remotely sensed imagery to model the environment of the playa. She reported her results in the form of maps using GIS' cartographic output capabilities. While she did not solve completely the mystery of the sliding rocks, she was able to disprove that winter ice caused the rocks to slide along together in rafts and that there are wind gusts strong enough to move the biggest rock on the playa.

Will it be a paper map, a graphic file posted to the Internet, or a video?

The answers to these questions will guide the cartographer in the design process. The design process can be broken down into stages. In the first stage of map design, imagination rules. What map type, size and shape, basic layout, and data will be used? The second stage is more practical and consists of making a specific plan. Based on the decisions made in the first stage, the symbols, line weights, colors, and text for the map are chosen. By the end of this stage, there should be a fairly clear plan for the map. During the third stage, details and specifications are finalized to account for the production method to be used. The actual software, hardware, and methods to be used must all be taken into consideration.

What makes a good map? Working in a digital environment, a mapmaker can change and test various designs easily. The map is a good one when it communicates the intended information, is pleasing to look at, and encourages map readers to ask thoughtful questions.

New Technologies

Mapping technology has gone from manual to magnetic, then to mechanical, optical, photochemical, and electronic methods. All of these methods have overlapped one another and each may still be used in some map-making processes. There have been recent advances in magnetic, optical, and most of all, electronic technologies.

All components of mapping systems—data collection, hardware, software, data storage, analysis, and graphical output tools—have been changing rapidly. Collecting location data, like mapping in general, has been more accessible to more people. The development of the Global Positioning System (GPS), an array of satellites orbiting Earth, gives anyone with a GPS receiver access to location information, day or night, anywhere in the world. GPS receivers are also found in planes, passenger cars, and even in the backpacks of hikers.

Satellites also have helped people to collect data about the world from space. Orbiting satellites collect images using visible light, infrared energy, and other parts of the electromagnetic spectrum. Active sensing systems send out radar signals and create images based on the return of the signal. The entire world can be seen easily with weather satellites, and other specialized satellite imagery can be used to count the trees in a yard.

These great resources of data are all stored and maintained as binary, computer-readable information. Developments in laser technology provide large amounts of storage space on media such as optical disks and compact disks. Advances in magnetic technology also provide massive storage capability in the form of tape storage, hard drives, and cloud storage. This is especially important for saving the large databases used for mapping.

Computer hardware and software continue to become more powerful and less expensive. Software continues to be developed to serve the specialized needs that mapping requires. Just as word processing software can format a paper, check spelling and grammar, draw pictures and shapes, import tables and graphics, and perform dozens of other functions, specialized software executes maps. The most common software used for mapping is called Geographic Information System (GIS) software. These systems provide tools for data input and for analysis and modeling of real-world spatial data, and provide cartographic tools for designing and producing maps.

Karen A. Mulcahy

UNITED STATES

The United States of America is the world's fourth-largest country in area, at 3.79 million square miles (9.83 million sq. km.). It is the third-largest country in population, with approximately 332 million people in 2020. Its forty-eight contiguous states are located in North America between Canada and Mexico, with Alaska in northwestern North America and Hawaii in the Pacific Ocean. In addition, the United States has possessions in the Caribbean Sea and the Pacific Ocean.

Topography

From the Atlantic Coastal Plain, the Appalachian Mountains rise, running from southern Canada to northern Alabama. To the east, between the Appalachians and the coastal plain, is the Piedmont, a region of rolling hills and plateaus. West of the Appalachians are more plateaus and rolling hills and the valleys of rivers such as the Ohio, Tennessee, and Mississippi. Farther to the west are the Great Plains, a vast grassland extending from Can-

Yosemite Valley from Wawona Tunnel view. (Mark J. Miller)

THE UNITED STATES

ada to Texas. The plains give way to the Rocky Mountains, one of North America's great mountain chains, which stretches from Canada down to central New Mexico. Mount Elbert at 14,440 feet (4,401 meters) is the tallest peak in the Rockies and in Colorado. To the west of the Rockies is the Great Basin, a desert area with inland drainage. The western edge of the Great Basin abuts the Cascade Range and the Sierra Nevada. West of the Cascades and Sierra Nevada are valleys, then coastal ranges.

Alaska is home to Denali (formerly Mount Mc-Kinley), the tallest mountain in North America at 20,310 feet (6,190 meters). The Hawaiian Islands are of volcanic origin, rising from the sea floor—what geologists refer to as a hot spot. In fact, the west coast of the United States, including Alaska, is part of what is known as the Pacific Ring of Fire. This name refers to the fact that there, the earth's tectonic plates converge, or slip past each other, along this ring, resulting in earthquakes and volcanic activity.

Climate and Vegetation

The primary factors of climate are temperature and precipitation, which are also the primary factors in vegetational patterns. The general precipitation pattern in the United States decreases from east to west, although it rises abruptly along the West Coast due to the coastal ranges. Average annual precipitation for Norfolk, Virginia, is 45 inches (114 centimeters); for St. Louis, Missouri, 37.5 inches (95 centimeters); for Denver, Colorado, 15 inches (38 centimeters); and for Reno, Nevada, 7.5 inches (19 centimeters). For San Francisco, it rises back up to 20 inches (51 centimeters).

Temperatures decrease to the north, although coastal locations are generally warmer than more inland locations at the same latitude as a result of the moderating effects of the oceans. For example, the normal January maximum temperature in Houston, Texas, is 61 degrees Fahrenheit (16.1 degrees Celsius); in Omaha, Nebraska, 31 degrees Fahrenheit (0.6 degrees Celsius); in Bismarck, North Dakota,

20 degrees Fahrenheit (6.7 degrees Celsius). San Francisco, California, has a normal January maximum almost 27 degrees Fahrenheit (14 degrees Celsius) warmer than that of Omaha, although both are at approximately the same latitude. The same situation exists for Seattle, Washington, and Bismarck. San Francisco and Seattle are warmer in the winter and cooler in the summer than Omaha and Bismark thanks to their coastal locations.

Because of the relationship between plant life and climate, the vegetational pattern is similar to the climate pattern. The lush forests of the east give way to the Great Plains where too little precipitation falls to support trees. Beyond the Great Plains are the high deserts of the Great Basin and the deserts of the Southwest. The Great Plains are grassland, while the deserts have much sparser vegetation. The one exception is the Sonoran Desert of southern Arizona which, because of its rainfall pattern, supports the greenest desert in the world. To the west of the deserts, more forests are located. They are fed by precipitation deriving from the moist winds coming in from the Pacific Ocean and rising up over the mountains, leading to plentiful rain and snow.

Exploration and Early Settlement
It is believed the first inhabitants of the Americas migrated from Asia across the Bering Strait approximately 30,000 years ago. These people scattered throughout North, Central, and South America. In the United States, they now are referred to as Native Americans. Many scholars believe the Vikings reached the coast of present-day Canada, and possibly the present-day United States, as early as 1000 CE. In 1492, Christopher Columbus, an Italian explorer sailing for Spain, was the first European of his time to visit the Americas, when he landed in the Bahamas, Cuba, and Hispaniola. John Cabot, an Italian-English explorer, first visited the North American mainland in 1497.

St. Augustine, in present-day Florida, was established by the Spanish in 1565 and Jamestown, in present-day Virginia, by the English in 1607. In 1620 the Pilgrims arrived from England via the Netherlands seeking religious freedom and established Plymouth Colony in present-day Massachu-

PERCEPTUAL AND VERNACULAR REGIONS OF THE UNITED STATES

Terms such as "the South," "the Southwest," and "the Midwest" are used and understood in daily life in the United States. Some geographers study the usage of these terms to try to understand how and why people apply them to certain places. For example, research shows that almost all Americans identify Alabama and Mississippi as in the South, Arizona and New Mexico as in the Southwest, and Nebraska and Iowa as in the Midwest. Where is Oklahoma, besides being central to these six states?

In a study, thirteen states were identified as being southern, with Oklahoma included next to last, eight states were identified as being in the Southwest, with Oklahoma the last to be included, and fourteen states were included in the Midwest, with Oklahoma included ninth. Therefore, Oklahoma is considered to be in all three regions, but on the margins of each. This is reflected in Oklahoma's settlement and culture. Oklahoma north of Interstate 40 was settled primarily from the Midwest and is similar to the Midwest in topography and vegetation. South of Interstate 40 and east of Interstate 35, settlement was primarily from the South (including the forced relocation of Native Americans along the Trail of Tears) and the land is similar to that of the South in topography and vegetation; to the west of Interstate 35 and in the Panhandle, Oklahoma is similar to the Southwest in topography and vegetation. In many respects, Oklahoma is the place where the South, the Southwest, and the Midwest merge. Texas, to the south, is seen more as southern and southwestern and is not seen as midwestern. Texas has stronger southern and southwestern influences and blends those two regions without the Midwestern influence.

setts. Later in the seventeenth century, France and England began a series of wars over their colonial holdings. The result in North America was the French and Indian War (1754-1763), in which the British were victorious. Colonists soon grew weary of being taxed by the British with no representation in governmental affairs, which led to the American Revolution (1775-1783) and the establishment of the United States of America. The Constitution became law in 1788; in 1789, George Washington became the first president.

Wheat field on Dutch flats near Mitchell, Nebraksa in 1910. (National Archives and Records Administration)

Westward Expansion

The first major US westward expansion was the Louisiana Purchase in 1803, which increased the size of the United States dramatically. In 1823 President James Monroe issued the Monroe Doctrine, which stated that European powers should not interfere with the newly established countries in the Americas. The United States used this policy in annexing Texas in 1845. This led to the Mexican-American War (1846-1848), won by the United States, and the establishment of the Rio Grande as the border between Texas and Mexico.

In 1846, the United States and Great Britain established the forty-ninth north parallel west to the Pacific Ocean as the border between the United States and Canada. In 1853, the United States purchased the area south of the Gila River in Arizona and New Mexico from Mexico to establish a southern rail link with the Pacific coast, in what is termed the Gadsden Purchase.

Waves of Immigration

There have been three primary waves of immigrants to the United States since European colonization. The first wave comprised two distinct groups and lasted until about 1870. One group of the first wave included the English, Germans, Scotch-Irish, and others from northern and western Europe. The other group in the first wave was made up of Africans forced to migrate as slaves. The second wave, which lasted approximately fifty years, consisted mostly of eastern and southern Europeans. In 1921 the United States adopted a quota system restricting immigration. In the 1960s, those restrictions were relaxed and the third wave began, dominated by Asians and Latin Americans.

Cultural Regions

The three primary waves of immigration were separated not only by time but also by space. Combined with early settlement patterns, the waves of immigration were instrumental in forging the regional cultures now present in the United States. The first wave, dominated by northern and western Europeans and Africans, went into the eastern United States. The English settled in New England and throughout the southern part of the country; the Scotch-Irish settled primarily in the Appalachian Mountains; many Germans settled in the mid-Atlantic region and the Midwest; Scandinavians settled primarily in Wisconsin, Minnesota, and the Dakotas. African slaves were sent to the South along the coastal plain and in the Mississippi River Valley—the primary areas for plantation agriculture.

In the second wave, southern and eastern Europeans moved into the mid-Atlantic states and then throughout the Midwest and Great Plains. In the third wave, Asians entered from the west rather than the east, settling primarily in the western United States; Latin Americans are found mostly in the Southwest.

The result is the cultural mosaic that includes such diverse regions as New England, the South, the Midwest, and the Southwest. New England was originally settled by the English and today reflects that strong influence. The South mixes English, Scotch-Irish, and African influences, and the Southwest mixes Latin American, Native American, and European influences.

Economy

The United States has the world's most advanced economy, with approximately 80 percent of its workers employed in service industries. This type of economy is termed "postindustrial." When the United States was founded, most workers were farmers. As the Industrial Revolution came to its shores, many people left farms for factories and the United States grew to be the industrial giant of the world. In the postindustrial economy, the United States has remained the world's industrial leader but with far fewer manufacturing employees. Capital-intensive manufacturing, relying on technology and mechanization, has replaced many manual-labor-intensive activities. For more than a century, the United States had the world's largest economy. In 2014, the US slipped behind China and now ranks as the second-largets economy in the world. In 2020, the US in the top dozen countries for highest income per capita.

Downtown Los Angeles with Mount Baldy in the background after a large snow storm. (Alek Leckszas)

The Statue of Liberty on Liberty Island. (Don Ramey Logan Jr)

Urbanization

The United States has continually become more urban. Every year, a greater percentage of its residents move to cities. In 2020, approximately 82.7 percent of US residents lived in metropolitan areas. The largest US metropolitan area was New York City, with a 2018 metropolitan population greater than 22 million; New York City's population was 8.3 million. The two other dominant U.S. metropolitan areas were Los Angeles, with 18.6 million residents, and Chicago, with 9.9 million. The ten largest cities in the US in 2019 were New York, Los Angeles, Chicago, Houston, Phoenix, Philadelphia, San Antonio, San Diego, Dallas, and San Jose.

James D. Lowry, Jr.

CANADA

Canada is the second-largest country in the world; only Russia has a larger area. Its 3.85 million square miles (9.98 million sq. km.) make it about the same size as Europe, and it covers six time zones. However, despite the country's great physical size, its population is comparatively small: somewhat more than 37.6 million people in 2020.

Canada is bounded by the Atlantic Ocean on the east, the Arctic Ocean on the north, the Pacific Ocean and Alaska on the west, and the Great Lakes and the United States on the south. The United States is its only contiguous neighbor. The country is made up of ten partly self-governing provinces and three territories. Ottawa is the capital city.

Canada's provinces and territories can be grouped into five major regions—the north, the Atlantic Provinces, Quebec, Ontario, and the west.

Residents of each region have a strong sense of regionalism, identifying themselves as different from people in the other regions of Canada.

The North

Canada has many zones in which distinct physical geographical features are found. The vast, cold, dry Arctic in the north—the region that includes most of Canada's territory above 60 degrees north latitude—has permanently frozen soil (permafrost), which makes life difficult. Vegetation in the Arctic is sparse. Lichens and mosses dominate and cling to the rocky surface in the barren landscape. Somewhat farther south, in the tundra zone, discontinuous permafrost enables the ground to thaw for a few brief weeks in the summer. The tree line is the region spanning the tundra, south of which trees can survive.

Sheep Slot Rapids on the Firth River in Canada's Ivvavik National Park. (Daniel Case)

CANADA

On the western edge of this northern Arctic region are the Rocky Mountains. The Yukon Territory has the highest mountain peak in the country—Mount Logan (19.551 feet/5,959). The Mackenzie River, Great Bear Lake, Great Slave Lake, and Hudson Bay are some of the exceptional physical features of the north. Settlements are sporadic, but can be found as far north as Baffin Island and Ellesmere Island.

The total population of Canada's Arctic region in 2020 was estimated at about 125,000 persons, residing in the Yukon, Northwest, and Nunavut territories, whose capital cities are Whitehorse, Yellowknife, and Iqaluit, respectively. This region was originally populated by the Inuit people (also called Eskimos), who dominate in the eastern half of the Arctic region. The north is rich in natural resources

and has mining operations, oil drilling in the Arctic Ocean, and other developments that extract natural resources from the ground.

Atlantic Canada

The Atlantic provinces officially include Newfoundland and Labrador and the three Maritime Provinces—Nova Scotia, New Brunswick, and Prince Edward Island. The first Europeans who came to North America were the Vikings who landed on the island of Newfoundland. In the year 1000, they established a settlement on the site of L'Anse aux Meadows on Newfoundland. Originally settled by the Beothuk people, who were later exterminated by the Europeans, Newfoundland and Labrador became Canada's tenth province in 1949.

The province of Nova Scotia was settled originally by the Micmac peoples. Nova Scotia and Newfoundland were sighted in 1497 by John Cabot, who claimed the region for England. In 1605, a fort was founded in the Annapolis Valley of Nova Scotia, named Port Royal by French explorer Samuel de Champlain. Port Royal holds the distinction of being the first European settlement north of Florida, the first French colony in North America, and Canada's oldest settlement. Nova Scotia is known for fishing, coal mining, and tourism, which brings many people to visit sights such as the Fortress of Louisbourg, Lunenberg, Halifax, and Cape Breton Island and highlands.

New Brunswick was originally part of Nova Scotia. After the American Revolution, thousands of British Loyalists emigrated from the United States, and many settled in New Brunswick. The city of Saint John, founded in 1785, is the oldest incorporated city in Canada. The capital of New Brunswick is Fredericton, located upriver from Saint John.

Noted natural attractions in the province are the Reversing Falls and the Bay of Fundy.

Charlottetown, the capital of Prince Edward Island, is where the Dominion of Canada was formed in 1867. The island is small, low in elevation, with predominantly red soil that is excellent for farming. Potatoes from the region are well recognized across Canada, their status as favored in Canada as Idaho potatoes are in the United States. The stories of Anne of Green Gables originated on this island. The ethnic makeup of Prince Edward Island is similar to that of the other Maritime Provinces, with many residents having English roots.

Quebec

The largest province in Canada is the most European in culture. This is a result of the overwhelming presence of the French language and culture, which has existed in the region since the St. Lawrence River was explored in 1534 by Jacques Cartier, followed by Samuel de Champlain and others. Lo-

Convention at Charlottetown, P.E.I., of Delegates from the Legislatures of Canada, New Brunswick, Nova Scotia, and Prince Edward Island to take into consideration the Union of the British North American Colonies. (Library and Archives Canada)

Parliament Hill, home of the federal government in Canada's capital city, Ottawa. (Coolcaesar)

cated almost completely within the Canadian Shield, the province is rich in natural resources, including minerals, forests, and rivers with hydroelectric potential. Only the St. Lawrence River Valley in the southern part of the province has land suitable for agriculture, and it is there that the majority of the population resides.

The province is dominated by Montreal, its financial center, and Quebec City, its political center. The majority of the population is of French background, with the only sizable English-speaking population being in Montreal. The people of Quebec are proud of their French heritage. Centuries of struggle for equal rights and recognition of a separate Francophone culture has led the province to change many of its laws on behalf of the French Movements to separate Quebec from the rest of Canada, on and off for the past half-century, and continue to this day.

Ontario

Ontario is the wealthiest and most populated province in Canada. Most of its people live in the southern part of the province, where Canada's largest city—Toronto—is found. Among the manufacturing industries in this region are the steel mills of Hamilton, the automobile manufacturing plants in Windsor and Oshawa, and the high-tech industries of Ottawa. Ottawa is also the nation's capital, and its large parliament buildings and other attractions bring many visitors to the area. Ottawa was chosen by Queen Victoria for the site of the capital over Montreal and Toronto.

The most southern point in Canada is found at Point Pelee and Pelee Island in Lake Erie. The climate and vegetation in this southern region of the province can be quite mild in the winter, with humid and often hot summers. Broadleaf forests dominate. Farther north, where the summers are warm to cool and winters are sometimes cold, vegetation changes to the boreal trees of the Canadian Shield. The Hudson Bay region in the far northern section of Ontario has Arctic conditions. Cities in the northern section of Ontario include Sudbury, which is well known for mining, Thunder Bay, and Sault Sainte Marie. Scenic attractions include Niagara Falls and Sibley Peninsula.

Western Canada

The remaining provinces of Canada constitute Western Canada—the four provinces west of Ontario to the Pacific Ocean. The majority of the south central part of this region is rich prairie farmland, but the far eastern, northeastern, and northern parts are dominated by the Canadian Shield. The coastal province of British Columbia, on the western edge of the region, is dominated by the Rocky Mountains, which range throughout the province. Climatic conditions vary in this area.

In the central Prairie Provinces, conditions usually involve warm to hot summers, with cold Arctic winds common in the winter. The region mostly experiences a dry continental climate. On the west coast, however, the maritime influence brings a great deal of moisture, and winters are generally cloudy with much precipitation. Winters are wet and mild, but summers can be quite warm and pleasant. Higher elevations experience cooler temperatures, and snow dominates in the interior of the province during the winter.

The Prairie Provinces of Manitoba, Saskatchewan, and Alberta were settled following the building of the Canadian Pacific and Canadian National Railroads at the end of the nineteenth century. At the beginning of the twentieth century, the region experienced a major boom in wheat farming and became a major wheat-growing area. The crop could be planted in the spring and ripened in the brief growing season, which can be as short as three months in the northern margin of the plowed land areas. Toward the west, the climate is drier and cattle ranching is dominant. Since the advent of mechanization, many farmers from the Prairie Provinces have moved away from the land, resulting in an outmigration of people from rural areas.

THE HEARTLAND-HINTERLAND CONCEPT

Geographers have promoted the "heartland-hinterland" concept to illustrate Canada's regional geography. Canada's heartland is southern Ontario and Quebec, where most of the manufacturing industries are located. The hinterland includes all provinces and regions that are farther away from the core of the country, from where raw materials are shipped to be manufactured.

Iron ore from mines in the Canadian Shield are shipped to the steel mills of Hamilton, where finished sheet-metal products are taken to nearby Oshawa for automobile manufacturing plants. Culturally, the heartland attracts artists and intellectuals to cities and universities. Financial headquarters are found in the downtown sections of Toronto and Montreal. In the heartland-hinterland model, the heartland has more power and influence, while the peripheral provinces and regions in the hinterland have less power and influence in a wide area of matters.

The major cities of the region include Winnipeg, Regina, Saskatoon, Calgary, and Edmonton. All those cities were influenced by the coming of the railroad and developed into major grain and meat markets. Three of the cities became provincial capitals. After the 1950s, the discovery of coal, oil, and natural gas in Alberta brought new industries and more people to the province. It has prospered because of these discoveries and today is one of the wealthiest Canadian provinces.

In the mountainous region of British Columbia are some localized lowland areas in which mining, farming, fruit-growing, and tourism are common. Fishing, especially for salmon, is also a major industry along the coast. The majority of the population is found in the Fraser River Valley in southern British Columbia and on Vancouver Island. The cities of Vancouver and Victoria dominate. Immigration to the Vancouver region is substantial; by 2016, more than half its people did not speak English as their first language. Nearly a third of the people who have immigrated to the Vancouver area have originated in Asia.

The Quartier Petit Champlain in Vieux-Québec is claimed to be the oldest commercial district in North America. (Wilfredor)

Culture

Canada is officially a multicultural country, which recognizes the many ethnic groups that have emigrated there. Atlantic Canada is dominated by descendants of earlier English, Irish, and Scottish settlers, and Quebec has the French. Ontario has a variety of ethnic groups, especially in cities such as Toronto. Italians, Poles, Greeks, Chinese, Scandinavians, and others are found throughout the province. Farther west, large population clusters of Germans, Ukrainians, and Scandinavians are found, while in British Columbia there is a growing concentration of East Indians and Asians.

Popular culture in Canada is similar to that in the United States. The favorite sport in the country is ice hockey, followed by Canadian football and baseball. Regional affiliation and loyalty of fans to sports teams is evident across the country as well, although the Montreal Canadiens and Toronto Maple Leafs maintain the strongest followings across the country thanks to their prominence in hockey history.

Mika Roinila

DISCUSSION QUESTIONS: UNITED STATES AND CANADA

Q1. How has the United States evolved from an agrarian society to one based on service industries? What is implied by the term "postindustrial"? How does the United States strike a balance between industrial and postindustrial?

Q2. In what ways has immigration shaped settlement patterns in the US and in Canada? What has been the role of indigenous peoples? How do the US and Canada embrace their diverse populations?

Q3. What commonalities in culture have emerged from the geographical location of the United States and Canada? What does the existence of a long and undefended border between the US and Canada imply about the two countries?

MEXICO

Shaped like a funnel that links North America with Central America, Mexico covers an area of almost 760,000 square miles (1.97 million sq. km.) making it the world's eighth-largest nation and the third-largest in Latin America, after Brazil and Argentina.

Physiography

The landmass of Mexico curves southeastward from 32 degrees north latitude at Tijuana, Baja California, to 15 degrees north latitude at Tapachula, Chiapas; and from longitude 117 degrees west at Tijuana to roughly 87 degrees west at Cancún. Mexico's northern border with the United States ex-

tends 1,954 miles (3,144 km.) from the Gulf of Mexico in the east to the Pacific Ocean in the west. In the east, more than half the US-Mexico border is formed by the Rio Bravo (Rio Grande in the United States). To the southeast, the Mexican states of Campeche and Chiapas border Guatemala, and the state of Quintana Roo borders Belize.

Landforms

Mexico is located at the intersection of four tectonic plates—the North American, the Caribbean, the Pacific, and the Cocos. As the plates move, they push up mountain ranges on the southern edge of the Mexican, or Central, Plateau. Some include ac-

Sierra Madre Occidental. (Christian Frausto Bernal)

MEXICO

tive volcanoes that erupt periodically. Earthquakes often occur; a 7.1-magnitude quake that struck Mexico City and nearby areas in 2017 killed hundreds of people and caused millions of dollars in damage.

Millions of years ago, the peninsula of Baja California in the northwest was attached to Mexico's mainland. As Baja California detached and moved northwest, the Gulf of California (called the Sea of Cortés in Mexico) opened. Many species of fish and animals live in these nutrient-rich waters. The Pacific gray whale migrates about 12,000 miles (19,300 km.) every year to calve in the Pacific lagoons of southern Baja California. Mexico has tried to curb commercial fishing in these waters to protect the marine life found here, but lack of funding has made enforcement of laws difficult.

Mountains dominate Mexico's landscape. The Sierra Madre Occidental, Mexico's largest range, extends along the western coast. It was here that the Spanish discovered some of the world's richest sil-

ver deposits, in the mid-sixteenth century. Mexico's second great mountain range, the Sierra Madre Oriental, runs parallel to the eastern coast, along the Gulf of Mexico.

The Mexican Plateau, a rugged central plateau that makes up most of Mexico, lies cradled between the Sierra Madres, or "mother ranges." The wide plains of the plateau average more than 6,000 feet (1,830 meters) above sea level. At the south end of the plateau lies the Valley of Mexico, where Mexico City, the capital, is located. The Valley of Mexico's year-round cool temperatures and rich, volcanic soils have attracted some of the densest populations in all of Latin America.

The Cordillera Neo-Volcanica south of Mexico City includes towering snowcapped volcanoes. The highest, Mount Orizaba, rises to 18,700 feet (5,700 meters) and is Mexico's tallest peak and the third-loftiest in North America. The Sierra Madre del Sur are low mountains along the southern Pacific coast.

Mexico narrows in the south to form the Isthmus of Tehuantepec, where the Pacific Ocean and the Gulf of Mexico are separated by only about 137 miles (220 km.). The Yucatán Peninsula of southeastern Mexico is the country's flattest region.

Regions of Mexico

The Central Plateau is Mexico's core region, with four-fifths of the population and several large cities, including the capital, Mexico City. It has vast plains and broad valleys at altitudes of around 6,550 feet (2,000 meters). Because of the higher elevations, the region's climate is pleasant, making it a desirable place to live. Mexico City, at about 7,350 feet (2,240 meters) above sea level, enjoys moderate temperatures.

Mexico City is a huge, busy, overpopulated, polluted metropolis and the center of government and commerce for Mexico. The oldest capital of the New World is rich in both indigenous and colonial history and is the home of one-quarter of the Mexican population. Built on the site of the ancient Aztec capital city, Tenochtitlán, Mexico City's foundation is the soft, unstable lake bed of Lake Texcoco. Mexico City has been sinking at an annual rate of approximately 6 inches (15 centimeters). The largest church in Latin America, Mexico City's cathedral, is situated there, in the world's largest Roman Catholic diocese.

An estimated twelve hundred rural migrants arrive in Mexico City daily seeking a better life. The city is encircled by *ciudades perdidas*, or "lost cities." Mexico City wrestles with severe water shortages, terrible pollution, uncontrollable traffic congestion, and the inability to provide even the most basic services, such as garbage pickup and disposal, to many of its inhabitants.

Mexico's Gulf Coast region—centered on the states of Veracruz and Tabasco—is one of the most rapidly developing regions of Mexico. Rich volcanic and alluvial soils make it one of the most productive agricultural zones in tropical Latin America. It has lush, tropical coastal plains and cooler inland mountains. The city of Veracruz is Mexico's leading port. This region was once home to three major pre-Columbian cultures—the Olmec "mother culture" of Mexico; the Totonacs of central Veracruz; and the Huastecs. Centuries later, the Spanish conquistadors, accompanied by Catholic priests, set out from here to conquer Mexico for "God, Gold, and Glory."

Farther south along the Gulf Coast lies the Yucatán Peninsula, Mexico's third region. This is a flat, low-lying area of approximately 70,000 square miles (181,000 sq. km.), with a 994-mile (1,600-kilometer) shoreline that borders both the Gulf of Mexico and the Caribbean Sea. It is composed of

Angel of Independence in Mexico City, Mexico. (Jptellezgiron)

27

limestone and coral and acts like a stone sponge, absorbing rain into the ground. The rain forms underground rivers that dissolve the soft limestone, causing the formation of sinkholes (cenotes), which the Maya used as water wells. There are no surface streams or rivers on the peninsula. Beautiful beaches and warm water have made tourist resorts in Cancún and Cozumel among the best in the world. Some of the finest archaeological sites in the Americas are found on the Yucatán Peninsula. They include Chichén Itzá, a well-preserved Maya site, and Uxmal, a late-Classical Maya temple site.

The South is Mexico's fourth region. This is the heartland of Mexico's indigenous population and is dominated by the cultures, traditions, way of life, and spiritual beliefs of the Olmecs, Zapotecs, Mixtecs, and Maya. With the exception of oil-rich Tabasco state, the South is Mexico's poorest region. It is decades behind central and northern Mexico in development and has fewer and smaller cities. In Oaxaca, the state with the largest concentration of indigenous people, a third of the population speak indigenous languages only.

Northern Mexico is the fifth major region of Mexico. It stretches from the beaches of Baja California to the marshes and islands of the Gulf of

CHICHÉN ITZÁ

The Maya, whose civilization is considered to have been one of the most advanced indigenous American cultures, flourished in Mexico between 250 and 900 C.E. Chichén Itzá is the best preserved of Mexico's Maya ruins, with temples, an observatory, and the largest ballcourt in Mexico. Kukulcan, or El Castillo, is the tallest, most imposing pyramid on the site. The Maya developed an advanced knowledge of astronomy, mathematics, and architecture. Although the Maya were long considered a peaceful race, it is now known that they used torture, mutilation, and human sacrifice for religious rituals and sporting events.

Mexico. Two great deserts, the Chihuahuan Desert—the largest in North America—and the Sonoran Desert, are located here. The United States-Mexico border delimits this region to the north and is a broad transition zone defined by a unique blend of Mexican and US languages, food, and music.

The border economy is dominated by maquiladora businesses—in which local workers assemble goods for US firms—and tourism. In the northeast, industrial towns like Monterrey,

A pyramid at Chichén Itzá. (Daniel Schwen)

Saint Prisca church in Taxco, Guerrero, Mexico. (Luidger)

Reynosa, and Matamoros bring jobs and foreign investment to the region. The sparsely populated arid plains of northern Mexico are devoted to cattle ranching and produce some of the finest beef in the Americas. In the arid northwest, irrigated agriculture and dams on the Colorado, the Sonora, and the Yaqui rivers make it possible for farmers to raise wheat, cotton, and vegetables on some of Mexico's most fertile land.

Human Geography

Mexico is the most populous Spanish-speaking nation in the world. *Mestizos* (people of mixed Spanish and native blood) form the majority of its population (60 percent), vastly outnumbering indigenous people, who are estimated at just over 13 million. The blending of Spanish and indigenous cultures makes Mexico's culture unique in Latin America.

Mexico's population in 2020 was estimated at 128 million, placing it tenth worldwide, behind China, India, and the United States, but ahead of Japan and Germany. Slightly more than 80 percent of all Mexicans live in urban areas, with one-quarter of the population living in Mexico City and the surrounding State of Mexico. About 82 percent of the population identify themselves as Roman Catholics; Mexico is the second-largest Roman Catholic nation in the world, after Brazil. Several preexisting indigenous spiritual traditions were absorbed by the Catholic faith, making it easier for the native people to embrace Catholicism. Mexico is reforming its economy and has experienced successful industrial growth. NAFTA was suspended by the United States-Mexico-Canada Agreement (USMCA) in July 2020.

Carol Ann Gillespie

GREENLAND AND THE ARCTIC

The region of Greenland and the Arctic, one of the most sparsely populated areas on Earth, extends from northern Alaska eastward through northern Canada to Greenland. Its southern edge is defined by the tree line that marks the boundary between the southern forests and the northern tundra. The entire region shares an Arctic environment characterized by long, cold winters and short, cool summers. Average daily temperatures for January range between -22 and -31 degrees Fahrenheit (-30 and -35 degrees Celsius). Average July daily temperatures range from 41 to 50 degrees Fahrenheit (5 to 10 degrees Celsius).

Year-round low temperatures, combined with annual total precipitation amounts of approximately 8 inches (200 millimeters), produce a cold, desert-like environment where only highly adapted species can survive. Because there is total darkness during the winter months and constant daylight during the summer, the area north of the Arctic Circle is often referred to as the Land of the Midnight Sun.

The only soils in this treeless region are poorly developed and underlain by permafrost—ground that remains permanently frozen. During the summer months at lower latitudes, the surface layers of soil thaw, allowing plant roots to penetrate only to the depth of the thawed ground (about 3.3 feet/1 meter). The vegetation consists of low, ground-level plant species, including sedges, mosses, lichens, and woody dwarf willow in lower latitudes.

Sermiligaaq, Greenland has more than 60 settlements. (Ray Swi-hymn)

GREENLAND AND THE ARCTIC REGION

This region is sparsely populated by Inuit (Eskimos) who trace their ancestry back to a common Thule culture dating back more than a millennium. They have developed a unique way of life that is specially adapted to the harsh Arctic environment.

The Arctic

The Arctic is best described by its terrestrial and marine ecological areas, or ecozones. There are three terrestrial ecozones in the Arctic; the Northern Arctic ecozone epitomizes the conditions that most people associate with the Arctic. This ecozone extends over most of the Arctic Islands. Vegetation is sparse and the region supports fewer than twenty species of mammals, among them caribou, musk ox, Arctic fox, polar bear, Arctic hare, and lemming.

The Southern Arctic terrestrial ecozone consists of rolling plains containing many lakes, with the occasional outcrop of bedrock of Precambrian Shield. Being slightly milder than the zone to the north, it has many wetland areas, which are home to a variety of sedges, mosses, and migrating waterfowl.

The Arctic Cordillera terrestrial ecozone is a mountainous region covering Eastern Baffin and Devon Islands and parts of Ellesmere and Bylot islands. These ice cap-covered mountains reach up to 6,500 feet (2,000 meters) in height. At lower elevations, there are some hardy plants; other than polar bears along the coasts, terrestrial mammals are largely absent.

The three Arctic marine ecozones differ in terms of ice conditions and marine life. The Arctic Basin marine ecozone is almost entirely covered by a massive, permanent ice cap, roughly centered on the North Pole, that covers the area from the Beaufort Sea, north and east to the northern tip of Greenland. The few animal species present include whales, seals, polar bears, and walruses that live along the edge of the ice pack.

The Arctic Archipelago marine ecozone, extending from Greenland to Alaska, consists of a maze of channels, straits, and fjords that run among the Arctic islands, where open water exists in some areas for two to three months every year. This region is best known from the adventures of early European explorers who came in search of the elusive Northwest Passage. This ecozone contains polynyas, areas of permanently open water, which contain significant concentrations of marine life, including whales, seals, polar bears, and Arctic cod. Humans are also attracted to these areas; the edges of polynyas have long served as locations for Inuit settlements. A dwindling number of Inuit, the traditional indigenous people of the Arctic, persist in their traditional hunting and fishing way of life in this region.

The Northwest Atlantic marine ecozone extends from the edge of Lancaster Sound and Greenland.

ALLIGATORS IN THE ARCTIC?

The Arctic was not always the cold and desolate place that it currently is. On Ellesmere Island in the far Arctic North, the fossilized remains of alligators, rhinoceros, tapirs, and tortoises—species found only in tropical forests—have been discovered. The remains of ancient redwood trees and the skull of an ancient flying lemur (today found only in Madagascar and southeast Asia) also indicate that the environment was once lush, warm, and moist. This evidence suggests that millions of years ago, landmasses in the Arctic region were physically connected to other landmasses. As a result of tectonic forces and continental drift, the area that is now the Arctic separated and drifted farther north and, though time, a different set of environmental conditions was established.

The warmer and shallower water is home to a variety of marine life, including twenty-two whale species and six species of seals.

Greenland

The largest island in the world, Greenland extends from Cape Morris Jesup in the north to Cape Farewell in the south. While politically part of Europe,

Glacier-fed (and carved) fjords in East Greenland, aka Østgrønland or Tunu, in the Sermersooq Kommune. (Doc Searls)

33

Map of the north Atlantic including some fantasy islands, c. 1570. (Abraham Ortelius)

geographically it is part of North America. It experiences the same cold, dry climatic characteristics as the Arctic, although there is greater variation in temperature and precipitation. Approximately 80 percent of the island is covered with the Greenland Ice Cap, a thick sheet of ice more than 10,000 feet (3,000 meters) deep. The ice flows through the coastal mountain valleys to the sea, forming dramatic glaciers, most notably the Ilulissat Icefjord. A small amount of open land exists along the southern and central western coasts, which have human settlement and sparse Arctic vegetation.

Both the Greenland and Arctic Inuit peoples (also known as Eskimos) trace their ancestry back to the Thule culture, found throughout the eastern Arctic and Greenland more than 1,000 years ago. Led by Erik the Red, the Vikings came into contact with the Thule on their arrival in Greenland. The

Vikings, the first Europeans to settle in Greenland, arrived in 982 and established two main settlements, the Eastern Settlement (near present-day Qaqortoq) and the Western Settlement (centered on Godthaab Fjord), which were abandoned before the fifteenth century. Starting in the eighteenth century, the Danish established outposts at Nuuk, Qasigiannguit, Narsaq, and Maniitsoq.

A self-governing administrative division of the Kingdom of Denmark, Greenland had 57,616 residents in 2020 living in small villages and towns along its west coast. In addition, the United States built and still operates the massive Thule Air Base on the northwest coast of the island. Its construction forced the resettlement of two Greenlandic villages to a location 60 miles (100 km.) north at Qaanaaq.

Catherine A. Hooey

THE CARIBBEAN

The Caribbean Sea is an extension of the western Atlantic Ocean that is bounded by Central and South America to the west and south, and the islands of the Antilles chain on the north and east. It is separated from another large body of water, the Gulf of Mexico, on the west by the Yucatán Channel, which runs from the north tip of Mexico's Yucatán Peninsula to the southern tip of Florida. The sea covers more than 965,000 square miles (2.5 million sq. km.).

The Caribbean islands, from the Bahamas to Trinidad and Tobago, form the core of the Caribbean region. They are usually divided into two major groupings—the Greater Antilles of Cuba, Hispaniola (the island shared by Haiti and the Dominican Republic), Puerto Rico, and Jamaica; and the Lesser Antilles, which contains two parallel chains of smaller islands. Islands in the inner arc, such as Montserrat and Martinique, have formed around volcanic peaks. Coral limestone islands, such as Barbados, form the outer arc. All the Caribbean islands lie in the Atlantic hurricane track from June through November.

In the Caribbean island realm, African and European influences have been historically strong, and plantation agriculture, tied to international markets, has dominated. Central America's Belize (formerly British Honduras) is sometimes grouped within the Caribbean because of its historical and cultural ties to the former British island colonies.

Johny Cay is one of the most beautiful cays of the San Andrés and Providencia Archipelago, Colombia. (Mario Carvajal)

THE CARIBBEAN

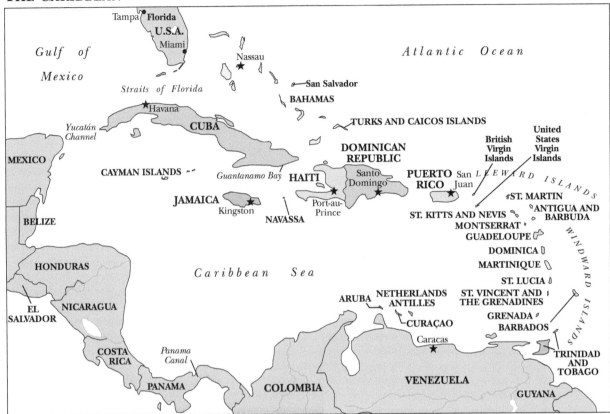

However, Hispanic migration from neighboring Central American countries has affected the demographic and cultural landscape of Belize.

The Caribbean region is rich in ethnic and cultural diversity. Peoples of African, European, and Asian heritage mingle in a tropical Caribbean environment. Many languages, cuisines, and religions are encountered. Nevertheless, the region as a whole has a distinctive regional character based on historical and cultural features and present economic circumstances.

Caribbean Commonalities

All the islands were once European colonies, whose indigenous Amerindian cultures were virtually eradicated. Spain took control of the Greater Antilles in the sixteenth century, followed by Great Britain, France, and the Netherlands, who carved up the Lesser Antilles among themselves. Britain took Jamaica from Spain in 1655, and France acquired what later became Haiti in 1697. For the past century, as many Caribbean countries gained political independence, the United States has played the major economic and geopolitical role in this strategic region. Puerto Rico became a US territory in 1898 and gained commonwealth status in 1952. The US Virgin Islands were purchased by the United States from Denmark in 1917.

Sugarcane, slavery, and plantation agriculture fashioned the economic and cultural landscape of the Ca-

Rose Hall, the estate house of a former sugar plantation, in Jamaica. (Urban Walnut)

The Gothic Cathedral of Santa María la Menor, Santo Domingo, Dominican Republic is the oldest cathedral in the Americas, built between 1514–1541. (Mario Roberto Durán Ortiz)

ribbean. Large-scale sugarcane plantations drove the physical transformation of the islands and a huge demand for slave labor. Slavery was abolished in 1838, and a new labor force arrived in the form of indentured laborers, chiefly from the Indian subcontinent and China. Today, the entire Caribbean is noted for its pronounced African-Caribbean cultural traits.

Finally, the Caribbean is fragmented, consisting of many small states with restricted economic resources and large, underemployed labor pools. Poverty is a feature of the Caribbean, whose people themselves are important exports. Caribbean economies rely heavily on tourism revenue.

Olwyn M. Blouet

DISCUSSION QUESTIONS:
MEXICO, GREENLAND AND THE ARCTIC, AND THE CARIBBEAN

Q1. How has Mexico's indigenous background combined with its Spanish heritage to create a unique culture? In what ways does the Spanish-infused Mexican culture differ from the Anglo-American culture of the US and Canada?

Q2. Where is Greenland? Why does it have stronger links to Europe than to North America? What geographical features account for Greenland's extremely small population?

Q3. What distinction did the Caribbean islands gain during the early European explorations? In what ways does the region's colonial heritage continue to express itself? What are the Caribbean's strongest cultural traits?

PHYSICAL GEOGRAPHY

Overview

Climate and Human Settlement

"Everyone talks about the weather," goes an old saying, "but nobody does anything about it." If everyone talks about the weather, it is because it is important to them—to how they feel and to how their bodies and minds function. There is plenty they can do about it, from going to a different location to creating an artificial indoor environment.

Climate

The term "climate" refers to average weather conditions over a long period of time and to the variations around that average from day to day or month to month. Temperature, air pressure, humidity, wind conditions, sunshine, and rainfall—all are important elements of climate and differ systematically with location. Temperatures tend to be higher near the equator and are so low in the polar regions that very few people live there. In any given region, temperatures are lower at higher altitudes. Areas close to large bodies of water have more stable temperatures. Rainfall depends on topography: The Pacific Coast of the United States receives a great deal of rain, but the nearby mountains prevent it from moving very far inland. Seasonal variations in temperature are larger in temperate zones.

Throughout human history, climate has affected where and how people live. People in technologically primitive cultures, lacking much protective clothing or housing, needed to live in mild climates, in environments favorable to hunting and gathering. As agricultural cultivation developed, populations located where soil fertility, topography, and climate were favorable to growing crops and raising livestock. Areas in the Middle East and near the Mediterranean Sea flourished before 1000 BCE.

Many equatorial areas were too hot and humid for human and animal health and comfort, and too infested with insect pests and diseases.

Improvements in technology allowed settlement to range more widely north and south. Sturdy houses and stables, internal heating, and warm clothing enabled people to survive and be active in long cold winters. Some peoples developed nomadic patterns, moving with herds of animals to adapt to seasonal variations.

A major challenge in the evolution of settled agriculture was to adapt production to climate and soil conditions. In North America, such crops as cotton, tobacco, rice, and sugarcane have relatively restricted areas of cultivation. Wheat, corn, and soybeans are more widely grown, but usually further north. Winter wheat is an ingenious adaptation to climate. It is sown and germinates in autumn, then matures and is harvested the following spring. Rice, which generally grows in standing water, requires special environmental conditions.

Tropical Problems

Some scholars argue that tropical climates encourage life to flourish but do not promote quality of life. In hot climates, people do not need much caloric intake to maintain body heat. Clothing and housing do not need to protect people from the cold. Where temperatures never fall below freezing, crops can be grown all year round. Large numbers of people can survive even where productivity is not high. However, hot, humid conditions are not favorable to human exertion nor (it is claimed) to mental, spiritual, and artistic creativity. Some tropical areas, such as South India, Bangladesh, Indonesia, and Central Africa, have developed large populations living at relatively low levels of income.

Slavery

Efforts to develop tropical regions played an important part in the rise of the slave trade after 1500 CE. Black Africans were kidnapped and forceably transported to work in hot, humid regions. The West Indian islands became an important location for slave labor, particularly in sugar production. On the North American continent, slave labor was important for producing rice, indigo, and tobacco in colonial times. All these were eclipsed by the enormous growth of cotton production in the early years of U.S. independence. It has been estimated that the forced migration of Africans to the Americas involved about 1,800 Africans per year from 1450 to 1600, 13,400 per year in the seventeenth century, and 55,000 per year from 1701 to 1810. Estimates vary wildly, but at least 12 million Africans were forced to migrate in this process.

European Migration

Migration of European peoples also accelerated after the discovery of the New World. They settled mainly in temperate-zone regions, particularly North America. Although Great Britain gained colonial dominion over India, the Netherlands over present-day Indonesia, and Belgium over a vast part of central Africa, few Europeans went to those places to live. However, many Chinese migrated throughout the Nanyang (South Sea) region, becoming commercial leaders in present-day Malaysia, Thailand, Indonesia, and the Philippines, despite the heat and humidity. British emigrants settled in Australia and New Zealand, Spanish and Italians in Argentina, Dutch (Boers) in South Africa—all temperate regions.

Climate and Economics

Most of the economic progress of the world between 1492 and 2000 occurred in the temperate zones, primarily in Europe and North America. Climatic conditions favored agricultural productivity. Some scholars believe that these areas had climatic conditions that were stimulating to intellectual and tech-

IRELAND'S POTATO FAMINE AND EUROPEAN EMIGRATION

Mass migration from Europe to North America began in the 1840s after a serious blight destroyed a large part of the potato crop in Ireland and other parts of Northern Europe. The weather played a part in the famine; during the autumns of 1845 and 1846 climatic conditions were ideal for spreading the potato blight. The major cause, however, was the blight itself, and the impact was severe on low-income farmers for whom the potato was the major food.

The famine and related political disturbances led to mass emigration from Ireland and from Germany. By 1850 there were nearly a million Irish and more than half a million Germans in the United States. Combined, these two groups made up more than two-thirds of the foreign-born U.S. population of 1850. The settlement patterns of each group were very different. Most Irish were so poor they had to work for wages in cities or in construction of canals and railroads. Many Germans took up farming in areas similar in climate and soil conditions to their homelands, moving to Wisconsin, Minnesota, and the Dakotas.

nological development. They argue that people are invigorated by seasonal variation in temperature, sunshine, rain, and snow. Storms—particularly thunderstorms—can be especially stimulating, as many parents of young children have observed for themselves.

Climate has contributed to the great economic productivity of the United States. This productivity has attracted a flow of immigrants, which averaged about 1 million a year from 1905 to 1914. Immigration approached that level again in the 1990s, as large numbers of Mexicans crossed the southern border of the United States, often coming for jobs as agricultural laborers in the hot conditions of the Southwest—a climate that made such work unattractive to many others.

Unpredictable climate variability was important in the peopling of North America. During the 1870s and 1880s, unusually favorable weather encouraged a large flow of migration into the grain-producing areas just west of the one-hundredth meridian. Then came severe drought and much agrarian distress. Between 1880 and 1890, the combined population of Kansas and Nebraska increased by about a million, an increase of 72 percent. During the 1890s, however, their combined population was virtually constant, indicating that a large out-migration was offsetting the natural increase. Much of the area reverted to pasture, as climate and soil conditions could not sustain the grain production that had attracted so many earlier settlers.

Climate variability can be a serious hazard. Freezing temperatures for more than a few hours during spring can seriously damage fruits and vegetables. A few days of heavy rain can produce serious flooding.

Recreation and Retirement

Whenever people have been able to separate decisions about where to live from decisions about where to work, they have gravitated toward pleasant climatic conditions. Vacationers head for Caribbean islands, Hawaii, the Crimea, the Mediterranean Coast, even the Baltic coast. "The mountains" and "the seashore" are attractive the world over. Paradoxically, some of these areas (the Caribbean, for instance) have monotonous weather year-round and thus have not attracted large inflows of permanent residents. Winter sports have created popular resorts such as Vail and Aspen in Colorado, and numerous older counterparts in New England. Large numbers of Americans have retired to the warm climates in Florida, California, and Arizona. These areas then attract

working-age adults who earn a living serving vacationers and retirees. Since these locations are uncomfortably hot in summer, their attractiveness for residence had to await the coming of air conditioning in the latter half of the twentieth century.

Human Impact on Climate

Climate interacts with pollution. Bad-smelling factories and refineries have long relied on the wind to disperse atmospheric pollutants. The city of Los Angeles, California, is uniquely vulnerable to atmospheric pollution because of its topography and wind currents. Government regulations of automobile emissions have had to be much more stringent there than in other areas to keep pollution under control.

Human activities have sometimes altered the climate. Development of a large city substitutes buildings and pavements for grass and trees, raising summer temperatures and changing patterns of water evaporation. Atmospheric pollutants have contributed to acid rain, which damages vegetation and pollutes water resources. Many observers have also blamed human activities for a trend toward global warming. Much of this has been blamed on carbon dioxide generated by combustion, particularly of fossil fuels. A widespread and continuing rise in temperatures is expected to raise water levels in the oceans as polar icecaps melt and change the relative attractiveness of many locations.

Paul B. Trescott

FLOOD CONTROL

Flood control presents one of the most daunting challenges humanity faces. The regions that human communities have generally found most desirable, for both agriculture and industry, have also been

the lands at greatest risk of experiencing devastating floods. Early civilization developed along river valleys and in coastal floodplains because those lands contained the most fertile, most easily irrigated soils for agriculture, combined with the convenience of water transportation.

The Nile River in North Africa, the Ganges River on the Indian subcontinent, and the Yangtze River in China all witnessed the emergence of civilizations that relied on those rivers for their growth. People learned quickly that residing in such areas meant living with the regular occurrence of life-threatening floods.

Knowledge that floods would come did not lead immediately to attempts to prevent them. For thousands of years, attempts at flood control were rare. The people living along river valleys and in floodplains often developed elaborate systems of irrigation canals to take advantage of the available water for agriculture and became adept at using rivers for transportation, but they did not try to control the river itself. For millennia, people viewed periodic flooding as inevitable, a force of nature over which they had no control. In Egypt, for example, early people learned how far out over the riverbanks the annual flooding of the Nile River would spread and accommodated their society to the river's seasonal patterns. Villagers built their homes on the edge of the desert, beyond the reach of the flood waters, while the land between the towns and the river became the area where farmers planted crops or grazed livestock.

In other regions of the world, buildings were placed on high foundations or built with two stories on the assumption that the local rivers would regularly overflow their banks. In Southeast Asian countries such as Thailand and Vietnam, it is common to see houses constructed on high wooden posts above the rivers' edge. The inhabitants have learned to allow for the water levels' seasonal changes.

Flood Control Structures

Eventually, societies began to try to control floods rather than merely attempting to survive them. Levees and dikes—earthen embankments constructed to prevent water from flowing into low-lying areas—were built to force river waters to remain within their channels rather than spilling out over a floodplain. Flood channels or canals that fill with water only during times of flooding, diverting water

away from populated areas, are also a common component of flood control systems. Areas that are particularly susceptible to flash floods have constructed numerous flood channels to prevent flooding in the city. For example, for much of the year, Southern California's Los Angeles River is a small stream flowing down the middle of an enormous, 20- to 30-foot-deep (6–9 meters) concrete-lined channel, but winter rains can fill its bed from bank to bank. Flood channels prevent the river from washing out neighborhoods and freeways.

Engineers designed dams with reservoirs to prevent annual rains or snowmelt entering the river upstream from running into populated areas. By the end of the twentieth century, extremely complex flood control systems of dams, dikes, levees, and flood channels were common. Patterns of flooding that had existed for thousands of years ended as civil engineers attempted to dominate natural forces.

The annual inundation of the Egyptian delta by the flood waters of the Nile River ceased in 1968 following construction of the 365-foot-high (111 meters) Aswan High Dam. The reservoir behind the 3,280-foot-long (1,000-meter) dam forms a lake almost ten miles (16 km.) wide and almost 300 miles (480 km.) long. Flood waters are now trapped behind the dam and released gradually over a year's time.

Environmental Concerns

Such high dams are increasingly being questioned as a viable solution for flood control. As human understanding of both hydrology and ecology have improved, the disruptive effects of flood control projects such as high dams, levees, and other engineering projects are being examined more closely.

Hydrologists and other scientists who study the behavior of water in rivers and soils have long known that vegetation and soil types in watersheds can have a profound effect on downstream flooding. The removal of forest cover through logging or clearing for agriculture can lead to severe flooding in the future. Often that flooding will occur many miles downstream from the logging activity. Devastating floods in the South Asian country of Bangladesh, for example, have been blamed in part on clear-cutting of forested hillsides in the Himalaya Mountains in India and Nepal. Monsoon rains that once were absorbed or slowed by forests now run quickly off mountainsides, causing rivers to reach unprecedented flood levels. Concerns about cause-and-effect relationships between logging and flood control in the mountains of the United States were one reason for the creation of the U.S. Forest Service in the nineteenth century.

In populated areas, even seemingly trivial events such as the construction of a shopping center parking lot can affect flood runoff. When thousands of square feet of land are paved, all the water from rain runs into storm drains rather than being absorbed slowly into the soil and then filtered through the watertable. Engineers have learned to include catch basins, either hidden underground or openly visible but disguised as landscaping features such as ponds, when planning a large paving project.

Wetlands and Flooding

Less well known than the influence of watersheds on flooding is the impact of wetlands along rivers. Many river systems are bordered by long stretches of marsh and bog. In the past, flood control agencies often allowed farmers to drain these areas for use in agriculture and then built levees and dikes to hold the river within a narrow channel. Scientists now know that these wetlands actually serve as giant sponges in the flood cycle. Flood waters coming down a river would spread out into wetlands and be held there, much like water is trapped in a sponge.

Draining wetlands not only removes these natural flood control areas but worsens flooding problems by allowing floodwater to precede downstream faster. Even if life-threatening or property-damaging floods do not occur, faster-flowing water significantly changes the ecology of the river system. Waterborne silt and debris will be carried farther. Trying to control floods on the Mississippi River has

had the unintended consequence of causing waterborne silt to be carried farther out into the Gulf of Mexico by the river, rather than its being deposited in the delta region. This, in turn, has led to the loss of shore land as ocean wave actions washes soil away, but no new alluvial deposits arrive to replace it.

In any river system, some species of aquatic life will disappear and others replace them as the speed of flow of the water affects water temperature and the amount of dissolved oxygen available for fish. Warm-water fish such as bass will be replaced by cold-water fish such as trout, or vice versa. Biologists estimate that more than twenty species of freshwater mussels have vanished from the Tennessee River since construction of a series of flood control and hydroelectric power generation dams have turned a fast-moving river into a series of slow-moving reservoirs.

Future of Flood Control

By the end of the twentieth century, engineers increasingly recognized the limitations of human interventions in flood control. Following devastating floods in the early 1990s in the Mississippi River drainage, the U.S. Army Corps of Engineers recommended that many towns that had stood right at the river's edge be moved to higher ground. That is, rather than trying to prevent a future flood, the Corps advised citizens to recognize that one would inevitably occur, and that they should remove themselves from its path. In the United States and a number of other countries, land that has been zoned as floodplains can no longer be developed for residential use. While there are many things humanity can do to help prevent floods, such as maintaining well-forested watersheds and preserving wetlands, true flood control is probably impossible. Dams, levees, and dikes can slow the water down, but eventually, the water always wins.

Nancy Farm Männikkö

ATMOSPHERIC POLLUTION

Pollution of the Earth's atmosphere comes from many sources. Some forces are natural, such as volcanoes and lightning-caused forest fires, but most sources of pollution are byproducts of industrial society. Atmospheric pollution cannot be confined by national boundaries; pollution generated in one country often spills over into another country, as is the case for acid deposition, or acid rain, generated in the midwestern states of the United States that affects lakes in Canada.

Major Air Pollutants

Each of eight major forms of air pollution has an impact on the atmosphere. Often two or more forms of pollution have a combined impact that exceeds the impact of the two acting separately. These eight forms are:

1. Suspended particulate matter: This is a mixture of solid particles and aerosols suspended in the air. These particles can have a harmful impact on human respiratory functions.

2. Carbon monoxide (CO): An invisible, colorless gas that is highly poisonous to air-breathing animals.

3. Nitrogen oxides: These include several forms of nitrogen-oxygen compounds that are converted to nitric acid in the atmosphere and are a major source of acid deposition.

4. Sulfur oxides, mainly sulfur dioxide: This sulfur-oxygen compound is converted to sulfuric acid in the atmosphere and is another source of acid deposition.

5. Volatile organic compounds: These include such materials as gasoline and organic cleaning solvents, which evaporate and enter the air in a vapor state. VOCs are a major source of ozone formation in the lower atmosphere.

6. Ozone and other petrochemical oxidants: Ground-level ozone is highly toxic to animals and plants. Ozone in the upper atmosphere, however, helps to shield living creatures from ultraviolet radiation.

7. Lead and other heavy metals: Generated by various industrial processes, lead is harmful to human health even at very low concentrations.

8. Air toxics and radon: Examples include cancer-causing agents, radioactive materials, or asbestos. Radon is a radioactive gas produced by natural processes in the earth.

All eight forms of pollution can have adverse effects on human, animal, and plant life. Some, such as lead, can have a very harmful effect over a small range. Others, such as sulfur and nitrogen oxides, can cross national boundaries as they enter the atmosphere and are carried many miles by prevailing wind currents. For example, the radioactive discharge from the explosion of the Chernobyl nuclear plant in the former Soviet Union in 1986 had harmful impacts in many countries. Atmospheric radiation generated by the explosion rapidly spread over much of the Northern Hemisphere, especially the countries of northern Europe.

Impacts of Atmospheric Pollution

Atmospheric pollution not only has a direct impact on the health of humans, animals, and plants but also affects life in more subtle, often long-term, ways. It also affects the economic well-being of people and nations and complicates political life.

Atmospheric pollution can kill quickly, as was the case with the killer smog, brought about by a temperature inversion, that struck London in 1952 and led to more than 4,000 pollution-related deaths. In the late 1990s, the atmosphere of Mexico City was so polluted from automobile exhausts and industrial pollution that sidewalk stands selling pure oxygen to people with breathing problems became thriving businesses. Many of the heavy metals and organic constituents of air pollution can cause cancer when people are exposed to large doses or for long periods of time. Exposure to radioactivity in the atmosphere can also increase the likelihood of cancer.

In some parts of Germany and Scandinavia in the 1990s, as well as places in southern Canada and the southern Appalachians in the United States, certain types of trees began dying. There are several possible reasons for this die-off of forests, but one potential culprit is acid deposition. As noted above, one byproduct of burning fossil fuels (for example, in coal-fired electric power plants) is the sulfur and nitrous oxides emitted from the smokestacks. Once in the atmosphere, these gases can be carried for many miles and produce sulfuric and nitric acids.

These acids combine with rain and snow to produce acidic precipitation. Acid deposition harms crops and forests and can make a lake so acidic that aquatic life cannot exist in it. Forests stressed by contact with acid deposition can become more susceptible to damage by insects and other pathogens. Ozone generated from automobile emissions also kills many plants and causes human respiratory problems in urban areas.

Air pollution also has an impact on the quality of life. Acid pollutants have damaged many monuments and building facades in urban areas in Europe and the United States. By the late 1990s, the distance that people could see in some regions, such as the Appalachians, was reduced drastically because of air pollution.

The economic impact of air pollution may not be as readily apparent as dying trees or someone with a respiratory ailment, but it is just as real. Crop damage reduces agricultural yield and helps to drive up the cost of food. The costs of repairing buildings or monuments damaged by acid rain are substantial. Increased health-care claims resulting from exposure to air pollution are hard to measure but are a cost to society nevertheless.

It is impossible to predict the potential for harm from rapid global warming arising from greenhouse gases and the destruction of the ozone layer by chlorofluorocarbons (CFCs), but it could be cata-

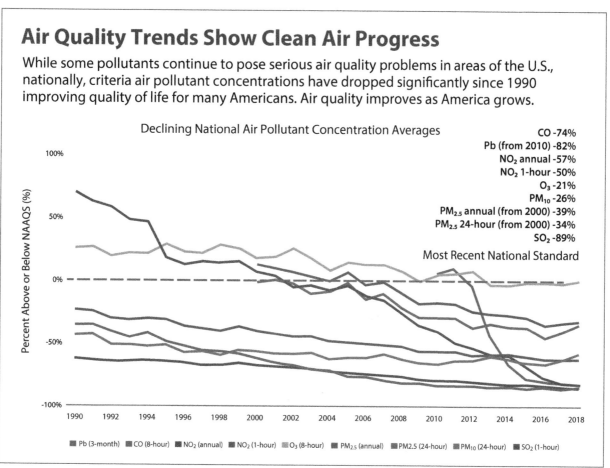

Air Quality Trends Show Clean Air Progress

While some pollutants continue to pose serious air quality problems in areas of the U.S., nationally, criteria air pollutant concentrations have dropped significantly since 1990 improving quality of life for many Americans. Air quality improves as America grows.

Declining National Air Pollutant Concentration Averages

CO -74%
Pb (from 2010) -82%
NO_2 annual -57%
NO_2 1-hour -50%
O_3 -21%
PM_{10} -26%
$PM_{2.5}$ annual (from 2000) -39%
$PM_{2.5}$ 24-hour (from 2000) -34%
SO_2 -89%

Most Recent National Standard

Percent Above or Below NAAQS (%)

Pb (3-month) CO (8-hour) NO_2 (annual) NO_2 (1-hour) O_3 (8-hour) $PM_{2.5}$ (annual) PM2.5 (24-hour) PM_{10} (24-hour) SO_2 (1-hour)

Source: U.S. Environmental Protection Agency, Our Nation's Air, Status and Trends Through 2018.

strophic. Rapid global warming would cause the sea level to rise because of the melting of the polar ice caps. Low-lying coastal areas would be flooded, or, in the case of Bangladesh, much of the country. Global warming would also change crop patterns for much of the world.

Solutions for Atmospheric Pollution

Although there is still some debate, especially among political leaders, most scientists recognize that air pollution is a problem that affects both the industrialized and less-industrialized world. In their rush to industrialize, many nations begin generating substantial amounts of air pollution; China's extensive use of coal-fired power plants is just one example.

The major industrial nations are the primary contributors to atmospheric pollution. North America, Europe, and East Asia produce 60 percent of the world's air pollution and 60 percent of its food supply. Because of their role in supplying food for many other nations, anything that damages their ability to grow crops hurts the rest of the world. In 2018, about 76 million tons of pollution were emitted into the atmosphere in the United States. These emissions mostly contribute to the formation of ozone and particles, the deposition of acids, and visibility impairment.

Many industrialized nations are making efforts to control air pollution, for example, the Clean Air Act of 1970 in the United States or the international Montreal Accord to curtail CFC production. Progress is slow and the costs of reducing air pollution are often high. Worldwide, bad outdoor air caused an estimated 4.2 million premature deaths in 2016, about 90 percent of them in low- and middle-income countries, according to the World Health Organization. Indoor smoke is an ongoing health threat to the 3 billion people who cook and heat their homes by burning biomass, kerosene, and coal.

In the year 2019 the record of the nations of the world in dealing with air pollution was a mixed one. There were some signs of progress, such as reduced automobile emissions and sulfur and nitrous oxides in industrialized nations, but acid deposition remains a problem in some areas. CFC production has been halted, but the impact of CFCs on the ozone layer will continue for many years. However, more nations are becoming aware of the health and economic impact of air pollution and are working to keep the problem from getting worse.

John M. Theilmann

DISEASE AND CLIMATE

Climate influences the spread and persistence of many diseases, such as tuberculosis and influenza, which thrive in cold climates, and malaria and encephalitis, which are limited by the warmth and humidity that sustains the mosquitoes carrying them. Because the earth is warming as a result of the generation of carbon dioxide and other "greenhouse gases" from the burning of fossil fuels, there is intensified scientific concern that warm-weather diseases will reemerge as a major health threat in the near future.

Scientific Findings

The question of whether the earth is warming as a result of human activity was settled in scientific circles in 1995, when the Second Assessment Report of the Intergovernmental Panel on Climate Change, a worldwide group of about 2,500 experts, was issued. The panel concluded that the earth's temperature had increased between 0.5 to 1.1 degrees Farenheit (0.3 to 0.6 degrees Celsius) since reliable worldwide records first became available in the late nineteenth century. Furthermore, the intensity of warming had increased over time. By the 1990s, the temperature was rising at the most rapid rate in at least 10,000 years.

The Intergovernmental Panel concluded that human activity—the increased generation of carbon dioxide and other "greenhouse gases"—is responsible for the accelerating rise in global temperatures. The amount of carbon dioxide in the atmosphere has risen nearly every year because of increased use of fossil fuels by ever-larger human populations experiencing higher living standards.

In 1998, Paul Epstein of the Harvard School of Public Health described the spread of malaria and dengue fever to higher altitudes in tropical areas of the earth as a result of warmer temperatures. Rising winter temperatures have allowed disease-bearing insects to survive in areas that could not support them previously. According to Epstein, frequent flooding, which is associated with warmer temperatures, also promotes the growth of fungus and provides excellent breeding grounds for large numbers of mosquitoes. Some experts cite the flooding caused by Hurricane Floyd and other storms in North Carolina during 1999 as an example of how global warming promotes conditions ideal for the spread of diseases imported from the Tropics.

Heat, Humidity, and Disease

During the middle 1990s, an explosion of termites, mosquitoes, and cockroaches hit New Orleans, following an unprecedented five years without frost. At the same time, dengue fever spread from Mexico across the border into Texas for the first time since records have been kept. Dengue fever, like malaria, is carried by a mosquito that is limited by temperature and humidity. Colombia was experiencing plagues of mosquitoes and outbreaks of the diseases they carry, including dengue fever and encephalitis, triggered by a record heat wave followed by heavy rains. In 1997 Italy also had an outbreak of malaria. An outbreak of zika in 2015–16, related to a virus spread by mosquitoes, raised concerns re-

garding the safety of athletes and spectators at the 2016 Summer Olympics in Rio de Janeiro and led to travel warnings and recommendations to delay getting pregnant for those living or traveling in areas where the mosquitoes are active.

The global temperature is undeniably rising. According to the National Oceanic and Atmospheric Administration, July 2019, was the hottest month since reliable worldwide records have been kept, or about 150 years. The previous record had been set in July 2017.

The rising incidence of some respiratory diseases may be related to a warmer, more humid environment. The American Lung Association reported that more than 5,600 people died of asthma in the United States during 1995, a 45.3 percent increase in mortality over ten years, and a 75 percent increase since 1980. Roughly a third of those cases occurred in children under the age of eighteen. Asthma is now one of the leading diseases among the young. Since 1980, there has been a 160 percent increase in asthma in children under the age of five.

Heat Waves and Health

A study by the Sierra Club found that air pollution, which will be enhanced by global warming, could be responsible for many human health problems, including respiratory diseases such as asthma, bronchitis, and pneumonia.

According to Joel Schwartz, an epidemiologist at Harvard University, air pollution concentrations in the late 1990s were responsible for 70,000 early deaths per year and more than 100,000 excess hospitalizations for heart and lung disease in the United States. Global warming could cause these numbers to increase 10 to 20 percent in the United States, with significantly greater increases in countries that are more polluted to begin with, according to Schwartz.

Studies indicate that global warming will directly kill hundreds of Americans from exposure to extreme heat during summer months. The U.S. Centers for Disease Control and Prevention have found that between 1979 and 2014, the death rate as a direct result of exposure to heat (underlying cause of death) generally hovered around 0.5 to 1 deaths per million people, with spikes in certain years). Overall, a total of more than 9,000 Americans have died from heat-related causes since 1979, according to death certificates. Heat waves can double or triple the overall death rates in large cities. The death toll in the United States from a heat wave during July 1999 surpassed 200 people. As many as 600 people died in Chicago alone during the 1990s due to heat waves. The elderly and very young have been most at risk.

Respiratory illness is only part of the picture. The Sierra Club study indicated that rising heat and humidity would broaden the range of tropical diseases, resulting in increasing illness and death from diseases such as malaria, cholera, and dengue fever, whose range will spread as mosquitoes and other disease vectors migrate.

The effects of El Niño in the 1990s indicate how sensitive diseases can be to changes in climate. A study conducted by Harvard University showed that warming waters in the Pacific Ocean likely contrib-

uted to the severe outbreak of cholera that led to thousands of deaths in Latin American countries. Since 1981, the number of cases of dengue fever has risen significantly in South America and has begun to spread into the United States. According to health experts cited by the Sierra Club study, the outbreak of dengue near Texas shows the risks that a warming climate might pose. Epstein and the Sierra Club study concur that if tropical weather expands, tropical diseases will expand.

In many regions of the world, malaria is already resistant to the least expensive, most widely distributed drugs. According to the World Health Organization (WHO), there were 219 million cases of malaria globally in 2017 and 435,000 malaria deaths, representing a decrease in malaria cases and deaths rates of 18 percent and 28 percent since 2010. Of those deaths, 403,000 (approximately 93 percent) were in the WHO African Region.

Bruce E. Johansen

PHYSIOGRAPHY OF
NORTH AMERICA

The North American region covers much of the Western Hemisphere, ranging about 5,000 miles (8,000 km.) north and south—from southern Mexico and the Caribbean to the near-polar lands of Canada and Greenland—and about 9,300 miles (15,000 km.) east and west—from the Aleutian Islands of Alaska to the eastern coast of Greenland. The majority of this landmass is situated within an area that is politically divided into Canada, the United States, and Mexico. It is naturally bounded by the Pacific, Arctic, and Atlantic oceans.

The southernmost country, Mexico, lies only 14 degrees north of the equator, and northernmost Greenland is 84 degrees north latitude. Longitudinally, the region spans about one-quarter of the globe, from the easternmost tip of Greenland, at longitude 13 degrees west (close to the longitude of the western coast of Ireland in Europe), to Attu Is-

Wonder Lake and Denali. (AlbertHerring)

PHYSICAL GEOGRAPHY OF NORTH AMERICA

land, the westernmost island in Alaska's Aleutian chain, at longitude 173 degrees east (across the International Date Line).

Volcanism and Diastrophism

Physiography concerns the character of land surfaces, from the gently rolling plains and lower hills to mountains and more spectacular scenery. However, discussions of the physiography of a continent or a region often begin with the higher and more rugged features because they are more noticeable as defining characteristics, and they provide natural boundaries to set one physiographic region apart from another. These dramatic, mountainous landscapes are created through one of two physical processes that deform Earth's surface: volcanism and diastrophism.

Volcanism is a process involving molten rock material that cools to form volcanic landforms. The term is derived from Vulcan, the Roman god of fire. Volcanic landforms are created when magma (molten rock) moves within Earth's crust and comes out onto the surface as lava (surface magma), or buckles and warps overlying rock without exiting the surface. All landforms created in this fashion are composed of igneous rock—rock made from magma or lava that cools and turns solid.

Surface volcanic features—volcanoes, cinder cones, spatter cones, lava flows—are immediately visible as landforms. The Cascade Mountain range, which runs from northern California to southern British Columbia in Canada, contains many volcanic peaks created from surface volcanic activity. Although internal igneous rocks form under the surface of the crust, they may become visible at the surface when the near-surface igneous masses are exposed, as overlying bedrock is weathered and eroded away. The core rock of several North American mountain ranges is igneous.

The Black Hills of South Dakota are a localized area of mountains produced from such internal volcanism; the Sierra Nevada in California and several of the ranges in the Rocky Mountain chain were pushed up and given surface relief by this type of in-

---

Volcanic National Parks

Because volcanic landforms are so dramatic, many have been preserved within national parks and national monuments. Mexico's Iztaccíhuatl and Popocatépetl (Izta-Popo) National Park, located southeast of Mexico City, contains two volcanoes over 17,000 feet (5,300 meters) high. Popocatépetl is one of Mexico's most active volcanos, producing periodic explosions resulting in evacuations of nearby towns and restricted access to the park. Mount Saint Helens, the last volcano to erupt in the continental United States in the twentieth century, is designated by the U.S. National Park Service a volcanic monument. However, many other volcanoes and volcanic features are preserved within U.S. park sites.

NATIONAL PARKS AND MONUMENTS AT EXTINCT VOLCANOES

Park or monument	State
Aniakchak National Monument	Alaska
Bandelier National Monument	New Mexico
Capulin Volcano National Monument	New Mexico
Crater Lake National Park	Oregon
Denali National Park	Alaska
Haleakala National Park	Hawaii
Lassen Volcanic National Park	California
Marianas Trench Marine National Monument	Guam, N.M.I.
Mount Rainier National Park	Washington
Newberry National Volcanic National Monument	Oregon
North Cascades National Park	Washington
Pinnacles National Park	California
Rio Grande del Norte National Monument	New Mexico

NATIONAL PARKS AND MONUMENTS AT AREAS
OF CINDER CONES, LAVA FLOWS, VOLCANIC NECKS,
OR GEOTHERMAL HOT SPRINGS AND GEYSERS

Park or monument	State
Chiricahua National Monument	Arizona
Craters of the Moon National Monument	Idaho
Devils Postpile National Monument	California
Devil's Tower National Monument	Wyoming
El Malpais National Monument	New Mexico
Lava Beds National Monument	California
National Park of American Samoa	American Samoa
Petroglyph National Monument	New Mexico
Sunset Crater Volcano National Monument	Arizona
Wrangell-St. Elias National Park	Alaska
Yellowstone National Park	Wyoming

---

trusive volcanic activity. Many volcanic landforms are geologically recent features. Still today, active volcanoes continue to change form and spew lava.

Diastrophism

Diastrophism involves the folding, faulting, tilting, warping, and uplifting of Earth's crust. The resulting deformations produce mountains, hills, ridges, and valleys. Death Valley, the lowest point of elevation in North America, was created when deep faults (cracks in Earth's crust) allowed a large block of the crust to sink downward. Most of Nevada is covered by a series of parallel, fault-block mountains produced when a similar cracking of the surface throughout the region induced widespread subsidence and uplift of large blocks of crust.

The ranges of the Appalachian Mountain chain in eastern North America were produced when adjacent blocks of continental crust were pushed together. This compression folded and buckled the surface into a rugged landscape, which has since been weathered and eroded into a series of ridges and valleys that comform to the different rock beds (resistant or more easily eroded) that make up the Appalachians.

A number of ranges in the Rocky Mountain chain were created or altered through diastrophism. Many individual ridges are the product of upturned and tilted rock beds. The ridges stand high above adjacent valleys because they are made of rock that is more resistant to erosion and has not worn down as quickly as the more easily eroded rock of the valleys.

Landforms and Physiographic Regions

North America is one of the most diverse continents in terms of physiography, resources, climate, and vegetation. Although the total area of North America is similar to that of Africa, its variety of landforms and environments is much greater because of the more diverse geological history and the greater expanse of latitudes covered by this region. North America contains high mountains, low mountains and rolling hills, canyons and plateaus, vast plains, continental and alpine glaciers, and great rivers and lakes. This diversity of physiography across a broad swath of latitudes has produced temperate-zone rain forests, subtropical forests, conifer and deciduous forests, grasslands, deserts, and Arctic and alpine tundra. North America can be regionally divided into six distinct physiographic regions.

Western Highlands

The Western Highlands contain the highest mountain ranges and most spectacular mountain scenery in North America. This region rises from the flatness of the interior plains to areas of steep slopes and rugged terrain; it is the source of water for most

Annapolis Rock Overlook, along the Appalachian Trail in South Mountain State Park, Maryland. (Patorjk)

of the west's rivers. It includes the highest slopes of the continental divide; the mountains of the Far West (in California, Oregon, Washington, and British Colombia); and the rugged, irregular mountains and plateaus of the intermountain region (the Columbia Plateau, Colorado Plateau, and the Basin and Range country of Nevada).

The Western Highlands spans the length of the western margin of the continent, from northern Alaska to central Mexico. The principal mountain ranges of this large highland region include the Sierra Nevada of California; the Great Basin Ranges of Nevada and western Utah; the Cascade Mountains of California, Oregon, Washington, and British Columbia; the Brooks Range of Alaska; and the Rocky Mountains, which run from central Alaska to Mexico.

The Rocky Mountain chain comprises several individual mountain ranges that intertwine along this western spine of North America. Ranges include the Alaska, St. Elias, and Wrangell mountains of Alaska; Mackenzie and Continental ranges of Yukon, British Columbia, and Alberta; Bitterroot and Flathead range and Salmon River Mountains of Idaho and Montana; the Tetons and Wind River range of Wyoming; the Wasatch and Uinta mountains of Utah; the Sawatch and San Juan Mountains of Colorado; the Sangre de Cristo Mountains of Colorado and New Mexico; the Sacramento Mountains of New Mexico; the Davis Mountains of western Texas; and the Sierra Madre Oriental and Sierra Madre Occidental of Mexico.

Because of the spectacular, beautiful appearance of the Western Highland landscapes, because lands here remained unsettled for so long and because much of it is owned by the governments of the United States, Canada, or Mexico, many scenic areas have been set aside as national parks and monu-

Valley of the Ten Peaks and Moraine Lake, Banff National Park, Canada. (Gorgo)

Fall Break, Buffalo National River, Arkansas. (OakleyOriginals)

ments. Some of the largest and most famous national parks in all three of these countries are found within the Western Highlands physiographic province.

Appalachian Highlands

This physiographic province includes the ranges and plateaus of eastern North America, a highland region trending northeast-southwest across much of southeastern Canada and the eastern United States. The Appalachian Highlands region manifests itself as a highly folded, ridge-and-valley landscape. These mountains are considerably lower in elevation than the Rocky Mountains, largely because they formed long before the Rocky Mountains and have been greatly worn down through millions of years of weathering and erosion.

Individual mountain ranges that collectively make up the Appalachian chain include the Long Range Mountains of Newfoundland, Laurentian Mountains of Quebec, White Mountains of New Hampshire, Green Mountains of Vermont, Adirondack and Catskill mountains of New York, Allegheny Mountains of Pennsylvania and West Virginia, Blue Ridge Mountains of Virginia, Great Smoky Mountains of Tennessee and North Carolina, Ozark Mountains of Missouri and Arkansas, and the Ouachita Mountains of Arkansas and southeastern Oklahoma.

The Ozarks and Ouachitas are separated from the principal Appalachian chain by more than 300 miles (480 km.) of flat terrain across the Mississippi River valley, but these latter two ranges are geologically, physiographically, and even culturally part of the Appalachian realm. The mountain ranges are punctuated here and there by broad, flat-to-undulating upland regions known as plateaus, such as the Allegheny Plateau in western New York and

Pennsylvania, the Cumberland Plateau in Tennessee, and the Ozark Plateau in Missouri.

Gulf and Atlantic Coastal Plains

Descending from the Appalachian Mountains toward the Atlantic Ocean, the landscape gradually slopes toward the gently rolling, low platform of the Piedmont region and finally levels off to a flat coastal plain nearer the ocean. This entire region is underlain by relatively unconsolidated marine sediments that extend out to sea to constitute the continental shelf. This low-lying, sandy, forested region of eastern North America (sometimes called the flatwoods) runs from the mid-Atlantic states to coastal Florida. It is crossed in several places by a winding maze of creeks and broadly meandering rivers that end near the ocean in bays, estuaries, and swampy wetlands.

The largest and most famous of these swamps is the Dismal Swamp along the Virginia-North Carolina border, and the Okefenokee Swamp in southern Georgia-northern Florida. Low-lying,

sandy islands lie just off shore along most of the coastline. Many of these islands are quite long and form barriers to the sea, especially along the northern Florida and North Carolina coasts. Some barrier islands are wide enough for urban development, but most are narrow and remain lightly inhabited or uninhabited.

On Florida's western coast, similar physiography is seen along the Gulf of Mexico coastal plain, stretching from western Florida to eastern Mexico. There are barrier islands, flat, sandy plains, and poorly drained rivers that end in wetlands. The largest swamp along the Gulf coastal plain is in the Mississippi River delta region, the so-called "bayou" region of Louisiana. The southern Atlantic and Gulf coastal plains are subtropical in climate and vegetation.

Central Plains

Between the highlands of the Rocky Mountains and the highlands of the Appalachian Mountains lies the continental interior; a broad, mostly even land-

Lake Drummond at Great Dismal Swamp National Wildlife Refuge in Virginia. (Rebecca Wynn/USFWS)

scape in the United States and Canada that occupies roughly one-fourth of the continental landmass. This physical province is often thought of as flat and featureless, yet most of the surface here is gently rolling, frequently dissected by streams and rivers. Wind and water have etched the plains to create eroded hills, such as the central Texas Hill Country, and "badlands" in parts of the Dakotas, Kansas, and Oklahoma.

The continent's largest area of sand dunes—a maze of dunes and ridges that rises to several hundred feet (more than 100 meters)—lies in a huge, rolling, grassland-covered region known as the Sand Hills, which covers almost one-third of Nebraska. High hills and low mountains interrupt the sloping plains in southwestern Oklahoma (the Wichita Mountains) and southwestern South Dakota (the Black Hills). The Black Hills are mountains, rising to more than 7,000 feet (2,130 meters).

The western edge of the Central Plains, where the Front Range of the Rockies abruptly rises, is at an elevation of 4,000 to 5,000 feet (1,215 to 1,520 meters). From here, the land surface gradually descends to the east at an average slope of 10 feet per mile (1.9 meters per km.). Much of the land surface in this physiographic province is composed of unconsolidated, or recently consolidated, materials transported and spread from the Rocky Mountains by the east-flowing rivers. The northern portion of this physical province has been scraped by numerous glacial advances, and continental glacial sediments—sand and silt—cover much of the northern plains. The North American prairie, the native grassland region of the continental interior, covers most of this physiographic region of the Central Plains; scattered trees and forests are more common in the eastern portion of this province.

Canadian Shield

The eastern half of Canada, most of Greenland, and a portion of the north central United States is part of this largest physiographic province in North America: an area of ancient crystalline rock covering 1.1 million square miles (2.8 million sq. km.). A "shield" is a geological term referring to a large, stable block of Earth's crust that has been unaffected

---

HIGHEST AND LOWEST

The highest average elevations in North America run along the western side of the continent, from Alaska to Mexico. The lower elevations are found along the coastal plains and in the center of the continent. The highest mountain in North America is Denali at 20,320 feet (6,194 meters), located in the Alaska Range. The highest peaks in Canada and the continental United States are also found in the western highlands. Mexico's highest peak is a volcano in the eastern part of the country, Pico de Orizaba (Citlaltéptl), which rises to more than 18,700 feet (5,700 meters). Two other massive volcanoes near Orizaba exceed 17,000 feet (5,300 meters). Although these volcanoes lie only 18 degrees north of the equator, they are capped by glacial ice throughout the year because of cold temperatures at such extreme altitudes.

The lowest point of land in North America is in the western highlands region, surrounded by high mountains: southeastern California's Death Valley, at an altitude of 284 feet (87 meters) below sea level. This is a deep graben valley, produced when parallel faults—or fractures in the earth's crust—allowed this block of crust to settle downward between the mountains of the Panamint Range and Amargosa Range.

---

by mountain-building for a long time. These ancient masses of crust typically form the core or central nucleus of each of the continents.

The Canadian Shield, sometimes called the Laurentian Shield, exemplifies this type of stable rock mass at the core on the North American continent. It is composed of Precambrian-age crystalline rock, 1 billion to 3 billion years old, and is covered with forest and tundra. The shield mass is twice as large as appears at the surface; to the west and south of the exposed shield, the rock is overlain by more recent geologic deposits. Therefore, the shield is found only at some depth beneath the surface in western Canada and in the eastern and central United States. Long ago, wide areas of this rock mass were severely compressed and contorted, and in these disturbed zones a large variety of minerals has been found. This is the most important region of non-engery minerals in Canada, producing sig-

Exposed rock part of the Canadian shield is common in northern Manitoba, Canada. (Ethan Sahagun)

nificant quantities of gold, nickel, copper, uranium, and iron ore.

The present surface of the shield bears little resemblance to the ancient disturbances of rock compression, because the surface has been repeatedly eroded as glaciers have expanded and retreated with each ice age. The result is a gently rolling surface covered with glacially scoured, rounded hills, a thin or nonexistent soil cover, a chaotic surface-water drainage system, swamps and muskegs, and thousands of lakes, large and small, created when water filled in the numerous depressions that were gouged out by glacial erosion after the continental glacier melted. Some of these lakes have disappeared over time, leaving behind clay-filled beds and offering some fertile ground in the otherwise soil-bare region of the shield. Along the eastern margins of the shield are several mountain ranges and hills that reach elevations exceeding 4,000 feet (1,200 meters), but elsewhere across the region, the topography is more gently undulating.

Dale R. Lightfoot

HYDROLOGY OF NORTH AMERICA

Hydrology is the study of water's properties, distribution, and circulation. Any study of North America's hydrology must deal with all the water on the continent, including the rivers, lakes, glaciers, wetlands, and groundwater. Hydrology has had a great impact on the human settlement and economics of this continent.

Excluding water locked in the Antarctic and Greenland ice caps, Canada holds about 20 percent of the world's freshwater. Almost 8 percent of Canada's surface (290,000 square miles/750,000 sq. km.) is covered by lakes and rivers. By contrast, lakes and rivers cover slightly more than 2 percent of the United States and an even smaller percentage of Mexico. The differences in coverage are largely due to each nation's geographical location and the resulting differences in precipitation. Another important factor is the degree of development and change that the hydrology of the land has undergone with human settlement.

Water Cycle

Water continually moves through a cycle called the hydrologic cycle. It first reaches the ground surface

Confluence of the Wisconsin and Mississippi Rivers, viewed from Wyalusing State Park in Wisconsin. (Cindy)

MAJOR WATERSHEDS OF NORTH AMERICA

The Tule Valley watershed and the House Range (Notch Peak) are part of the Great Basin's Great Salt Lake hydrologic unit. (Qfl247)

through precipitation. It then either evaporates or transpires back into the air, sinks into the ground, or flows overland as runoff until it collects and eventually empties into lakes or streams. The streams can be tributaries to larger rivers, which continue flowing downhill until the water empties into the oceans. Water from the oceans then evaporates, and some of this moist air blows over the land and precipitates. In this way, North America's freshwater is continually renewed. However, the amount of precipitation and water flowing in streams varies by season and from year to year. On average, the continental United States receives approximately 4 million cubic miles (16 million cubic km.) of precipitation each year, which eventually makes its way back through the water cycle.

Hydrology and Topography

North America can be divided into water drainage basins, also called watersheds. Drainage basins are the total land area that a river and its tributaries drain. The largest drainage basin in North America is the Mississippi River system, which drains a land area of 1,250,000 square miles (3,250,000 sq. km.). This is the third-largest drainage basin in the world, after South America's Amazon River and Central Africa's Congo River.

Areas of high elevation, such as mountain ranges, separate drainage basins from each other and are called drainage divides. In North America, the continental divide formed by the Rocky Mountains helps determine which way water generally flows. Near the East Coast, the Appalachian Range also acts as a drainage divide. Water that falls on the western side of the Rocky Mountains eventually empties into the Pacific Ocean. Water that falls between the Rocky Mountains and Appalachians and south of the Great Lakes eventually reaches the Gulf of Mexico. Water that falls on the eastern side of the Appalachians or in the Great Lakes drainage basin generally flows toward the East Coast and into the Atlantic Ocean. Water from a large part of central Canada flows into the Hudson Bay.

Water in north central Canada generally empties into the Arctic Ocean. In Mexico, water on the western side of the Continental Divide flows into the Pacific Ocean or Gulf of California. Water on the eastern side eventually flows into the Gulf of Mexico.

Rivers

The Mississippi River, one of the longest in North America, has tributaries from thirty-two states and two Canadian provinces. Its drainage basin covers about 40 percent of the continental United States as well as parts of Canada. The river flows for 2,348 miles (3,778 km.) and eventually empties into the Gulf of Mexico. It has an average flow rate of approximately 4.8 million gallons (18,200 cubic meters) per second, the largest in North America. The Mississippi River is a vital part of the United States' economy, because it provides water for irrigation and industry in ten states as well as an excellent and reliable means of transporting goods from the northern and central states to the rest of the world.

River flow rates in North America vary dramatically with season, rainfall, and snowmelt. For example, Canada's Saskatchewan River has a maximum

flow rate that is almost sixty times larger than its smallest recorded flow rate. The river with the largest average flow rate in Canada is the St. Lawrence River, which flows from the Great Lakes. The longest river in Canada is the Mackenzie River, which empties into the Arctic Ocean after flowing through the Northwest Territories. Its drainage basin is the second largest in North America and the twelfth largest in the world (580,000 square miles/ 1,500,000 sq. km.).

Rivers in Mexico flow generally eastward into the Gulf of Mexico or westward into the Pacific Ocean or the Gulf of California. The Rio Grande flows south and southeast for 1,800 miles (2,900 km.) from the San Juan Mountains in southern Colorado, becoming part of the border between Mexico and the United States. The Rio Grijalva and Rio Usumacinta have the largest flow rates (3.70 trillion cubic feet/105 billion cubic meters) per year of all the rivers in Mexico. Together, they account for almost one-third of the total flow of all Mexico's rivers combined. They are each about 440 miles (700 km.) long and flow into the Gulf of Mexico.

Lakes

The five Great Lakes—Erie, Huron, Michigan, Ontario, and Superior—lie in the central region of North America, mostly along the border between Canada and the United States. Lake Michigan is the only one of the Great Lakes entirely within the United States. These lakes were instrumental in enabling early European colonists to penetrate rapidly into the center of the continent. In 2017, approximately 30 percent of Canadians and 10 percent of U.S. citizens lived in the region of the Great Lakes—more than 44.5 million people in all. The Great Lakes region is important to 25 percent of Canada's agriculture and 7 percent of the United States' agriculture. Industry in Canada and the United States also is heavily dependent on the Great Lakes region.

Terra MODIS image of the Great Lakes, January 27, 2005, showing ice beginning to build up around the shores of each of the lakes, with snow on the ground. (NASA)

Great egret at Chapala Lake, Mexico's largest freshwater lake. (Sambit 1982)

These five lakes extend more than 850 miles (1,370 km.) from east to west. They hold roughly 5,500 cubic miles (23,000 cubic km.) of water, which is about one-quarter of all the freshwater held in the world's lakes. The only lake system equal to this volume is Lake Baikal in Russia. The Great Lakes together cover 94,950 square miles (246,000 sq. km.) and are the largest inland bodies of water in the world. Individually, three of the six largest inland water bodies by surface area are Lakes Superior, Huron, and Michigan. Lake Superior is the largest of the five Great Lakes, in both area and depth. It could easily hold the water of the other four lakes combined. The drainage basin for the Great Lakes totals about 298,500 square miles (773,000 sq. km.).

Canada contains more lake surface area than any other country in the world, including more than 500 lakes larger than 40 square miles (100 sq. km.). Other large lake systems in Canada include the Great Bear Lake (12,275 square miles/31,792 sq. km.) and Great Slave Lake (11,170 square miles/28,930 sq. km.), both located in the Northwest Territories. By area, these two lakes are the ninth- and tenth-largest inland surface water bodies in the world. Lake Winnipeg in Manitoba Province

is the twelfth-largest inland surface water body in the world by area, but with a maximum depth of only 60 feet (18 meters), its total water volume is relatively smaller.

Mexico has few large lakes because of its dry climate. Its largest lake is Lake Chapala, between the states of Jalisco and Michoacán. It is 50 miles (80 km.) long and has a surface area of 420 square miles (1,080 sq. km.).

Not all lakes in North America contain freshwater. The Great Salt Lake in Utah is even saltier than the oceans. It covers an area of about 1,700 square miles (4,400 sq. km.), just a fraction of its size 15,000 years ago. Back then, the Great Salt Lake was part of a lake fourteen times larger, one which geologists call Lake Bonneville. Over time, evaporation and a drier climate reduced the size of the lake and increased the concentration of salts, resulting in today's Great Salt Lake.

In contrast to the saline water of the Great Salt Lake, a large lake with some of the world's purest water is Lake Tahoe, on the border between California and Nevada in the Sierra Nevada range. Surrounded by tall mountains, Tahoe is a young lake fed by runoff and snowmelt. Its water is so clean that it needs virtually no treatment to meet drinking water standards and is so clear that a person can see to depths of more than 100 feet (30 meters) from the surface.

Groundwater

An enormous amount of freshwater, called groundwater, lies beneath Earth's surface. There is about twice as much groundwater in North America as the water in all its lakes and rivers. In the United States, approximately 38 percent of the population depends upon groundwater for the majority of its drinking water. In many areas, groundwater is pumped up to the surface through wells for use. But most groundwater flows slowly into lakes and rivers. In this way, groundwater can recharge rivers and keep them flowing even during droughts. Nearly all the people of Florida get their drinking water from groundwater. One of the most important aquifers in that state is the Biscayne Aquifer, which supplies water to most of southern Florida.

The High Plains Aquifer, also called the Ogallala Aquifer, spans the midwestern and southwestern states of South Dakota, Wyoming, Nebraska, Colorado, Kansas, Oklahoma, New Mexico, and Texas. This aquifer once held more freshwater than all the lakes and rivers of the world combined. It is especially important to agriculture, because it provides almost one-third of the groundwater used for agriculture in the United States. In 2020, the rate of water recharging this important aquifer was much less than the rate at which it was being pumped out, raising worries that this aquifer is gradually being depleted.

Overpumping of groundwater has caused problems in other parts of North America. In California's San Joaquin Valley, overpumping to support the rich agriculture in the region has caused the land in some areas to sink by as much as 28 feet (8.5 meters) since the 1920s. In the Houston-Galveston area of Texas, the land has sunk more than 4 feet (1.2 meters), making the area more susceptible to seasonal flooding. Another dramatic case is Mexico City, which relies almost entirely upon its aquifer for its water supply. The growth in the city's population

Aerial photo of the California Aqueduct at the Interstate 205 crossing, just east of Interstate 580 junction. (Ikluft)

Sawgrass prairie in Everglades National Park. (Moni3)

has resulted in rapid depletion of this water supply resource at the same time that industrial pollution has lowered the water's quality.

Wetlands

Swamps, marshes, bogs, and fens make up the wetlands, which are an important part of North America's hydrology. They not only are important ecologically but also help purify water, recharge groundwater aquifers, and act as holding areas for flood waters to reduce flood damage. The extent of wetlands in North America has changed dramatically. Before colonization by Europeans, the United States had approximately 350,000 square miles (900,000 sq. km.) of wetlands, but it had less than half this amount in 2020. Alaska contains most of the country's wetlands. The Everglades in Florida is another large wetlands area.

Canada has not experienced the same magnitude of wetlands reduction, except in more populated areas such as near the Great Lakes. Canada contains approximately one-fourth of all the world's wetlands, covering about 13 percent of Canada's surface area. The wetlands along the shores of the Great Lakes have been reduced by more than 50 percent since the nineteenth century, as towns and agriculture developed along their shores.

Glaciers and Ice

In 2020, glaciers and ice sheets covered about 10 percent of the world's land surface and held about 68.7 percent of the world's freshwater. Most of this is in Antarctica and Greenland. In the northern latitudes of North America, much of the water is frozen. Greenland is covered by 660,000 square miles (1,700,000 sq. km.) of ice. After Antarctica and Greenland, Canada has the largest ice-covered area (57.660 square miles/150,000 sq. km.). The United States has 29,000 square miles (75,000 sq. km.) of area covered by ice, most of which is in Alaska. Ice often has existed in an area for a long time, as opposed to the relatively short time water spends in rivers and lakes. Ice in some of Canada's glaciers is more than 100,000 years old.

Human Impact on North American Hydrology

Approximately 7 percent of the area of the United States is covered by bodies of water; however, the vast majority of the human population lives near these water bodies. Bodies of water provide ready sources of accessible water for drinking, industry, and irrigation. People have sought to control bodies of water to benefit themselves. When a dam is built, a special kind of lake, called a reservoir, is created behind it. The reservoir can be a reliable source of water during times of drought. Dams, together with levees, also are useful in controlling floods and thereby preventing damage to communities. The downside of this type of control of hydrology is that it can affect the amount of water that flows as well as the water's usefulness.

One example is the Colorado River. At about 1,400 miles (2,330 km.) in length, it is the longest river west of the Rocky Mountains. The Colorado has many large dams along it, including the Glen Canyon and Hoover dams. As a result of the enormous reservoirs behind these dams, large amounts of water are lost to evaporation. It is estimated that up to 10 percent of the water in the Colorado River that comes into Lake Powell, the reservoir behind the Glen Canyon Dam, evaporates. Losses from this and other reservoirs increase the salinity of the water. Water is also taken from the river for irrigation of crops. By the time the Colorado River empties into the Gulf of California, the flow has been drastically reduced, and the water is often too saline to be used for irrigation, industry, or drinking water.

All the major rivers in the continental United States have been dammed, diverted, or drained, except the Yellowstone River, which flows freely for its entire 671-mile (1,080-km.) length. It rises in Wyoming and flows through Montana before emptying into the Missouri River.

Thomas R. MacDonald

DISCUSSION QUESTIONS: PHYSIOGRAPHY AND HYDROLOGY OF NORTH AMERICA

Q1. What mountains constitute the Western Highlands? When did they come into existence? How do their characteristics differ as they pass from Canada to the United States to Mexico? What is the continental divide?

Q2. What is the Canadian Shield? How do geographers describe the shield's origin? How does the shield's surface appearance belie its true size?

Q3. How have the geographical features of North America helped to define the location of the continent's most important watersheds? How does the flow of rivers in Canada differ from that in the United States? How can Mexico's rivers be best described?

PHYSIOGRAPHY OF THE CARIBBEAN

The Caribbean region comprises the Caribbean Sea and the many islands that surround it. The region's east-west extent is 1,800 miles (2,900 km.) from the Yucatán Peninsula to Barbados. Its north-south extent is 760 miles (1,200 km.) from the Bahamas to South America. Island groups include the Greater Antilles (Cuba, Hispaniola, Jamaica, and Puerto Rico), the Lesser Antilles, and numerous other small islands.

Geologic History

The oldest rocks of the Caribbean are Permian (200 million years old), but the oldest landscapes date only to the Cretaceous period (70 million years old). Two major mountain-building episodes created the Caribbean landscapes of today. In the first, the North American and South American plates separated and moved westward. A part of the Pacific plate was forced northeastward between the larger plates, becoming the Caribbean plate. This created a subduction zone that led to volcanism and diastrophism in the Greater Antilles. This episode was followed by a period of tectonic inactivity, during which the lands eroded and thick limestone deposits were laid on the seafloor.

The second tectonic episode began in the Pliocene epoch and continues today. The Caribbean plate is now moving eastward relative to surround-

Pico Duarte is the highest point in the Dominican Republic, the Island of Hispaniola, and the entire Caribbean. (Adrian Michael)

ing plates. Thus, the Greater Antilles are along a lateral fault zone, whereas the Lesser Antilles have formed along a new subduction zone. The Atlantic plate is moving westward 1 to 2 inches (3 to 4 centimeters) per year, being subducted beneath the Caribbean plate. Volcanic eruptions are the result.

Besides volcanism, the region suffers also from earthquakes around its periphery, especially along the lateral fault in the Greater Antilles and the subduction zone to the east. Violent earthquakes have occurred, destroying the Jamaican cities of Port Royal (1692) and Kingston (1902) and Haiti's Port-au-Prince (2010).

Physiographic Regions
The landforms of the Caribbean region are the result of two tectonic episodes and periods of erosion and deposition. Three regions are identified: the Greater Antilles; the High Islands of the Lesser Antilles; and the Low Islands of the Lesser Antilles, the Bahamas, and offshore of South America.

The Greater Antilles, formed early in the geologic history of the Caribbean region, were produced by combinations of volcanism, folding, faulting, and uplift. Evidence of these events is found in the varied rocks of the islands, which include lavas, granites, metamorphic rocks, and limestone. Currently, this region has been relatively quiet, dominated by denudation due to wave and stream erosion and solution weathering.

The highest mountains in the Caribbean region are in the Greater Antilles. They include Pico Duarte (10,417 feet/3,175 meters) in the Domini-

Plymouth, the former capital city and major port of Montserrat, on 12 July 1997, after pyroclastic flows burned much of what was not covered in ash. (USGS)

> ### VOLCANIC ERUPTIONS ON MONTSERRAT
> The Caribbean island of Montserrat became newsworthy in 1995 when a volcano in the Soufriére Hills began to erupt. This stratovolcano, which had been dormant since the 17th century, began a series of violent eruptions after years of seismic activity. The eruptions destroyed the capital city of Plymouth, burying it under more than 12 meters (39 ft) of mud. Residents of the island were encouraged to move to the north, away from the eruptions, or to leave the island altogether. Following a series of nearly 80 magmatic explosions occurring between August and October of 1997, most of the island's eleven thousand inhabitants had left, and the southern two-thirds of the island was declared off limits, as the volcanic activity continued. A partial evaluation of the island was mandated following a 2007 eruption. The volcano is currently monitored by the Montserrat Volcano Observatory.

can Republic, Blue Mountain of Jamaica (7,402 feet/2,256 meters), Cerro de Punta in Puerto Rico (4,390 feet/1,338 meters), and Pico Turquino in the Sierra Maestra of Cuba (6,578 feet/2,005 meters). Land elevations extend below sea level on Hispaniola's Lago Enriquillo (at -114 feet/-35 meters). The mountains owe their existence to the uplift of folded and faulted blocks whose rise ended in the late Tertiary period.

As a result of the combination of high relief and high annual rainfall, denudation rates are high and erosion features are prominent. Rivers have carved deep, steep-sided valleys along which landslides frequently occur. Human removal of native forests has accelerated denudation. Streams deposit their loads on coastal lowlands or in the sea where the material is reworked by waves.

Limestone plateaus containing karst landforms are also common. Limestone, rich in soluble calcium carbonate, is readily weathered into karst features, such as cones and sinkholes. Cone karst is dominated by steep-sided hills, also called *mogotes* or towers, which are remnants of limestone rock after millions of years of solution weathering. The Cockpit Country of Jamaica is a good example of cone karst. It is so rugged that few roads exist there. Sinkholes are common, ranging from small ponds to huge, steep-sided collapse sinks such as the vertical holes up to 200 feet (60 meters) deep at the Rio Camuy caves in Puerto Rico. Karst terrain is associated with underground drainage, which forms caves and subsurface stream passages. Surface streams are usually absent here.

The coastlines of the Greater Antilles are mostly rugged, with cliffs cut by wave erosion. In areas of low relief and low wave energy, extensive beaches and coastal mangrove swamps are found. Coral reefs occur offshore, especially on the windward sides of islands. Reefs are sensitive to sediment pollution, so they seldom occur near stream mouths.

Where streams occur, they tend to be short and steep. Perennial streams, common on the rainy windward sides of these islands, have more even flow regimes. Intermittent streams, typical of the drier lee slopes, are subject to flash flooding, often destroying bridges, roads, and other property.

The islands of the Lesser Antilles are made up of two island arcs, the Inner Arc and the Outer Arc. High islands of the Inner Arc include Saba and St. Kitts in the north, through Nevis, Montserrat, western Guadeloupe, Dominica, Martinique, St. Lucia, St. Vincent, and Grenada in the south. All these islands were formed from relatively recent volcanic eruptions and contain considerable local relief. There have been many volcanic eruptions in these islands in recorded history.

Effects of Volcanoes

Because the primary island landforms are volcanoes, slopes are steep and rugged. On older volcanoes, deep ravines have formed and landslides are common during heavy rains. Eroded volcanoes soon lose their conical shapes. Relatively little flat land is available for agriculture, limiting population densities throughout history.

Other landforms related to volcanoes are steep rock towers, such as those of St. Lucia and similar features on Martinique, formed when thick lava plugged volcanic craters. After millions of years of erosion of the weaker surrounding rocks, the plugs have been exposed and stand in relief. Also related to volcanism are the yellow sulfur springs (*soufrières* in French), found in St. Lucia, Dominica, and Montserrat. Toxic fumes from these springs have killed most of the surrounding vegetation.

Volcanoes of the Lesser Antilles are composite or stratovolcanoes, large mountains built of layers of ash and lava. Eruptions tend to be explosive, raining layers of ash over everything. Hot clouds of ash, called *nuées ardentes*, accompany many eruptions, flowing downhill at great speeds, killing and destroying everything in their paths. The most recent

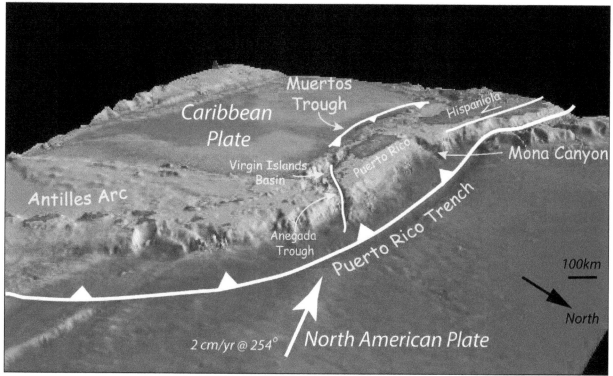

Bathymetry of the northeast corner of the Caribbean Plate showing the major faults and plate boundaries. (NOAA)

notable one erupted in 1995, when Montserrat's long-dormant Soufriere Hills volcano erupted explosively. A series of later eruptions in 1996 and 1997 destroyed the capital city of Plymouth. In 2010, the island's airport was destroyed by still more volcanic activity. By 2020, although the volcano had become less active, most of Montserrat has been rendered uninhabitable.

The high volcanic islands induce substantial rainfall, which runs off in steep, short streams. The smaller high islands tend to lack lee rain shadows, and the uplands are cloaked in lush rain forests.

Effects of Low Relief

The many low islands of the Caribbean region lack significant relief, not rising more than 660 feet (200 meters) above sea level. Notable are the Bahamas, the Outer Arc of the Lesser Antilles, and Caribbean islands offshore of South America. Although such islands have varying geologic histories, they all have low relief and a dominance of surface limestone. One significant result of low relief is low annual rainfall. These islands receive less rainfall because they lack the extensive surfaces necessary to produce afternoon convection, and they lack the relief necessary to produce mountain rainfall. They depend on stray thundershowers, mostly from easterly waves and occasional hurricanes. For example, Inagua in the Bahamas receives only 27 inches (685 millimeters) of rain per year, and Bonaire, off the coast of South America, receives only 20 inches (510 millimeters).

Water

This comparative scarcity of rainfall means that freshwater at the surface is rare. If it exists at all, fresh groundwater occurs in thin subsurface layers called lenses floating on top of the denser salt water. Overuse of such groundwater results in loss of the resource, because freshwater recharge is very slow. Because of the lack of rainfall, the low islands rarely have soils good enough for agriculture, and they lack the tropical forests of larger islands. These difficult environments were largely avoided by colonizers, and they still suffer from low development, except in a few cases, such as the Bahamas and the Cayman Islands, where tourism and banking have enabled people to thrive.

P. Gary White

Climatology of North America

Climates of North America are diverse, ranging from extremely dry to extremely wet, from bitterly cold to oppressively hot. Most of North America's habitable regions are located within the Northern Hemisphere midlatitudes, between 23.5 degrees north (the Tropic of Cancer) and approximately 60 degrees north. Midlatitude climates are zones of transition between the warm tropics and the cold polar regions. Pronounced seasonality exists with regard to weather variables as a wide array of individual weather events occurs.

Complicating any discussion of climate is the impact of global warming. Atmospheric scientists are recording year-after-year warmer temperatures for many locales in North America. The weather has turned stormier as well, with many regions experiencing more large-scale weather events than ever before. Efforts to mitigate global warming have had only limited success at best.

North American Weather

North America has some of the most violent weather in the world. On no other continent is there a regular mix of such vastly different air. This is the result of the unique arrangement of mountains, which are aligned roughly north to south, allowing cold and warm air to mix freely. Warm, moist air from the Gulf of Mexico moves northward into the

Fall foliage at Missisquoi National Wildlife Refuge, Vermont. (Ken Sturm/USFWS)

CLIMATIC REGIONS OF NORTH AMERICA

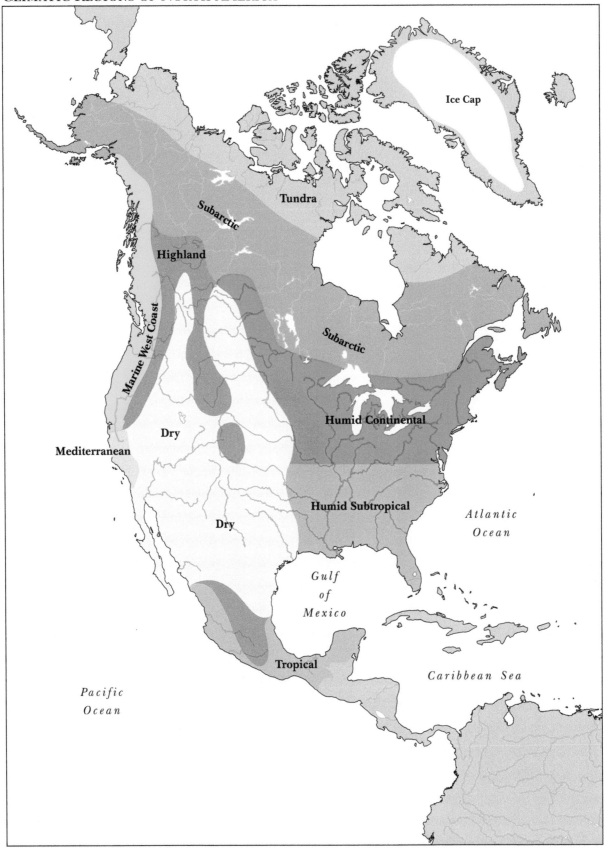

central portion of the continent, where it mixes with cold, dry air emanating from northern Canada, resulting in volatile weather.

Most significant weather is associated with migratory midlatitude, or frontal, cyclones. These storms are the largest on Earth, with diameters between 1,000 and 2,000 miles (1,600 and 3,200 km.). These storm systems have a low-pressure core in which surface air converges and rises, creating clouds and precipitation. From the low-pressure core extends a cold front, with heavy precipitation producing cumulonimbus clouds, and a warm front with stratus clouds, also exist. Warm-front precipitation is normally light and widespread.

Virtually every type of severe weather can be found in relation to frontal cyclones, especially along the cold front. Heavy rain, heavy snow, sleet, freezing rain, hail, high winds, lightning, and tornadoes all can develop, depending upon specific conditions. Because of the frequency and magnitude of these systems, the United States leads the world in the number of tornadoes. Tornadoes usually develop in the large cumulonimbus clouds that accompany a cold front or in advance of a cold front in a squall line. Over 750 tornadoes per year are confirmed in the United States alone. Southern portions of Canada are also subject to tornadoes.

Dry Climates

Dry climates are a feature of the southwestern regions of the continent. Dry climates, which may be classified as subtropical desert, hot desert, subtropical steppe, and hot steppe, occur when net evaporation rates exceed net precipitation totals. In North America, hot desert climates occur only in the southwestern United States and northern Mexico. Wetter steppe climates surround the true desert core, which is largely centered on the state of Arizona.

North America's dry region is induced by a variety of climatic factors acting in concert. First, the region is located at approximately 30 degrees north latitude in an area of subsiding air, as dictated by the general circulation of the atmosphere. Subsiding air causes clear skies because it discourages clouds and precipitation. Second, the desert region is located along the western portion of the continent

near cold ocean currents. The cold California current chills overlying air so it resists rising.

A third factor is the phenomenon known as rain shadow. A rain shadow occurs when warm, moist air traverses a mountain range. As the air ascends the windward side of the mountains, clouds and precipitation occur. Air flow descends over the opposite, or lee, side, where subsidence limits clouds and precipitation. All three factors combine to ensure the predominance of sinking air in the southwestern United States.

The true desert of North America is classified as a subtropical desert, although northern portions may be classified as midlatitude deserts. Although precipitation totals—about 0 to 10 inches (0 to 250 millimeters)—are fairly uniform throughout the year, the most intense precipitation typically occurs during the warm months and is associated with hot, rising air stimulated by high surface temperatures. Precipitation totals increase slightly in association with this summer "monsoon" of the desert southwest.

During other times of the year, rain and snow are generated by migratory frontal cyclones. However, precipitation totals are frequently low, especially during cooler times. Precipitation varies greatly from year to year, since many years may pass with very little precipitation, only to be followed by a year with precipitation far above normal. Monthly or yearly precipitation averages may, therefore, be somewhat misleading and unrepresentative.

The highest natural air temperature ever recorded in the Western Hemisphere—134 degrees Fahrenheit (56.7 degrees Celsius)— was measured at Death Valley, California, on July 10, 1913. However, temperatures in subtropical deserts fluctuate considerably, both daily and seasonally. Daytime summer temperatures can be exceedingly high, while nighttime temperatures can be quite cool because of low humidity levels. Average monthly temperatures can be near 100 degrees Fahrenheit (38 degrees Celsius) for some locations, as daily maximum temperatures frequently climb to 109 degrees Fahrenheit (43 degrees Celsius). Temperatures during the winter months are much cooler, with average monthly temperatures in January and February near 50 degrees Fahrenheit (10 degrees Celsius).

Sonoran Desert terrain near Tucson. (Santryl)

The vast differences between summer and winter result in high annual temperature ranges.

More northerly regions are classified as midlatitude deserts, which are similar to subtropical deserts with regard to precipitation but typically have greater annual temperature ranges than subtropical deserts. While summer temperatures are similar in both regions, winter temperatures are colder in midlatitude deserts than in subtropical deserts.

The true desert of North America is bounded on all sides by the wetter steppe climate transition zone. As in the true desert, annual precipitation varies greatly. However, the annual totals of 10 to 30 inches (250 to 760 millimeters) exceed those of a true desert. Because of higher precipitation totals, steppe grasses proliferate, becoming more widespread, thicker, and taller away from the true desert core. The grasslands ultimately culminate at the boundary of an adjacent climate zone.

Precipitation totals typically reach a maximum during winter for steppe locations on the poleward side of the true desert. Precipitation in these locations is triggered by the passage of migratory midlatitude storm systems. On the equator side of the true desert, precipitation totals maximize during the summer season when thunderstorm activity is greater.

Temperatures in steppe deserts are similar to those in true deserts, except that they vary less by season. Therefore, temperature ranges are smaller than in true deserts. The steppe regions are also reflective of latitude in that poleward locations experience cooler overall temperatures than locations farther equatorward.

Mediterranean Climates

In North America, the Mediterranean climate is found only in a coastal area in the state of California. This climate type is characterized by a distinct summer dry period and a wet winter. The summer dry is induced by the presence of a semipermanent high-pressure cell that dominates over the Pacific Ocean. During the warm season, this Hawaiian High becomes particularly well established and exerts a subsiding force on the atmosphere, which prohibits cloud and precipitation processes. During winter, the high weakens, allowing migratory storms to pass through the region.

The frequency and magnitude of these storm systems is highly variable. Some winters are predominantly dry, while during others, copious precipitation falls. Precipitation generally increases with latitude for this climate type. This is because even during fairly dry winters, northern areas are affected by frontal cyclones passing on the poleward side of the region. Total annual precipitation is approximately 15 inches (380 millimeters).

Temperatures for the region are mild through the cool months and mild to hot during warm months. In general, temperatures are milder near the coast and become more extreme farther inland. Summertime coastal temperatures are typically 68 to 86 degrees Fahrenheit (20 to 30 degrees Celsius), while inland locations can have daily highs near 100 degrees Fahrenheit (38 degrees Celsius). During cool months, inland temperatures can dip below freezing, while coastal locations, moderated by the ocean, remain cool with temperatures in the range of 50 to 61 degrees Fahrenheit (10 to 16 degrees Celsius).

Marine West Coast

North of the Mediterranean climate along the Pacific Coast lies the marine west coast climate region. This region extends along the coastal margins of the continent, from northern California through Alaska. It is typified by mild temperatures through the seasons and a fairly even distribution of precipitation through the months. Temperatures are mild for the relatively high latitude of the region.

The moderation of the nearby ocean and the prevalence of frequent low-level cloud cover associated with migratory frontal cyclones ensure mild temperatures.

Although the number of days of precipitation may be high, annual precipitation totals are low, especially near the coast. Much precipitation is generated by vertically limited stratus clouds, which produce only light drizzle. Inland, in higher elevations, the precipitation is much greater: In some locations, it exceeds 100 inches (2.540 millimeters) per year, similar to tropical rain forest totals. A slight autumn precipitation maximum occurs, and a slight minimum occurs during the summer months. In general, there is a fairly even distribution of precip- itation year round.

Even with the relatively high latitudes of this climate region, extreme cold is rare, as is snowfall in lower elevations. Average temperatures are above freezing throughout the year. Snow typically occurs only in higher elevations, where average temperatures are much lower. Temperature ranges are fairly small, mainly because of the moderating influence of the ocean. Temperature ranges increase farther

A day of big surf in La Jolla, California. (Bengt Nyman)

inland, where the moderating effect of the ocean diminishes. Maximum monthly temperatures rarely exceed 75 degrees Fahrenheit (24 degrees Celsius) as maximum monthly averages are usually between 50 and 55 degrees Fahrenheit (10 and 13 degrees Celsius). Monthly temperature averages during winter are usually about 32 degrees Fahrenheit (0 degrees Celsius), as daily winter minimums rarely dip below -20 degrees Fahrenheit (-29 degrees Celsius).

Humid Subtropical

Humid subtropical climates have relatively mild winters and long, hot summers. Much of this is related to the maritime influence exerted on the area by the warm waters of the Gulf of Mexico and the southwest North Atlantic Ocean. The southeastern portion of the United States falls into this climate classification.

Hurricane Wilma, 2005. (NASA)

HURRICANE DORIAN

In 2019, Hurricane Dorian, an extremely powerful and devastating Category 5 Atlantic hurricane, became the most intense tropical cyclone on record to strike the Bahamas, and one of the most powerful hurricanes recorded in the Atlantic Ocean, with winds peaking at 185 miles (295 km.) per hour. Damage to the islands left most structures flattened and 70,000 people homeless. After it ravaged through the Bahamas, Dorian proceeded up the coasts of the southeastern US and Atlantic Canada, leaving behind considerable damage.

Average cold-month temperatures are typically between 39 and 60 degrees Fahrenheit (4 and 16 degrees Celsius), but freezing temperatures, although unexpected, are not uncommon. The lowest recorded temperature for southeastern Louisiana, for example, was approximately 10 degrees Fahrenheit (-12 degrees Celsius), recorded in 1989. The coldest month in this climate usually is January (approximately 50 degrees Fahrenheit/10 degrees Celsius, on average), the time of most frequent freezes. Extremely cold temperatures are relatively uncommon.

Summer months are normally hot, with average temperatures around 79 to 82 degrees Fahrenheit (26 to 28 degrees Celsius) with rather high relative humidity levels. The average relative humidity on a July morning is typically between 85 and 95 percent. High humidity in conjunction with high temperature produces sultry, oppressive weather with low cooling power. Temperature maximums during the summer months typically approach 95 degrees Fahrenheit (35 degrees Celsius), with temperatures reaching as high as 102 degrees Fahrenheit (39 degrees Celsius).

Precipitation is relatively evenly distributed throughout the year in this area, averaging 40 to 60 inches (1,020 to 1,520 millimeters). Heavy precipitation is fairly common, generally associated with summer convective thunderstorms. Such

Cumulus clouds hover above a yellowish prairie at Badlands National Park, South Dakota. (Wing-Chi Poon)

thunderstorms occur almost daily throughout the summer and early autumn months in the southeastern United States as a result of diurnal heating of the land surface.

Tropical cyclones also add to the summer precipitation. North Atlantic tropical cyclones are most common during the late summer. The tropical cyclone is an important component of the precipitation regime of the humid subtropical climate region during the summer and autumn months and may be looked upon as a benefit to the precipitation regime, especially during early autumn.

The driest months of the year are typically October and November, when a transition occurs between precipitation-forcing mechanisms. Frontal cyclones are a common feature in this region during the winter and spring, as the polar jet stream reaches its closest position to the equator (near 30 degrees north). The frequency of frontal cyclones also increases during these times, ensuring a relatively even distribution of precipitation through the region from month to month.

Humid Continental

Humid continental climates are confined to the central to northeastern portion of the continent. Most of the inhabited portions of the United States and Canada lie within this climate region. It is similar to the humid subtropical climates, except that precipitation totals are less and temperatures are lower.

Precipitation is fairly well-distributed throughout the year, but a slight maximum is perceptible during the summer months. Precipitation is largely controlled by migratory frontal cyclone passages during all seasons except summer. Summer precipitation is predominantly convective, initiated by surface heating, although some summer precipitation stems from frontal passages. These precipitation events are usually associated with cooler air migrating down from northern Canada. Such occurrences spell relief from fairly high summer temperatures. Annual precipitation totals are usually between 20 and 40 inches (510 and 1,020 millimeters).

Summer temperatures range from fairly warm to hot in the continental interior. The continental landmass itself causes extreme seasonal fluctuations, because the regions are far removed from the moderating effects of large bodies of water, which tie up available energy through evaporation processes and heat storage. In summer months, when high solar angles are prevalent, the land heats quickly.

During winter, the opposite occurs: Low solar angles predominate, so there is little surface heating. The absence of a large body of water augments warming or chilling of the surface and the overlying air, depending on the season, so seasonal extremes are typical. Many locations in the humid continental climate region have monthly summertime averages of 81 to 84 degrees Fahrenheit (27 to 29 degrees Celsius), while monthly winter averages range from 30 to 41 degrees Fahrenheit (-1 to +5 degrees Celsius).

Subarctic

In North America, subarctic climates exist north of humid continental climate regions, covering more than half of Alaska and Canada. The region is typically referred to as the boreal forest, because it is dominated by extensive coniferous forest. Summer temperatures are similar to, but cooler than, those of the humid continental climate region. Winter temperatures, however, are much colder than those of the adjacent climate region. Below-freezing average monthly temperatures are quite common. In some locations, average temperatures can be below freezing for up to seven months.

Extreme seasonality of temperatures is the norm, as monthly average temperature ranges may be as high as 70 to 100 degrees Fahrenheit (21 to 38 degrees Celsius). The transition seasons, autumn and spring, are relatively short in comparison with the length of summer and winter, with the latter season extremely long. Precipitation is usually very low, because the cold air temperatures prohibit the storage of large amounts of moisture. Total precipitation is typically between 5 and 20 inches (120 and 500 millimeters), with the majority falling during the warm months.

Group of Inuit people building an igloo. (Library of Congress Prints and Photographs Division)

Tundra

Tundra regions exist poleward of the subarctic climate region. The tundra climate type is named for tundra vegetation, which consists of low-growing flowering plants, shrubs, small trees, lichens, and mosses. Typically, the mean monthly temperature of the warmest month does not exceed 50 degrees Fahrenheit (10 degrees Celsius). The region lies around or poleward of 60 degrees latitude. Because of the high latitude, winter temperatures are extremely cold as the Sun shines for only a few hours each day. Summer temperatures are only slightly above freezing on average, even though there is virtually continuous daylight. The constant cold causes permafrost, constantly frozen subsoil. In winter, all soil is frozen, but during the summer months, the top layers of soil thaw. This creates a highly saturated condition, as liquid water cannot penetrate the frozen ground below.

Precipitation amounts are relatively low, typically around 5 inches (130 millimeters) per year. Precipitation is distributed somewhat evenly through the year, but a summertime maximum is prevalent. This is a result of the extreme cold of winter, when air has little moisture-carrying capacity, thus essentially creating a cold desert.

Ice Cap

Greenland is the only location in North America, or the Northern Hemisphere, that is classified as an ice cap region. Ice covers the ground throughout the year there, as mean monthly temperatures remain below freezing. Cold, dense air forms above the chilled Greenland ice cap surface and spills toward lower elevations, ensuring frigid temperatures even along the coast. The region receives little precipitation, as a result of the extremely cold temperatures. The precipitation that does occur is mainly limited to late summer and autumn, when temperatures are highest. Precipitation is typically less than 5 inches (130 millimeters) per year.

Highland

High elevations are typically included within other climate regions. The climates of mountainous regions are driven by elevation and may not adequately represent any particular climate zone. Highland climates are driven by vertical zonation, which refers to the progression of climatic zones with increasing elevation. The progression of climate zones with height largely mimics the progression of climate zones with increasing latitude. Rapid changes in slope, orientation, and elevation make specific categorization of these regions problematic. Temperature and precipitation regimes change according to these factors. Individual locations may be greatly different from one another even though they may be in close proximity to each other. Highland climates largely exist along the Rocky Mountain spine that runs from Alaska through Canada and the western United States to Mexico.

Anthony J. Vega

CLIMATOLOGY OF THE CARIBBEAN

The Caribbean extends from Trinidad in the south (10 degrees north latitude) to the northern Bahamas (27 degrees north latitude). With the exception of the northern Bahamas, it lies entirely within the tropics. Such a location means that temperatures on its islands are generally warm year-round, with no major seasonal differences. A typical average annual temperature near sea level is 80 degrees Fahrenheit (27 degrees Celsius), and the coolest month averages above 64 degrees Fahrenheit (18 degrees Celsius).

Climate Controls

Caribbean climates are governed by several factors, including latitude, trade winds, maritime influences, mountains, and nearness to North America. The low latitude of the region causes temperatures to be high all year because of the high Sun angles. However, the latitudinal extent of the Caribbean, about 17 degrees, means that there are temperature-related differences from north to south. Temperature varies more by season in the Bahamas than it does southward.

Crane Beach is situated on the south east coast of Barbados. (Johnmartindavies)

A road in the Roseau (Dominica) area is littered with structural debris, damaged vegetation and downed power poles and lines after Hurricane Maria. (Roosevelt Skerrit)

Prevailing winds throughout the Caribbean region are the northeasterly trade winds. These are tropical winds off the ocean, carrying considerable heat and moisture. Typical trade winds have relative humidities of 70 percent or more. Although this alone is not sufficient for precipitation, all that is needed is some triggering mechanism to cause clouds and rainfall. Common triggers are forced rise of the trade winds over mountains (orographic effect), convectional rise of air above heated land surfaces, and the passage of low pressure, often in the form of easterly waves.

The fact that most of the islands of the Caribbean are mountainous has two important consequences. First, because air tends to be cooler at higher altitudes, at a rate of 3.5 Fahrenheit degrees per 1,000 feet (6 Celsius degrees per 1,000 meters), the uplands have cooler climates than at sea level. Although most people of the region live near sea level, in coastal plains and port cities, the wealthy

often live in upland suburbs where temperatures are less oppressive.

A second consequence of mountains involves their effects on the trade winds. Mountains force the trade winds to rise, and in the process, the moist air cools and clouds and rain may result. Most of the rain falls on the northeast, windward sides of mountains. Windward slopes receive the highest annual rainfall totals in the region, sometimes more than 200 inches (5,000 millimeters). Lee slopes, to the southwest of the mountains, are typically drier, with annual rainfall less than half that of the windward slopes.

The surface of the Caribbean region is mostly water, so that temperatures are moderated, being neither as hot nor as cold as at the midlatitudes. In addition, the chief ocean currents of the Caribbean arrive from the tropical Atlantic Ocean. These waters help maintain the tropical temperatures and protect the islands from northern climatic influences.

The Greater Antilles and the Bahamas are relatively close to the continent of North America, so in winter they are influenced by cold air masses from the north. Although a cold air mass is warmed on its journey across the warm Gulf of Mexico and Caribbean waters, it can still depress temperatures by up to 18 Fahrenheit degrees (10 Celsius degrees) below normal in the Bahamas and Cuba.

Climate Types

The Caribbean region has three distinct climate types, defined primarily by rainfall patterns: tropical wet, tropical wet and dry, and tropical dry.

The tropical wet climate is hot all year and receives more than 80 inches (2,000 millimeters) of precipitation per year. There is no significant dry season. Substantial rainfall occurs in every month, although summer months are wettest. Rain is

chiefly orographic. This climate is found on the windward sides of the Greater Antilles and almost all of the Lesser Antilles.

The tropical wet and dry climate is also hot all year. Rain falls chiefly as afternoon convectional thunderstorms. This climate receives two-thirds of its annual rainfall in summer; a four- to six-month dry season is centered on winter. Drought is possible, especially from March to May. This climate occurs on the lee slopes of the Greater Antilles.

The tropical dry climate is found chiefly in the rain shadows to the lee of the highest mountains of the region, especially the highlands of Hispaniola. Annual precipitation totals less than 20 inches (500 millimeters). Smaller pockets of dry climate also are found on the lee sides of Jamaica and Puerto Rico. Additionally, small islands that lack relief, and thus lack the triggers of convection and orographic ef-

Signs denoting past hurricanes, Rum Point, Grand Cayman Island. (Lhb1239)

fect, are usually dry. These include the Bahamas, Turks and Caicos, Cayman Islands, St. Martin, and the Netherlands Antilles off the north coast of South America.

Easterly Waves and Hurricanes

Caribbean climates are affected by easterly waves and hurricanes. Easterly waves are elongated (northeast to southwest) troughs of low pressure that ride the trade winds, mostly in summer. As these troughs move westward, they trigger showers and thundershowers, accounting for much of the total rainfall received by many islands, especially those lacking relief.

In the summer and fall months, for reasons not entirely understood, certain easterly waves may develop into hurricanes; strong rotary cyclonic storms known for high winds, heavy precipitation, and coastal flooding. Most islands are affected by these storms once every three to ten years. Among the devastating effects of hurricanes in the Caribbean are serious floods, major landslides, wind damage to crops and forests, and damage to infrastructure, including tourist facilities.

Influences of Climate

Climate influences many aspects of life in the Caribbean, including agriculture and health. Because there is no winter here, agricultural possibilities are good, depending largely on annual precipitation and altitude. Tropical wet areas are ideal for growing cocoa, bananas, coconuts, nutmeg, and citrus, which are major exports for some islands. Islands with extensive uplands can grow vegetables and flowers, often associated with the midlatitudes. Drier leeward locales are suitable for a variety of crops, including sugarcane, fiber crops, and tobacco, when irrigation is available.

Climate affects health in various ways. As in all humid tropical locales, the lack of frost and abundance of heat and moisture promote a variety of tropical diseases that affect humans, animals, and plants. Yellow fever, malaria, dengue fever, cholera, and typhoid once ravaged the people of the Caribbean, although most have been controlled, except for periodic outbreaks. The extent to which heat affects human performance is still being studied and debated.

P. Gary White

DISCUSSION QUESTIONS: CLIMATOLOGY

Q1. How do geographical factors contribute to North America's comparatively frequent violent weather? Why do tornadoes occur so often in the United States? Why is the Caribbean region especially subject to hurricanes?

Q2. What are the prevailing climate types in North America? Why is the area that experiences a Mediterranean climate so limited? Where in North America are the coldest climates experienced?

Q3. What role does the Caribbean's tropical location play in the region's climate? How do geographical features define the climate on individual islands? Why are some Caribbean islands directly influenced by North American weather systems?

BIOGEOGRAPHY AND NATURAL RESOURCES

Overview

Minerals

Mineral resources make up all the nonliving matter found in the earth, its atmosphere, and its waters that are useful to humankind. The great ages of history are classified by the resources that were exploited. First came the Stone Age, when flint was used to make tools and weapons. The Bronze Age followed; it was a time when metals such as copper and tin began to be extracted and used. Finally came the Iron Age, the time of steel and other ferrous alloys that required higher temperatures and more sophisticated metallurgy.

Metals, however, are not the whole story—economic progress also requires fossil fuels such as coal, oil, natural gas, tar sands, or oil shale as energy sources. Beyond metals and fuels, there are a host of mineral resources that make modern life possible: building stone, salt, atmospheric gases (oxygen, nitrogen), fertilizer minerals (phosphates, nitrates, and potash), sulfur, quartz, clay, asbestos, and diamonds are some examples.

Mining and Prospecting

Exploitation of mineral resources begins with the discovery and recognition of the value of the deposits. To be economically viable, the mineral must be salable at a price greater than the cost of its extraction, and great care is taken to determine the probable size of a deposit and the labor involved in isolating it before operations begin. Iron, aluminum, copper, lead, and zinc occur as mineral ores that are mined, then subjected to chemical processes to separate the metal from the other elements (usually oxygen or sulfur) that are bonded to the metal in the ore.

Some deposits of gold or platinum are found in elemental (native) form as nuggets or powder and may be isolated by alluvial mining—using running water to wash away low-density impurities, leaving the dense metal behind. Most metal ores, however, are obtained only after extensive digging and blasting and the use of large-scale earthmoving equipment. Surface mining or strip mining is far simpler and safer than underground mining.

Safety and Environmental Considerations

Underground mines can extend as far as a mile into the earth and are subject to cave-ins, water leakage, and dangerous gases that can explode or suffocate miners. Safety is an overriding issue in deep mines, and there is legislation in many countries designed to regulate mine safety and to enforce practices that reduce hazards to the miners from breathing dust or gases.

In the past, mining often was conducted without regard to the effects on the environment. In economically advanced countries such as the United States, this is now seen as unacceptable. Mines are expected to be filled in, not just abandoned after they are worked out, and care must be taken that rivers and streams are not contaminated with mine wastes.

Iron, Steel, and Coal

Iron ore and coal are essential for the manufacture of steel, the most important structural metal. Both raw materials occur in many geographic regions. Before the mid-nineteenth century, iron was smelted in the eastern United States—New Jersey, New York, and Massachusetts—but then huge hematite deposits were discovered near Duluth, Minnesota, on Lake Superior. The ore traveled by ship to steel mills in northwest Indiana and northeast Illinois, and coal came from Illinois or Ohio. Steel also was made in Pittsburgh and Bethlehem in Pennsylvania, and in Birmingham, Alabama.

After World War II, the U.S. steel industry was slow to modernize its facilities, and after 1970 it had great difficulty producing steel at a price that could compete with imports from countries such as Japan, Korea, and Brazil. In Europe, the German steel industry centered in the Ruhr River valley in cities such as Essen and Düsseldorf. In Russia, iron ore is mined in the Urals, in the Crimea, and at Krivoi Rog in Ukraine. Elsewhere in Europe, the French "minette" ores of Alsace-Lorraine, the Swedish magnetite deposits near Kiruna, and the British hematite deposits in Lancashire are all significant. Hematite is also found in Labrador, Canada, near the Quebec border.

Coal is widely distributed on earth. In the United States, Kentucky, West Virginia, and Pennsylvania are known for their coal mines, but coal is also found in Illinois, Indiana, Ohio, Montana, and other states. Much of the anthracite (hard coal) is taken from underground mines, where networks of tunnels are dug through the coal seam, and the coal is loosened by blasting, use of digging machines, or human labor. A huge deposit of brown coal is mined at the Yallourn open pit mine west of Melbourne, Australia. In Germany, the mines are near Garsdorf in Nord-Rhein/Westfalen, and in the United Kingdom, coal is mined in Wales. South Africa has coal and is a leader in manufacture of liquid fuels from coal. There is coal in Antarctica, but it cannot yet be mined profitably. China and Japan both have coal mines, as does Russia.

Aluminum

Aluminum is the most important structural metal after iron. It is extremely abundant in the earth's crust, but the only readily extractable ore is bauxite, a hydrated oxide usually contaminated with iron and silica. Bauxite was originally found in France but also exists in many other places in Europe, as well as in Australia, India, China, the former Soviet Union, Indonesia, Malaysia, Suriname, and Jamaica.

Much of the bauxite in the United States comes from Arkansas. After purification, the bauxite is combined with the mineral cryolite at high temperature and subjected to electrolysis between carbon electrodes (the Hall-Héroult process), yielding pure aluminum. Because of the enormous electrical energy requirements of the Hall-Héroult method, aluminum can be made economically only where cheap power (preferably hydroelectric) is available. This means that the bauxite often must be shipped long distances—Jamaican bauxite comes to the United States for electrolysis, for example.

Copper, Silver, and Gold

These coinage metals have been known and used since antiquity. Copper came from Cyprus and takes its name from the name of the island. Copper ores include oxides or sulfides (cuprite, bornite, covellite, and others). Not enough native copper occurs to be commercially significant. Mines in Bingham, Utah, and Ely, Nevada, are major sources in the United States. The El Teniente mine in Chile is the world's largest underground copper mine, and major amounts of copper also come from Canada, the former Soviet Union, and the Katanga region mines in Congo-Kinshasa and Zambia.

Silver often occurs native, as well as in combination with other metals, including lead, copper, and gold. Famous silver mines in the United States include those near Virginia City (the Comstock lode) and Tonopah, Nevada, and Coeur d'Alene, Idaho. Silver has been mined in the past in Bolivia (Potosi mines), Peru (Cerro de Pasco mines), Mexico, and Ontario and British Columbia in Canada.

Gold occurs native as gold dust or nuggets, sometimes with silver as a natural alloy called electrum. Other gold minerals include selenides and tellurides. Small amounts of gold are present in sea water, but attempts to isolate gold economically from this source have so far failed. Famous gold rushes occurred in California and Colorado in the United States, Canada's Yukon, and Alaska's Klondike region. Major gold-producing countries include South Africa, Siberia, Ghana (once called the Gold Coast), the Philippines, Australia, and Canada.

THE EXXON VALDEZ OIL SPILL

On March 24, 1989, the tanker Exxon Valdez, with a cargo of 53 million gallons of crude oil, ran aground on Bligh Reef in Prince William Sound, Alaska. Approximately 11 million gallons of oil were released into the water, in the worst environmental disaster of this type recorded to date. Despite immediate and lengthy efforts to contain and clean up the spill, there was extensive damage to wildlife, including aquatic birds, seals, and fish. Lawsuits and calls for new regulatory legislation on tankers continued a decade later. Such regrettable incidents as these are the almost inevitable result of attempting to transport the huge oil supplies demanded in the industrialized world.

Petroleum and Natural Gas

Petroleum has been found on every continent except Antarctica, with 600,000 producing wells in 100 different countries. In the United States, petroleum was originally discovered in Pennsylvania, with more important discoveries being made later in west Texas, Oklahoma, California, and Alaska. New wells are often drilled offshore, for example in the Gulf of Mexico or the North Sea. The United States depends heavily on oil imported from Mexico, South America, Saudi Arabia and the Persian Gulf states, and Canada.

Over the years, the price of oil has varied dramatically, particularly due to the attempts of the Organization of Petroleum Exporting Countries (OPEC) to limit production and drive up prices. In Europe, oil is produced in Azerbaijan near the Caspian Sea, where a pipeline is planned to carry the crude to the Mediterranean port of Ceyhan, in Turkey. In Africa, there are oil wells in Gabon, Libya, and Nigeria; in the Persian Gulf region, oil is found in Kuwait, Qatar, Iran, and Iraq. Much crude oil travels in huge tankers to Europe, Japan, and the United States, but some supplies refineries in Saudi Arabia at Abadan. Tankers must exit the Persian Gulf through the narrow Gulf of Hormuz, which thus assumes great strategic importance.

After oil was discovered on the shores of the Beaufort Sea in northern Alaska (the so-called North Slope) in the 1960s, a pipeline was built across Alaska, ending at the port of Valdez. The pipeline is heated to keep the oil liquid in cold weather and elevated to prevent its melting through the permanently frozen ground (permafrost) that supports it. From Valdez, tankers reach Japan or California.

Drilling activities occasionally result in discovery of natural gas, which is valued as a low-pollution fuel. Vast fields of gas exist in Siberia, and gas is piped to Western Europe through a pipeline. Algerian gas is shipped in the liquid state in ships equipped with refrigeration equipment to maintain the low temperatures needed. Britain and Northern Europe benefit from gas produced in the North Sea, between Norway and Scotland.

Shale oil, a plentiful but difficult-to-exploit fossil fuel, exists in enormous amounts near Rifle, Colorado. A form of oil-bearing rock, the shale must be crushed and heated to recover the oil, a more expensive proposition than drilling conventional oil wells. In spite of ingenious schemes such as burning the shale oil in place, this resource is likely to remain largely unused until conventional petroleum is used up. A similar resource exists in Alberta, Canada, where the Athabasca tar sands are exploited for heavy oils.

John R. Phillips

RENEWABLE RESOURCES

Most renewable resources are living resources, such as plants, animals, and their products. With careful management, human societies can harvest such resources for their own use without imperiling future supplies. However, human history has seen many instances of resource mismanagement that has led to the virtual destruction of valuable resources.

Forests

Forests are large tracts of land supporting growths of trees and perhaps some underbrush or shrubs. Trees constitute probably the earth s most valuable, versatile, and easily grown renewable resource. When they are harvested intelligently, their natural environments continue to replace them. However, if a harvest is beyond the environment's ability to restore the resource that had been present, new and different plants and animals will take over the area. This phenomenon has been demonstrated many times in overused forests and grasslands that reverted to scrubby brushlands. In the worst cases, the abused lands degenerated into barren deserts.

The forest resources of the earth range from the tropical rainforests with their huge trees and broad diversity of species to the dry savannas featuring scattered trees separated by broad grasslands. Cold, subarctic lands support dense growths of spruces and firs, while moderate temperature regimes produce a variety of pines and hardwoods such as oak and ash. The forests of the world cover about 30 percent of the land surface, as compared with the oceans, which cover about 70 percent of the global surface.

Harvested wood, cut in the forest and hauled away to be processed, is termed roundwood. Globally, the cut of roundwood for all uses amounts to about 130.6 billion cubic feet (3.7 billion cubic meters). Slightly more than half of the harvested wood is used for fuel, including charcoal.

Roundwood that is not used for fuel is described as industrial wood and used to produce lumber, veneer for fine furniture, and pulp for paper prod-

ucts. Some industrial wood is chipped to produce such products as subflooring and sheathing board for home and other building construction. Most roundwood harvested in Africa, South America, and Asia is used for fuel. In contrast, roundwood harvested in North America, Europe, and the former Soviet Union generally is produced for industrial use.

It is easy to consider forests only in the sense of the useful wood they produce. However, many forests also yield valuable resources such as rubber, edible nuts, and what the U.S. Forest Service calls special forest products. These include ferns, mosses, and lichens for the florist trade, wild edible mushrooms such as morels and matsutakes for domestic markets and for export, and mistletoe and pine cones for Christmas decorations.

There is growing interest among the industrialized nations of the world in a unique group of forest products for use in the treatment of human disease. Most of them grow in the tropical rainforests. These medicinal plants have long been known and used by shamans (traditional healers). Hundreds of pharmaceutical drugs, first used by shamans, have been derived from plants, many gathered in tropical rainforests. The drugs include quinine, from the bark of the cinchona tree, long used to combat malaria, and the alkaloid drug reserpine. Reserpine, derived from the roots of a group of tropical trees and shrubs, is used to treat high blood pressure (hypertension) and as a mild tranquilizer. It has been estimated that 25 percent of all prescriptions dispensed in the United States contain ingredients derived from tropical rainforest plants. The value of the finished pharmaceuticals is estimated at US$6.25 billion per year.

Scientists screening tropical rainforest plants for additional useful medical compounds have drawn on the knowledge and experience of the shamans. In this way, the scientists seek to reduce the search time and costs involved in screening potentially useful plants. Researchers hope that somewhere in the dense tropical foliage are plant products that could treat, or perhaps cure, diseases such as cancer or AIDS.

Many as-yet undiscovered medicinal plants may be lost forever as a consequence of deforestation of large tracts of equatorial land. The trees are cut down or burned in place and the forest converted to grassland for raising cattle. The tropical soils cannot support grasses without the input of large amounts of fertilizer. The destruction of the forests also causes flooding, leaving standing pools of water and breeding areas for mosquitoes, which can spread disease.

Marine Resources

When renewable marine resources such as fish and shellfish are harvested or used, they continue to reproduce in their environment, as happens in forests and with other living natural resources. However, like overharvested forests, if the marine resource is overfished—that is, harvested beyond its ability to reproduce—new, perhaps undesirable, kinds of marine organisms will occupy the area. This has happened to a number of marine fishes, particularly the Atlantic cod.

When the first Europeans reached the shores of what is now New England in the early seventeenth century, they encountered vast schools of cod in the local ocean waters. The cod were so plentiful they could be caught in baskets lowered into the water from a boat.

At the height of the New England cod fishery, in the 1970s, efficient, motor-driven trawlers were able to catch about 32,000 tons. The catch began to decline that year, mostly as a result of the impact of fifteen different nations fishing on the cod stocks. As a result of overfishing, rough species such as dogfish and skates constitute 70 percent of the fish in the local waters. Experts on fisheries management decided that fishing for cod had to be stopped.

The decline of the cod was attributed to two causes: a worldwide demand for more fish as food and great changes in the technology of fishing. The

technique of fishing progressed from a lone fisher with a baited hook and line, to small steam-powered boats towing large nets, to huge diesel-powered trawlers towing monster nets that could cover a football field. Some of the largest trawlers were floating factories. The cod could be skinned, the edible parts cut and quick-frozen for market ashore, and the skin, scales, and bones cooked and ground for animal feed and oil. A lone fisher was lucky to be able to catch 1,000 pounds (455 kilograms) in one day. In contrast, the largest trawlers were capable of catching and processing 200 tons per day.

In the 1990s, the world ocean population of swordfish had declined dramatically. With a worldwide distribution, these large members of the billfish family have been eagerly sought after as a food fish. Because swordfish have a habit of basking at the surface, fishermen learned to sneak up on the swordfish and harpoon them. Fishermen began to catch swordfish with fishing lines 25 to 40 miles (40 to 65 kilometers) long. Baited hooks hung at intervals on the main line successfully caught many swordfish, as well as tuna and large sharks. Whereas the harpoon fisher took only the largest (thus most valuable) swordfish, the longline gear was indiscriminate, catching and killing many swordfish too small for the market, as well as sea turtles and dolphins

As a result of the catching and killing of both sexually mature and immature swordfish, the reproductive capacity of the species was greatly reduced. Harpoons killed mostly the large, mature adults that had spawned several times. Longlines took all sizes of swordfish, including the small ones that had not yet reached sexual maturity and spawned. The decline of the swordfish population was quickly obvious in the reduced landings. But things have changed remarkably, thanks to a 1999 international plan that rebuilt this stock several years ahead of schedule. Today, North Atlantic swordfish is one of the most sustainable seafood choices.

Albert C. Jensen

93

NONRENEWABLE RESOURCES

Nonrenewable resources are useful raw materials that exist in fixed quantities in nature and cannot be replaced. They differ from renewable resources, such as trees and fish, which can be replaced if managed correctly. Most nonrenewable resources are minerals—inorganic and organic substances that exhibit consistent chemical composition and properties. Minerals are found naturally in the earth's crust or dissolved in seawater. Of roughly 2,000 different minerals, about 100 are sources of raw materials that are needed for human activities. Where useful minerals are found in sufficiently high concentrations—that is, as ores—they can be mined as profitable commercial products.

Economic nonrenewable resources can be divided into four general categories: metallic (hardrock) minerals, which are the source of metals such as iron, gold, and copper; fuel minerals, which include petroleum (oil), natural gas, coal, and uranium; industrial (soft rock) minerals, which provide materials like sulfur, talc, and potassium; and construction materials, such as sand and gravel.

Nonrenewable resources are required as direct or indirect parts of all the products that humans use. For example, metals are necessary in industrial sectors such as construction, transportation equipment, electrical equipment and electronics, and consumer durable goods—long-lasting products such as refrigerators and stoves. Fuel minerals provide energy for transportation, heating, and electrical power. Industrial minerals provide ingredients needed in products ranging from baby powder to fertilizer to the space shuttle. Construction materials are used in roads and buildings.

Location

When minerals have naturally combined together (aggregated) they are called rocks. The three general rock categories are igneous, sedimentary, and metamorphic. Igneous rocks are created by the cooling of molten material (magma). Sedimentary rocks are caused when weathering, erosion, trans-portation, and compaction or cementation act on existing rocks.

Metamorphic rocks are created when the other two types of rock are changed by heat and pressure. The availability of nonrenewable resources from these rocks varies greatly, because it depends not only on the natural distribution of the rocks but also on people's ability to discover and process them. It is difficult to find rock formations that are covered by the ocean, material left by glaciers, or a rainforest. As a result, nonrenewable resources are distributed unevenly throughout the world.

Some nonrenewable resources, such as construction materials, are found easily around the world and are available almost everywhere. Other nonrenewable resources can only be exploited profitably when the useful minerals have an unusually high concentration compared with their average concentration in the earth's crust. These high concentrations are caused by rare geological events and are difficult to find. For example, an exceptionally rare nonrenewable resource like platinum is produced in only a few limited areas.

No one country or region is self-sufficient in providing all the nonrenewable resources it needs, but some regions have many more nonrenewable resources than others. Minerals can be found in all types of rocks, but some types of rocks are more likely to have economic concentrations than others. Metallic minerals often are associated with shields (blocks) of old igneous (Precambrian) rocks. Important shield areas near the earth's surface are found in Canada, Siberia, Scandinavia, and Eastern Europe. Another important shield was split by the movement of the continents, and pieces of it can be found in Brazil, Africa, and Australia.

Similar rock types are in the mountain formations in Western Europe, Central Asia, the Pacific coast of the Americas, and Southeast Asia. Minerals for construction and industry are found in all three types of rocks and are widely and randomly distributed among the regions of the world.

The fuel minerals—petroleum and natural gas—are unique in that they occur in liquid and gaseous states in the rocks. These resources must be captured and collected within a rock site. Such a site needs source rock to provide the resource, a rock type that allows the resource to collect, and another surrounding rock type that traps the resource. Sedimentary rock basins are particularly good sites for fuel collection. Important fuel-producing regions are the Middle East, the Americas, and Asia.

Impact on Human Settlement

Nonrenewable resources have always provided raw materials for human economic development, from the flint used in early stone tools to the silicon used in the sophisticated chips in personal computers. Whole eras of human history and development have been linked with the nonrenewable resources that were key to the period and its events. For example, early human culture eras were called the Stone, Bronze, and Iron Ages.

Political conflicts and wars have occurred over who owns and controls nonrenewable resources and their trade. One example is the Persian Gulf War of 1991. Many nations, including the United States, fought against Iraq over control of petroleum production and reserves in the Middle East.

Since the actual production sites often are not attractive places for human settlement and the output is transportable, these sites are seldom important population centers. There are some exceptions, such as Johannesburg, South Africa, which grew up almost solely because of the gold found there. However, because it is necessary to protect and work the production sites, towns always spring up near the sites. Examples of such towns can be found near the quarries used to provide the material for the great monuments of ancient Egypt and in the Rocky Mountains of North America near gold and silver mines. These towns existed because of the nonrenewable resources nearby and the needs of the people exploiting them; once the resource was gone, the towns often were abandoned, creating "ghost towns," or had to find new purposes, such as tourism.

More important to human settlement is the control of the trade routes for nonrenewable resources. Such controlling sites often became regions of great wealth and political power as the residents taxed the products that passed through their community and provided the necessary services and protection for the traveling traders. Just one example of this type of development is the great cities of wealth and culture that arose along the trade routes of the Sahara Desert and West Africa like Timbuktu (in present-day Mali) and Kumasi (in present-day Ghana) based on the trade of resources like gold and salt.

Even with modern transportation systems, ownership of nonrenewable resources and control of their trade is still an important factor in generating national wealth and economic development. Modern examples include Saudi Arabia's oil resources, Egypt's control of the Suez Canal, South Africa's gold, Chile's copper, Turkey's control over the Bosporus Strait, Indonesia's metals and oil, and China's rare earth element.

Gary A. Campbell

NATURAL RESOURCES OF NORTH AMERICA

Natural resources are materials found naturally in the environment that are used or valued by people as a resource. They provide the raw materials—wood pulp, iron, water, petroleum, soil, and food—that are transformed by refining and manufacturing to make things that are used in daily life. Renewable resources can regenerate themselves and will remain available if they are not used faster than they can regrow or regenerate. These include water, fish, timber, and soil. Nonrenewable resources, which cannot be replaced after they are used, include oil, coal, natural gas, iron, and other minerals.

The United States and Canada contain the greatest concentration of natural wealth in the world. Domestically produced mineral, timber, and agricultural resources that are sold locally and overseas have created enormous economic wealth. The countries of North America also import natural resources from other countries. The United States has only about 4.25 percent of the world's population, but as the world's leading manufacturing country, it consumes about 24 percent of the energy resources annually consumed in the world. Some of the resources that the United States consumes are im-

ported, mainly raw materials for use by industry, but much is mined, cut, refined, and produced in the country. Canada and Mexico also have abundant renewable and nonrenewable resources and contribute to the outstanding resource bounty of the North American continent.

Renewable Resources: Water

The humid eastern portions of North America receive much more rainfall annually than the semiarid interior and the more arid western portions of the continent. However, even the arid regions may have freshwater resources available, either from surface rivers or from underground water in aquifers (porous rock beds into which water seeps and slowly flows underground).

There is no water shortage in North America as a whole, but rapidly increasing demand in some places has caused localized problems. The agricultural western portions of the United States, which provide much of the country's fruits and vegetables, depend almost entirely on irrigation for their production. Many of the surface and subsurface sources of water used there for irrigation, as well as for the region's

The McArthur River Uranium Mine, in northern Saskatchewan, Canada, is the world's largest high-grade uranium deposit. (Turgan)

97

SELECTED RESOURCES OF NORTH AMERICA

growing industries and population, are being used faster than they are replenished. Part of the problem is that 60 percent of the land area of the United States (the west and southwest) receives only about 25 percent of the country's annual precipitation.

Many underground sources of water, like the vast Ogallala Aquifer, have for decades been used at a faster rate than that at which water seeps back in. The western US water table has been declining by 3 feet (1 meter) each year in some places. Above-ground water sources are also endangered. The US Water Resources Council has found water shortages to be imminent in the Rio Grande basin and the lower Colorado River basin. The Colorado River, which is drained for irrigation by seven arid and semiarid states, has become little more than a trickle of water laden with salts by the time it reaches Mexico. The Missouri River basin also has begun to experience water shortages. The eastern United States is not exempt from water shortages, especially in the heavily urbanized and industrialized Northeast. In this more humid area of the continent, the problem is usually one of water quality, rather than quantity.

The danger of absolute water shortage in Canada as a whole is remote; however, local shortages occur in parts of the country. The semiarid western prairie region, an area with low precipitation and high evaporation of moisture, is most susceptible to shortages. Some areas face shortages because local facilities for water purification, storage, and distribution are lacking. Additional factors are that 60 percent of Canada's surface runoff is carried by rivers that flow north toward the Arctic Ocean and away from the more populated area near the US border, and more than one-third of Canada's average annual precipitation is in the form of snow, which is not available in winter and can be overabundant in the spring.

Forest Products

Canada and the United States rank first and second, respectively, in world export of forest products, together accounting for one-third of these

---

THE OGALLALA AQUIFER

The Ogallala Aquifer—an enormous underground reservoir of water-soaked sand, silt, and gravel stretching from South Dakota to western Texas—has for decades been pumped faster than it can be replenished by rainfall and surface runoff. As a result, the aquifer's water levels have declined by an estimated 2-10 feet per year, around 9% of the total volume, since the 1940s.

The entire Ogallala Aquifer system underlies about 174,000 square miles of eight states. Because human activity and land surface saturation rates vary across the high plains that overlie this vast area, some areas have been drained more rapidly, while others have remained unchanged.

The Ogallala aquifer varies in thickness from 1,000 feet (300 meters) in Nebraska to just a few feet in parts of Texas. Most of the excessive pumping from this aquifer is due to widespread crop irrigation. The withdrawal has been most excessive where irrigation is heavy but the aquifer is not deep. In parts of Texas, declines have exceeded 100 feet (30 meters) over the last half of the twentieth century.

As the water table declines, the cost of pumping water from greater depths increases the cost to farmers irrigating their crops, which gradually forces farmers to abandon the practice of irrigation. This has already happened in parts of Nebraska, Kansas, Oklahoma, Colorado, and Texas. If the aquifer were completely drained, it could take six thousand years to refill naturally.

---

exports. Because of variations in terrain and climate across the continent, different types of forests are found in various regions. Half of Canada is covered with trees, more than any other country except Russia. Most Canadian forests are coniferous, with softwood, cone-bearing needleleaf trees such as pine, spruce, and fir. Much of the forest in British Colombia—about half of all timber harvested in Canada—is used for lumber products. Some lumber and most of the pulpwood (used for paper products) comes from the boreal forest in Ontario and Quebec. The boreal forest is the largest forest in

Taiga forest in the boreal forest region in Quebec, Canada. (peupleloup)

Canada and remains the least accessible. Canadian hardwoods (broadleaf trees such as birch, maple, and beech) are in relatively short supply. Canada has far fewer acres dedicated to hardwood forest than the United States has, and these forests are found mostly in southern Ontario and southern Quebec, near the US border.

The most important forest resources of the United States are the northeast hardwoods, which provide most of the hardwood lumber for furniture, paneling, and cabinets produced in the country; the western timberlands of the mountain and coastal forests between the Pacific coast and the Rocky Mountains, which account for most of the construction lumber produced in the country; and the southeast mixed forests, which are harvested mostly for pulpwood, although the region also produces some hardwood and softwood lumber.

Fish and Aquatic Products

In 2018, the United States and Canada ranked fifth and twenty-fifth, respectively, in the world in total fish catch. In the United States, commercial fishing is concentrated in four fishing regions: the Atlantic (off the coast of the New England, mid-Atlantic, and southern states), which accounts for 37 percent of the national catch by value; the Gulf of Mexico, with 16 percent; the Alaska region with 32 percent; and the Pacific coast with 12 percent. Inland sources account for fish and other aquatic creatures caught in freshwater rivers and lakes, or raised and harvested in hatcheries.

The majority of the commercial fish harvest in Canada comes from the northwest Atlantic region, around the eastern Maritime Provinces. The remainder is collected off the western coast of British Columbia and from inland sources. The Gulf of Mexico, the Caribbean Sea, the Pacific Ocean, and the Sea of Cortés offer abundant tropical and sub-tropical fishing grounds for Mexico.

In Canada and the United States, the most important fishing grounds historically have been around the Grand Banks area of the northwest Atlantic (the New England and Maritime Provinces region). But over the years, overfishing in the Grand Banks led to an alarming depletion of the cod population. In July 1992, the Canadian government enacted a ban on cod fishing. The cod moratorium had an immediate and devastating effect, especially on the economy of Newfoundland and Labrador, where upwards of 50,000 people lost their livelihood. In the years since, those areas long dependent on cod fishing refocused their efforts on crab, shrimp, lobster, and salmon with much success. And beyond fishing, the Maritime Provinces were, to a large extent, compelled to diversify their economies, and have done so very well. The moratorium remains larges in place, although a very limited amount of cod fishing is now allowed. Today, the main fishing states and provinces areAlaska and British Columbia (salmon, crab, lobster), California (lobster, salmon), Texas-Louisiana (shrimp), Massachusetts-Connecticut-Rhode Island (lobster, shrimp, tuna), and Nova Scotia and Newfoundland and Labrador (lobster and salmon).

Fertile Soil

An essential resource, soil is potentially renewable because new soil is created through the weathering and decomposition of bedrock beneath the surface, even as topsoil is being washed away from the surface. The most fertile soils and the largest farms in North America are found in the Central Plains of Canada and the United States, and in the midwestern United States. This region produces most of the continent's wheat, corn, sorghum, milo, and other grains, and annually produces a huge surplus that is exported to countries all over the world.

Fishing boats and lobster traps in Salvage, Newfoundland. (Sallyledrew)

OVERFISHING IN ATLANTIC CANADA

The Canadian government has appealed to the world community to try to save what remains of its once-rich Atlantic cod fishery. The international Law of the Sea bans commercial fishing within 200 nautical miles off shore, but Canada has claimed a 300-nautical-mile Exclusive Economic Zone maritime boundary to curtail foreign commercial fishing trawlers.

At one time, large schools of northern cod were treated as an inexhaustible resource. However, unsustainable fishing practices led to a dramatic fall to an estimated 1% of northern cod stocks from 1980s levels and finally brought action in 1992 when the Canadian government announced a moratorium on northern cod fishing. A similar reduction in Atlantic salmon fishing began with closure of commercial Atlantic salmon fisheries in 1984 and a full moratorium on all commercial salmon fisheries in eastern Canada by 2000. In both cases the Canadian government paid fishermen not to catch fish in order to preserve salmon stocks for future commercial and sport fishing.

The Canadian government has also tried to convince commercial cod and salmon fishermen to fish for the more plentiful crabs, scallops, and Arctic char. However, foreign "factory ships," mostly trawlers from Russia and Japan, continued capturing migratory cod that swam into international waters beyond Canada's Exclusive Economic Zone. Today, Fisheries and Oceans Canada oversees conservation and sustainable development efforts through the Canadian Coast Guard, Freshwater Fish Marketing Corporation and the Canadian Hydrographic Service. Northern cod stock has reached about 25% of the levels seen in the '80s, however scientists do not know when the fish stocks will be restored.

The eastern United States and Canada have relatively smaller farms, which produce fruits, vegetables, cotton, tobacco, peanuts, and soybeans. Some valleys of the US intermountain west are productive where irrigated.

Volcanically derived soils found in parts of Mexico are highly productive and, because of the tropical and subtropical climates of Mexico, can be farmed throughout the year. The United States is the most agriculturally productive country in the world and produces far more food each year than is consumed within the country. Some of the food surplus is stored, but most is sold to other countries or distributed for free through food- aid programs.

Active Permian Basin pumpjack east of Andrews, Texas. (Zorin09)

Nonrenewable Resources: Oil

North America is one of the world's largest producers of oil in 2019. However, North America has about 12.5 percent of the world's proven oil reserves, while the Arabian Peninsula has 28 percent. Only about 3 percent of the continent's oil reserves are found in Mexico, mostly around the Gulf Coast.

Most oil in the United States and Canada is produced in northern Alaska, the Texas-Louisiana Gulf coastal region, the midcontinent district from eastern Kansas to western Oklahoma and Texas, and along the Front Range of the Canadian Rocky Mountains, in central and northern Alberta. Lesser oil fields are found in southern California, west central Appalachia, central Alabama, the northern Great Plains, and the central and southern Rocky Mountains.

During the 1970s the United States and Canada began exploiting their Arctic territories—Alaska, the Yukon, and the Northwest Territories—more heavily. In 1977 the Trans-Alaska pipeline was opened, pumping oil from Alaska's North Slope 800 miles (1,290 km.) south to the warm-water port of Valdez, Alaska, where it is shipped off by tankers. The pipeline's maximum throughput occurred in 1988. By 2019, the pipeline was running at only one-third capacity and providing just 2.6 percent of US oil needs.

The United States and Canada also have great oil reserves not counted as proven reserves. This oil, which is found within sandstone or shale bedrock, is now being extracted via a controversial method called hydraulic fracking. The amount of oil stored in these alternative deposits is said to rival the proven reserves of the Arabian Peninsula.

Natural Gas

North America is one of the world's largest producers of natural gas, with about 8.6 percent of the world's reserves. Natural gas previously was used only near the areas where it was produced, but since a process was discovered to liquefy it, it can be transported more easily. The distribution of natural gas follows the pattern of oilfields, because oil and natural gas are usually found in similar geologic formations.

Coal

Coal was important in the rise of North America as an industrial power and was the chief source of energy for industrial and electrical production until the 1960s, when oil began to be more widely used. Today, the United States is the world's third-largest producer of coal (after China and India).

The US has the world's largest coal reserves, but nonetheless has seen coal consumption fall precipitously in recent years. In 2019, coal produced approximately 23.5 percent of US electricity, a 71 percent drop in just five years and the lowest level since 1979. The decline in coal consumption is the result of both the closing of coal-fired power plants and the decrease in demand due to greater competition from natural gas and from renewable energy sources. Environmental concerns have also played a role in coal-plant capacity reductions and shut-

downs. Coal-plant closings peaked in 2015, at least in part because stricter emissions standards took effect that year. The rate of coal plants going offline remnained high through the rest of the decade.

In 2020, the US had three main coal-producing regions: the Western, the Appalachian, and the Illinois Basin regions. The western coal region produces more than half of the country's coal, virtually all of it extracted from surface mines. Wyoming, the leading coal state, is the source of about 40 percent of all US coal. A single mine in Wyoming, the North Antelope Rochelle mine, produces more coal than all of West Virginia, the largest coal-producing state in the Appalachian region, which also includes Ohio, Pennsylvania, Tennessee, and Virginia. Slightly more than one-quarter of all US coal comes from this region, mostly from underground mines. The Illinois Basin region accounts for about 7 percent of the country's coal production.

Renewable Energy Resources

North American has been a world leader in the utilization of renewable energy. In just the period be-

Coastal exposure of the Point Aconi Seam (bituminous coal; Pennsylvanian) exposed at Point Aconi, Nova Scotia. (M.C. Rygel)

The San Gorgonio Pass wind farm in Riverside County, California. (Erik Wilde)

tween 2000 and 2018, the use of renewable energy in the United States has more than doubled, and it now provides more than 17 percent of the country's primary energy supply. In Canada, renewable energy is also the source of 17 percent of the country's primary energy, and 67 percent of its electricity generation. Until recently, hydropower accounted for more than half of the renewable energy sources applied to electricity generation in both countries. It still does in Canada, where many provinces derive 90 percent or more of their electricity from hydroplants. But in the US, 2019 figures show that wind power surpassed hydropower as the leading renewable energy source. In Canada, wind power was responsible for about 4.6 percent of electricty generation. Solar power, though becoming more affordable, still only accounts for 1.5 percent of the electricity generation in the US, and even less (0.51 percent) in Canada.

In 2019, geothermal energy sources produced about 0.4 percent of the US's utility-scale electricity. Seven geothermal power plants are currently online in seven states. Neither Canada nor Mexico operate geothermal power plants.

In Mexico, renewable sources supply 22 percent of the country's electricity-generation capacity. In 2013, this 22-percent figure broke down to 18.1 percent hydropower, 2.5 percent wind power, and 0.1 percent solar. Solar power holds great potential in Mexico, where vast areas of the country receive close to the maximum hours of sunlight on an annual basis. The Mexican government has put forth ambitious plans to increase the solar power capacity dramatically in the coming decade.

Iron Ore

Iron ore is the primary ingredient in steel, which is necessary for industrial construction, railroads, highways, commercial buildings, and other features of developed countries. The United States and Canada are both major producers and consumers of iron ore, most of which comes from Minnesota, Michigan, and Utah, central Alabama, and the Canadian provinces of Quebec and Newfoundland and Labrador. The United States and Canada are the world's eighth and ninth largest producers, respectively, of iron ore.

Other Minerals

Many other nonenergy minerals are mined and refined in North America, including gold, silver, copper, nickel, uranium, cadmium, mercury, lead, zinc, and molybdenum. The continent's minerals are consumed by the diverse manufacturing industries of North America and sold to other countries. North America's most important regions for nonenergy minerals are the Canadian Shield in eastern Canada; the volcanic and metamorphosed mountainous areas, including the western Rocky

Mountain region; and the Appalachian Mountain region of the United States.

Human Geography of Natural Resources

Abundant resources helped to build and develop the continent's waterway, railroad, and highway transportation networks; transportation, in turn, helped to open up the coal and oil fields, and to exploit and distribute the abundant wealth of fuels, metals, timber, and other resources. Because natural resources are so involved in the development of landscapes and the progress of society, there is a close relationship between resources and human settlement patterns, economic growth, and industry. These variations in the human geography of the continent result, in part, from spatial variations in the location of natural resources.

Resources and Settlement

One of the most formidable barriers to the settlement of North America was the lack of water in the

Cave of the Crystals or Giant Crystal Cave is a cave connected to the Naica Mine, in Naica, Chihuahua, Mexico. (Alexander Van Driessche)

semiarid and arid lands of the west. Although the technology to secure more water resources now exists, allowing for dense settlement and even agriculture on arid lands, water supply still influences human settlement. Water must be found locally or transported, via canal or pipe, before new neighborhoods can be built or towns and cities can expand. Many settlements in North America began as sites of mining activity, lumber camps, or fishing villages, later growing into larger towns and cities.

Resources and Industry

Industry developed at sites across the continent that had favorable access to sources of energy needed for industrial power. The first manufacturing industries in the new United States of America—iron works and textile mills—were sited along the edge of the New England upland and the Appalachian Piedmont. There, rapidly running rivers could be diverted to turn water wheels, whose energy could move the gears and levers of industry. More diverse types of refining and manufacturing facilities later were sited around the coal fields of the eastern and western Appalachian Mountains in Pennsylvania, and still later, around the earliest oil and gas fields in western Pennsylvania.

As the settlement frontier pushed westward, new energy resource deposits were discovered and tapped, and new industries sprang up around the coalfields in Indiana, Illinois, and eastern Michigan. The principal industrial region of the United States and Canada emerged around Great Lakes port set-tlements that were ideally situated between the Appalachian coal fields, the source of energy for industry and northern Minnesota and southern Ontario, the source of iron ore, the raw material for steel and related industrial products. Newer industrial regions emerged around the oil fields of Oklahoma, Texas, and southern California, and the Gulf Coastal Plain in Louisiana, Texas, and Mexico. Later, the energy provided by hydropower in the Pacific Northwest encouraged industry in that region.

Future of Resources in North America

The North American continent has vast forests, good agricultural land, and extensive deposits of industrial and energy resources. These resources are generally concentrated in large enough quantities to make long-term extraction economically feasible. Most of the sites that were historically the richest in raw materials have continued to support timber, agricultural, drilling, or mining operations. The continental and offshore mineral storehouse still contains outstanding resource reserves that have attracted future exploration and exploitation as well as substantial environmental concerns. Resource development is focused to a significant extent on the humid conifer forests of the southeastern United States, the oil and gas supplies in the Arctic, oil shale deposits in Colorado and Wyoming, and the tar sands in the Canadian province of Alberta.

Dale R. Lightfoot

NATURAL RESOURCES OF THE CARIBBEAN

The natural resources of the Caribbean are severely limited. In terms of nonrenewable resources, the Greater Antilles is richer in minerals than the Lesser Antilles. Cuba has the most diversified mineral resources, including nickel, iron ore, chrome, cobalt, copper, petroleum, manganese, and marble. Nickel, used in the production of military hardware, is Cuba's most important mineral. In 2019, Cuba was the world's ninth-leading nickel producer.

Jamaica and Guyana have extensive deposits of bauxite, the ore used for aluminum production. Since 1990, however, only Jamaican mines have operated. Some bauxite is mined and processed locally into alumina and sold mainly to the United States and Canada. Most bauxite refining occurs in North America because processing it requires considerable energy. Jamaica was once the world's largest supplier of bauxite, with several foreign companies operating on the island. Now Jamaica must compete with high-grade ores from Australia, China, Brazil, and other countries. Jamaica also mines gypsum, marble, sand, gravel, and industrial lime.

Haiti's mineral resources, including bauxite, copper, gold, iron, silver, sulfur, and tin, are underdeveloped. Its bauxite mine closed in the 1980's. Haiti is the poorest country in the Western Hemisphere, and its high population growth rate has resulted in more people than the weak resource base can provide for. Its economy has yet to recover from the damage caused by a catastrophic earthquake in 2010.

The Dominican Republic produces small quantities of gold, copper, and platinum. Amber and dore (an alloy of silver and gold) are also produced and are major exports. The Dominican Republic is also noted for having the world's largest known salt deposits. Small amounts of manganese, lead, copper, and zinc are found throughout most of the Caribbean islands.

Energy Resources

The Caribbean region is devoid of most energy resources, except in Trinidad and Tobago. Trinidad and Tobago is the largest oil and gas producer in the Caribbean and among the top five exporters of liquefied natural gas in the world. Barbados and Cuba also produce small amounts of petroleum, but not enough to export.

Renewable energy plays a minor but growing role in the Caribbean. Solar energy holds the greatest potential, with many areas receiving upwards of 200 days of sunshine per year. Despite many proposals

Pointe-a-Pierre Oil Refinery, San Fernando, Trinidad. (Christianwelsh)

and ambitious goals, the region has yet to embrace solar energy on a meaningful basis. More promising is wind power, with Jamaica, Aruba, and the Dominican Republic, among other Caribbean locales, having built extensive wind farms. Much slower to catch on is geothermal energy, with only the French island of Guadeloupe using it, though St. Kitts and Nevis have plans for building a facility.

Firewood is an important fuel source in the Caribbean region, especially among the rural poor. Even though it is looked upon as old-fashioned and inefficient, many residents also convert it into charcoal for use in cooking foods.

Water Resources

Precipitation varies from island to island. As a rule, low-lying islands such as Aruba, Bonaire, and Curaçao do not receive much rainfall, because the moisture-laden air does not rise high enough to cool and produce rainfall. Therefore, they receive less than 25 inches (635 millimeters) of rain each year. This causes severe water shortages, which affect industry and the tourist trade. Aruba, Grand Cayman, and Trinidad and Tobago, among other islands, have built desalinization plants to produce or supplement their water supplies. In general, agriculture is limited by lack of water. Antigua, Barbuda, and Anguilla are flat, semiarid, limestone islands that of-

ten experience drought conditions. They frequently must import water to meet their needs.

Other Caribbean islands have high central volcanic cores. The Windward Islands are especially mountainous. These volcanic peaks and mountains force moisture-laden winds to rise, cool, and drop their precipitation on the islands. Water is generally in sufficient supply for tourism and agriculture.

Precipitation varies markedly within the islands themselves. Heavier rains generally fall on the northeast (windward) sides of the islands and lighter rain on the southwest (leeward) sides. Montego Bay, Jamaica, receives an average of 51 inches (1,295 millimeters) of rainfall each year. Kingston, on Jamaica's southern coast, receives only 31 inches (790 millimeters) yearly because precipitation is blocked by the mountains. Irrigation is necessary for agriculture in southern Jamaica. This type of variation in rainfall is found in many Caribbean islands.

While the islands of the Caribbean region are limited in natural resources, they are rich in natural beauty, pleasant climates, and spectacular sunsets. Their sandy beaches, coral reefs, and luxury resorts draw millions of visitors each year. By far, the tourist trade is the most valuable source of revenue the Caribbean region possesses.

Carol Ann Gillespie

DISCUSSION QUESTIONS: NATURAL RESOURCES

Q1. How has North America's mineral wealth contributed to the continent's development? How does the continent's production of natural resources compare to its consumption of resources? For what natural resources are the Caribbean islands most noted?

Q2. How has North America's abundance of fresh water impacted agriculture? What role do underground water sources play? Where do arid parts of the continent derive the water supplies needed for irrigation?

Q3. Where are North America's primary fossil fuel resources located? Why has the use of coal been eclipsed by other energy sources? To what extent have the countries of North America and the Caribbean embraced the use of alternative energy sources?

NORTH AMERICAN FLORA

The world's major biomes are all represented in the diverse vegetation of North America. Forest is the native flora of almost half of mainland Canada and the United States. Before European settlement, the forest was nearly continuous over the eastern, and much of the northern, part of the continent. Grasses covered a large part of the continental interior. Desert vegetation is native in the Southwest, tundra in the far north. Over much of the continent, human activity has virtually eliminated native vegetation.

Coniferous Forests of the West

Along the Pacific coast, from Alaska to northern California, evergreen, coniferous forest grows luxuriantly, watered by moisture-laden winds blowing inland from the ocean. This lowland forest includes some of the largest and longest-lived trees in the world. North of California, characteristic trees include Sitka spruce, western hemlock, and western red cedar. Douglas fir, one of the major timber species in North America, is also common. The northwest coastal coniferous forest is sometimes called

Two giant sequoias, Sequoia National Park. The right-hand tree bears a large fire scar at its base. (Aronlevin)

Scenic view of Jacques-Cartier River valley from Andante mountain, Jacques-Cartier National Park, Quebec, Canada. (Cephas)

temperate rain forest because, in its lushness, it resembles the tropical rain forests. Many of the trees of the coastal forest have been cut for timber.

In California, the dominant coastal conifer species is the coast redwood. The tallest tree in the world, coast redwoods can reach 330 feet (100 meters) and live 2,000 years.

Coniferous forest also grows along the Cascade Mountains and the Sierra Nevada. Trees of the Cascades include mountain hemlock and subalpine fir at high elevations, and western hemlock, western red cedar, and firs somewhat lower. Sierra Nevada forests include pines, mountain hemlock, and red fir at high elevations; and red and white firs, pines, and Douglas fir somewhat lower. Ponderosa pine is dominant at low mountain elevations in both of these Pacific ranges.

The giant sequoia, long thought to be the largest living organism on Earth, grows in scattered groves in the Sierra Nevada. (The largest organism actually may be a very old tree root-rot fungus that covers about 1,500 acres [600 hectares] in Washington

State.) Although shorter than the coast redwood, the giant sequoia is larger in trunk diameter and bulk. It can reach 260 feet (80 meters) tall and 30 feet (10 meters) in circumference.

Coniferous forest also dominates the Rocky Mountains and some mountainous areas of Mexico. In the Rockies, Englemann spruce and subalpine fir grow at high elevations, and Douglas fir, lodgepole pine, and white fir somewhat lower. Ponderosa pine grows throughout the Rockies at low elevations and is a dominant tree in western North America.

Boreal Coniferous Forest

Just south of the Arctic tundra in North America is a broad belt of boreal, coniferous, evergreen forest. It is often called taiga, the Russian name for similar forest growing in northern Eurasia. However, in large areas of northeastern Siberia, the dominant tree is larch, which is deciduous, shedding its leaves each autumn, whereas the North American taiga is mostly evergreen. Taiga is the most extensive coniferous forest in North America, covering nearly 30 percent of the land area north of Mexico. It grows across Alaska and Canada and southward into the northern Great Lakes states and New England. White spruce and balsam fir dominate much of the Canadian taiga. Taiga has little shrub or understory vegetation, but the ground is commonly carpeted with small plants.

Eastern Deciduous Forest

A forest of mainly broad-leaved, deciduous trees is the native vegetation of much of eastern North America. Narrow fingers of this forest, growing along rivers, penetrate westward into the interior grasslands. European settlers cut most of the eastern forest, but second-growth forest now covers considerable areas. The plants are closely related genetically to plants of the temperate deciduous forests in Europe and Asia. In contrast, the plants of other biomes in North America are generally not closely related to the plants that occur in the same biomes elsewhere in the world, although they look similar.

In the eastern deciduous forest, maple and oak are widespread—maples especially in the north, oaks in the south. There are major subdivisions within the forest. These include oak and hickory forest in Illinois, Missouri, Arkansas, and eastern Texas, and also in the east—Pennsylvania, Virginia, and West Virginia—where oak and chestnut forest formerly predominated; beech and maple forest in Michigan, Indiana, and Ohio; and maple and basswood forest in Wisconsin and Minnesota. The forest in parts of Michigan, Wisconsin, Minnesota, and New England contains not only deciduous trees but also evergreen conifers, including pines and hemlock. Vast native pine stands in the Great Lakes states have been cut for lumber.

Plant diseases have changed the composition of the eastern forest. American chestnut was once an important tree, but has now nearly disappeared due to an introduced fungal disease. Dutch elm disease has similarly devastated American elms.

Other Forests

The southeastern United States, excluding the Florida peninsula, once supported open stands of pine and also mixed evergreen and deciduous forest. The mixed forest included a variety of pines, evergreen species of oaks, and deciduous trees. In much of Florida, the native vegetation is a mixture of deciduous and evergreen trees that are subtropical rather than temperate. In many parts of the Southeast, people have replaced the native vegetation with fast-growing species of pines for timber production.

In Mexico, tropical rain forest and savanna are prominent on the west coast, in the south and east, and in the Yucatán. On the south coasts of Mexico and Florida, swamps of mangrove trees are common.

Central Grasslands

The central plains of North America, a wide swath from the Texas coast north to Saskatchewan, Canada, were once a type of vast grassland called the prairie. The climate there is too dry to support trees, except along rivers. Fire is important for maintaining grasslands. From

A glimpse of the southern Great Plains in southern Oklahoma. (Billy Hathorn)

west to east, there is a transition from the more desertlike short-grass prairie (the Great Plains), through the mixed-grass prairie, to the moister, richer, tall-grass prairie. This change is related to an increase in rainfall from west to east. Grasses shorter than 1.5 feet (0.5 meter) dominate the short-grass prairie. In the tall-grass prairie, some grasses grow to more than 10 feet (3 meters). Colorful wildflowers brighten the prairie landscape.

Grassland soil is the most fertile in North America. Instead of wild prairie grasses, this land now supports agriculture and the domesticated grasses corn and wheat. The tall-grass prairie, which had the best soil in all the grassland, has been almost entirely converted to growing corn. Much of the grassland that escaped the plow is now grazed by cattle, which has disturbed the land and aided the spread of invasive, nonnative plants.

Other outlying grasslands occur in western North America. Between the eastern deciduous forest and the prairie is savanna, a grassland with scattered deciduous trees, mainly oaks. Savanna also occurs over much of eastern Mexico and southern Florida.

Scrub and Desert

In the semiarid and arid West, the natural vegetation is grass and shrubs. Over a large part of California, this takes the form of a fire adapted scrub community called chaparral, in which evergreen, often spiny shrubs form dense thickets. The cli-

West Saguaro National Park around Sombrero Mountain near Tucson, Arizona. (WClarke)

mate, with rainy, mild winters and hot, dry summers, is like that around the Mediterranean Sea, where a similar kind of vegetation, called maquis, has evolved. However, chaparral and maquis vegetation are not closely related genetically. Human beings have greatly altered the chaparral through the grazing of livestock and other disturbances.

The North American deserts, which are located between the Rocky Mountains and the Sierra Nevada, cover less than 5 percent of the continent. Shrubs are the predominant vegetation, although there are also many species of annuals. Desert plants are sparsely distributed, and many have either small leaves or none at all. Cacti and other succulent plants are common.

In southern California, Arizona, New Mexico, west Texas, and northwestern Mexico, there are three distinct deserts. These are all hot deserts, like Africa's Sahara. The Sonoran Desert stretches from southern California to western Arizona and south into Mexico. A characteristic plant of the Sonoran is the giant saguaro cactus. To the east of the Sonoran,

---

INVASIVE NONNATIVE PLANTS

Approximately 50,000 plant species growing in North America originated in other continents. Although the large majority of these nonnative plants cause no problems, an estimated 4,300 are invasive, contributing to the displacement of native plants. About half of these invasive plants were brought to North America to beautify gardens, and their spread threatens North American ecosystems. Hawaii, Florida, the Great Lakes, and the west coast are North American regions most affected by invasive nonnative plants.

in west Texas and New Mexico, is the Chihuahuan Desert, where a common plant is the agave, or century plant. North of the Sonoran Desert, in southeastern California, southern Nevada, and northwestern Arizona, is the Mojave Desert, where the Joshua tree, a tree-like lily, is a well-known plant. It can reach 50 feet (15 meters) in height. Creosote bush is common in all three deserts. To the north, the Mojave Desert grades into the Great Basin Desert, which is a large and bleak cold desert. The dominant plant of the Great Basin Desert is the sagebrush. Plant diversity there is less than that of the hot American deserts.

Deserts dominated by grasses rather than shrubs once occurred at high elevations near the Sonoran and Chihuahuan deserts. Much of this area has been overtaken by such desert scrub species as creosote bush, mesquite, and tarbush. Grazing by cattle may have been a factor in this change. Desert is very fragile. Even one pass with a heavy vehicle causes lasting damage.

Tundra

Tundra vegetation grows to the northern limits of plant growth, above the Arctic Circle, in Canada. The flora consists of only about 600 plant species.

In contrast, tropical regions that are smaller in area support tens of thousands of plant species. Tundra is dominated by grasses, sedges, mosses, and lichens. Some shrubby plants also grow there. Most tundra plants are perennials. During the short Arctic growing season, many of these plants produce brightly colored flowers. Like desert, tundra is exceptionally fragile, and it takes many years for disturbed tundra to recover.

Tundra also occurs southward, on mountaintops, from southern Alaska into the Rocky Mountains, the Cascades, and the Sierra Nevada. This alpine tundra grows at elevations too high for mountain coniferous forest.

Coastal Vegetation

Along the coasts of the Atlantic Ocean and the Gulf of Mexico, the soil is saturated with water and is very salty. Tides regularly inundate low-lying vegetation. The plants in these salt marsh areas are not diverse, consisting mainly of grasses and rushes. Marshes are a vital breeding ground and nursery for fish and shellfish, and play an important role in absorbing and purifying water from the land. Coastal marshes are being lost to development at an alarming rate.

Jane F. Hill

NORTH AMERICAN FAUNA

The wildlife of North America can be grouped within two large regions: the Nearctic realm, which covers most of North America from the Arctic to northern Mexico; and the Neotropical realm, which covers southern Mexico and all of Central America. Species in the Nearctic region are similar to those of Eurasia and North Africa, and originally reached North America from Eurasia by passing over the Bering Strait land bridge that once connected Siberia and Alaska, about 60 million years ago. Species in the Neotropical realm are distinctly different from Nearctic wildlife, and reached Central America and Mexico by gradual movement up the isthmus of Panama from South America.

North America's fauna also can be grouped by regions that reflect such climatic influences as latitude, the position of mountain ranges and oceans, and the plants and trees that grow in that area (grasslands, desert, forest, or tundra). Generally, Arctic animals are found to the far north and on the highest slopes of mountains. As one goes farther south, or farther down the sides of mountains, the animals will be those of the forest, grasslands, or desert environments.

Arctic Tundra

Animals of the far north are similar to those found in Eurasia, and are well-adapted to their cold, treeless environment. Many of these animals evolved from Ice Age species as the glaciers that once covered North America slowly retreated northward. They are large in size and thickly furred, allowing

American bison. (United States Department of Agriculture)

Habitats and Selected Vertebrates of North America

them to maximize conservation of body heat. Large herbivores such as musk oxen and caribou graze on grasses, lichens, and mosses and are, in turn, a food source for polar bears and Arctic wolves. Smaller predators such as the Arctic white fox feed on Arctic hares and small rodents such as voles, lemmings, or the Arctic ground squirrels that subsist upon the small shrubs, berries, and grass seeds of the tundra.

Seals and whales proliferate in the Arctic seas. Birdlife in the Arctic tundra is nearly absent in the winter months, with the exception of willow ptarmigans and snowy owls. In the three to four months of summer, several bird species use the region as a breeding ground where young can be hatched and fed with the abundant insect life that emerges during the long and warmer days of summer. Among these migratory species are many varieties of waterfowl, including Canada geese, snow geese, whooping swans, trumpeter swans, phalaropes, plovers, and Arctic terns.

Boreal Forests

Farther south, below the tree line, coniferous northern forests of fir, spruce, cedar, hemlock, and pine provide shelter and food for the moose, mule deer, and snowshoe hares that browse on the needles of these trees. Squirrels, chipmunks, and porcupines also thrive in the forests of the far north. Predators of this region include martens, fishers, lynx, wolves, weasels, red foxes, and wolverines. Ponds, rivers,

ADAPTING TO EXTREMES

The extremes of climate in North America have caused many animals to adapt in interesting ways. In the far north, animals such as polar bears, musk oxen, and caribou tend to be large in size, so that their external skin area is minimized in relation to their internal portions, allowing the least possible heat loss to the cold air. Many arctic animals have both inner and outer layers of fur, trapping their body heat in the insulating layer. The feet of arctic animals are thickly furred and quite broad, allowing the animals to move over snow without sinking into it.

Animals of the southern deserts have equally effective ways of adapting to the extreme dryness and heat of their environment. They collect and store food when it is abundant after a rare rainfall and then survive long periods of drought on the surplus. They obtain water from the vegetation they eat, rather than from lakes or streams. Many desert animals are pale in color, reflecting back the heat of the sun and blending in with the surrounding landscape. Desert animals are largely nocturnal, moving about only in the cool of the night, after the moist dew falls.

marshes, and swamps are common in this habitat, and provide homes for beavers, muskrats, river otters, minks, and such common fish species as whitefish, perch, pickerel, and pike.

Bird species native to the boreal forest include jays (Canada, blue, and gray), thrushes, finches, nuthatches, loons, osprey, ravens, and crows. A wide variety of songbirds (warblers) and some hummingbirds nest in the boreal woods in the summer months.

Hardwood Forests

The eastern half of North America, from southern Canada to Florida, was once thickly covered by deciduous forests of maple, oak, beech, ash, sycamore, hickory, and other trees that shed their leaves in the winter. Where this forest remains, it provides food for a wide variety of mammals, including black bears, red foxes, racoons, red and gray squirrels, minks, muskrats, jackrabbits, cottontail rabbits, ground-

Blue jay (Cyanocitta cristata) in Algonquin Provincial Park, Canada. (Mdf)

Mexican Spider Monkey (Ateles geoffroyi vellerosus), is a subspecies of Geoffroy's Spider Monkey. (Alex Lee)

hogs, chipmunks, mice, moles, bobcats, skunks, ermine, opossums, and porcupines. Fish species common to the rivers of eastern and southeastern North America include catfish, suckers, and gar. Several species of salamanders and turtles live in the marshes, streams, and rivers of the deciduous forests, especially in the Appalachians. Waterfowl such as kingfishers, herons, ducks, and grebes also live along the waterways of the hardwood forest region. White-tailed deer, black bears, coyote, and bobcats have adapted to life in many suburban areas.

Grasslands
The prairies in the center of North America provide abundant grasses and other herbaceous plants for many small mammals and grazing animals. Jackrabbits, badgers, prairie chickens, and small rodents such as pocket gophers, prairie dogs, and

Richardson's ground squirrels, feed on grass and roots, as do such larger herbivores as pronghorn antelopes and the American bison. The numerous small rodents and the openness of the grasslands provide optimal habitat conditions for such raptors as owls, hawks, and falcons. Waterfowl nest in the many seasonal watering holes (sloughs) that dot the prairies, although farm drainage and extended periods of drought have greatly reduced this habitat. Seeds and insects are plentiful in the grasslands, supporting such bird species as grouse, quail, partridge, and finches.

Rocky Mountains
The high mountains that run along the western side of North America are inhabited by a number of unique species. Bighorn sheep, Rocky Mountain goats, mule deer, and elk graze on the grasses of the

foothills and slopes of the mountains. Kodiak bears, grizzly bears, and mountain lions prey upon the grazing animals, and bald and golden eagles subsist upon the ground squirrels, marmots, voles, shrews, and pikas that live in the grasses and scattered forests of the lower mountain slopes. Dipper birds feed from the fast-running mountain streams and are found nowhere else on the continent.

Southwestern Deserts

A large number of animals have adapted to the lack of vegetation and water that exists in much of the southwestern United States and Mexico. Kangaroo rats, pocket mice, jackrabbits, armadillos, peccaries, ring-tailed cats, and ground squirrels all survive in this hostile environment. Predators include bobcats, desert foxes, badgers, and coyotes.

The most prolific forms of wildlife in the arid deserts of southernmost North America are the reptiles, including many different species of lizards, rattlesnakes, toads, and iguanas. Because they can only be active when the outside air temperature provides warmth for basic body functions, reptiles and amphibians are quite rare in the most northern parts of the continent but thrive in the arid and hot southern parts of North America. Roadrunners are a major bird predator of these reptiles, along with eagles and hawks.

Neotropical Forests

The tropical rain forests of Central America and southern Mexico have an astonishing variety of wildlife. The forest canopy has abundant bird life in the form of macaws, parrots, turkey vultures, and flycatchers. Monkeys of many varieties (spider, howler, squirrel, and capuchin) as well as sloths and tamarinds also live in the canopy, feeding on the many fruiting trees. On the floor of the forest, many varieties of ants, spiders, beetles, and chiggers provide food for smaller predators such as anteaters and various species of bats. A number of animals indigenous to South America have adapted to and thrive in Central America, including tapirs, capybaras, pacas, jaguars, ocelots, and agoutis.

An olive ridley sea turtle nesting on Escobilla Beach, Oaxaca, Mexico. (Claudio Giovenzana)

ENDANGERED NORTH AMERICAN MAMMALS

Common Name	Scientific Name	Range
Bat, Florida bonneted	Eumops floridanus	Southeastern U.S.A. (Florida)
Bat, gray	Myotis grisescens	Midwestern U.S.A.
Bat, Hawaiian hoary	Lasiurus cinereus semotus	Pacific U.S.A. (Hawaiian)
Bat, Indiana	Myotis sodalis	Eastern and Midwestern U.S.A.
Bat, little Mariana fruit	Pteropus tokudae	Pacific U.S.A. (Guam)
Bat, Mexican long-nosed	Leptonycteris nivalis	Southwestern U.S.A. (Texas), Mexico
Bat, Ozark big-eared	Corynorhinus townsendii ingens	Southwestern U.S.A.
Bat, Pacific sheath-tailed	Emballonura semicaudata rotensis	Pacific (Northern Mariana Islands, American Samoa)
Bat, Virginia big-eared	Corynorhinus townsendii virginianus	Northeastern U.S.A.
Caribou, Woodland	Rangifer tarandus caribou	Pacific U.S.A. (Idaho, Washington), Canada
Deer, Key	Odocoileus virginianus clavium	Southeastern U.S.A. (Florida)
Ferret, black-footed	Mustela nigripes	Mountain Prairie U.S.A., Mexico
Fisher	Pekania pennanti	Pacific Southwestern U.S.A.
Fox, San Joaquin kit	Vulpes macrotis mutica	Pacific Southwestern U.S.A. (California)
Jaguar	Panthera onca	Southwestern U.S.A. (Arizona, New Mexico)
Jaguarundi, Gulf Coast	Herpailurus yagouaroundi cacomitli	Southwestern U.S.A. (Texas)
Jaguarundi, Sinaloan	Herpailurus yagouaroundi tolteca	Mexico
Kangaroo rat, Fresno	Dipodomys nitratoides exilis	Pacific Southwestern U.S.A. (California)
Kangaroo rat, Giant	Dipodomys ingens	Pacific Southwestern U.S.A. (California)
Kangaroo rat, Morro Bay	Dipodomys heermanni morroensis	Pacific Southwestern U.S.A. (California)
Kangaroo rat, San Bernardino	Dipodomys merriami parvus	Pacific Southwestern U.S.A. (California)
Kangaroo rat, Stephen's	Dipodomys stephensi	Pacific Southwestern U.S.A. (California)
Kangaroo rat, Tipton	Dipodomys nitratoides nitratoides	Pacific Southwestern U.S.A. (California)
Margay	Leopardus wiedii	Mexico
Mountain beaver, Point Arena	Aplodontia rufa nigra	Pacific Southwestern U.S.A. (California)
Mouse, Alabama beach	Peromyscus polionotus ammobates	Southeastern U.S.A. (Alabama)
Mouse, Anastasia Island beach	Peromyscus polionotus phasma	Southeastern U.S.A. (Florida)
Mouse, Choctawhatchee beach	Peromyscus polionotus allophrys	Southeastern U.S.A. (Florida)
Mouse, Key Largo cotton	Peromyscus gossypinus allapaticola	Southeastern U.S.A. (Florida)
Mouse, N.M. meadow jumping	Zapus hudsonius luteus	Southwestern U.S.A.
Mouse, Pacific pocke	Perognathus longimembris pacificus	Pacific Southwestern U.S.A. (California)
Mouse, Perdido Key beach	Peromyscus polionotus trissyllepsis	Southeastern U.S.A. (Alabama, Florida)
Mouse, salt marsh harvest	Reithrodontomys raviventris	Pacific Southwestern U.S.A. (California)
Mouse, St. Andrew Beach	Peromyscus polionotus peninsularis	Southeastern U.S.A. (Florida)
Ocelot	Leopardus pardalis	Southwestern U.S.A.
Panther, Florida	Leopardus pardalis	Southeastern U.S.A. (Florida)
Pronghorn, Sonoran	Antilocapra americana sonoriensis	Southwestern U.S.A. (Arizona), Mexico
Rabbit, Columbia Basin Pygmy	Brachylagus idahoensis	Pacific U.S.A. (Washington)
Rabbit, Lower Keys marsh	Sylvilagus palustris hefneri	Southeastern U.S.A. (Florida)
Rabbit, riparian brush	Sylvilagus bachmani riparius	Pacific Southwestern U.S.A. (California)
Rice rat, Silver	Oryzomys palustris natator	Southeastern U.S.A. (Florida)
Seal, Hawaiian monk	Monachus schauinslandi	National Marine Fisheries Service
Sea lion, Steller	Eumetopias jubatus	Western U.S.A.
Seal, Ringed	Phoca hispida ladogensis	National Marine Fisheries Service
Sheep, Peninsular bighorn	Ovis canadensis nelson	Pacific Southwestern U.S.A. (California)
Sheep, Sierra Nevada bighorn	Ovis canadensis sierra	Pacific Southwestern U.S.A. (California)
Shrew, Buena Vista Lake ornate	Sorex ornatus relictus	Pacific Southwestern U.S.A. (California)
Squirrel, Carolina northern flying	Glaucomys sabrinus coloratus	Southeastern U.S.A.
Squirrel, Mount Graham red	Tamiasciurus hudsonicus grahamensis	Southwestern U.S.A. (Arizona)
Vole, Amargosa	Microtus californicus scirpensis	Pacific Southwestern U.S.A. (California)
Vole, Florida salt marsh	Microtus pennsylvanicus dukecampbelli	Southeastern U.S.A. (Florida)
Whale, beluga	Delphinapterus leucas	National Marine Fisheries Service
Whale, blue	Balaenoptera musculus	National Marine Fisheries Service
Whale, bowhead	Balaena mysticetus	National Marine Fisheries Service
Whale, false killer	Pseudorca crassidens	National Marine Fisheries Service
Whale, finback	Balaenoptera physalus	National Marine Fisheries Service
Whale, killer	Orcinus orca	National Marine Fisheries Service
Whale, North Atlantic right	Eubalaena glacialis	National Marine Fisheries Service
Whale, North Pacific right	Eubalaena japonica	National Marine Fisheries Service
Whale, sei	Balaenoptera borealis	National Marine Fisheries Service
Whale, sperm	Physeter catodon (=macrocephalus)	National Marine Fisheries Service
Wolf, gray	Canis lupus	Mountain Prairie U.S.A.
Wolf, Mexican	Canis lupus baileyi	Southwestern U.S.A. (Arizona, New Mexico)
Wolf, red	Canis rufus	Southeastern U.S.A. (Florida)
Woodrat, Key Largo	Neotoma floridana smalli	Southeastern U.S.A. (Florida)
Woodrat, riparian	Neotoma fuscipes riparia	Pacific Southwestern U.S.A. (California)

Source: U.S. Fish and Wildlife Service, U.S. Department of the Interior.

Coastal Regions

The beaches, shores, lagoons, and marshes that line the North American continent are home to many kinds of animals that feed upon the ocean life or the intertidal plants and animals that live in the midzone between fresh and salt water. Many different kinds of migratory waterfowl exploit the small crustaceans and mollusks that live at the water's edge, including sandpipers, stilts, curlews, and flamingos. Seals, sea otters, and walruses are found on both coasts in the north of the continent, while Steller's sea lions and California sea lions are found only on the Pacific coast. The lagoons of the southern Atlantic coast and the Gulf of Mexico are home to manatees, alligators, pelicans, egrets, and spoonbills.

Helen Salmon

CARIBBEAN FLORA AND FAUNA

Most of the Caribbean region has been affected by human activities. Deforestation began with the development of the sugarcane culture in the seventeenth century. When forests are cut for farmland, soil erosion and depletion often occur. Jamaica, Haiti, and many of the smaller islands have suffered acute ecological degradation. Thorn scrub and grasses have replaced native forests that were cleared for farming. This new vegetation does not protect the ground from the Sun and provides little protection against moisture loss during drought. Livestock grazing also has contributed to ecological degradation.

These complex and fragile island ecosystems are finally being appreciated and protected. Most islands recognize that they must balance development with protection of the natural environment. Many have established active conservation societies and national wildlife trusts for this purpose.

Flora

The Caribbean region is noted for its diverse and varied vegetation. Flowers thrive in the moist, tropical environments found on many islands. Hibiscus, bougainvillea, and orchids are just a few of the endless varieties found here.

The only rain forest left in the Caribbean Islands is a small area on Guadeloupe. However, many of the islands still have large stands of good secondary forest that are being harvested selectively by commercial lumber companies. Gommier, balata, and blue mahoe are some of the valuable species of trees

Hiking trail in the Morne Trois Pitons National Park on the island of Dominica. (Aneil Lutchman)

cut commercially. Such tropical hardwoods as cedar, mahogany, palms, and balsa also grow on many islands. Martinique has some of the largest tracts of forest (rain forest, cloud forest, and dry woodland) left in the Caribbean.

While the Caribbean rain forest is not as diverse as those of Central and South America, it still supports numerous plant species. Several of these plants are endemic to the region. For example, Jamaica has more than 3,000 species with 800 endemic to Jamaica. Orchids and bromeliads are stunning examples of climbing and hanging vegetation in the rain forests of the Caribbean. Huge tree ferns, giant elephant ear plants, figs, and balsam trees are also found in these tropical island rain forests.

Plantations of commercial timber (blue mahoe, Caribbean pine, teak, and mahogany) have been established in many locations. These plantations reduce pressure on natural forest, help protect watersheds and soil, and provide valuable wildlife habitat. Much of the original forests of the Caribbean have been cut down to make room for sugar plantations and for use as fuel. Haiti, once covered with luxurious forests, is now on the verge of total deforestation. Soil erosion and desertification have turned the country into an ecological disaster. St. Kitts, on the other hand, is one of the few places in the world where the forest is actually expanding. It provides abundant habitat for exotic vines, wild orchids, and candlewoods.

Closer to the coasts, dry scrub woodland often predominates. Some trees lose their leaves during the dry season. The turpentine tree is common to

Guadeloupe National Park. (Ofol)

the dry scrub forests. It is sometimes called the tourist tree because of its red, peeling bark (like the skin of sunburned tourists). Tree bark and leaves are often sold in the marketplaces and used for herbal remedies and bush medicine.

The marshy coastal waters of many of the islands have dense mangrove swamps. Mangroves grow roots that are thick and stand above the water line. These wetlands provide habitats for the manatee and American crocodile and for huge numbers of migratory birds and resident birds such as the egret and heron.

Dominica, one of the Windward Islands, is known as the Nature Island of the Caribbean. Most of the southern part of the island (17,000 acres/6,900 hectares) has been designated the Morne Trois Pitons National Park, which became a World Heritage Site in 1998. Elfin forests of dense vegetation and low-growing plants cover the highest volcanic peaks and receive an abundance of rain. The trees, stunted by the wind, have leaves adapted with drip tips to cope with the excess moisture. Lower down, the slopes are covered with rain forest. Measures are

Coquí frog at La Finca Caribe, Vieques, Puerto Rico. (Cathybwl)

being taken to protect this forest from being replaced for farmland or for economic gain, thereby preserving it as a valuable water source and a unique laboratory of scientific research. On many Caribbean islands, tourism is a vital source of income, awareness of the need for conservation is increasing, and forests are being protected.

Fauna

Mammals are not good colonizers of small islands, so there is a scarcity of species in the Caribbean. Most of the more commonly seen species (opossum, agouti, mongoose, and some monkeys) were introduced by humans. Bats are the exception to this rule: They migrated to the Caribbean Islands on their own. The mongoose was imported by colonists to kill snakes and rats on sugarcane estates. It never achieved its purpose because it hunts during the day, and rats are nocturnal. Unfortunately, the mongoose has con-

tributed to the extinction of many species of ground-nesting birds, lizards, green iguanas, brown snakes, and red-legged tortoises.

The green vervet monkeys of Barbados, Grenada, and St. Kitts and Nevis were introduced from West Africa by the French during the seventeenth century. Many of them are exported for use in medical research because they are relatively free of disease.

Many islands have their own endemic species of birds, such as the Grenada dove and the Guadeloupe woodpecker. The islands are important stepping-stones in the migration of birds through the Americas. As a result, this region is a popular area for ornithologists and is home to several important nature reserves. Many species of parrots, macaws, and hummingbirds thrive in this region.

Trinidad and Tobago, close to the South American mainland, have no endemic species of their own, yet together they have more species of birds—

Antillean or Caribbean manatee. (Chris Muenzer)

123

ENDANGERED CARIBBEAN SPECIES

Common Name	Scientific Name	Range
Anole, giant Culebra	*Anolis roosevelti*	Puerto Rico
Blackbird, yellow shouldered	*Agelaius xanthomus*	Puerto Rico
Boa, Puerto Rican	*Epicrates inornatus*	Puerto Rico
Boa, Virgin Islands tree	*Epicrates monensis granti*	Puerto Rico, Virgin Islands
Frog, Puerto Rican wetland	*Eleutherodactylus Juanariveroi*	Puerto Rico
Hawk, broad-winged Puerto Rican	*Buteo platypterus brunnescens*	Puerto Rico
Hawk, sharp-shinned Puerto Rican	*Accipiter striatus venator*	Puerto Rico
Iguana, Jamaican	*Cyclura collei*	Jamaica
Nightjar, Puerto Rican	*Caprimulgus noctitherus*	Puerto Rico
Lizard, St. Croix ground	*Ameiva polops*	Virgin Islands
Parrot, Bahaman or Cuban	*Amazona leucocephala*	Bahamas, Cayman Islands, Cuba
Parrot, Puerto Rican	*Amazona vittata*	Puerto Rico
Pigeon, Puerto Rican plain	*Columba inornata wetmorei*	Puerto Rico
Sea Turtle, Hawksbill	*Eretmochelys imbricate*	Puerto Rico, Virgin Islands
Sea Turtle, Leatherback	*Dermochelys coriacea*	Puerto Rico, Virgin Islands*

Note: * this species has an extensive range
Source: U.S. Fish and Wildlife Service, U.S. Department of the Interior.

485 species confirmed as of March 2019—than any other Caribbean island. There are also hundreds of recorded species of butterflies. The Asa Wright Center is one of the most important bird-watching sites in the world.

Lizards and geckos are found on most of the Caribbean islands. There are many species of snakes, turtles, and iguanas scattered throughout the islands. Many are restricted to one island; Jamaica has at least twenty-seven endemic species. Most of the reptiles found in the Caribbean region are harmless to humans, although snakes and geckos are superstitiously feared by the locals. The fer de lance snake is venomous, but its bite is seldom fatal. It lives in isolated areas of dry scrubland and river valleys. Iguanas are found in declining numbers on many islands. These herbivores spend their time hiding in trees and low scrub and trying to avoid contact with humans. The Antilles iguana and the rare, harmless racer snake are in grave danger of extinction in the Caribbean region. Scientists are actively engaged in several projects to learn about and protect these endangered species.

Marine turtles—including the loggerhead, leatherback, hawksbill, and green turtles—are found throughout Caribbean waters. The females come ashore on isolated sandy beaches between the months of May and August to lay eggs. There are programs in place to protect the turtles from poachers on many of the islands. The iguana, marine turtle, and many other species are now protected on reserves in the Caribbean region.

Amphibians such as toads and frogs are common in a variety of shapes and colors. The Cuban pygmy frog is called the world's smallest frog; Dominica's crapaud, or mountain chicken, is one of the largest.

Carol Ann Gillespie

DISCUSSION QUESTIONS: FLORA AND FAUNA

Q1. How do geographers classify North America's biomes? Which biome is the most widespread? How does plant life vary among the biomes?

Q2. What general statements can be made about the transition of North American animal life from the Arctic southward into Mexico? What animal species are most characteristic of the various North American biomes? Where are reptiles most common?

Q3. Why are the island ecosystems of the Caribbean considered so fragile? Which types of mammals are most abundant in the Caribbean? What impact have introduced species had on the native wildlife?

HUMAN GEOGRAPHY

OVERVIEW

THE HUMAN ENVIRONMENT

No person lives in a vacuum. Every human being and community is surrounded by a world of external influences with which it interacts and by which it is affected. In turn, humans influence and change their environments: sometimes intentionally, sometimes not, and sometimes with effects that are harmful to these environments, and, in turn, to humans themselves. Humans have always shaped the world in which they live, but developments over the past few centuries have greatly enhanced this capacity.

Many people feel a sense of alarm about the consequences of widespread adoption of modern technology, including artificial intelligence (AI) and accelerating human population growth in the world. Travel and transportation among the world's regions have been made surer, safer, and faster, and global communication is virtually instantaneous. The human environment is no longer a matter of local physical, biological, or social conditions, or even of merely national or regional concerns—the postmodern world has become a true global community.

Students of human geography divide the human environment into three broad areas: the physical, biological, and social environments. The study of ecology describes and analyzes the interactions of biological forms (mainly plants and animals) and seeks to uncover the optimal means of species cooperation, or symbiosis. Everything that humans do affects life and the physical world around them, and this world provides potentials for and constraints on how humans can live.

As people acquired and shared ever-more knowledge about the world, their abilities to alter and

shape it increased. Humans have always had a direct impact on Earth. Even 10,000 years ago, Neolithic people cut down trees, scratched the earth's surface with simple plows, and replaced diverse plant forms with single crops. From this basic agricultural technology grew more complex human communities, and people were freed from the need to hunt and gather. The alteration of the local ecosystems could have deleterious effects, however, as gardens turned eventually to deserts in places like North Africa and what later became Iraq. Those who kept herds of animals grazed them in areas rich in grasses, and animal fertilizer helped keep them rich. If the area was overgrazed, however, destroying important ground cover, the herders moved on, leaving a perfect setup for erosion and even desertification. Today, people have an even greater ability to alter their environments than did Neolithic people, and ecologists and other scientists as well as citizens and politicians are increasingly concerned about the negative effects of modern alterations.

The Physical Environment

The earth's biosphere is made up of the atmosphere—the mass of air surrounding the earth; the hydrosphere—bodies of water; and the lithosphere—the outer portion of the earth's crust. Each of these, alone and working together, affect human life and human communities.

Climate and weather at their most extreme can make human habitation impossible, or at least extremely uncomfortable. Desert and polar climates do not have the liquid water, vegetation, and animal life necessary to sustain human existence. Humans can adapt to a range of climates, however. Mild variations can be addressed simply, with clothing and shelter. Local droughts, tornadoes, hurricanes, heavy winds, lightning, and hail can have devastating effects even in the most comfortable of climates. Excess rain can be drained away to make habitable land, and arid areas can be irrigated. Most people live in temperate zones where weather extremes are rare or dealt with by technological adaptation. Heating and, more recently, air conditioning can create healthy microclimates, whatever the external conditions. Food can be grown and then transported across long distances to supply populations throughout the year.

The hydrosphere affects the atmosphere in countless ways, and provides the water necessary for human and other life. Bodies of water provide plants and animals for food, transportation routes, and aesthetic pleasure to people, and often serve to flush away waste products. People locate near water sources for all of these reasons, but sometimes suffer from sudden shifts in the water level, as in tidal waves (tsunamis) or flooding. Encroachment of salt water into freshwater bodies (salination) is a problem that can have natural or human causes.

The lithosphere provides the solid, generally dry surface on which people usually live. It has been shaped by the atmosphere (especially wind and rain that erode rocks into soil) and the hydrosphere (for example, alluvial deposits and beach erosion). It serves as the base for much plant life and for most agriculture. People have tapped its mineral deposits and reshaped it in many places; it also reshapes itself through, for example, earthquakes and volcanic eruption. Its great variations—including vegetation—draw or repel people, who exploit or enjoy them for reasons as varied as recreation, military defense, or farming.

The Biological Environment

Humans share the earth with over 8 million different species of plants, animals, and microorganisms—of which only about 2 million have been identified and named. As part of the natural food chain, people rely upon other life-forms for nourishment.

Through perhaps the first 99 percent of human history, people harvested the bounty of nature in its native setting, by hunting and gathering. Domestication of plants and animals, beginning about 10,000 years ago, provided humans a more stable and reliable food supply, revolutionizing human communities. Being omnivores, people can use a wide variety of plants and animals for food, and they have come to control or manage most important food sources through herding, agriculture, or mechanized harvesting. Which plants and animals are chosen as food, and thus which are cultivated, bred, or exploited, are matters of human culture, not, at least in the modern world, of necessity.

Huge increases in human population worldwide have, however, put tremendous strains on provision of adequate nourishment. Areas poorly endowed with foodstuffs or that suffer disastrous droughts or blights may benefit from the importation of food in the short run, but cannot sustain high populations fostered by medical advances and cultural considerations.

Human beings themselves are also hosts to myriad organisms, such as fungi, viruses, bacteria, eyelash mites, worms, and lice. While people usually can coexist with these organisms, at times they are destructive and even fatal to the human organism. Public health and medical efforts have eradicated some of humankind's biological enemies, but others remain, or are evolving, and continue to baffle modern science.

The presence of these enemies to health once played a major role in locating human habitations to avoid so-called "bad air" (*mal-aria*) and the breeding grounds of tsetse flies or other pests. The use of pesticides and draining of marshy grounds have alleviated a good deal of human suffering. Human efforts can also control or eliminate biological threats to the plants and animals used for food, clothing, and other purposes.

Social Environments

Human reproduction and the nurturing of young require cooperation among people. Over time, people gathered in groups that were diverse in age if not in other qualities, and the development of towns and cities eventually created an environment in which otherwise unrelated people interacted on intimate and constructive levels. Specialization, or division of labor, created a higher level of material wealth and culture and ensured interpersonal reliance.

The pooling of labor—both voluntary and forced—allowed for the creation of artificial living environments that defied the elements and met human needs for sustenance. Some seemingly basic human drives of exclusivity and territoriality may be responsible for interpersonal friction, violence and, at the extreme, war. Physical differences, such as size, skin, or hair color, and cultural differences, including language, religion, and customs, have often divided humans or communities. Even within close quarters such as cities, people often separate themselves along lines of perceived differences. Human social identity comes from shared characteristics, but which things are seen as shared, and which as differentiating, is arbitrary.

People can affect their social environment for good and ill through trade and war, cooperation and bigotry, altruism and greed. While people still are somewhat at the mercy of the biological and physical environments, technological developments have balanced the human relationship with these. Negative effects of human interaction, however, often offset the positive gains. People can seed clouds for rain, but also pollute the atmosphere around large cities, create acid rain, and perhaps contribute to global warming.

Human actions can direct water to where it is needed, but people also drain freshwater bodies and increase salination, pollute streams, lakes, and oceans, and encourage flooding by modifying riverbeds. People have terraced mountainsides and irrigated them to create gardens in mountains and deserts, but also lose about 24 billion metric tons of soil to erosion and 30 million acres (12 million hectares) of grazing land to desertification each year. These negative effects not only jeopardize other species of terrestrial life, but also humans' ability to live comfortably, or perhaps at all.

Globalization

Humankind's ability to affect its natural environments has increased enormously in the wake of the Industrial Revolution. The harnessing of steam, chemical, electrical, and atomic energy has enabled people to transform life on a global scale. Economically, the Western world still to dominates global markets despite effort of China to capture the crown, and computer and satellite technology have made even remote parts of the globe reliant on Western information and products. Efficient transportation of goods and people over huge distances has eliminated physical barriers to travel and commerce. The power and influence of multinational corporations and national corporations in international markets continues to grow.

Human environmental problems also have a global scope: Extreme weather, changes in ocean temperatures and sea level rise, global warming, and the spread of disease by travelers have become planetary concerns. International agencies seek to deal with such matters, and also social and political concerns once left to nations or colonial powers, such as population growth, the provision of justice, or environmental destruction within a country. Pessimists warn of horrendous trends in population and ecological damage, and further deterioration of human life and its environments. Optimists dismiss negative reports as exaggerated and alarmist, or expect further technological advances to mitigate the negative effects of human action.

Joseph P. Byrne

POPULATION GROWTH AND DISTRIBUTION

The population of the world has been growing steadily for thousands of years and has grown more in some places than in others. On November 2019, the total population of the earth had reached 7.7 billion people. The population of the United States in August 2019 was approximately 329.45 million. India's population in November 2019 was 1.37 billion, making it the world's second most populous country. China's population was about 1.45 billion—about 1 in 5 people on the planet.

How Populations Are Counted

The U.S. Constitution requires that a census, or enumeration, of the population of the United States be conducted every ten years. The U.S. Census Bureau mails out millions of census forms and pays thousands of people (enumerators) to count people that did not fill out their census forms. This task cost about US$5.6 billion in the year 2010, and estimates for the 2020 census have risen to over US$15 billion. Despite this great effort, millions of people are probably not counted in every U.S. census. Moreover, many countries have much less money to spend on censuses and more people to count. Therefore, information about the population of many poor or less-developed countries is even less accurate than that for the population of the United States.

Counting how many people were alive a hundred, a thousand, or hundreds of thousands of years ago is even more difficult. Estimates are made from archaeological findings, which include human skeletons, ruins of ancient buildings, and evidence of ancient agricultural practices. Historical records of births, deaths, taxes paid, and other information are also used. Although it is not possible to estimate the global population 1,000 years ago with great accuracy, it is a fascinating topic, and many people have participated in estimating the total population of the planet through the ages.

History of Human Population Growth

Ancient ancestors of humans, known as hominids, were alive in Africa and Europe around 1 million years ago. It is believed that modern humans *(Homo sapiens sapiens)* coexisted with the Neanderthals *(Homo sapiens neandertalensis)* about 100,000 years ago. By 8000 BCE (10,000 years ago) fully modern humans numbered around 8 million. If the presence of archaic *Homo sapiens* is accepted as the beginning of the human population 1 million years ago, then the first 990,000 years of human existence are characterized by a very low population growth rate (15 persons per million per year).

Around 10,000 years ago, humans began a practice that dramatically changed their growth rate: planting food crops. This shift in human history, called the Agricultural Revolution, paved the way for the development of cities, government, and civilizations. Before the Agricultural Revolution, there were no governments to count people. The earliest censuses were conducted less than 10,000 years ago in the ancient civilizations of Egypt, Babylon, China, Palestine, and Rome. For this reason, historical estimates of the earth's total population are difficult to make. However, there is no argument that human numbers have increased dramatically in the past 10,000 years. The dramatic changes in the growth rates of the human population are typically attributed to three significant epochs of human cultural evolution: the Agricultural, Industrial, and Green Revolutions.

Before the Agricultural Revolution, the size of the human population was probably fewer than 10 million people, who survived primarily by hunting and gathering. After plant and animal species were domesticated, the human population increased its growth rate. By about 5000 BCE, gains in food production caused by the Agricultural Revolution meant that the planet could support about 50 million people. For the next several thousand years, the human population continued to grow at a rate of about 0.03 percent per year. By the first year of the common era, the planet's population numbered about 300 million.

At the end of the Middle Ages, the human population numbered about 400 million. As people lived in densely populated cities, the effects of disease increased. Starting in 1348 and continuing to 1650, the human population was subjected to massive declines caused by the bubonic plague—the Black Death. At its peak in about 1400, the Black Death may have killed 25 percent of Europe's population in just over fifty years. By the end of the last great plague in 1650, the human population numbered 600 million.

The Industrial Revolution began between 1650 and 1750. Since then, the growth of the human population has increased greatly. In just under 300 years, the earth's population went from 0.5 billion to 7.7 billion people, and the annual rate of increase went from 0.1 percent to 1.1 percent. This population growth was not because people were having more babies, but because more babies lived to become adults and the average adult lived a longer life.

The Green Revolution occurred in the 1960s. The development of various vaccines and antibiotics in the twentieth century and the spread of their use to most of the world after World War II caused big drops in the death rate, increasing population growth rates. Feeding this growing population has presented a challenge. This third revolution is called the Green Revolution because of the technology used to increase the amount of food produced by farms. However, the Green Revolution was really a combination of improvements in health care, medicine, and sanitation, in addition to an increase in food production.

Geography of Human Population Growth

The present-day human race traces its lineage to Africa. Humans migrated from Africa to the Middle East, Europe, Asia, and eventually to Australia, North and South America, and the Pacific Islands. It is believed that during the last Ice Age, the world's sea levels were lower because much of the world's water was trapped in ice sheets. This lower sea level created land bridges that facilitated many of the major human migrations across the world.

Patterns of human settlement are not random. People generally avoid living in deserts because they lack water. Few humans are found above the Arctic Circle because of that region's severely cold climate. Environmental factors, such as the availability of water and food and the livability of climate, influence where humans choose to live. How much these factors influence the evolution and development of human societies is a subject of debate.

The domestication of plants and animals that resulted from the Agricultural Revolution did not take place everywhere on the earth. In many parts of the world, humans remained as hunter-gatherers while agriculture developed in other parts of the world. Eventually, the agriculturalists outbred the hunter-gatherers, and few hunter-gatherers remain in the twenty-first century. Early agricultural sites have been found in many places, including Central and South America, Southeast Asia and China, and along the Tigris and Euphrates Rivers in what is now Iraq. The practice of agriculture spread from these areas throughout most of the world.

By the time Christopher Columbus reached the Americas in the late fifteenth century, there were millions of Native Americans living in towns and villages and practicing agriculture. Most of them died from diseases that were brought by European colonists. Colonization, disease, and war are major mechanisms that have changed the composition and distribution of the world's population in the last 300 years.

The last few centuries also produced another change in the geography of the human population. During this period, the concentration of industry in urban areas and the efficiency gains of modern agricultural machinery caused large numbers of people to move from rural areas to cities to find jobs. From 1900 to 2020 the percentage of people living in cities went from 14 percent to just about 55 percent. Demographers estimate that by the year 2025, more than 68 percent of the earth's population will live in cities. Scientists estimate that the human population will continue to increase until the year

2050, at which time it will level out at between 8 and 15 billion.

Earth's Carrying Capacity

Many people are concerned that the earth cannot grow enough food or provide enough other resources to support 15 billion people. There is great debate about the concept of the earth's carrying capacity—the maximum human population that the earth can support indefinitely. Answers to questions about the earth's carrying capacity must account for variations in human behavior. For example, the earth could support more bicycle-riding vegetarians than car-driving carnivores. Questions about carrying capacity and the environmental impacts of the human race on the planet are fundamental to the United Nations' goals of sustainable development. Dealing with these questions will be one of the major challenges of the twenty-first century.

Paul C. Sutton

GLOBAL URBANIZATION

Urbanization is the process of building and living in cities. Although the human impulse to live in groups, sharing a "home base" probably dates back to cave-dweller times or before. The creation of towns and cities with a few hundred to many thousands to millions of inhabitants required several other developments.

Foremost of these was the invention of agriculture. Tilling crops requires a permanent living place near the cultivated land. The first agricultural villages were small. Jarmo, a village site from c. 7000 BCE, located in the Zagros Mountains of present-day Iran, appears to have had only twenty to twenty-five houses. Still, farmers' crops and livestock provided a food surplus that could be stored in the village or traded for other goods. Surplus food also meant surplus time, enabling some people to specialize in producing other useful items, or to engage in less tangible things like religious rituals or recordkeeping.

Given these conditions, it took people with foresight and political talents to lead the process of city formation. Once in cities, however, the inhabitants found many benefits. Walls and guards provided more security than the open country. Cities had regular markets where local craftspeople and traveling merchants displayed a variety of goods. City governments often provided amenities like primitive street lighting and sanitary facilities. The faster pace of life, and the exchange of ideas from diverse people interacting, made city life more interesting and speeded up the processes of social change and invention. Writing, law, and money all evolved in the earliest cities.

Ancient and Medieval Cities

Cities seem to have appeared almost simultaneously, around 3500 BCE, in three separate regions. In the Fertile Crescent, a wide curve of land stretching from the Persian gulf to the northwest Mediterranean Sea, the cities of Ur, Akkad, and Babylon rose, flourished, and succeeded one another. In Egypt, a connected chain of cities grew, soon unified by a ruler using Memphis, just south of the Nile River's delta, as his strategic and ceremonial base. On the Indian subcontinent, Mohenjo-Daro and Harappa oversaw about a hundred smaller towns in the Indus River valley. Similar developments took place about a thousand years later in northern China.

These first city sites were in the valleys of great river systems, where rich alluvial soil boosted large-scale food production. The rivers served as a "water highway" for ships carrying commodities and luxury items to and from the cities. They also furnished water for drinking, irrigation, and waste disposal. Even the rivers' rampages promoted civilization, as making flood control and irrigation systems required practical engineering, an organized workforce, and ongoing political authority to direct them.

Eurasia was still full of peoples who were not urbanized, however, and who lived by herding, pirating, or raiding. Early cities declined or disappeared, in some cases destroyed by invasions from such forces around 1200 BCE. Afterward, the cities of Greece became newly important. Their surrounding land was poor, but their access to the sea was an advantage. Greek cities prospered from fishing and trade. They also developed a new idea, the city-state, run by and for its citizens.

Rome, the Greek cities' successor to power, reached a new level of urbanization. Its rise owed more to historical accident and its citizens' political and military talents than to location, but some geographical features are salient. In some ways, the fertile coastal plain of Latium was an ideal site for a great city, central to both the Italian peninsula and the Mediterranean Sea. There, the Tiber River becomes navigible and crossable.

In other ways, Rome's site was far from ideal. Its lower areas were swampy and mosquito-ridden. The seven hills, with their sacred sites later filled with public buildings and luxury houses, imposed a crazy-quilt pattern on the city's growth. Romans built cities with a simple rectangular plan all over Europe and the Middle East, but their home city grew in a less rational way.

At its peak, Rome had a million residents, a population no other city reached before nineteenth century London. It provided facilities found in modern cities: a piped water supply, a sewage disposal system, a police force, public buildings, entertainment districts, shops, inns, restaurants, and taverns. The streets were crowded and noisy; to control traffic, wheeled wagons could make deliveries only at night. Fire and building collapse were constant risks in the cheaply built apartment structures that housed the city's poorer residents. Still, few wanted to live anywhere but in Rome, their world's preeminent city.

In the Early Middle Ages after the western Roman Empire collapsed, feudalism, based on land holdings, eclipsed urban life. Cities never disappeared, but their populations and services declined drastically. Urban life still flourished for another millenium in the eastern capital of Constantinople. When Islam spread across the Middle East, it caused the growth of new cities, centered around a mosque and a marketplace.

In the twelfth and thirteenth centuries, life revived in Western Europe. As in the Islamic cities, the driving forces were both religious—the building of cathedrals—and commercial—merchants and artisans expanding the reach of their activities. Medieval cities were usually walled, with narrow, twisting streets and a lack of basic sanitary measures, but they drew ambitious people and innovative forces together. Italy's cities revived the concept of the city-state with its outward reach. Venice sent its merchant fleet all over the known world. Farther north, Paris and Bologna hosted the first universities. The feudal system slowly gave way to nation-states ruled by one king.

Modern Cities

Modern cities differ from earlier ones because of changes wrought by technology, but most of today's cities arose before the Industrial Revolution. Until the early nineteenth century, travel within a city was by foot or on horse, which limited street widths and city sizes. The first effect of railroads was to shorten travel time between cities. This helped country residents moving to the cities, and speeded raw materials going into and manufactured goods coming out of the factories that increasingly dotted urban areas. Rail transit soon caused the growth of a suburban ring. Prosperous city workers could live in more spacious homes outside the city and ride rail lines to work every day. This pattern was common in London and New York City.

Factories, the lifeblood of the Industrial Revolution, were built in pockets of existing cities. Smaller cities like Glasgow, Scotland, and Pittsburgh, Pennsylvania, grew as ironworking industries, using nearby or easily transported coal and ore resources, built large foundries there. Neither industrialists nor city authorities worried about where the people working there would live. Workers took whatever housing they could find in tenements or subdivided old mansions.

Beginning in the 1880s, metal-framed construction made taller buildings possible. These skyscrapers towered over stately three- to eight-story structures of an earlier period. Because this technology enabled expensive central-city ground space to house many profitable office suites, up through the 1930s, city cores became quite compacted. Many people believed such skyward growth was the wave of the future and warned that city streets were becoming sunless, dangerous canyons.

Automobiles kept these predictions from fully coming true. As car ownership became widespread, more roads were built or widened to carry the traffic. Urban areas began to decentralize. The car, like rail transit before it, allowed people to flee the urban core for suburban living. Because roads could be built almost anywhere, built-up areas around cities came to resemble large patches filling a circle, rather than the spokes-of-a-wheel pattern introduced by rail lines. Cities born during the automotive age tend to have an indistinct city center, surrounded by large areas of diffuse urban development. The prime example is Los Angeles: It has a small downtown area, but a consolidated metropolitan area of about 34,000 square miles (88,000 sq. km.).

Almost everywhere, urban sprawl has created satellite cities with major manufacturing, office, and shopping nodes. These cause an increasing portion of daily travel within metropolitan areas to be between one edge city and another, rather than to and from downtown. Since these journeys have an almost limitless variety of start points and destinations within the urban region, mass transit is only a partial solution to highway crowding and air pollution problems.

The above trends typify the so-called developed world, especially the United States. Many cities in poor nations have grown even more rapidly but with a different mix of patterns and problems. However, the basic pattern can be detected around the globe, as urban dwellers seek to better their own cir-

URBANIZATION AND DEVELOPING NATIONS

The urban population, or number of people living in cities, in North America accounts for about 75 percent of its total population. In Europe, about 90 percent of the population lives in cities. In developing countries, the urban population is often less than 30 percent. The term "urbanization" refers to the rate of population growth of cities. Urbanization mainly results from people moving to cities from elsewhere. In developing countries, the urbanization rate is very high compared to those of North America or Europe. The high rate of urbanization of these countries makes it difficult for their governments to provide housing, water, sewers, jobs, schools, and other services for their fast-growing urban populations.

cumstances. Today, 55 percent of the world's population lives in urban areas, and that percentage is expected to rise to 68 percent by 2050. Projections show that urbanization combined with the overall growth of the world's population could add another 2.5 billion people to urban areas by 2050, with close to 90 percent of this increase taking place in Asia and Africa, according to a United Nations data set published in May 2018.

Megacities and the Future

In the year 2019 the world had thirty-three megacities, defined as urban areas with a population of 10 million or more. The largest was Tokyo, with an estimated 37.5 million people in 2018, predicted to grow to around 37 million by 2030. Second-largest was Delhi, with more than 28.5 million in 2018 and predicted to grow to around 38.94 million by 2030. Megacities in the United States include New York-Newark with a population of 18.8 million and Los Angeles at 12.5.

Megacities profoundly affect the air, weather, and terrain of their surrounding territory. Smog is a feature of urban life almost everywhere, but is worse where the exhaust from millions of cars mixes with industrial pollution. Some megacities have slowed the problem by regulating combustion technology; none have solved it. Huge expanses of soil pre-

URBAN HEAT ISLANDS

Large cities have distinctly different climates from the rural areas that surround them. The most important climatic characteristic of a city is the urban heat island, a concentration of relatively warmer temperatures, especially at nighttime. Large cities are frequently at least 11 degrees Fahrenheit (6 degrees Celsius) warmer than the surrounding countryside.

The urban heat island results from several factors. Primary among these are human activities, such as heating homes and operating factories and vehicles, that produce and release large quantities of energy to the atmosphere. Most of these activities involve the burning of fossil fuels such as oil, gas, and coal. A second factor is the abundance of heat-absorbing urban materials, such as brick, concrete, and asphalt. A third factor is the surface dryness of a city. Urban surface materials normally absorb little water and therefore quickly dry out after a storm. In contrast, the evaporation of moisture from wet soil and vegetation in rural areas uses a large quantity of solar energy—often more than is converted directly to heat—resulting in cooler air temperatures and higher relative humidities.

empted by buildings and pavements can turn heavy rains into floods almost instantly, and the ambient heat in large cities stays several degrees higher than in comparable rural areas. Recent engineering studies suggest that megacities create instability in the ground beneath, compressing and undermining it.

How will cities evolve? Barring an unforeseen technological or social breakthrough, the current growth and problems will probably continue. The process of megapolis—metropolitan areas blending together along the corridors between them—is well underway in many areas. Predictions that the computer will so change the nature of work as to cause massive population shifts away from cities have not been proven correct. Despite its drawbacks, increasing numbers of people are drawn to urban life, seeking the economic opportunities and wider social world that cities offer.

Emily Alward

PEOPLE OF NORTH AMERICA

In the late fifteenth century Christopher Columbus sailed across the Atlantic Ocean in search of Asia. When he reached land, he thought he had arrived in India. Therefore, he called the people he encountered "Indians." Soon, European explorers realized that they had found a world that was new to them, and the origin of its inhabitants aroused considerable interest.

After the Spanish priest José de Acosta studied these inhabitants in the late sixteenth century, he concluded that not only were the people of the Americas like human beings elsewhere, but also the plants and animals were similar to those found in other parts of the world. For example, North America and Eurasia (Europe and Asia) both had oak trees, pine trees, deer, rabbits, bears, and wolves, among other common species.

Acosta initially thought that the animals might have arrived in America by boat, along with humans. However, many of the animals were dangerous, and it is unlikely that people would have brought such creatures with them deliberately. Both animals and humans, Acosta therefore suggested, must have arrived in North America by a land connection. North America was not attached to Eurasia in the Atlantic so, Acosta reasoned, human beings must have come to the continent by a connection with Asia in the northern Pacific. More than 400 years later, most scholars still agreed with this.

Archaeological Evidence

In Europe during the second half of the nineteenth century, many early human remains were found. It was learned that early people, especially the so-called Neanderthals, had a long history in Eurasia. These people lived during the Pleistocene epoch, or Ice Age, and they hunted such now-extinct mammals as the mammoth, giant bison, ground sloth, and woolly rhinoceros.

Human settlers arrived in the Americas long after they had become established in Eurasia. For many years, it was believed that the first humans arrived in the Americas long after the end of the Ice Age, in fact, not much earlier than the beginning of the common, or Christian, era. This idea was shattered in the 1920's and 1930's, when stone projectile points were found embedded in the remains of extinct Pleistocene mammals in the southwestern

Three Native American women in Warm Springs Indian Reservation, Wasco County, Oregon. (1902) (Library of Congress Archives)

BERING STRAIT MIGRATIONS

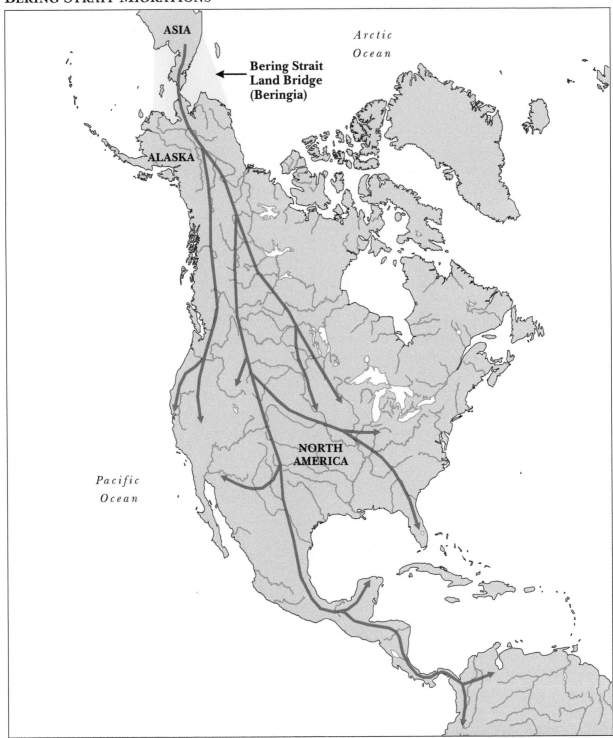

United States. Early people in the Americas also had blades, flakes, choppers, scrapers, knives, hammerstones, and bone tools. These people often are referred to as the Llano culture, after the Llano Estacado of New Mexico and Texas where many of the artifacts were found.

The oldest artifacts were part of the so-called Clovis Culture, named from materials found near Clovis, New Mexico. Clovis remains date from about 10,000 BCE, which is late in the Pleistocene epoch. These people are sometimes referred to as "big game hunters" after the large animals that they pursued. They are also known as Paleo-Indians since they are believed to be the ancestors of the present Native Americans.

Archaeologists generally agree that the Clovis date of 10,000 BCE is the earliest date for which there is firm evidence of humankind in North America. Many people argue that humans were in North America far earlier; however, the archaeological establishment stands by the Clovis date of 12,000 years ago. Archaeologists concede that humans might have been in North America much earlier, but argue that conclusive proof is lacking.

Although the specific time of the first immigrants' arrival in North America is in dispute, it is generally accepted that they came from Asia. Asia and North America now are separated by the narrow Bering Strait. When sea levels dropped during the Ice Age, however, the Bering Strait vanished and left a broad land bridge that connected Asia and North America. This land bridge, called Beringia by scientists, allowed animals and human beings to cross freely between the continents. When melting glaciers caused the sea to rise, Beringia was flooded and the Bering Strait re-formed. This happened a number of times during the Pleistocene epoch. Once in North America, migrants from Asia might have moved southward through an ice-free corridor east of the Rocky Mountains. Some authorities believe that early people followed the coast to lower latitudes.

Other Evidence

Scholars in fields other than archaeology have studied early human beings in North America, and some of their work supports the views of archaeologists. There is an interesting argument from the field of linguistics (the study of languages). The traditional view has been that there are about sixty Native American language families in North America and 100 in South America. One study, however, asserts that there are only three language families in both continents. The Eskimo-Aleut family, spoken in Alaska and northern Canada, arrived there 4,000

THE PLEISTOCENE EPOCH

The Pleistocene epoch, or Ice Age, lasted from about 1.8 million to ten thousand years before the present. It is thought that approximately every forty thousand years, the amount of radiation from the sun that is received at approximately 65 degrees north and south latitude changes. When sun's radiation level decreases, the climate becomes colder and glaciers advance; when it increases, the climate becomes warmer and glaciers retreat. As glaciers advance, they absorb a great deal of water and cause the sea level to drop hundreds of feet. Melting ice during warm interglacial periods causes sea levels to rise. Early migrants to North America lived under these changing conditions during the Pleistocene Epoch.

---

to 5,000 years ago. Speakers of languages in the Na-Dene family of western Canada and the southwestern United States migrated from Asia somewhat earlier. All remaining languages of North and South America are placed in the Amerindian family, which arrived from Asia more than 12,000 years ago. These three families are related to language families of Eurasia. Therefore, this study implies that there were three major migrations from Eurasia.

Geneticists who study physical features acquired from heredity have examined remains of early people in the Americas. Africa was the original home of humankind, and from there humans colonized the rest of the world. By analyzing genetic differences among populations, one can estimate when various groups became separated from one another. One study suggests that Asians began settling North America about 30,000 years ago.

A genetic study based on DNA, the basic material that transmits the hereditary pattern of all living things, also supports the likelihood of three migrations from Asia to North America. The earliest one, called the first Paleo-Indian migration, crossed Beringia from 20,000 to 40,000 years ago. After arriving in the Americas, these people diversified. They correspond to the Amerindians.

The Na-Dene speakers of western Canada and the US Southwest followed the same route to North America 5,000 to 10,000 years ago. A third migration occurred from 6,000 to 12,000 years ago. These people were not related genetically to the population of Siberia, that part of Asia from which the other migrations seem to have originated. Instead, this group is more like the populations of East Asia and Polynesia. Rather than crossing Beringia, this second wave of immigrants might have reached America by crossing the Pacific Ocean by boat. The genetic studies tend to reinforce the linguistic and archaeological research.

Ongoing Controversies

Although the majority of archaeologists believe that there is no firm evidence for the presence of humans in North America more than 12,000 years ago, some disagree. Research from Monte Verde in southern Chile reveals that the site was inhabited at least 1,000 years before Clovis. Early remains of a settlement there were found preserved in a peat bog. The site appears to have been occupied by twenty to thirty people for a year or so. Among the artifacts recovered were the remains of shelters, animals hides, food (including some mastodon meat), fires, and a child's footprint. A number of prominent archaeologists who examined the site believe that it is genuine. If human beings were in South America more than 1,000a years before Clovis, others must have passed through North America considerably earlier.

Early human artifacts have been found in other coastal areas, and it has been suggested that people followed the coast as they moved south. The interior route east of the Rocky Mountains in Canada would have been forbidding, but the coast was not. There, people might have constructed primitive craft and used them to journey down the coast. Coastal areas are appealing because they have a relatively mild climate and provide abundant food in

Pyramid of the Magician, a Maya site in Uxmal, Mexico. (Rob Young)

the form of fish, shellfish, sea mammals, and the like.

Materials that have been found in California have led some researchers to believe that humans were there long before Clovis. A number of skeletons from the San Diego area have been dated, the oldest at about 48,000 years old. However, the reliability of the dating technique employed in this study—protein racemization—is questioned by many authorities. Charcoal from fires, bones, shells, and primitive stone tools also have been found in the San Diego area. Examination of the weathering of the stone artifacts and the age of the surface and soil in which they were found has led some researchers to claim that humans might have been in that area 100,000 years ago. Critics, however, argue that the stone artifacts were fashioned by nature, and there is no precise dating for the finds. Primitive, highly weathered stone artifacts also have been found at various sites in the California desert, particularly the Calico Mountains. Again, some scholars think they might be 100,000 or older, but others insist that the so-called artifacts were shaped by natural processes.

The Dating Controversy

The traditional view has been that the first humans who arrived in North America 12,000 or so years ago were thoroughly modern. All humans of recent times are classified as *Homo sapiens sapiens*. Earlier peoples included *Homo erectus* and more primitive varieties of *Homo sapiens*. For example, the famed Neanderthal people of Eurasia are *Homo sapiens*, but not *Homo sapiens sapiens*. Because the early migrants to the Americas arrived from Asia, they and their present Native American descendants are believed to be Mongoloid (Asian) in race. This has been challenged. If the first migrants arrived in North America 100,000 years ago as some believe, they could not have been *Homo sapiens sapiens*, since that subspecies of humans did not ex-

ist then. Some very primitive skulls have been found in Nebraska, Minnesota, and Brazil that are obviously extremely old. They display the prominent brow ridges of the Neanderthal people of Europe, although these might be defined as Neanderthal-like and not classic Neanderthal.

Negritoids, related to the small, dark-skinned people who still occupy remote sections of Africa, Asia, and the western Pacific, might have been in North and South America. If they were in Asia, they could have reached America by way of Beringia. The Botocudo tribe of South America is very negritoid in appearance. If they were in South America, they could have been in North America as well.

There might have been an Australoid presence in North America. Australoids are related to the indigenous people of Australia. The native peoples of California shared many traits with the Australian Aborigines, including certain physical characteristics, male pubertal rites from which women were driven away with a device known as a bull roarer, and hunting with throwing spears and curved throwing sticks. The ancestors of the Australian aborigines came from Asia, and some might have migrated to North America as well as to Australia. There may have also been a "Europoid" presence in

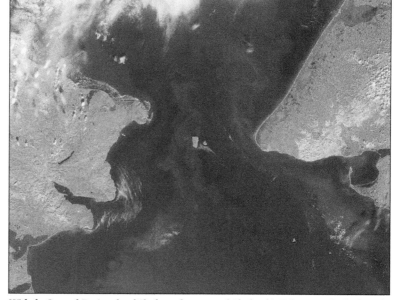

With the Seward Peninsula of Alaska to the east, and Chukotskiy Poluostrovof Siberia to the west, the Bering Strait separates the United States and the Russian Federation by only 90 kilometers. (NASA)

North America—people somewhat similar to those of Europe who were present in East Asia in the remote past. The Ainu of Japan are a contemporary example.

How the First People Came

How could these people have gotten to North America? The earliest migrants could have arrived by way of Beringia, as did others. People related to modern Polynesians might have reached North or South America by boat. Certain elements of the Native American population are genetically related to East Asians and Polynesians, and might have entered the Americas by way of the Pacific Ocean, rather than via Beringia.

In addition, some Europeans might have migrated across the Atlantic Ocean to North America. This does not mean that modern Native Americans are not of ultimate Asian origin. It appears that Asians did enter North America, and their direct descendants are the present Native Americans. However, the ancestors of modern Native Americans might have been neither the first nor the only people to enter North America.

Dispersion Throughout the Hemisphere

Once humans arrived in North America, they spread out across the continent. By the end of the Pleistocene epoch, about 10,000 years ago, many large mammals had become extinct, including the mammoth, mastodon, giant bison, camel, and ground sloth. It has been suggested that the "big game hunters" contributed to the extinction of Pleistocene fauna. Modern human beings are not the first people to have had an adverse impact on the environment.

As time passed, other animals were hunted and plants were gathered and eaten. Stone weapons were still used, and an instrument called an atlatl was employed to propel large darts, or spears. These devices gave extra leverage, which allowed spears to be thrown with great force. The bow-and-arrow was a rather late invention.

Early Agriculture

Southern Mexico and Central America became major centers of plant domestication. From there, some crops were diffused to North America and were grown as far north as the upper St. Lawrence River Valley in eastern Canada. The most important were maize (corn), beans, and squash. Tobacco was also grown in North America, but it came from South America and not Mesoamerica. The bottle gourd was cultivated and used as a water container and for other purposes. The bottle gourd is a native of Africa, and its presence in the Americas thousands of years ago is a mystery. Acorns, nuts, and the like were placed on a *metate*, a stone slab, and ground with a *mano*, a hand-held stone. The soil was prepared with digging sticks and hoes. Livestock were absent; the only domesticated animal was the dog.

Population Estimates

Recent estimates of the population in 1492 in what is now the United States and Canada vary from as few as under 4 million to nearly 10 million. The actual number was probably closer to the lower figure. California and a few other regions had relatively large populations. Agriculture was absent there, but abundant resources allowed for a sizable number of inhabitants. The interior Southwest supported a moderate population among the Pueblos and other sedentary tribes. The Southeast was rather populous and had permanent towns. A few other areas, generally coastal, supported substantial populations. Vast stretches of North America, especially the arid and very cold realms, had few inhabitants.

The Vikings were in North America about 1000 CE, and the continent might have been visited by ancient Chinese, Phoenicians, Carthaginians, and Romans, among others. These early contacts, if they took place, had little or no significance for the peopling of North America. The vast transformation of North America came with the arrival of Europeans after 1492.

John A. Milbauer

People of the Caribbean

The first human inhabitants of the islands of the Caribbean were Caribs and Arawaks. Christopher Columbus arrived in 1492 and was followed by many Spaniards who exploited the Native Americans for labor. Danish, Dutch, English, and French explorers, planters, and settlers came later. European diseases decimated the Arawaks and Caribs. As a result, the Europeans turned to African laborers, sometimes as indentured servants, but more often and increasingly as slaves brought in captivity, mostly from the West African coast and Angola. Chinese and Indian indentured servants were eventually brought to some islands.

Native Americans

The earliest evidence of human colonization of the Caribbean is found in Cuba, Haiti, and the Dominican Republic, where sites have been dated to around 3500 to 4000 BCE. These migrants probably came from the Yucatán peninsula. There is evidence of subsequent migration, the most far-reaching of which occurred between 500 and 250 BCE. It seems to have come from the Orinoco drainage and the river systems of South America's northeast coast, mostly from present-day Venezuela.

When Europeans arrived, they found the Caribbean to be densely inhabited by diverse indigenous

Reconstruction of a Taíno village in Cuba. (Michal Zalewski)

groups. They came to view the Caribbean as having two kinds of people, Caribs and Tainos (Arawaks), but it probably was more ethnically complex than that. From 1492 to 1660, European diseases and the effects of forced labor decimated and almost eliminated the native peoples of the Caribbean. It has been estimated that the native population of Hispaniola (modern-day Dominican Republic and Haiti) was well in excess of 3 million in 1496. By 1508 the population was only 100,000. This was the general pattern for the tropical Caribbean islands and the coasts of the Americas. Carib peoples still exist, though, on parts of the islands of Dominica and St. Vincent.

Sugar and Settlers

Europe's demand for sugar drove most aspects of settlement in much of the Caribbean. The Dutch supplied much of the initial workforce and knowledge and were responsible for the firm implantation of sugar and slavery in the Caribbean. By 1640 the English had 52,000 whites on their islands of Barbados, Nevis, and St. Kitts. This contrasts sharply with 22,000 at the same time in the settlements of New England. By the end of the 1650's there were 15,000 white Frenchmen in Martinique and Guadeloupe. At first the whites grew tobacco, but a shift to sugar began in 1645, and sugar was dominant by the 1670s. As a result, African slaves became the majority of the British islands' population. In Barbados, there were more than 50,000 African slaves by the end of the 1600s, while the white population had declined to about 17,000. Similar changes occurred in the French islands, but at a slower pace.

In Jamaica, sugar developed more slowly, but by the 1680s, the island's sugar industry started to expand. Slaves began arriving at the rate of more than 36,000 per year in the 1680s. By 1768 the slave population in Jamaica had reached 167,000 while the white population had grown to only 18,000. St. Domingue (Haiti), a French possession, developed in a manner similar to that of Jamaica. By 1740 St. Domingue had 117,00 slaves, about 13,000 free persons of color, and a similar number of white Frenchmen. In 1791 the slaves rose in revolt and eliminated the white population, and modern-day Haiti came into being.

In the late eighteenth century, the estimated slave populations in the West Indies were as follows: French West Indies, including Haiti, 575,000; British West Indies, 467,000; the Spanish West Indies, 80,000. In the early 1860s the slave population in Cuba was 370,553 and in Puerto Rico, 41,738. The slave population altered Cuba's culture for some time, so that it resembled Haiti. Subsequent immigration from Spain in the late nineteenth century shifted Cuba's culture back to being distinctly Spanish.

United States Virgin Islands

In 1666 Denmark occupied St. Thomas. Five years later, a colony was founded there to supply the mother country with cotton, indigo, and other products. Slaves from Africa were first introduced to St. Thomas in 1673 to work the cane fields. The first regular consignment of slaves did not arrive until 1681. In 1684 the Danes claimed neighboring St. John and colonized the islands with planters from St. Thomas in 1717. In 1733 they purchased St. Croix from the French. By 1742 there were 1,900 African slaves on St. Croix alone. The United States purchased the islands from Denmark in 1917.

Indentured Servants

The emancipation of slaves throughout the West Indies from 1834 to 1838 resulted in a severe labor shortage on the sugar plantations. This need was met by the subsidized immigration of laborers from India and China. British Guiana (Guyana) and Trinidad received the majority of the indentured laborers. During the 1850s and 1860s, 142,000 Chinese went to Cuba. Many of the Chinese came from Canton.

Indentured servirtude was banned in the United States in 1865 by the Thirteenth Amendment to the US Constitution. The British banned indentured servitude in their Caribbean possessions in 1917. The practice was declared prohibited worldwide by the Universal Declaration of Human Rights, adopted by the United Nations in 1948.

Dana P. McDermott

POPULATION DISTRIBUTION OF NORTH AMERICA

Patterns of settlement in modern North America are the result of more than 1,000 years of influence and change. Europeans, Africans, and Asians all left their mark on a continent that had been dominated by different indigenous nations. By 2020, North America was home to more than 500 million people of different cultures dispersed throughout Greenland, Mexico, Canada, and the United States.

The United States had the largest group—about 333 million. Mexico had the second-largest popula-tion, approximately 128 million people. Canada was much smaller, with around 37 million residents. The island of Greenland had about 58,000 inhabit-ants, although its total land area is bigger than Mex-ico. In terms of gross distribution, almost seven of every 10 North Americans lived in the United States.

Canada and the United States are sometimes col-lectively called Anglo-America, reflecting their common British heritage. Mexico, by contrast, has

A typical day in the Mexico City subway. (maykonrod)

POPULATION DENSITIES IN NORTH AMERICA
(BASED ON 2018 ESTIMATES)

Fewer than 1 person/sq. mi.

1–10 persons/sq. mi.

11–50 persons/sq. mi.

51–100 persons/sq. mi.

101–500 persons/sq. mi.

501–1,000 persons/sq. mi.

More than 1,000 persons/sq. mi.

a strong Spanish heritage. Greenland is actually part of the Kingdom of Denmark; however, most Greenlanders are Inuit.

Geographic Patterns of Modern Settlement

The population in North America is not spread evenly; it is quite clustered. Most North Americans live in cities and their surrounding suburbs, or in smaller towns. In 2019, just over 80 percent lived in urban areas—a big change from 1800, when only 5 percent lived in a city.

In Greenland, almost all the population lives in the towns and villages along the western coast. In contrast, much of Mexico's population is found in a wide band extending west-to-east across the middle third of the country. Toward the east coast is Mexico City, home to nearly 9 million people in 2015. This city was the biggest in North America in 2019 and the world's fifth-largest urban agglomeration.

In the United States, about two-thirds of the cities with more than 1 million people are located east of the Missouri River. Rural (farming) areas there also have more people per square mile than in the western half of the United States. The largest urban concentration is in the area that stretches from Washington, D.C., to Boston, Massachusetts, including New York (8.3 million people), Washington, D.C. (6.2 million), Boston (4.6 million), Baltimore (2.8 million), and Philadelphia (1.58 million).

In Canada, almost 90 percent of the population lives within 100 miles (160 km.) of the United States border. Nearly half of all Canadians are concentrated in southern Ontario and along Quebec's St. Lawrence River Valley. This area, extending from Windsor (near Detroit, Michigan) to Quebec City is nicknamed "The Corridor" because of its linear shape and the fact that so many people live along it. The biggest cities on that corridor are Toronto (6.4

Caribana is a festival celebrating Caribbean culture and traditions. Held each summer in Toronto, Canada, it is North America's largest street festival. (Loozrboy)

million people), Montreal (3.5 million), and the capital city, Ottawa (just under 1 million).

Along the west coast of Anglo-America are several big cities, extending from San Diego (2.9 million) and the Los Angeles area (13.1 million) in California and Seattle (3.0 million) in Washington, to Vancouver, British Columbia (2.2 million) in Canada. Beyond the coastal area, much of the western and northern parts of North America are sparsely settled. This is mostly due to the nature of the landscape and the climate.

Impact of European Settlement

The earliest discovery of North America by foreign explorers is believed to have been by Norse seafarers known as Vikings. Erik the Red established a Viking colony in Greenland in the 980s and his son Leif Erikson, became the first European to land in North America, in 1000. The Vikings built a colony in the northernmost tip of Newfoundland and Labrador on the east coast of Canada, but it was later abandoned.

Around 1600 Europeans began building settlements in North America again, and many of those became modern major cities. However, Native Americans, Eskimos, and Inuit already had built communities, and had lived there for at least 12,000 years before the Europeans arrived. These peoples lived throughout North America, from the Arctic Ocean to the southern border of Mexico. Contact with European diseases and the disruption of traditional lifestyles and homelands dramatically reduced their populations.

By the seventeenth century, a number of European countries were beginning to explore and settle North America, drawn by the great variety of raw materials and resources there—fish, furs, minerals, lumber, and fertile soil for crops.

The clusters of European colonies, mainly along the Atlantic coast of North America, resulted in different cultural regions. The French were in the northern areas; English Puritans dominated southern New England, around Boston; the Dutch, and later the English, were around New York City; and English Quakers settled around Philadelphia. Both England and France established plantations in the southeastern region, the English in Virginia and South Carolina, and the French along the Gulf of Mexico. The Southeast also became the destination for many Africans, brought across the Atlantic Ocean as slaves to work on the plantations. These large farms grew crops such as cotton and tobacco, which were in great demand in Europe.

The Spanish domination of Mexico was based out of the colonial capital of Mexico City, which was the main population center of New Spain. For decades after Mexico gained independence in 1821, the country was politically unstable. In contrast to Canada and the United States, Mexico lost half its territory in wars. These losses, combined with the legacy of Spanish colonial control and its internal land confiscation, created a weakened economy and burgeoning poverty that continued into the twenty-first century.

Distributing Population: The Ability to Move

In North America, overland travel by wagon was difficult because there was no road system such as the network that exists today. It was much easier to move about by boat on the rivers and along the coast. Because of this, the biggest population growth occurred in cities and surrounding areas that were located along the northeast Atlantic coast, the St. Lawrence River, and the Gulf of Mexico. In Anglo-America, New York was the biggest city, with about 100,000 people. Settlements were also built along rivers, such as the Mississippi, that linked the US Midwest with the larger cities of the Southeast. River ports such as St. Louis and New Orleans grew quickly because of their location on major waterways.

Boat traffic going down midwestern rivers did not help businesses and cities in the Northeast. Because of competition for trade, northeastern interests built connected river and canal systems. New York business people paid for the construction of the Erie Canal in 1825, which linked the Hudson River at New York City to Lake Erie at Buffalo. A second canal system went from Philadelphia and Baltimore to Pittsburgh and the Ohio Valley; there, Cincinnati and Louisville became important port cities.

Those new transportation routes meant that much of the better agricultural lands within the eastern half of United States and the Great Lakes

The ceremony for the driving of the golden spike at Promontory Summit, Utah on May 10, 1869 celebrating the completion of the First Transcontinental Railroad. (Andrew J. Russell - Yale University Libraries)

area in Canada could be settled, because farmers could get their crops to markets in the bigger cities. As a result, most of the US Midwest and Southeast were settled by the mid-nineteenth century. Many of the new settlers came from Europe and Britain, and a large slave population was forcibly brought from West Africa.

In Canada, a population of fewer than 300,000 Native Americans and Europeans in 1760 grew to 3.2 million by 1860. In the United States over this same 100-year period, the population grew from about 2.5 million to more than 30 million. New York City alone had around 1 million people. However, most people at that time did not live in cities: In the United States, only 25 percent were city-dwellers; in Canada, it was less than 12 percent.

By the middle of the nineteenth century, the Industrial Revolution had begun to cause great

change, not only in where people lived and what their jobs were, but also in where population growth went in North America. An important invention that changed the population distribution in North America was the railroad. Much of this new technology came from England, and it was especially influential in Anglo-America. The impact of rail in Mexico was minor, and in Greenland it was nonexistent. With the new railroad technology, anywhere a track could be laid meant easy movement. This opened up much more of the United States, in particular, for people to settle.

Cities, towns, farms, and mines were built throughout the western half of the United States. The discovery of gold near San Francisco in 1849 helped speed up the building of a rail line to the West Coast. Rail reached the Pacific Ocean in 1869. In Canada, the transcontinental railway was completed in 1885. It went from the East Coast to the West Coast near the present-day Canada-United States border. It strongly influenced where population settled in the western half of Canada. The number of people who live west of the Missouri and Mississippi rivers and east of the Rocky Mountains is much less than live in the eastern half of North America, but without the railroad, these regions might not have seen much settlement until the twentieth century.

By 1920 most of Anglo-America's remaining agricultural lands were settled. Population in the United States reached just more than 106 million, of which 51 percent lived in urban areas. In Canada, there were about 8.5 million people (37 percent urban). New York City reached 4.75 million, making it the biggest city in either country. In Mexico, population continued to grow, but poverty and inequality were major factors in limiting population distribution and suppressing development. In the relatively short time since Europeans built settlements along the Atlantic coast and Gulf of Mexico, much of North America was changed by settlers, both urban and rural. It was one of the quickest transformations of a landscape in the history of the human population.

North Americans on the Road

The first half of the twentieth century saw the invention of the car, the building of highways, more factories and businesses in cities, and farms growing more food with fewer farm workers because of new machines. Urbanization increased. Between 1920 and 1960, the sprawling corridor on the eastern US seaboard and Canada's "Corridor" grew in particular. By 1970, 70 percent of the people in North America lived in cities, and urban growth continued.

In 2020, Mexico, the United States, and Canada were 80.7 percent, 82.7 percent, and 81.6 percent urban, respectively. One reason that cities can continue to grow outward is because increasing numbers of people own cars. In Mexico City, there were almost 5 million cars in 2017. In the United States, there were more than 273.6 million car registrations in 2018—about 1 car for every 1.2 persons.

This means individuals can drive to work and do not have to live close to their place of employment. The car and the highway system have created sprawling suburbs. Trucks enable industries and businesses also to move to the suburbs. Fewer people and fewer jobs are in the inner city or in the downtown areas; more farmland is being converted into suburbs.

The New Destinations

In the United States in 2017, about 14 percent of the population moved each year. With the car and truck, this becomes quite easy. The moving of people from region to region is called migration. In the US, the most prominent trend has been for people to move southward and westward. The fastest-growing U.S. states have been in the south and southwest, including Florida, Texas, North Carolina, Arizona, and Colorado, with New York, Illinois and California losing the most people to domestic migration. The fastest-growing Canadian provinces have been British Columbia, Nova Scotia, and Ontario.

The cities in the US South, part of the so-called Sunbelt, have also been destinations for many people. Large numbers of people began moving there from the Midwest and Northeast. In part, this is because of the growth of high-technology businesses that have been established there, such as petrochemicals, aeronautic industries, computer-related businesses, and the space program. At the same time, some of the old factories in the Midwest and Northeast became obsolete and closed. The West Coast continued to be a popular destination, fueled in part by the growth in computer industries and the entertainment business.

Migration is a serious problem in Mexico. There, the geographic pattern involves the movement of rural populations to cities—primarily Mexico City. Rural-to-urban domestic migration has caused high rates of urbanization in a country that already experiences high population growth. The city struggles to provide enough new housing, new jobs, new schools, clean water, and sewerage capacity for its nearly 9 million people.

Timothy J. Bailey

POPULATION DISTRIBUTION OF THE CARIBBEAN

With a total population of about 43 million people in 2020, the Caribbean region has roughly the same number of people as California and Oregon combined. However, the Caribbean's people are unevenly distributed among the many islands of its independent nations and colonial dependencies. This uneven distribution is closely linked to the centers of production of the region's main agricultural products: sugar, coffee, and tobacco, as well as the locations of port cities largely developed to export these products.

The Caribbean's most populous nation occupies its largest island: Cuba, whose 2020 population was about 11 million people. With an area of 42,803 square miles (110,860 sq. km.), Cuba has a population density of about 260 residents per square mile (100 per sq. km.). Its total area is similar to that of the US state of Tennessee, but Tennessee has about

POPULATION DENSITIES OF CARIBBEAN COUNTRIES
(BASED ON 2018 ESTIMATES)

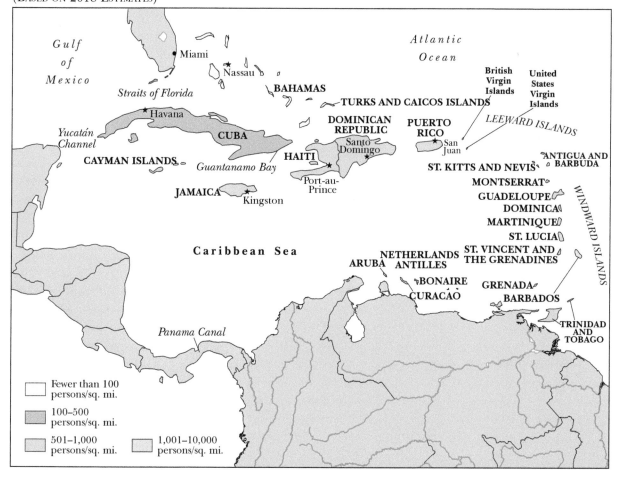

60 percent the number of people and is thus 60 percent as crowded.

Population Densities

The average population density for the Caribbean region as a whole is roughly 500 people per square mile (193 per sq. km.). By contrast, the United States has an average population density of about 88 people per square mile (34 per sq. km.). However, while the Caribbean is the most densely populated region in the Western Hemisphere, its countries show a wide range of individual population densities. The region's most densely populated island nation is Barbados. It had about 294,500 people living on its 166 square miles (430 sq. km.) in 2020. Its population density of about 1,775 people per square mile (685 people per sq. km.) is nearly three times that of the Caribbean's next most densely populated nation, St. Vincent and Grenadines. The country with the lowest population density is the Bahamas, with about 63 people per square mile (24 people per sq. km.).

The Caribbean's second-largest country, in both area and population, is the Dominican Republic, with 10.5 million inhabitants on 18,792 square miles (48,670 sq. km.). Its population density is about 560 people per square mile (216 people per sq. km.). The Spanish-speaking Dominican Republic shares the island of Hispaniola with French- and Creole-speaking Haiti. With 10,714 square miles (27,750 sq. km.), Haiti has only 57 percent of its neighbor's area, but its 2020 population of just over 11 million makes Haiti nearly twice as crowded.

No other Caribbean islands approach Cuba, the Dominican Republic, and Haiti in the size of their populations. Puerto Rico, a self-governing commonwealth of the United States, is next biggest, with a 2020 population of 3,189,068. Jamaica has 2,808,570 people on an island a tenth the size of Cuba. Trinidad and Tobago is the only other nation with a population of over 1 million, with about 1,208,789 people on a small group of islands covering just under 2,000 square miles (5,180 sq.km.). Haiti, Jamaica, and Trinidad and Tobago all have population densities of the same order of magnitude—more than double that of Cuba.

Among the Caribbean's other island nations and colonial dependencies, the next largest in population are French-ruled Guadeloupe (395,700) and Martinique (376,480), the Bahamas (337,721), Barbados (294,560), St. Lucia (166,487), Grenada (113,094), the US-ruled Virgin Islands (106,9770, and St. Vincent and the Grenadines (101,390). All remaining countries and territories had populations of less than 100,000 in 2020.

Major Population Clusters

The Caribbean has four major population concentrations of people. The largest is found on the central and southern portions of the island of Hispaniola (divided into Haiti and the Dominican Republic), roughly connecting the cities of Port-au-Prince and Santo Domingo. The three remaining large clusters include northwest Cuba (including the city of Havana), Puerto Rico (including San Juan), and Jamaica (including Kingston).

While country population totals and population clusters are highest on the larger Greater Antilles islands of Cuba, Hispaniola, Puerto Rico, and Jamaica, high densities can be found on many of the smaller islands of the Lesser Antilles.

Kristopher D. White

DISCUSSION QUESTIONS: PEOPLE AND POPULATION DISTRIBUTION

Q1. Where is Beringia? When did it exist? What role did it play in populating the Americas? What routes do paleontologists believe these earliest people followed once they arrived in North America?

Q2. Who are the indigenous people of the Caribbean? Which groups did Europeans encounter when they first arrived? What was the role of slavery during the Caribbean's colonial period? What is indentured servitude, and how was it applied in the Caribbean?

Q3. When did the Vikings conduct explorations in Greenland and North America? What became of the settlements they established? When did the next big wave of European exploration begin?

Culture Regions of North America

The culture regions of North America extend from the Arctic Circle, through Canada and the United States, and south to Mexico and include many people, languages, and lifestyles.

Greenland

Although citizens of Greenland are also Danish citizens, Greenland is a part of North America. Connected by a submarine ridge to North America,

Greenland lies only 16 miles (26 km.) from the Canadian island of Ellesmere. About two-thirds of Greenland's approximately 836,000 square miles (2.18 million sq. km.) lies within the Arctic Circle and 79 percent of Greenland's land area is covered by thick sheets of ice. These facts indicate the types of cultural groups and activities that could flourish in Greenland. Less than 1 percent of the island is used for agriculture. At its southernmost point, the

Nuuk, Greenland's main road Aqqusinersuaq with Hotel Hans Egede on the right. (Oliver Schauf)

island provides pasture for sheep and reindeer, which are raised for meat, wool, and milk.

About 90 percent of Greenland's 57,616 people are of Inuit heritage, although over the years there has been an intermixing of Inuit and Europeans, mostly Danes, who have emigrated to the island. The Inuit of Greenland, like those of Canada and Alaska, have had a traditional lifestyle adapted to a cold climate where there is little vegetation, and where people must seek much of their livelihood from nature itself. Many people fish the coastal waters and rivers. Greenland's chief exports are canned and frozen cod, prawns, and other marine life, as well as zinc and lead from the island's mines.

Greenland has neither metropolitan areas nor superhighways. Of its 161 localities, only about 88 have populations of more than 100. Nuuk, the largest city and capital, has a population of only about 18,000 (2020). Near the coastal areas are some highways, but much of the year, over much of the land, Greenlanders must travel on iceways.

While Greenlanders have relatively modern lifestyles, they still adhere to many of the values and beliefs that were brought to Greenland by the Thule culture in 900 CE. The Inuit culture is now found in its pure form only in the Thule District; however, Greenlandic, a three-dialect form of the Inuit language, is spoken throughout Greenland. Inuit folktales and folk arts are part of the culture of Greenland. Dogs remain important to the Inuit, both for transportation and as companions in a cold, isolated climate. Each year Greenland hosts the Arctic Circle Race, an international race for cross-country skiers across 100 miles (160 km.) at the Arctic Circle. Greenlanders follow the visiting skiers in dogsleds.

Greenland's culture is also influenced by Scandinavian culture. According to Icelandic and Greenlandic sagas, Erik the Red settled in Greenland around 985. To encourage other Vikings to settle on the island, he named it Greenland to make prospective settlers think it was a land where people could live prosperously. In 1000 Erik's son, Leif Erikson, introduced Christianity to Greenland. His mother built the first Christian church in Greenland, at Brattahild. Instruction in modern Green-

THE INFLUENCES OF GEOGRAPHY, HISTORY, AND TECHNOLOGY

The present patterns of population distribution in North America are the result of myriad factors. However, the primary determinants are different for Greenland, Mexico, and Canada and the United States. Geography has played a particularly critical role in Greenland, where the physical landscape effectively restricted settlement to the coastal areas. The rugged landscape has limited the influence of historical colonization and the impacts of transportation improvements.

History explains the continuing legacy of Spanish conquest in Mexico. Ongoing problems of land ownership, poverty, and exploitation have strongly influenced the mass migration to Mexico City, and the limited resources available to much of the population to change their situation.

Technology has played a huge role in the exploration and settlement of Canada and United States. The railroad and automobile, more than any other technologies, have shaped population distribution in Anglo-America.

land's schools is conducted in the Greenlandic language, but administrative affairs and most media coverage are in Danish.

Nunavut

Just to the west of Greenland lies Nunavut, a Canadian administrative unit that was created in 1999 out of the eastern portion of the Northwest Territories. Nunavut is made up of the traditional lands of the Canadian Inuit, whose language, Inuktitut, is the official language of the government. Most of Nunavut's residents identify themselves as Inuit (83.6 percent). The Inuit of Nunavut are closely related to the people of Greenland and are also descendants of members of the Thule culture. While the popular culture of North America has come to Nunavut, as many as 40 percent of its adults do not participate in a wage-based economy. A substantial number follow the traditional cultural beliefs and practice the traditional arts of the Inuit.

The Inuit value the land, sea, animals, weather, human goodness, and sharing. Older people are respected and regarded as cultural experts. Younger

A ceremony commemorating the establishment of Nunavut, Canada, April 1999. (Ansgar Walk)

Inuit look to older people for guidance about hunting, the weather, astronomy, cultural beliefs, and ethics. While most modern-day Inuit live permanently in their villages, their ancestors were nomadic, moving from place to place to follow game. Many of the remaining elderly lived a nomadic lifestyle in their early years. They have shared their stories with others in their families and villages to keep alive the traditional Inuit lifestyle. In Nunavut, some of the Inuit elders broadcast their stories on the radio in the Inuktitut language.

Canada's Atlantic Provinces

South of Nunavut lie Canada's Atlantic Provinces: Newfoundland and Labrador, Nova Scotia, Prince Edward Island, and New Brunswick. Although some Inuit and Native Canadians from the Micmac group live in Newfoundland and Labrador, 95 percent of the people there are descendants of immigrants from southwestern Ireland or southwestern England. This heritage, and the fact that Newfoundland and Labrador is somewhat isolated, have given the culture of the province a special flavor. The English spoken in Newfoundland and Labrador is much like the English of seventeenth century West County England. The local citizens also cling to Irish traditions. At public gatherings,

bards and minstrels often do presentations in an Irish manner.

About 58 percent of the people of Newfoundland and Labrador live in urban areas. They engage in fishing, processing of seafood, and mining, particularly the extraction of iron ore. Emigration to other parts of Canada and to the United States has been high for several decades.

In 1605 people from France settled at Port Royal in presen-day Nova Scotia. Port Royal became the seat of the French Colony of Acadia. Port Royal, now known as Annapolis Royal, was the first permanent North American settlement of Europeans north of Florida. The French lived and flourished in this area early in the seventeenth century, but, with the Treaty of Paris of 1763, France had to cede its North American lands to Great Britain. Most of the Acadian French refused to swear loyalty to the British government and were deported. Many went to New Orleans, where their culture became known as Cajun. In 2016, about 2.6 percent of the people of Nova Scotia traced their ancestry to the original Acadians who settled there. These people are often bilingual and generally live on farms as their ancestors did.

About 80 percent of the citizens of Nova Scotia are descendants of immigrants from the British Isles. They practice the occupations of their ancestors, fishing for lobsters, oysters, clams, and scallops, or dairy farming. Many are of Scottish heritage, and there is a strong Scottish cultural influence within the province.

Around Halifax and Shelburne live many descendants of the West Indian slaves who left boats docked in this area during the eighteenth and nineteenth centuries. There are about 21,000 First Nations people in Nova Scotia, some of whom live on government-administered lands. About 90 percent of the people of Nova Scotia speak English exclusively; about 10 percent are bilingual, speaking English and French.

Although Prince Edward Island has a rural atmosphere, it is the most densely populated Canadian province. The areas around Charlottetown are urbanized. The majority of the island's inhabitants are descendants of immigrants from the British Isles. Another several thousand people are descendants of Acadians who were not deported in the eighteenth century; they often are French speakers. About 1.3 percent of its citizens are Chinese Canadian. More than one-third of the citizens of Prince Edward Island are Roman Catholics.

New Brunswick is the only Canadian province that has both English and French as its official languages. The majority of its residents are descended from citizens of the British Isles: Some of their ancestors were Loyalists during the American Revolution; others were immigrants from England, Ireland, and Scotland during the nineteenth and early twentieth centuries. There are also French-speaking descendants of Acadians who were not deported in the eighteenth century; African Canadians, mostly descendants of loyalist slaves; Native Canadians living on government-administered lands; and more recent Asian immigrants.

Quebec and Ontario

About 78 percent of the people of Quebec are French-speaking, and French is the official language of the province. While people of French heritage, known as Quebecois, dominate the culture of Quebec, people of English heritage dominate the economy. This leads

to much tension within Quebec. Many French speakers have attempted to establish their own personal identities in an economy dominated by a culture different from their own. In Quebec, 75 percent of the people live in urban areas. Special singers (*chansonniers*) present music that identifies with the Quebeçois search for political and cultural identity. Radio and television programs are broadcast in both French and English, and there are several newspapers in each language.

Ontario is the political and industrial center of Canada. English is spoken there, since relatively few French ever settled in this province. While Ontario is the industrial center of Canada and its citizens make up about 52 percent of the Canadian workforce, many Ontario residents cross the border to work or attend schools in the United States.

About 27,000 Cree Indians live in northern Quebec and another 36,000 live in Ontario, mostly on government-administered land. Other Native Canadians live in rural areas of southern Ontario. Large numbers of immigrants from Asia have settled in Toronto.

Toronto, Ontario and Montreal, Quebec are major centers for the Canadian performing arts. Their respective theaters, symphonies, and ballet companies rival each other. These two cities are also important in the sporting life of Canada. Ice hockey is the national sport, and both Toronto and Montreal have major-league ice hockey teams.

The Montreal Canadiens at the Bell Centre. (Acrocynus)

Canada's Prairie Provinces

The central provinces of Canada—Manitoba, Saskatchewan, and Alberta—are known as the Prairie Provinces. English is the official language, but all these provinces have many citizens of German, Scandinavian, Ukrainian, and French ancestry. These people have kept their heritages alive through ethnic festivals at which they wear traditional costumes, sing music of their ancestral lands, and prepare and eat the foods of those homelands. Alberta's recent immigrants include those who came there to work in the rich petroleum fields near Edmonton.

Abundant natural resources are found in the Prairie Provinces. Alberta and Saskatchewan are also prime farming areas. Huge wheat farms and cattle ranches dot the landscape there. Some areas of these provinces have a cowboy culture complete with rodeos and western wear.

The Northwest Territories and the Yukon

There are two large, sparsely populated areas in northern Canada: the Northwest Territories and the Yukon. Major towns in these areas are Yellowknife, capital of the Northwest Territories; Whitehorse, capital of the Yukon Territory; and Dawson City, Hay River, Fort Smith, and Inuvik. About 40 percent of the people are of European or mixed European heritage. Some are descendants of people who came searching for gold; others are transients who have come for a variety of employment reasons and do not plan to settle there. Many transients work in the gold and diamond mines of the Yellowknife area.

Hill-side mining during the Klondike Gold Rush, c.1899. (John Scudder McLain—Alaska and the Klondike)

The people of European descent tend to live in settlements along the Mackenzie River, although a few live in the far northern coastal regions. One such settlement is Inuvik, the administrative center for the western Arctic. Inuvik is literally the "end of the road." People who live there are at the end of the Dempster Highway that connects the western Arctic to Whitehorse, Yukon.

About 11 percent of the people are western Inuit, who live mostly in the far north and east of the Northwest Territory. The western Inuit have many of the same customs and beliefs as other Inuit groups. About 36 percent of the people of these territories are First Nations Native Canadians. They comprise several tribes who speak languages from the Athabascan group. They live near Fort Smith and along the Mackenzie River. Many of them were originally fur traders and, as demand for that occupation has declined, have had difficulty adapting to a wage-based economy.

British Columbia

The Canadian province of British Columbia is a cultural mix. More than 50 percent of its inhabitants are descendants of people from the British Isles, and many towns and streets are named for places in England and for members of British royalty. However, Vancouver has one of the largest Chinese populations outside of China, Taiwan, and Hong Kong. There are also people of Scandinavian, German, and Ukrainian heritage in the urban areas of the province. Most of the people of the province live in the Vancouver-Victoria region in the southwestern corner of British Columbia. Groups of people also live in the north, along the Pacific coast, and in the sparsely populated interior mountainous areas. These include the northern Inuit. The Athabascan live in the northern part of the province. Other Native Canadian groups live along the coast, where they have traditionally made their living by catching fish.

The United States

About 17 percent of the people of the United States are of a British or Irish heritage; about 13 percent are descendants of people from Germany; about 12

Little Italy, Manhattan, New York, c.1900. (Library of Congress)

percent are of African-American heritage; about 5 percent are of Italian heritage. The remaining U.S inhabitants represent nearly all the languages and cultural groups on Earth. These cultural groups tend to be clustered in various areas. The public policies of the United States, while encouraging people to assimilate and adapt to the prevailing culture, also encourage them to preserve their cultural heritage.

Northeastern United States

The northeastern United States includes New York and the six New England states: Maine, New Hampshire, Vermont, Massachusetts, Rhode Island, and Connecticut. This area was originally settled by Europeans, mainly from England, the Netherlands, and France. In 2020, in such densely populated metropolitan areas as New York City and Boston, there are many people of Asian, African,

and Latino heritage. While most citizens of the northeastern part of the United States, like the nation as a whole, are of Christian heritage, there are also many adherents to Jewish and Muslim faiths in the metropolitan areas.

To a certain extent, the US culture is held together by basic values: patriotism, belief in freedom, and love of family. The northeastern part of the United States is where many of these values were secured, since the Revolutionary War began there.

US culture is supplemented and enriched by the ethnic neighborhoods, festivals, and foods that are found in the urban areas of the northeast. People of Chinese, Indian, Korean, Arabic, and African-American bring diversity to the urban centers in this area. The northeastern part of the United States also is home to world-famous seats of learning, including Harvard University, the Massachusetts Institute of Technology, and Yale University.

Middle Atlantic States

The Middle Atlantic states of Pennsylvania, New Jersey, Delaware, and Maryland are highly industrialized areas with large metropolitan centers. Their citizens are generally of European heritage, but, particularly in the urban areas, there are many citizens of Asian and African backgrounds.

In the Philadelphia-Lancaster area of Pennsylvania live the Amish, an Anabaptist religious group who immigrated to the United States from Germany. The Amish have led lives largely isolated from the other citizens of the United States, shunning modern conveniences such as automobiles and electricity, and have been self-sufficient farmers.

Southeastern States

The southeastern states—Virginia, North Carolina, South Carolina, Georgia, and Florida—began as agricultural states. While there is still much farming there, by 2020, some of these states, particularly North Carolina and Georgia, had become heavily industrialized.

Large urban centers in the southeast include Charlotte, North Carolina; Atlanta, Georgia; and Miami, Florida. Nevertheless, many of the people of the southeast live in small towns or in rural areas where life is more leisurely paced than in large cities. More than 50 percent of the US African-American population lives in the Southeast. In the industrialized north most African Americans live in urban areas; in the Southeast, by contrast, many live in rural areas on small farms. The African Americans of the Southeast trace their heritage to slaves who worked in the fields and homes of the European planters who settled there in the seventeenth and eighteenth centuries.

In the western mountains of North Carolina live Native Americans, the Eastern Cherokee, who have

Amish farm near Morristown, New York. (ilamont)

Miners posing with their guns in Eskdale, West Virginia during the Paint-Cabin Creek strike from 1912-1913. (Register-Herald Reporter)

an Indian reservation there. Members of another Native American group, the Seminole, live in Florida, where they fish and raise cattle. While most members of this group live a lifestyle similar to that of the mainstream US culture, some maintain the folkways of their ancestors.

Mexicans in the United States

Many recent Mexican immigrants had settled in central North Carolina to work in the meat-packing industry. These immigrants brought with them Mexican culture, foods, music, and other additions to the cultural landscape of North Carolina.

Many senior citizens from northern states escape to Florida for the winter; many have emigrated there to live year-round. The US Census Bureau predicts that by 2030, 32.5 percent of Florida's population will be over 60 years old.

Cuban Americans are another cultural group of the Southeast. They originally fled Cuba after Fidel Castro became that country's president, and have more or less preserved their culture as it existed when they left Cuba in the late 1950s and early 1960s. The Cuban Americans of southern Florida have exerted substantial influence, especially in Florida politics.

Southern Appalachian States

The southern Appalachian states of West Virginia, western Virginia, Kentucky, and Tennessee were settled by people of European heritage who crossed the Appalachian Mountains in the eighteenth and early nineteenth centuries. These people were often Scotch-Irish subsistence farmers. They brought to the mountains their dialect and love of song, dance, and storytelling. Remnants of the dialect, the songs, and the stories remain to this day, in the isolated mountain areas.

Early in the nineteenth century, German and Swiss immigrants traveled up the Mississippi River to the Ohio River and up the Ohio into the Appalachians in order to work in the processing of iron ore mined in the area. At the end of the nineteenth century, coal mining began in the Appalachians, and other groups began to move into the interior of the Appalachians, joining the Scotch-Irish who had made a living on small farms. Immigrants from Poland, Wales, Italy, Austria, and England, and African Americans from southern areas, moved into the Appalachians to become coal miners. Because of the hard life that they endured in the coalfields of Appalachia, the miners of Appalachia became a special cultural group, sharing beliefs and a unique way of life. by 2020, most of the coal mines had closed, disrupting a long-standing social contruct.

Southern Appalachia has a rural culture dominated by love of family and love of church. Fundamental Christianity has many dedicated followers there. In the southern Appalachians, country music was born and is produced. Nashville, Tennessee, home of the Grand Ole Opry, is the country music capital of the world.

Lower South

The southern states of Alabama, Mississippi, and Louisiana have much in common with the southeastern states. Their nineteenth century historical background includes large plantations, owned by European settlers and worked by African-American slaves. In 2020, these states had large rural populations composed of African Americans and descendants of the people who ran the plantations. There are still large farms producing cotton, sugarcane, and rice. These farms are now worked by laborers from all cultural groups.

Unlike other southern states, which were settled by the English, Louisiana's original European settlers were Spanish and French; therefore, New Orleans and areas surrounding it have many French streets and place names. The famous New Orleans festival of Mardi Gras is a French celebration. The Creoles, who generally live in New Orleans and sur-

Mardi Gras Day, New Orleans: Krewe of Kosmic Debris revelers on Frenchmen Street. (Infrogmation)

Untitled. *Ansel Adams, 1941. Taken near Canyon de Chelly, Arizona. (Ansel Adams - U.S. National Archives and Records Administration)*

rounding areas, are descendants of the original Spanish and French settlers. Some members of the Creole group are descendants of both the original European settlers and African Americans. Creoles have a distinct English dialect that incorporates some aspects of the language of their French ancestors. Some Creoles are bilingual, speaking both English and French. Up until 1868, all official documents of the Louisiana state government were written in both French and English.

Between 1763 and 1770, the British exiled about 8,000 French settlers from Acadia, which is now Nova Scotia. These settlers, called Cajuns, went to Louisiana, where their descendants now live in the bayou areas around New Orleans. Cajuns speak a distinct English dialect that has a French influence. Many of the descendants of the Acadians are bilin-

gual, speaking both French and English. About 30 percent of the people in the lower South are of African-American heritage. Some of the African Americans living in Louisiana speak a special French dialect as well.

Midwest

The Midwest comprises the states that make up the interior of the United States. While there are many urban centers in the Midwest—including Chicago, Illinois; Minneapolis-St. Paul, Minnesota; Omaha, Nebraska; and St. Louis, Missouri—the region is most characterized by vast stretches of flat land planted in corn to the south and wheat to the north. The Midwest was originally settled by people of several European countries, including Ireland, Great Britain, Germany, and Scandinavia.

Today the descendants of these early settlers give cultural variety to an area that the casual observer may see as homogeneous. In Iowa and Wisconsin, many descendants of German settlers keep their cultural heritage alive by staging German festivals such as Oktoberfest, by cooking and eating the foods of their ancestors, and by performing German music and dances. In parts of Ohio and Indiana, there are members of the Amish and Mennonite religious groups. Their minimalist lifestyles are noteworthy when compared to the complex lives that many other Americans live.

In Minnesota and eastern North Dakota, many people of Norwegian and Swedish heritage live. About 10,000 Chippewa Native Americans also live in Minnesota, some on the White Earth Reservation. Other Chippewa live in Michigan, some on reservations. Many Sioux live on the Pine Ridge and Rosebud reservations in South Dakota; others live on the Standing Rock Reservation that straddles the borders of North Dakota and South Dakota.

The Southwest

The southwestern states of Texas, New Mexico, and Arizona have a cultural mix different from that of the East and the Midwest. The first Europeans to live in this area were Spanish, and their influence can be seen in the place names, architecture, foods, language, and lifestyle. Immigration from Mexico in recent years reinforced the Spanish influences in this area. The blend of the United States and Mexican cultures in such Texas cities as San Antonio and El Paso and such Mexican cities as Juarez has led to the formation of the Tex-Mex culture, characterized by foods, dress, music, and dialect combining Spanish and US cultures.

Native Americans also live throughout the Southwest, on reservations, in towns and cities, and on their own farmland. Many Navajo live in Arizona; members of the Laguna Pueblo tribe live in New Mexico. The jewelry and pottery of the Native Americans of this area are prized throughout the world.

Rocky Mountain States

The Rocky Mountain area of the United States has densely populated urban areas such as Denver, Colorado; Salt Lake City, Utah; and Las Vegas, Nevada. But much of this area is made up of sparsely populated mountains and deserts. Throughout the region are huge sheep and cattle ranches.

One distinctive group found in the Rocky Mountain States is the Mormons, members of the Church of Jesus Christ of Latter-day Saints. This group left the eastern part of the United States in the middle part of the nineteenth century and moved to the Great Salt Lake Desert area in what is now the state of Utah. It was a barren land that no one else wanted. Through their knowledge of land management and irrigation, the Mormons prospered there. Thousands of Mormons still live in areas around Salt Lake City and Ogden, Utah.

Many Native American groups, including the Blackfoot in Montana and the Cheyenne in Wyoming, live in the Rocky Mountain states, some on reservations.

Pacific States

The Pacific states of California, Oregon, and Washington have a cultural diversity that is not found to such an extent in other parts of the United States. Since the middle of the eighteenth century, people from Asia have come to the United States, although at times in American history both state governments and the federal government tried to hamper this immigration.

Men lounging outside a saloon and a Chinese laundry, Salt Lake City, Utah, 1910.

Many people who can trace their ancestry to China, Taiwan, Japan, South Korea, Vietnam, and other Asian countries live along the Pacific Coast. Some of these people have formed ethnic communities in such large cities as Los Angeles and San Francisco. Here they speak the languages of their homelands, eat their native foods, and practice a lifestyle similar to the one they left behind in Asia. People of Asian ancestry generally prefer to live in culturally diverse communities and embrace the mainstream US culture.

In Southern California, there are many immigrants from Mexico, as well as descendants of people who immigrated from Mexico in past decades. They have brought the culture of Mexico to California, imbuing California with foods, music, and a special atmosphere it would not otherwise have.

Throughout the Pacific Northwest there are Native American groups. Some of these people, such as the Coeur d'Alene, live on Indian reservations where their native culture is kept alive through the work of tribal councils and the knowledge handed down from older people to younger. Other Native Americans live in the towns, cities, and rural areas of this region.

Mexico

A variety of societies ruled Mexico before Spain began occupying the region in 1519. The Olmecs, Maya, and Aztecs all developed brilliant civilizations noted for their advanced arts and sciences, monumental architecture, and large-scale political systems. Mestizos (people of mixed Spanish and native blood) form the majority of Mexico's population. In their efforts to create a new society in

San Lorenzo Monument 3 (also known as Colossal Head 3). Museo de Antropología de Xalapa, Veracruz, Mexico. (Maribel Ponce Ixba)

Mexico, the Spanish government and its partner, the Roman Catholic Church, permitted a measure of continuity with the indigenous past. Local officials administered Mexico through the indigenous institutions, wherever such existing customs and practices did not conflict with the demands of church and state. Racial mixing fostered cultural mixing, producing a blending of Spanish and indigenous culture that is unique in Latin America.

About 80 percent of Mexicans live in cities. Mexico has one of the largest Roman Catholic populations in the world. About 82 percent of its nearly 130 million people classify themselves Catholics—despite the government's long-standing twentieth history of anticlericalism. The absorption of indigenous spiritual traditions into Catholic practices has made it easier for native peoples to embrace Catholicism.

Annita Marie Ward

Culture Regions of the Caribbean

The culture regions of the Caribbean islands, formed in the wake of the European conquest that began with explorer Christopher Columbus in 1492, marked a new phenomenon in the history of Western civilization—a relationship between masters and slaves that resulted not in the imposition of one culture on top of another but in the constant transfusion of numerous traditions derived from Europe, Africa, and the indigenous peoples of America.

The Spanish, English, French, and Dutch colonists created a world whose crux was the sugar plantation. African and native slaves and their offspring, known as "mulattoes" and "mestizos," kept their own customs alive while superficially accepting the domination of the whites. Out of this admixture arose a distinct Caribbean identity, neither black nor white, neither native nor European. The arrival of immigrants from the Indian subcontinent in the nineteenth century added still another component.

Culture of the Planter Class

Europeans carved the Caribbean into rival spheres of influence: Cuba, Puerto Rico, and the Dominican Republic formed the Hispanic bloc; Haiti, Guadeloupe, and Martinique composed the French zone; Jamaica, Barbados, Trinidad, and the Windward and Leeward Islands made up the English contingent; and the Dutch Antilles rounded out the picture.

Whether Latin or Anglo-Saxon, the conquerors shared certain cultural characteristics, related to the need to protect the sugar plantation complex, that they passed on to their descendants: a monolithic religion (Catholicism, Anglicanism, or Dutch Protestantism) preaching defined notions of social status; military, political, and religious power invested in a planter class that identified more closely with its metropolitan motherland than with the New World; and a formal ban on mixing of the races that was often ignored.

All the Caribbean islands had black and mixed-ancestry majority populations by the nineteenth century, and the culture of the masters could not remain undiluted. People of mixed ancestry attained a degree of social acceptance then unimaginable in the United States.

Culture of the Indigenous

The most devastating legacy of the European conquest was genocide, the eradication of the Carib, Arawak, and other indigenous peoples who occupied the islands until the end of the fifteenth century. The Indoamerican culture, however, was not entirely wiped out. It survives in place names such as Guanabacoa in Cuba and the straw-patched hut called a *bohio*, which shelters farm laborers. Some

In the 18th century, Sint Eustatius was the most important Dutch island in the Caribbean. (Walter Hellebrand)

Caribbean Indian gods, transformed into saints, have entered the folk religions of the islands. A mestizo culture, borrowing freely from Europe, Indoamerica, and Africa, arose in the sixteenth century. Aspects of that culture that still survive include the white-laced *guayabera* shirt worn in Puerto Rico, and the smoking of tobacco, originally an indigenous crop harvested by African slaves and sold to the Europeans. The music of the Caribbean, merengue and dances such as the mambo, are probably the mestizo product best known in the United States.

The African heritage of the Caribbean is preserved in its purest form in Haiti, which makes up half the island of Hispaniola. The religion called voodoo, which involves ritual magic and the physical possession of believers by spirits, can be traced directly to the West Coast of Africa. The Haitian people, nearly all descended from slaves imported by the French and Spanish, developed their own language, Creole, that preserves a largely African vocabulary while using French grammar.

The other half of Hispaniola, the Dominican Republic, is the best example of a Caribbean mulatto culture, blending African with European. Race mixing has gone on for so long and to such an extent that almost every Dominican might be classified as a "mulatto." African languages are no longer spoken in the Caribbean, but Dominican Spanish carries its own Creole inflection. Voodoo there blends almost seamlessly with Roman Catholicism, producing a cult of saints to whom believers attribute magical powers without invoking their African origin.

Anglo-Caribbean Culture

Two elements make the English-speaking Caribbean islands stand out from the rest. One is an Anglophile (English-loving) elite that, regardless of skin color, turns its eyes to London and tries to recreate the glory days of the British Empire in miniature form, whether by playing the game of cricket or

Slaves planting sugar cane, Antigua, 1823. (William Clark)

by strictly adhering to mainstream British Protestantism. This fixation with the colonial metropolis has resulted in a more disparaging attitude toward race mixing, although a not insignificant portion of the population is mulatto. The poems of the Nobel Prize in Literature author Derek Walcott from St. Lucia, cannily reconstruct the dilemma of race and culture in the English-speaking islands. An African-American culture survives there largely in opposition to the elite culture, as evidenced by the continued popularity of reggae music and religious sects such as the Rastafarians in Jamaica.

Another unique feature of the Anglo-Caribbean is the sizeable presence of generations of women and men descended from migrants from preindependence India, including the areas that are now Pakistan and Bangladesh. Torn between nostalgia for a homeland most have never known and a language and set of customs imposed by the British, East Indians in the Caribbean have produced their own subculture, mining elements of the Hindu, Muslim, and English elements from their heritage. In the novels and essays of Trinidad-born author V. S. Naipaul, the contradictions of East Indian identity in the New World are beautifully explored.

Julio César Pino

166

EXPLORATION OF NORTH AMERICA

The first explorers of North America were the peoples who crossed from Asia to present-day Alaska several thousand years ago. They became the original population of both North and South America. Their cultures developed and thrived for many centuries, completely unknown to the peoples of Europe. Finally, Europeans began to discover America, although accounts of the early expeditions are shrouded in myth and legend.

Early Legends and Explorations

Legendary accounts begin in the sixth century with the supposed voyage of St. Brendan, Bishop of Clonfert in Ireland. Although Brendan is credited with reaching the western limits of the known world, written accounts of his exploits did not appear until the tenth century. It is impossible to know what those limits were, either the Canary Islands off the coast of Africa or the east coast of North America.

The next legend is that of the Seven Cities. When Muslim Arabs conquered Spain in 711 CE, it was said that seven Christian bishops and their congregations fled west across the Atlantic Ocean and founded seven cities. This story was passed down orally for centuries. After Spanish maps of North

Leif Eriksson Discovers America *by Hans Dahl (1849-1937).*

VOYAGES OF COLUMBUS, 1492-1502

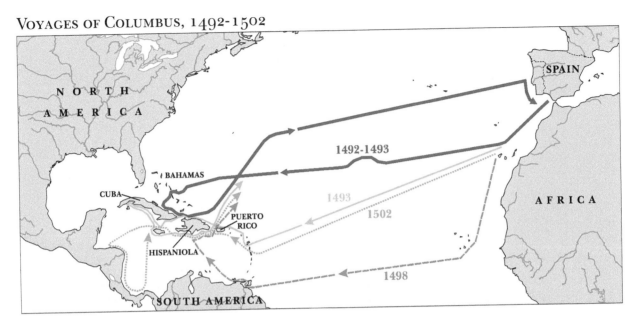

America began to appear in the sixteenth century, the seven cities were identified as the pueblos of New Mexico.

The last major legend is the account of the Welsh adventurer Madoc in the late twelfth century. Written accounts about 400 years later, after the colonization of America by Europeans, claimed that Madoc and a large group of settlers in ten ships landed on the east coast of North America. Later reports of "white Indians" who spoke the Welsh language in Tennessee, Kentucky, and on the upper Missouri River have been linked to the adventures of Madoc.

Historically, European exploration of North America began in the ninth and tenth centuries, when Norse adventurers from Norway discovered and colonized Greenland. The colony, established by Erik the Red in 986 CE, survived for about 500 years before vanishing for unknown reasons. About the year 1000, Leif Erikson, the son of Erik, led a party south and west of Greenland. After landing on Baffin Island and in Labrador, his party established the colony of Vinland, probably in what is now Newfoundland. Either tragedy or a return to Greenland ended this settlement after about twenty years, but the forests of Labrador continued to provide timber for Greenland for many years. Accounts of Vinland written in the thirteenth century provided little aid to later Europeans.

Spanish Discoveries

In 1477 Christopher Columbus, a native of Genoa, Italy, joined an English trading expedition to Iceland. Any plans he might have had to get to Asia by crossing the North Pole ended after he experienced the harsh conditions of this voyage. Columbus then developed his grand design of reaching Asia by sailing west across the Atlantic Ocean, but he miscalculated the size of the world. He thought that a journey of 2,400 miles (3,860 km.) would take him to Japan. The actual distance is about 10,600 miles (17,055 km.), with the American continents blocking the path. In 1492 Columbus secured the financial backing of the Spanish monarchs, Ferdinand and Isabella, and sailed west. After his landing on San Salvador in the Bahamas, he made three more trips, still with the hope of reaching Asia. Columbus's voyages opened the New World to further exploration and colonization, and began the modern era.

In 1513 Vasco Nuñez de Balboa crossed the Isthmus of Panama—the connecting link between North and South America—and discovered the vast Pacific Ocean that further separated Europeans from Asia. The first Spanish attempt to explore the North American mainland was by Juan Ponce de León, who had sailed with Columbus on his second journey and had discovered Puerto Rico. León made two trips, in 1513 and in 1521, to Florida, which he first thought was an island rather than a peninsula. He explored both coasts, naming the peninsula for the fact that he landed first on Easter Sunday ("Pascua de Florida"—the Feast of Flowers) and for the abundant and beautiful vegetation of the area. León was killed by a poison-tipped arrow in 1521.

In 1526 Lucas Vazquez de Ayllon established the first Spanish settlement in the New World, near the mouth of the Savannah River in Georgia. Hardships including native hostility and disease soon led to the abandonment of this colony.

Early Interior Explorations

The next step for the Europeans was to investigate the size, shape, and human geography of the New World. In 1524 Giovanni da Verrazzano, an Italian, was commissioned by King Francis I of France to find a Northwest Passage through or around North America to Asia. Verrazzano became the first European to sail into New York Bay. In 1534 Jacques Cartier, also commissioned by Francis I, discovered the St. Lawrence River, immediately raising hopes

> ### PONCE DE LÉON AND THE FOUNTAIN OF YOUTH
>
> The Spanish explorer Juan Ponce de Léon was motivated by legends about a fountain of youth in which elderly people bathing themselves became young again. Legends of such fountains had existed in Europe for many centuries. In Florida in 1513, Léon and his men sampled every brook, spring, and puddle they could find, but found no such fountain.

of a Northwest Passage. Cartier travelled about 1,000 miles (1,600 km.) into the Appalachian mountains and valleys. He eventually reached the present-day sites of Quebec and Montreal, but he was stopped at the Lachine Rapids south of Montreal, far short of his goal of reaching the Pacific Ocean.

The next expedition was by the Spanish explorer, Hernando de Soto. In 1539 de Soto was made the governor of Florida, which at the time meant the entire southeastern corner of North America. Although he was authorized to explore and colonize

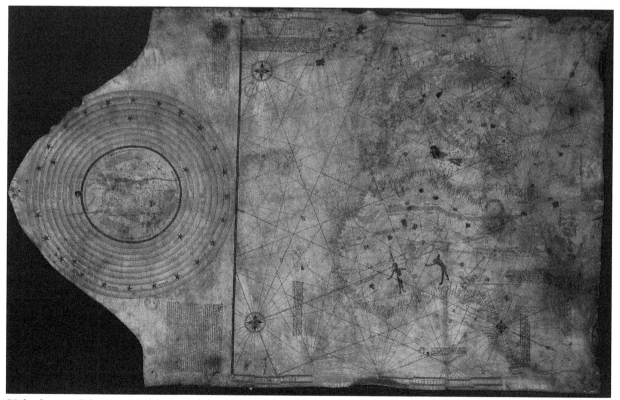

"Columbus map," drawn c. 1490 in the Lisbon workshop of Bartolomeo and Christopher Columbus. (Gallica Digital Library)

MAJOR VOYAGES TO THE NEW WORLD AFTER COLUMBUS

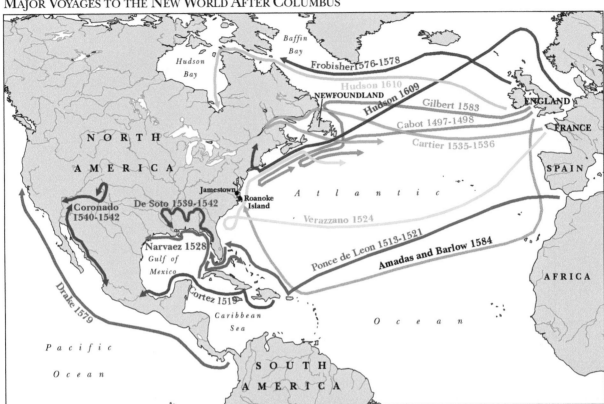

the area, De Soto's major goal was to find gold. He began his exploration at Tampa Bay with about 600 men and traveled up the Atlantic Coastal Plain. As far north as Ayllon's abandoned settlement, de Soto faced opposition from the native inhabitants. Moving inland, he entered the southern section of the Appalachians. Here he made contact with people who had not yet met the Europeans and initially were friendly to the Spanish.

After crossing the Appalachian Mountains, de Soto discovered the Hiwassee River, a tributary of the Tennessee River and part of the vast Mississippi River system. Now facing native hostility, he continued west into the central lowlands. He died in 1541, after discovering the Mississippi River near Memphis, Tennessee, and he was buried in the river. Although de Soto had found the legendary town of Chisca, where Memphis now stands, he found no gold.

Another Spaniard, Francisco Coronado, began an expedition north from Mexico at about the same time as de Soto's. Like de Soto, he was searching for gold. With Coronado were 336 soldiers and several hundred Native Americans. In 1540, the group entered the desert basin of Arizona, followed the Colorado River, and became the first Europeans to view the magnificent Grand Canyon. After crossing the Rocky Mountains, Coronado found himself on the Great Plains of North America. Here the Europeans saw for the first time the huge herds of bison, numbering in the millions, that roamed the plains. Coronado traveled as far east as Kansas before turning back to Mexico. Like de Soto, he had found no gold.

Search for the Northwest Passage

In 1607 the search for a Northwest Passage was renewed by Henry Hudson from England. In the most determined search yet, Hudson sailed up the west coast of Greenland as far as 80 degrees 27 minutes north latitude, far north of the Arctic Circle, before realizing that there was no hope of crossing the Arctic Ocean. Still, he believed that only a narrow isthmus separated the Atlantic and Pacific oceans. On another voyage in 1609, this time financed by the Dutch East India Company, Hudson discovered the Hudson River in New York Bay. After sailing up the river about 150 miles (240 km.), he realized that this route, like all earlier routes, did not reach the Pacific Ocean.

In 1608 Samuel de Champlain of France founded Quebec. The following year, he discovered the lake in upstate New York to which he gave his name, and he later mapped Lake Huron. Champlain did more than any explorer to that time to make this area of North America known to Europeans.

The next phase of the search for a Northwest Passage came when the French began exploring the Mississippi River-Great Lakes section of the central lowlands. In 1634, Jean Nicolet reached Green Bay on the west side of Lake Michigan. This exploration involved the rich fur trade that the French were developing, but it also included the possibility that the Mississippi River might be a route to the Pacific Ocean. In 1666 Father Claude Dahlon, a Roman Catholic priest and a trained geographer, ascended the Fox River and portaged to the Wisconsin River, but did not reach the Mississippi River.

The journey of Louis Joliet and Jacques Marquette, beginning in 1673, followed Dahlon's route and reached the mighty Mississippi River. They descended the river as far as the mouth of a tributary, the Arkansas River, before confirming their growing realization that the Mississippi flowed south to the Gulf of Mexico rather than west to the

Sacajawea

A member of the Shoshone tribe, Sacajawea was born in about 1786 in present-day Idaho. After being captured by an enemy tribe, she was sold to a French fur trader and became his wife. In the winter of 1804-1805, Lewis and Clark met Sacajawea and her husband in North Dakota and hired them as guides for their expedition. Carrying her infant son on her back, Sacajawea, with her knowledge of the area and of the tribal languages, became a valuable asset to the expedition, especially when they met the Shoshone. Sacajawea died about 1812.

Pacific Ocean. In 1682 another French adventurer, Rene-Robert Cavelier de La Salle, followed the same route and reached the Gulf of Mexico at present-day New Orleans.

Later Explorations

English exploration of the Atlantic Coastal Plain began at Jamestown, Virginia, in 1607. By 1700 they had moved west to the fall line on the eastern slope of the Appalachian Mountains. After taking over French claims to North America in 1763, the English explored and settled the Mississippi River-Great Lakes and Ozark sections of the central lowlands. A major step toward exploring the unknown regions of this territory came in 1775, when Daniel Boone built the Wilderness Road through Cumberland Gap in

Map of Lewis and Clark's expedition: It changed mapping of northwest America by providing the first accurate depiction of the relationship of the sources of the Columbia and Missouri Rivers, and the Rocky Mountains. Published c. 1814.

George Caleb Bingham's Daniel Boone Escorting Settlers through the Cumberland Gap *(1851–52) is a famous depiction of Boone. (George Caleb Bingham, The Bridgeman Art Library)*

the southern Appalachians. This opened up the valleys of the Tennessee, Kentucky, and Ohio rivers all the way to the Mississippi River.

In 1803 the Louisiana Purchase added the vast Great Plains west of the Mississippi River to the new United States. In 1803, US President Thomas Jefferson authorized Meriwether Lewis and William Clark to explore this territory. Their famous expedition left St. Louis, Missouri, in May 1804, and returned in September 1806. They had followed the Missouri River to its source, crossed the Continental Divide in the Rocky Mountains, and followed the Columbia River to the Pacific Ocean. Journals kept by every member of this expedition provided valuable information about the physical geography, animal and plant life, and human geography of the area.

Many other expeditions explored the west. In 1806 Zebulon Pike, searching for the source of the Arkansas River, discovered the peak in Colorado that bears his name. Sylvester Pattie and his son explored the intermontane region west of the Rockies and the Gila River area of the southwest. In 1832 Benjamin Bonneville went through the South Pass of the Rockies in Wyoming and paved the way for the famous Oregon Trail. John C. Frémont and others explored the Great Basin around the Great Salt Lake and then across the Sierra Nevada to the Pacific coastlands of California. Future exploration and development revealed that North America was rich in natural resources. Lumber, iron ore, coal, copper, silver, and the gold that the early Spanish had sought, were all found in abundance. To these were added the rich oil deposits in many areas of North America.

Glenn L. Swygart

EXPLORATION OF THE CARIBBEAN

The European explorers who first sailed to the Caribbean did not expect to discover it. At first they assumed that the sea and its islands were merely way stations on their Atlantic passage to Asia. Gradually, however, the fact that the Caribbean was a partly enclosed sea itself became apparent.

The Caribbean's indigenous people included the Caribs (from whom the name of the sea was later derived) and the Arawaks. The Arawak umbrella also covers local groups such as the Taino and the Ciboney. It is not known exactly when the Caribs and Arawaks arrived in the Caribbean basin. Historians generally believe the Arawak arrived around 300 CE, the Caribs a millennium later.

When Christopher Columbus sailed out of Palos, Spain, in August, 1492, most learned Europeans already knew that the world was round. However, they believed—and correctly—that Asia was about 15,000 miles (24,000 km.) to the west of Europe's farthest western reaches. They therefore believed it was not worth the effort to sail west, instead of making the long overland trek from west to east. Columbus, a sailor from Genoa who was employed by the king and queen of Spain, thought that the world was smaller than it actually is. As a consequence, he thought that the east-to-west distance from Europe to Asia was considerably shorter than was actually the case. Finding evidence in various geographical texts to back him up, he persuaded the Spanish monarchs to finance his voyage.

Columbus's expedition included three ships: *Nina*, *Pinta* (captained by Martin Alonzo Pinzón), and *Santa Maria*. The ships stopped at Gomera in the Canary Islands on September 9. This was their last landfall before they reached the New World. After a voyage of slightly longer than four weeks, Columbus's expedition reached landfall on an island that Columbus called San Salvador (24 degrees north latitude, 75 degrees west longitude). Its Arawak name was Guanahani; it is now called Watling's Island. It is part of the Bahamian chain on the east central edge of the island group, just north of the Tropic of Cancer. Two days later, the expedition discovered the nearby island now called Rum Cay. The Spaniards encountered local plant and animal spe-

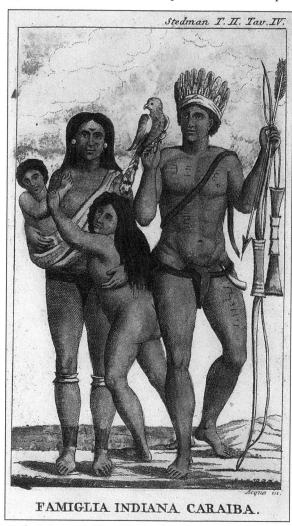

Carib (Kalina or Galibi) indian family after a painting by John Gabriel Stedman. (John Gabriel Stedman)

cies and also native human populations whom they did not antagonize.

Two Major Islands

The small Bahamian islands, which Columbus believed to lie off the coast of Malaysia, were insufficient finds for Columbus's purposes. Columbus heard from the Arawak people of a large island to the south called Colba (present-day Cuba). From October 20 to December 3, the expedition explored the northeast coast of Cuba. It then moved on to Hispaniola, Columbus giving the island its name (Española in Spanish), which means "Little Spain." Thus the two biggest islands in the Caribbean, the anchors of the Greater Antilles, had been found.

The expedition returned triumphantly to Spain in early 1493. Immediately, plans for a second expedition were made; it left Spain in late September 1493, and made landfall in the Caribbean on November 3. This time, Columbus and crew landed to the south and east of their previous destination, on the island of Dominica. From Dominica, the fleet sailed northwest toward Cuba, sighting many of the Lesser Antilles, stretching in a northwesterly chain from Dominica, and also the much larger island of Puerto Rico. Once Columbus had returned to Cuba, he led a small expedition that probed to the south and discovered Jamaica. Columbus's third voyage, in 1497, was devoted to finding the American mainland, although Trinidad was sighted on the way to Venezuela. At the first Caribbean landfall of his fourth and final voyage in 1502, Columbus's ships discovered Martinique.

After Columbus's return to Spain and his death in 1506, the rule of Cuba was entrusted to Nicolás de Ovando. In 1506 Ovando sent his talented lieutenant Juan Ponce de León to search for gold on Puerto Rico (called Borinquen by its indigenous people). In June, Ponce de León landed at Añasco Bay on the west coast of the island, but soon the native Tainos resisted. The Tainos were eventually subjected to forced labor.

Later Expeditions

Not all islands were explored in the first wave, and some seem to have never been inhabited even by indigenous peoples. Some of the Lesser Antilles, of little immediate interest to trade- and gold-hungry

NAMING THE ISLANDS

The names first given to the Caribbean islands by Columbus and other early explorers may be confusing to persons used to present-day nomenclature. The Spanish explorers tended either to retain native place names (for example, Cuba) or to name places with religious meanings, such as the names of saints or other religious words (for example, Trinidad, which means "trinity"). Many of the names given to islands by Columbus have changed. The island Columbus named after Saint Martin of Tours is not the present French/Dutch island Saint Martin/Sint Maarten, but the formerly British-ruled Isle of Nevis. San Jorge, just north of Nevis, is now familiar as Saint Kitts. Although an explorer may have given an island a name, time may not always have retained it.

explorers, were not extensively probed. Barbados was only sighted by Portuguese explorers on their way to Brazil in the early 1500s, and Europeans did not set foot upon it until the English arrived there in 1627.

Outlying islands such as Bermuda were also late to be explored. In 1505, the first European landfall in Bermuda was made by Juan de Bermudez,$Ide Bermudez, Juana Spanish explorer for whom the island is named. Bermuda was only permanently settled by the English in the early seventeenth century. Some islands close to the South American coast (Bonaire, Curaçao, and Aruba) were discovered by Alonso de Ojeda in 1499, but they were deemed useless other than as a source of slaves.

Once the Pacific Ocean was discovered by Vasco Nuñez de Balboa in 1513, the Spanish concentrated most of their efforts on exploring that side of the hemispheric landmass as well as exploring and settling the areas to the south. Although the American continent narrowed at the Panama isthmus, there was no point of direct access to the Pacific in the area of the Caribbean latitudes. People could voyage to the Caribbean from the east, but they could not voyage westward from there. The geographic orientation of the modern Caribbean was thus toward Europe, Africa, and North America—not Asia, to which its first European explorers aspired to journey.

Nicholas Birns

POLITICAL GEOGRAPHY OF NORTH AMERICA

North America comprises mainland Canada, Mexico, and the United States; offshore islands and island groups belonging to these countries; the islands of Greenland, Saint Pierre and Miquelon, and Bermuda, which are located in the North Atlantic and belong to, respectively, Denmark, France, and the United Kingdom.

Central America is sometimes classified as a geographical part of North America but is usually considered a separate subregion of the Americas; it is closely associated—both culturally and politically—with the continent of South America. That region is discussed in Volume 1, "South and Central America," of World Geography. This chapter only discusses the three countries of mainland North America and the island countries and territories of the Caribbean, or the West Indies.

Canada

Modern Canada started as a French possession, New France, centered in what became the province of Quebec and Quebec City. The first English colonies in the region were in Newfoundland, Nova Scotia, and the Hudson Bay drainage basin. Conflict between Britain and France dominated its history in the seventeenth century. The conflict culminated

Senate Chamber, Centre Block of the Parliament of Canada. (Mightydrake)

POLITICAL DIVISIONS WITHIN CANADA, UNITED STATES, AND MEXICO

Key to States
(shown by numbers on map)

1 Aguascalientes	6 Tlaxcala
2 Guanajuato	7 Distrito Federal
3 Querétaro	8 Morelos
4 Hidalgo	9 Puebla
5 Mexico	

in the Seven Years' War, known in North America as the French and Indian War (1754-1763), in which Britain took Canada from France. Britain then possessed a colony populated almost wholly by non-English subjects, the French-speaking Québécois. Scots and the Loyalists fleeing the American Revolution dominated the subsequent settlement of Canada by English speakers. There also were armed conflicts between Canada and the United States. During the American Revolution and the War of 1812, US forces invaded Canada in unsuccessful attempts to conquer it.

Revolts against the British Crown in 1837-1838 pushed Canada toward the process of the Canadian Confederation, which eventually resulted in the formation of the Dominion of Canada in 1867, with the British monarch as the formal head of state, or sovereign. With the passage of the Statute of Westminster (1931) by the Parliament of the United Kingdom, Canada became an independent state, with the reigning British monarch acting (with respect to Canada) as essentially a titular head of state. Thus, Canada became an independent constitutional monarchy, symbolizing the collective heritage of Canada and the United Kingdom. Canada developed a parliamentary government with the prime minister as the head of government.

Mexico

Modern Mexico began with Hernán Cortés's conquest (1519-1521) of the Aztec Empire and its capital city, Tenochtitlán, located on the site of modern Mexico City. Mexico developed into a rich Spanish possession with a vast territory that included present-day California, Nevada, Utah, Arizona, New Mexico, Texas, and parts of Colorado and Oklahoma. In 1810 Father Miguel Hidalgo y Costilla issued a manifesto calling for the end of Spanish rule. Although the Hidalgo Revolt was unsuccessful, it ignited the Mexican War of Independence, which lasted until 1821 and resulted in the establishment of an independent Mexican state.

In 1846 the Mexican-American War began, in which Mexico lost roughly one-third of the territory it had in 1821. Causes of the war were rooted in the expansionist policy of the United States based on

the doctrine of Manifest Destiny, that is, the belief that the United States was destined to expand across North America. In 1861 three European powers (France, Spain, and the United Kingdom) occupied Mexico. After Spain and the United Kingdom withdrew, France invited Austrian archduke Maximilian, a brother of Emperor Franz Joseph of Austria-Hungary, to rule as Mexico's titular emperor. The forces of Benito Juárez, exiled president of Mexico, overthrew Maximilian in 1867 and restored the republic. Dictators subsequently ruled Mexico until 1910, when a revolution broke out led by military commanders such as Pancho Villa and Emiliano Zapata. For much of the twentieth century, Mexico was ruled by the Institutional Revolutionary Party (Partido Revolucionario Institucional —PRI), which was formed as a result of the Mexican Revolution.

United States of America

The United States began as thirteen British colonies in North America that declared independence in 1776. The country underwent great expansion, subsequent wars with Great Britain (1812-1814) and Mexico (1846-1848), and the American Civil War (1861-1865). The Civil War settled major issues with regard to slavery, which was abolished, and supremacy of federal law over state law. The country expanded farther, all the way to the Pacific and, with the Spanish-American War in 1898, entered the international stage as a world power. The United States is a constitutional democratic republic in which some decisions are made by direct democratic processes and some others, by democratically elected representatives of the people.

Governmental Structures

Canada has a federal system of government. Federal legislative authority is vested in the Parliament of Canada, which consists of the sovereign (represented by the governor-general), the Senate, and the House of Commons. Legislative and executive authority is divided between the central government and the country's ten provinces and three territories, whose powers are similar in scope to those of the fifty US states. The Parliament of Canada is

The National Palace, symbolic seat of the executive branch in Mexico City, Mexico.

assigned authority over such matters as control of the armed forces, the regulation of trade and commerce, and currency. The Governor-General who represents the Crown is appointed by the reigning monarch of the Commonwealth (acting as the monarch of Canada and not as the British monarch) upon the advice of the Canadian government. In order to form a government, the Governor-General calls upon the leader of the party winning the most seats in Parliament in a general election. That party leader becomes the prime minister and generally chooses to form the cabinet from among party colleagues who have been elected to Parliament.

Mexico is a federal republic comprising thirty-one states and the Federal District, which is similar to the United States' District of Columbia. The official name of the country is the Estados Unidos Mexicanos (United Mexican States). Governmental powers are divided among executive, legislative, and judicial branches, but in practice, the president has strong control. The PRI was the dominant political party for more than seventy years but is now one of four major parties alternating in power. The president is popularly elected and can serve only one six-year term. The president is empowered to select a cabinet as well as the governor of the Federal District, the attorney general, ambassadors, high-rank-

ing military officers, and supreme court justices, who serve life terms. At the state and local levels, governors, unicameral (one-house) legislatures and mayors are elected by popular vote. Governors serve for six years, deputies for three. States can levy taxes and have all powers not delegated to the federal government, but in fact, they have relatively little revenue-generating potential or political power. The military has long been an apolitical force in the country, but its high-ranking officers are appointed by the president and serve at his or her discretion.

The United States is also a federal republic, comprising fifty states, the District of Columbia, and five permanently inhabited territories: Guam, Puerto Rico, the US Virgin Islands, the Northern Mariana Islands, and American Samoa. Governmental powers are, as in Mexico, divided among the executive, legislative, and judicial branches. The executive and legislative branches are almost equally strong, but the US Supreme Court has the final say in many matters, especially those of constitutional importance. Lower federal courts, the US district courts, and the US circuit courts of appeals can also make rulings on constitutionality and can even find the president, for example, in contempt of court and possibly subject to criminal proceedings.

The fifty states are generally autonomous and have considerable law-making and revenue-generating power. Pursuant to the US Constitution, any powers not expressly granted to the federal government are reserved to the states. The US Constitution also recognizes Native American tribes as distinct governments. Consequently, tribal governments of indigenous American nations have significant self-governing powers, similar to those of state governments. The governors of the states are popularly elected, as are the state legislatures. Almost all state legislatures are bicameral (two-house). The US military is apolitical. The president is commander-in-chief of all US armed forces, but can only appoint or nominate the Joint Chiefs of Staff, who are subject to Senate approval.

Immigration and Drug-Trafficking Issues

Migration from Mexico to the United States has been one of the greatest waves of immigration in US history. Many have come as legal immigrants or with temporary working papers, but many more have just crossed the border in search of work or for a better life. Since the early 1980s, this migration flow has been supplemented, and later surpassed, by the flow of migrants into Mexico from Central American countries. These people, fleeing civil wars, chronic poverty, and violence in Honduras, El Salvador, and Guatemala, started using Mexico as a route into the United States. Many have been injured or even died in the attempt to cross through other countries.

Migration of Mexicans into the United States is most often tied to Mexico's economic hardships and underemployment; it is also stimulated by the demand for a low-skilled workforce on the part of labor-intensive sectors of the US economy. The border crossing issue has been further complicated by the activities of Mexican drug cartels and organized crime syndicates, which became heavily invested in illicit drug smuggling and human trafficking between Latin America and the United States.

The first barrier on the 1,954-mile (3,144-km.)-long US-Mexico border was built by the United

Mexico–United States barrier at the border of Tijuana, Mexico and San Diego, USA. The crosses represent migrants who died in the crossing attempt. Some identified, some not. A surveillance tower is in the background. (Tomascastelazo)

States between 1909 and 1911 and by Mexico around 1920. The border barriers were extended in the 1920s, 1940s, and the 1990s when the first major walled structures were constructed. Significant additions and enhancements were made after the US Congress passed the Secure Fence Act of 2006, especially between 2017 and 2020.

Internal Divisions

Both Mexico and Canada have had long-standing internal cultural and political strife. In Canada, the province of Quebec has remained strongly attached to the French language and distinct French-Canadian culture. The Québécois firmly retain their identity and customs. Many have strived for an independent Quebec. This has put great strain on the Canadian governmental structure. Many efforts have been made to encourage Quebec to stay in the Canadian state, including placing the French language on an equal footing with English. The loss of Quebec would geographically isolate the Atlantic provinces of Newfoundland and Labrador, New Brunswick, Nova Scotia, and Prince Edward Island (the latter three also known as the Maritime Provinces) from the rest of Canada. The problem of Quebec's political status has remained unresolved to this day.

In Canada, much like in the United States, indigenous peoples have been at the bottom of the social and economic scale with high rates of poverty, unemployment, health issues, and other social problems. But unlike in the United States, self-government rights of indigenous peoples in Canada are limited by the longstanding Indian Act, which is a source of discontent among Canada's First Nations. Aggressive, forced assimilation in the past had disrupted native communities. In a landmark 2008 court settlement, the Canadian government formally apologized to former students of aboriginal boarding schools and established the Truth and Reconciliation Commission of Canada (TRC). In 2015, the commission published its report that described forced removal of aboriginal children for schooling as cultural genocide. Since the 2010s, the federal government has begun to consistently address the problems rooted in the legacy of neglect

Cover of They Came for the Children. *(Truth and Reconciliation Commission of Canada)*

and failed policies felt by generations of indigenous Canadians.

Mexican society is sharply divided by income and educational level. Although a middle class has grown in the cities, the principal division is between the wealthy, well-educated elite and the urban and rural poor. From the 1940s through the 1970s, Mexico enjoyed sustained economic growth and political stability. At the close of the 1970s, a petroleum boom produced growth rates of 8 to 9 percent. It seemed that Mexico could be headed toward the status of an industrialized country, but there were serious flaws in the country's development model. Highly inequitable income distribution left 30 to 50 percent of the population mired in poverty, particularly in the countryside.

In 1994 there was an uprising of Mayan peasants in Chiapas in southern Mexico. While the rebels, known as Zapatistas, never posed a threat to the regime, they did focus attention on Mexico's economic and political injustices. Economic and fiscal crises of the 1980s and 1990s as well as corruption, power abuses and impunity, activities of gangs and drug cartels, and low productivity have all impeded the country's socioeconomic development well into the twenty-first century.

In the United States, the Puerto Rican Nationalist Party revolts of the 1950s were the culmination of a coordinated armed struggle for full independence of the island commonwealth from the United States. Since the second half of the twentieth century, the independence movement has not been widely supported by Puerto Ricans. However, several referendums about gaining statehood for Puerto Rico have been unsuccessful.

The United States faces a high level of income disparity between a small group of the very rich and the rest of the population, especially the large segment of the population that lives below the poverty line, slightly above the poverty line, and "paycheck to paycheck," lacking stable or sufficient income to cope with any disruption in employment. While the problem has not been as acute as in Mexico or the Caribbean, the gap between the two groups has been

widening since the 1980s. The latter group, which historically included large percentages of African Americans, Native Americans, and Hispanics as well as poor whites, has been growing while the middle class shrinks. According to the authoritative Pew Research Center, the share of American adults who live in middle-income households has decreased from 61 percent in 1971 to 51 percent in 2019.

Many factors have contributed to the process, including the loss of manufacturing jobs due to globalization, the decline of labor unions, the eroding value of the minimum wage, skill-based technological change, growth in compensation for high-level management at the expense of labor wages and salaries, anti-labor policies favoring corporations, and slower expansion of the welfare state in the United States in comparison to other Western industrialized nations. The growth of income disparity has been accompanied by the ideological polarization of the American society, evidenced by ultraconservative and ultra-left grassroots movements, ex-

Political cartoon from the Progressive Era, when wealth concentration was similar to that of the present, shows how the concentration of wealth in a few hands leads to the extinguishing of individualism, initiative, ambition, untainted success, and independence. (Library of Congress)

treme partisanship in Congress, the politicization of mainstream media, and a sharp divide along generational lines.

Intergovernmental Treaties and Organizations

The North American Free Trade Agreement (NAFTA) went into effect in January 1994 and lowered or eliminated tariffs among its three members, Canada, Mexico, and the United States. For a quarter of a century, NAFTA was the largest trading bloc in the world by gross domestic product (GDP). However, NAFTA was criticized in both the United States and Mexico. Americans complained that little benefit has come to the United States and jobs have been lost to Canada and Mexico. Multinational companies moved factories to Mexico to take advantage of the cheaper labor and access to US markets. At the same time, Mexico lost over 1 million farm jobs as imported corn and other grains became cheaper than locally produced ones after the removal of tariffs. Because of these and other disadvantages, in 2017 Mexico and Canada have agreed to the proposal of the US administration to renegotiate NAFTA. The new treaty, called the United States-Mexico-Canada Agreement (USMCA), was ratified by all three countries and took effect on July 1, 2020.

The United States and Canada, which share the longest land border between any two countries, have a close political, military, and law-enforcement relationship. A principal instrument of the US-Canada cooperation on transboundary issues is the International Joint Commission, established under the Boundary Waters Treaty of 1909. Both countries were among the founding members of the North Atlantic Treaty Organization (NATO), the world's largest peacetime military alliance. The air forces of the United States and Canada operate under joint command in accordance with a 1957 pact that established the binational North American Air Defense Command (NORAD), later renamed North American Aerospace Defense Command. It conducts aerospace warning, aerospace control, and maritime warning in the defense of North America against any possible threat by aircraft, missiles, or space vehicles.

Canada, Cuba, the Dominican Republic, Haiti, Mexico, and the United States were among the founding members of the United Nations. Mexico and the United States were also founding members of the Organization of American States (OAS); Canada and the Caribbean nations joined the OAS later. Canada, the United States, and Mexico are members of the Asia-Pacific Economic Cooperation (APEC).

Internationally, all sovereign countries of mainland North America and the Caribbean are members in good standing of almost every major international organization. In the twenty-first century, bilateral relations among North American nations have not been marred by military conflicts or notable territorial disputes.

Dana P. McDermott

POLITICAL GEOGRAPHY OF THE CARIBBEAN

The Caribbean

The island region of the Caribbean, or the West Indies, has thirteen sovereign states and seventeen non-sovereign territories. Sovereign Caribbean nations are Antigua and Barbuda, the Bahamas, Barbados, Cuba, Dominica, the Dominican Republic, Grenada, Haiti, Jamaica, Saint Kitts and Nevis, Saint Lucia, Saint Vincent and the Grenadines, and Trinidad and Tobago.

The non-sovereign territories in the Caribbean include the US Virgin Islands and Puerto Rico, which are organized (that is, self-governing) unincorporated territories of the United States. Anguilla, British Virgin Islands, Turks and Caicos, Cayman Islands, and Montserrat are the British overseas territories. The islands of Bonaire, Sint Eustatius, and Saba are special municipalities within the Netherlands, while the islands of Aruba and Curaçao and part of the island of Saint Martin (known in Dutch as Sint Maarten) enjoy the status of constituent countries of the Netherlands. The other part of the island of Saint Martin as well as the islands of Guadeloupe, Martinique, and Saint Barthélemy belong to the French system of overseas collectivities. (The Caribbean island region, or the West Indies, is distinct from the larger Caribbean region, which also includes parts of Mexico, Central America, and South America that are bordering the Caribbean Sea.)

Independence

Europeans started occupying various parts of the Caribbean soon after Christopher Columbus first set foot in the region. Piracy thrived in the Caribbean in the seventeenth and eighteenth centuries because of the proximity of the many islands to major shipping lanes. By the end of the eighteenth century, practically all islands were claimed by various European colonial powers. Hundreds of thousands of slaves were brought from Africa to work the sugar plantations established by the colonists.

The first Caribbean nation to gain independence was Haiti, in 1804, after a thirteen-year uprising that ended both the slavery and colonial rule in the French part of the island of Hispaniola. The Haitian Revolution has remained the only successful

Parliament of the Bahamas, located in downtown Nassau. (UpstateNYer)

slave uprising in history. The Spanish part of Hispaniola, occupied by Haitian forces, became independent as the Dominican Republic in 1844. Cuba achieved independence in 1902, as a result of the Spanish-American War. The rest of the now-sovereign Caribbean nations—all former British possessions—attained independence between 1962 and 1983. Movements toward independence in the remaining dependent territories in the Caribbean have not acquired sufficient majority support of their populations.

Since gaining independence, Cuba, Haiti, the Dominican Republic, and several other Caribbean nations have experienced a string of governments dominated by military dictators and corrupt politicians. In Cuba, decades of repression and corruption led to a 1959 communist revolution headed by Fidel Castro. Under Castro's authoritarian rule, supported by the Soviet Union, Cuba repelled the 1961 CIA-led Bay of Pigs Invasion. In 1962, Cuba was the focus of the Cuban Missile Crisis, a major confrontation between the United States and the Soviet Union initiated by Soviet ballistic missile deployment on the island. A 1979 left-wing coup in Grenada brought to power the People's Revolutionary Government, which was toppled by the US military and its Caribbean allies. The US invasion of Grenada was the only time when US and Cuban soldiers (the latter were engaged in construction works in Grenada) fought each other directly.

Like Canada, most of the former British colonies in the Caribbean that became independent nations are technically constitutional monarchies, in which the current British monarch formally serves as the reigning monarch of each of these countries. These include Antigua and Barbuda, the Bahamas, Barbados, Grenada, Jamaica, Saint Kitts and Nevis, Saint Lucia, and Saint Vincent and the Grenadines. Haiti, Dominica, the Dominican Republic, and Trinidad and Tobago have the republican form of government. Nominally, Cuba is also a republic, but that country has been under authoritarian one-party rule since 1959.

The Caribbean island nations vary in their systems of government. Cuba has a one-party rule, while Haiti has almost thirty political parties. The

Cuban Prime Minister, Fidel Castro, arriving at MATS Terminal, Washington D.C. (1959). (Library of Congress)

two-party systems of Jamaica and the Bahamas are similar to that of the British government. Although Antigua and Barbuda and Trinidad and Tobago also have more than one political party, only one has remained in power for most of these countries' histories as independent nations.

French dependencies have their own elected governments but also elect their representatives to the French parliament, the National Assembly. (Similarly, residents of Greenland elect two representatives to Denmark's parliament.) The US, British, and Dutch possessions also have their own elected governments and are considered citizens of the controlling states, but they do not have voting representatives in the parliaments of the controlling states.

Local Issues

In 1982, the governments of the Bahamas, the United States, and the United Kingdom (on behalf of the Turks and Caicos) signed a tripartite agreement known as the Operation Bahamas Turks and Caicos (OPBAT) with the aim to interdict illicit drug trafficking, particularly from Colombian drug cartels to the US mainland. Since then, the US Coast Guard has been actively assisting the above islands in the search and seizure of vessels suspected of involvement in drug trafficking. Gangs involved in transshipment of cocaine and hashish oil to the United States have also been active in Puerto Rico, Jamaica. Haiti, and the Dominican Republic.

Illicit migration of Cuban nationals to the United States via a short maritime crossing between Cuba and Florida as well as via overland routes has been a decades-long challenge for the United States. In 2017, the two countries signed a joint statement ending the so-called "wet-foot, dry-foot" policy by which Cuban nationals who reached US soil were permitted to stay. Cuban migration by sea has since dropped significantly, but land border crossings have continued.

Crime, poverty, and income disparity have long been the prevalent social problems in the Caribbean and have remained a pressing developmental challenge in that region. In the twenty-first century, Haiti has remained the poorest country in the Western Hemisphere and has continued to experience bouts of political instability. Jamaica has had one of the highest murder rates in the world. There have been indications that international drug cartels displaced from the continental Americas have been moving into the Caribbean and that this accounts, at least partially, for rapid the growth of levels of crime in recent decades. In some Caribbean states, organized crime has strong political influence.

The Caribbean countries all belong to the United Nations (UN) and to the Organization of American States (OAS). Most sovereign states and one British dependency, Montserrat, are members of the Caribbean Community (CARICOM), an intergovernmental organization established in 1973 with the aims of coordinating economic and foreign policies, development planning, and trade relations. CARICOM was instrumental in the creation of the Association of Caribbean States (ACS) in 1994. That organization, with a predominantly economic agenda, has a larger membership; besides the island nations, ACS includes countries of Central America and South America bordering the Caribbean Sea. A smaller regional organization, the Organization of Eastern Caribbean States (OECS), consists mostly of former British colonies that have retained close historical, cultural, and economic ties.

P. S. Ramsey

DISCUSSION QUESTIONS: POLITICAL GEOGRAPHY

Q1. How did European colonial legacies affect the evolution of governments in Canada, the United States, and Mexico? What caused Canada to transition from being a French colony to an English colony? How is today's relationship between Canada and the British crown best described?

Q2. How have the indigenous people of North America fared in countries dominated by European-derived cultures? In what ways is the experience of the Inuit and other native peoples of the Far North different from those of Native Americans in the continental US and Mexico?

Q3. How many independent nations are located in the Caribbean? How are their governments similar? How do they differ? What characterizes the relationship of the non-sovereign states to their parent countries?

URBANIZATION OF NORTH AMERICA

To geographers, urbanization is both a process and a state. The state or extent of urbanization for a region or country at a point in time can be measured by the ratio of city to rural dwellers, by the number of designated "urban areas," or by the proportion of land included in designated urban or metropolitan areas. As a historical process, urbanization is descriptive of the ways in which a society creates high-density population centers that serve multiple needs of the broader community. Urbanization also refers to the ways in which the relationships of cities to rural areas, and of cities to cities, change over time.

In a broader sense, urbanization also can describe the ways in which cities and urban life affect the society and culture of people, changing attitudes, norms, customs, and even language. Urbanization is a matter of interrelationships that develop in complex ways, and numerous factors affect these developments: population changes, technology, economic conditions, and governmental and legal actions. In dealing with a single urban area and its development, one finds that transportation opportunities, communication possibilities, regional economic development patterns, on-site resources (minerals or recreation facilities), and governmental decisions influence success or failure. In examining the process for a region or nation, many of the same factors apply.

Early America

When Europeans first began colonizing North America, only a few of the native inhabitants lived in urbanized cultures. The Mississippian culture centered on Cahokia, several tribes in the Southwest, and the peoples of Mexico's Tenochtitlán had developed population centers of relative sophistication, but the influence these had on their respective regions is unknown. Except for the Tenochtitlán, they did not inspire further urbanization. The Europeans who colonized North America came from societies that were becoming increasingly urbanized, and it was the European pattern of early urban development that would characterize growth in the New World.

European concepts of property rights, Christian institutions, the use of horses and wheeled vehicles, continuing contacts with the parent countries (especially England, Spain, and France), the economic needs or desires of those countries, and a frontier mentality that craved expansion and ignored native

View of Boston, the oldest surviving aerial photograph ever taken. October 13th, 1860.

187

MAJOR URBAN CENTERS IN NORTH AMERICA

GREENLAND

Fairbanks
Anchorage

CANADA

Edmonton
Vancouver
Saskatoon
Seattle
Calgary
Regina
Portland
Winnipeg
Butte
Quebec
Montreal
Halifax
Ottawa
Minneapolis
Toronto
Boston
Milwaukee
Sacramento
Chicago
Detroit
New York
San Francisco
Baltimore
Philadelphia
Salt Lake City
St. Louis
Cincinnati
Washington D.C.
Denver
Indianapolis
Las Vegas
UNITED STATES
Louisville
Los Angeles
San Diego
Phoenix
Oklahoma City
Atlanta
Dallas
Birmingham
Charleston
Savannah
San Antonio
New Orleans
Houston
Atlantic Ocean

Miami

Pacific Ocean

Monterrey
Gulf of Mexico

MEXICO

Guadalajara
Mexico City ★
Oaxaca

Caribbean Sea

The City of Chicago, Illinois is an example of the early American grid system of development. The grid is enforced even on uneven topography.

claims to land, were dominant factors in the early settling and urbanization of North America.

The most heavily urbanized region was the seaboard from Boston to Philadelphia. Together with Newport and New York, these four merchant-controlled cities regulated the flow of raw materials to England and the Caribbean, and of manufactured goods into the British colonies. The southern region of the future United States had only Charleston as a major port. Despite the later growth of Savannah and New Orleans, the US South remained relatively underurbanized until after the Civil War. In the Northeast, competition among cities and continuing immigration fueled both the growth of new cities as well as dissatisfactions with England, factors that led to the American Revolution.

In 1690, an estimated 10 percent of the American population lived in colonial cities. The 1790 census showed an urban population of 5.1 percent. Cities stagnated as trade with England fell off and migration inland increased. New York City, however, soon benefited from special agreements made with transatlantic shippers, and never looked back. By 1840 its population was 250,000, and it was the main port of entry for immigrants from Europe.

The early nineteenth century witnessed the growing population and urbanization of the American interior. Cities required transportation and communication links with other cities, and waterways provided the most efficient channels for both. Canals in Pennsylvania and New York opened up the Great Lakes to the eastern port cities.

With the exception of Chicago, Indianapolis, and Milwaukee, all the major urban centers of the upper Midwest began between 1790 and 1820. Although initially tied economically and socially to the urban Northeast, these towns of the Ohio Valley and lower

Great Lakes soon controlled the rural hinterland around them and created a distinctive frontier culture that played an important role in defining the US identity. These cities, like their eastern elders, were strongly federalist and democratic in outlook and supported national initiatives that aided the urban economies and expansion: national currency, tariffs, and federally supported transportation, especially railroads.

Until the early twentieth century, urbanization meant both the growth of older metropolitan areas and the creation of new ones. The harnessing of steam power revolutionized both processes. In 1840 only twelve US cities had more than 10,000 inhabitants; in 1860, 101 cities had exceeded that population threshold. Manufacturing that once required river waterpower could locate anywhere, including the cities, where financial and legal resources as well as cheap labor resided. Steamboats made the transatlantic voyage faster and cheaper,

and spurred immigration (as did European events such as the Irish potato famine of the later 1840s). Steamboats also enhanced river traffic, as they could defy the current and carry huge loads in both directions.

The development of the railroad had perhaps the widest influence. Railways increased access to otherwise isolated areas, shortened travel and shipping time, greatly reduced the costs of shipping, brought raw materials to manufacturing sites, created crossroads with river cities like St. Louis and Cincinnati, stimulated the founding and development of cities along its path, and eventually connected the east and west coasts and their cities.

By the 1820s, Baltimore was the United States' third-largest city (after New York and Philadelphia) thanks primarily to the Baltimore and Ohio Railroad. Chicago grew overnight into America's rail center, becoming the terminus for eleven trunk lines. This led to its dominance in meat-packing and

Rowhouses, Federal Hill neighborhood, Baltimore. (ktylerconk)

other regional manufacturing and service industries. By 1850 U.S. cities were connected by almost 9,000 miles (1,480 km.) of track, most of which were in the northern states. By 1860 that figure was 31,000 miles (50,000 km.); by 1870, 53,000 (85,000 km.).

The largely agricultural economy of the South did not depend on rail transport, and while the Midwest and Northeast were being sewn ever more tightly together, the South remained isolated and underdeveloped. Its lack of urban, industrial, and transportation infrastructure weighed heavily against it during the Civil War. Aggressive capitalization and development followed the war and helped integrate the South into the US mainstream. Atlanta was rebuilt as a major rail hub, and Birmingham, Alabama, emerged as a center of steel production.

Metropolitan Growth

During the 1840s, the urban proportion of the U.S. population doubled to about 11 percent, and by 1880 had risen to 28 percent. The form as well as the distribution of cities changed. New York City claimed 1 million inhabitants in 1860, and other eastern cities grew apace as a result of foreign and regional immigration. Cities became more densely populated in part because of the development and spread of tightly packed, inexpensive tenements. Elevators and steel-beam construction allowed cities to build upward, further increasing urban densities. Some housing units occupied older, industrial buildings, while specially built so-called dumbbell apartments (named for their characteristic shape) provided cramped but efficient residences. In parts of New York City, densities hit 100 persons per acre (250 persons per hectare).

Cities also spread out, as both horse-drawn and steam omnibuses carried workers and shoppers into and out of the city along regular routes. These routes connected outlying smaller towns and stimulated growth along them. Central portions of cities grew into "downtowns" or central business districts, with concentrations of manufacturing, services, shopping, and entertainment. People changed residences at increasing rates, in some cities as high as 30 percent per year. In the later nineteenth century,

wealthier people occupied the central portions of urban residential space, with the middle classes farther out, and the poor—along with noxious industries—farthest away.

With high rates of European immigration, ethnic enclaves developed in eastern and, later, midwestern cities. In Greek Towns or Little Italys, customs, traditions, and languages were preserved, and new immigrants found welcoming communities. Along the West Coast, and later farther east, Asian communities likewise developed, often in older neighborhoods. In many cities, the pattern of ethnic or racial supplanting is quite clear, with, for example, an Irish neighborhood giving way to Eastern European Jews, then to African Americans who migrated from the South, and, more recently, to Asians.

Such growth and development had their costs. The high concentration of people in small areas meant enormous daily demands for food, water, and sanitation services. Scientific farming meant increased crop yields in the urban hinterland and, combined with rail transport and mechanical refrigeration, served the rising demand efficiently. Provision of potable water often proved more of a problem, but inventive engineers, urban planners, and politicians seemed to manage. Larger-capacity sewers were developed to carry away waste, but air and water pollution, animal dung, and garbage remained major urban problems.

Social problems plagued the growing cities also. Industrialization and new immigrants needing assimilation provoked demand for education. Poverty, transience, and poor housing fostered disease, organized violence, crime, and social pathologies like illiteracy, alcoholism, and child and spousal abuse. Tensions in cities sometimes erupted into ethnic or racial violence, with Catholics giving way to African Americans as targeted victims.

Expansion and Reform

In the West, gold and other mineral strikes drew people to San Francisco, Butte (Montana), and Denver. Timber and Alaskan gold brought people to the Pacific Northwest. The railroad connected these coastal centers with the Midwest and East; by 1910, San Francisco and Los Angeles each had

Port of San Francisco in 1851. (Library of Congress)

more than 250,000 inhabitants, and Denver, Seattle, and Portland more than 200,000 each. Speculators attempted to prompt a land rush to Southern California in the 1880s, but more than 100 townships remained unoccupied.

The South and Southwest grew slowly, with larger centers like New Orleans and Louisville (Kentucky) reliant on connections to the Midwest. In 1910, Atlanta, the South's major inland rail center, had only 154,000 people, and industrial power Birmingham (Alabama) only 132,000. As in the Midwest and still in the East, migration from rural areas was important, but relatively few of the 23.5 million immigrants to America from 1880 to 1920 settled in the South; in 1910, 70 percent of urban Americans still lived in the upper Midwest and Northeast. The South and Southwest both profited enormously from the invention and marketing of air conditioning in the 1930's.

The late nineteenth and early twentieth centuries was the era of urban reform. Both urban political bosses and legitimate urban reformers carried out social and physical improvements that made life more tolerable, and greater attention was given to the physical layout and amenities of cities. From New York with its Central Park to Seattle and its "emerald necklace" of parks, a new breed of land use planners—with grand ideas of the City Beautiful movement and newly legalized tools for zoning and land use organization—sought to rationalize and aesthetically enhance urban space. Metropolitan comprehensive planning emerged, continuously clashing with property rights and the realities of capitalist land-use economics. By the twenty-first century, evolved into regional planning efforts and federally based programming efforts that affected every US city.

Approaching Modernity

In the early twentieth century, the automobile encouraged the spread outward from urban cores, and created the need for complex circulation systems.

This freed development from urban transit lines and fostered the further annexation of surrounding towns. City center and periphery became more closely tied, but the spatial separation of disparate population groups became ever greater. African-American migration from the South into northern cities peaked between 1910 and 1930, at the same time that foreign immigration dropped off. Territoriality and nativist and racist sentiments flared—sometimes violently—in cities in the 1920s, leading to the isolation of African Americans in specific neighborhoods of cities and towns.

During the Great Depression of the 1930's, rural-to-urban migration decreased, and the era of federal aid to cities began. The federal government's New Deal urban policies revolved around aiding the unemployed, and thousands of federally funded works projects repaired or enhanced the fabric of US cities. Slum clearance and public-housing projects eliminated some of the blight, and experimental planned communities for low-income families—like Greenbelt, Maryland (1937)—foreshadowed later, more extensive federal efforts.

The Tennessee Valley Authority and rural electrification movements put many rural people in touch with urban values and culture, especially through the radio, and later, television. Like no other inventions, those devices homogenized US culture around broadcasts that originated in huge cities and reflected their origins. Federal money also funded extensive road, bridge, and expressway construction, which more tightly connected US cities and opened the hinterland. In the 1950s, President Dwight D. Eisenhower's interstate highway system continued this trend and created the largest construction project in human history.

Postwar Developments

World War II increased migration to US cities, and federal aid to returning veterans helped millions buy their own homes. Tract housing projects developed on the fringes of large cities and were linked to them with fine roadways. New manufacturing and industries relocated to the suburbs, where land was relatively inexpensive and a younger, eager, increasingly well-educated population was concentrating. Efficient air travel and federal aid to airports revolutionized the movement of people and goods as railroads had done earlier. Inner cities stagnated and declined, while aesthetic and recre-

Wilson Dam, completed in 1924, was the first dam under the authority of TVA, created in 1933. (Tennessee Valley Authority)

View of The Pond and Midtown Manhattan from the Gapstow Bridge in Central Park. (Ajay Suresh)

ational considerations began to dictate the locations of people and businesses.

The flight of affluent whites from central cities, urban pollution, and inner-city problems of poverty, unemployment, poor housing, and failed urban renewal efforts plagued major cities by the 1960s. Many cities exploded in violent riots in the 1960s. The government responded to these problems with its Great Society programs. President Lyndon B. Johnson's War on Poverty sought to alleviate the economic roots of urban problems, and comprehensive and regional planning were revived to coordinate the federal efforts. These measures neither succeeded, nor were they eliminated. President Richard Nixon attempted to create a population flow toward rural areas. President Ronald Reagan emphasized revenue sharing and a reduction in federal programs.

Late twentieth century efforts to create transportation efficiencies resulted in more complex road systems but few successful mass-transit programs. These have connected cities within regions into megalopolises, such as the Northeast Corridor (Boston to Washington, D.C.), or Southern California between Los Angeles and San Diego. Efforts at cleaning up polluted water and air resources were more successful. Many inner cities underwent restoration or revival, both physical and economic, and neighborhoods once abandoned by the middle classes were gentrified by wealthier people seeking easier access to downtown. Energetic African Americans, Hispanics, and Asian immigrants revitalized hundreds of neighborhoods. Federal efforts in the 1980s and 1990s generally sought to subsidize private enterprise rather than supplant it.

By 2020, approximately 82 percent of Americans live in urban areas or their suburbs. For most metropolitan areas, the suburbs are growing at a somewhat brisker rate (16 percent in 2016) than the cities (13 percent). Nevertheless, several large US cities—New York, Los Angeles, Chicago, Boston, San Francisco, and Washington, D.C.—among others—have emerged as engines of innovation and job creation. Each of these cities is noted for a large and highly skilled workforce, its colleges and universities, the presence of diverse communities, and for having in place the necessary transportation infrastructure to ensure future growth.

Perhaps because of their success, many of these same cities are seeing skyrocketing costs of living, worrisome levels of homelessness, poorly performing schools, higher crime rates, and other problems that ultimately lower the overall quality of life. The increasing ability to work remotely, a trend greatly accelerated by the COVID-19 pandemic in 2020, could be the key that ultimately allows many city dwellers to relocate to suburban or rural communities.

Canada

Like the United States, both Canada and Mexico were settled by Europeans with a tradition of town life. Canada's major cities began as French fur-trading posts along rivers in the southern part of present-day Canada. They grew as defensive positions in the early nineteenth century against an agreesive United States, and as further resources (timber, wheat farming) were developed. By the 1920s, more than half the Canadian population was still rural, and no city exceeded 1 million inhabitants.

In 2020 most of Canada's population and population centers were located in a 745-mile (1,200-km.)-long corridor stretching from Windsor, Ontario, to Montreal, within 200 miles (320 km.) of the US border. Toronto, and to a lesser degree Montreal, dominate the region and, with Vancouver to the west, contain nearly 32 percent of the Canadian population of 37.6 million (2020 estimate).

Urbanization of the population occurred rapidly in Canada, largely as a result of mechanized and scientific farming, immigration, industrialization,

Chinatown gate, Vancouver, British Columbia, Canada. (Suman Chakrabarti)

San Pedro Garza García (also known as San Pedro or Garza García) is a city-municipality of the Mexican state of Nuevo León and part of the Monterrey Metropolitan area. It is a contemporary commercial suburb of the larger metropolitan city of Monterrey. (Rick González)

and the growth of the service sector of the urban economy. By 2020, more than 81 percent of Canadians lived in cities. Efficient mass transit systems rather than extensive roadways distinguish Canadian from US cities, but sprawling suburbs, malls, and vertical downtown development parallel US patterns of development.

Mexico

The Spanish ruled Mexico from small cities and towns, but urbanization is a recent phenomenon. Mexico's population rose from 35 million in 1960 to more than 128 million in 2020. Natural growth, along with migration from the countryside, accounts for the urban explosion. In 2020, 80 percent of the population lived in cities. Guadalajara has 1.4 million inhabitants; Monterrey, 1.1 million. In 2015, more than 85 percent of Mexico's population had piped water or flush toilets. All of Mexico's major cities suffer from congestion, poverty, and pollution. Largely uncontrolled suburban sprawl and annexation of surrounding towns characterize physical growth.

Joseph P. Byrne

URBANIZATION OF
THE CARIBBEAN

In 2018, more than 80 percent of the people of the Caribbean region were living in cities—a rate of urbanization exceeding that of the world as a whole. The largest cities of the region are the capitals of the Caribbean's biggest countries: Santo Domingo in the Dominican Republic; Port au Prince in Haiti; San Juan in Puerto Rico; Havana in Cuba; and Kingston in Jamaica. Each of these cities has populations of over 1 million. In 2020, the most urbanized Caribbean entities were Puerto Rico (93.6 percent), the Dominican Republic (82.5 percent), and Cuba (77 percent).

Social scientists link this increasing urbanization phenomenon to the processes of economic change experienced by the region. It is in great part the product of migration of the rural population to urban centers, particularly to capital cities. People move to large cities expecting to find better social services, increased job opportunities, higher incomes, better education, and other amenities.

Mechanization of agriculture also has contributed to encouraging people to leave the countryside for the cities. New technologies and advanced farming equipment have reduced the demand for agricul-

View of Santo Domingo, Dominican Republic from space, 2010. (NASA Astronauts)

Old San Juan, Puerto Rico.

tural workers. However, the decline in the number of rural residents has caused economic problems in islands where agriculture remains among the most important economic activities. Overall, the rural-to-urban migration has caused labor shortages in the countryside and contributed to an increase in the unemployment rates of the city. Shifting land use from crop production to less-intensive grazing is among the results of the lack of rural labor.

Some Caribbean cities are not prepared to absorb large groups of immigrants. They lack the infrastructure and support agencies necessary to deal with this influx. The contradiction is that most of these nations have designed development programs that favor cities and neglect the countryside. A major problem associated with massive rural-to-urban migration is the emergence of shantytowns. These squalid areas are part of the landscape in cities where government cannot meet the demands of uncontrolled urban growth.

Providing housing for fast-growing urban populations is challenging for Caribbean nations, which are among the poorest in the Western Hemisphere. The housing units that abound in their shantytowns are usually built of cardboard, scraps of wood, and various types of light metals. In more prosperous areas, structures are adapted according to local environmental conditions, and buildings are able to withstand the high winds of tropical storms. For example, apartment complexes in Havana and San Juan are made of concrete and cement. The homes of the Haitian, Dominican, and Puerto Rican middle and elite classes are built with these materials as well. In addition, shopping centers can be seen in some suburban communities of the most prosperous Caribbean countries, accentuating in many ways the contrasts within these cities.

Jose Javier Lopez

ECONOMIC GEOGRAPHY

OVERVIEW

TRADITIONAL AGRICULTURE

Two agricultural practices that are widespread among the world's traditional cultures, slash-and-burn and nomadism, share several common features. Both are ancient forms of agriculture, both involve farmers not remaining in a fixed location, and both can pose serious environmental threats if practiced in a nonsustainable fashion. The most significant difference between the two forms is that slash-and-burn generally is associated with raising field crops, while nomadism as a rule involves herding livestock.

Slash-and-Burn Agriculture

Farmers have practiced slash-and-burn agriculture, which is also referrred to as shifting cultivation or swidden agriculture, in almost every region of the world where the climate makes farming possible. Humans have practiced this method for about 12,000 years, ever since the Neolithic Revolution. Swidden agriculture once dominated agriculture in more temperate regions, such as northern Europe. It was, in fact, common in Finland and northern Russia well into the early decades of the twentieth century. Today, between 200 and 500 million people use slash-and-burn agriculture, roughly 7 percent of the world's population. It is most commonly practiced in areas where open land for farming is not readily available because of dense vegetation. These regions include central Africa, northern South America, and Southeast Asia

Slash-and-burn acquired its name from the practice of farmers who cleared land for planting crops by cutting down the trees or brush on the land and then burning the fallen timber on the site. The farmers literally slash and burn. The ashes of the burnt wood add minerals to the soil, which temporarily improves its fertility. Crops the first year following clearing and burning are generally the best crops the site will provide. Each year after that, the yield diminishes slightly as the fertility of the soil is depleted.

Farmers who practice swidden cultivation do not attempt to improve fertility by adding fertilizers such as animal manures but instead rely on the soil to replenish itself over time. When the yield from one site drops below acceptable levels, the farmers then clear another piece of land, burn the brush and other vegetation, and cultivate that site while leaving their previous field to lie fallow and its natural vegetation to return. This cycle will be repeated over and over, with some sites being allowed to lie fallow indefinitely while others may be revisited and farmed again in five, ten, or twenty years.

Farmers who practice shifting cultivation do not necessarily move their dwelling places as they change the fields they cultivate. In some geographic regions, farmers live in a central village and farm cooperatively, with the fields being alternately allowed to remain fallow, and the fields being farmed making a gradual circuit around the central village. In other cases, the village itself may move as new fields are cultivated. Anthropologists studying indigenous peoples in Amazonia, discovered that village garden sites were on a hundred-year cycle. Villagers farmed cooperatively, with the entire village working together to clear a garden site. That garden would be used for about five years, then a new site was cleared. When the garden moved an inconvenient distance from the village, about once every twenty years, the entire village would move to be closer to the new garden. Over a period of approximately 100 years, a village would make a circle through the forest, eventually ending up close to where it had been located long before any of the present villagers had been born.

In more temperate climates, individual farmers often owned and lived on the land on which they practiced swidden agriculture. Farmers in Finland, for example, would clear a portion of their land, burn the brush and other covering vegetation, grow grains for several years, and then allow that land to remain fallow for from five to twenty years. The individual farmer rotated cultivation around the land in a fashion similar to that practiced by whole villages in other areas, but did so as an individual rather than as part of a communal society.

Although slash-and-burn is frequently denounced as a cause of environmental degradation in tropical areas, the problem with shifting cultivation is not the practice itself but the length of the cycle. If the cycle of shifting cultivation is long enough, forests will grow back, the soil will regain its fertility, and minimal adverse effects will occur. In some regions, a piece of land may require as little as five years to regain its maximum fertility; in others, it may take 100 years. Problems arise when growing populations put pressure on traditional farmers to return to fallow land too soon. Crops are smaller than needed, leading to a vicious cycle in which the next strip of land is also farmed too soon, and each site yields less and less. As a result, more and more land must be cleared.

Nomadism

Nomadic peoples have no permanent homes. They earn their livings by raising herd animals, such as sheep, cattle, or horses, and they spend their lives following their herds from pasture to pasture with the seasons. Most nomadic animals tend to be hardy breeds of goats, sheep, or cattle that can withstand hardship and live on marginal lands. Traditional nomads rely on natural pasturage to support their herds and grow no grains or hay for themselves. If a drought occurs or a traditional pasturing site is unavailable, they can lose most of their herds to starvation.

THE HERITAGE SEED MOVEMENT

Modern hybrid seeds have increased yields and enabled the tremendous productivity of the modern mechanized farm. However, the widespread use of a few hybrid varieties has meant that almost all plants of a given species in a wide area are almost identical genetically. This loss of biodiversity, or, the range of genetic difference in a given species, means that a blight could wipe out an entire season's crop. Historical examples of blight include the nineteenth century Great Potato Famine of Ireland and the 1971 corn blight in the United States.

In response to the concern for biodiversity, there has been a movement in North America to preserve older forms of crops with different genes that would otherwise be lost to the gene pool. Nostalgia also motivates many people to keep alive the varieties of fruits and vegetables that their grandparents raised. Many older recipes do not taste the same with modern varieties of vegetables that have been optimized for commercial considerations such as transportability. Thus, raising heritage varieties also can be a way of continuing to enjoy the foods one's ancestors ate.

In many nomadic societies, the herd animal is almost the entire basis for sustaining the people. The animals are slaughtered for food, clothing is woven from the fibers of their hair, and cheese and yogurt may be made from milk. The animals may also be used for sustenance without being slaughtered. Nomads in Mongolia, for example, occasionally drink horses' blood, removing only a cup or two at a time from the animal. Nomads go where there is sufficient vegetation to feed their animals.

In mountainous regions, nomads often spend the summers high up on mountain meadows, returning to lower altitudes in the autumn when snow begins to fall. In desert regions, they move from oasis to oasis, going to the places where sufficient natural water exists to allow brush and grass to grow, allowing their animals to graze for a few days, weeks, or months, then moving on. In some cases, the pressure to move on comes not from the depletion of food for the animals but from the depletion of a water source, such as a spring or well. At many natural desert oases, a natural water seep or spring provides only enough water to support a nomadic group for a few days at a time.

In addition to true nomads—people who never live in one place permanently—a number of cultures have practiced seminomadic farming: The temperate months of the year, spring through fall, are spent following the herds on a long loop, sometimes hundreds of miles long, through traditional grazing areas; then the winter is spent in a permanent village.

Nomadism has been practiced for millennia, but there is strong pressure from several sources to eliminate it. Pressures generated by industrialized society are increasingly threatening the traditional cultures of nomadic societies, such as the Bedouin of the Arabian Peninsula. Traditional grazing areas are being fenced off or developed for other purposes. Environmentalists are also concerned about the ecological damage caused by nomadism.

Nomads generally measure their wealth by the number of animals they own and so will try to develop their herds to be as large as possible, well beyond the numbers required for simple sustainability. The herd animals eat increasingly large amounts of vegetation, which then has no opportunity to regenerate, and desertification may occur. Nomadism based on herding goats and sheep,

---

DESERTIFICATION

Desertification is the extension of desert conditions into new areas. Typically, this term refers to the expansion of deserts into adjacent nondesert areas, but it can also refer to the creation of a new desert. Land that is susceptible to prolonged drought is always in danger of losing its vegetative ground cover, thereby exposing its soil to wind. The wind carries away the smaller silt particles and leaves behind the larger sand particles, stripping the land of its fertility. This naturally occurring process is assisted in many areas by overgrazing.

In the African Sahel, south of the Sahara, the impact of desertification is acute. Recurring drought has reduced the vegetation available for cattle, but the need for cattle remains high to feed populations that continue to grow. The cattle eat the grass, the soil is exposed, and the area becomes less fertile and less able to support the population. The desert slowly encroaches, and the people must either move or die.

---

for example, has been blamed for the expansion of the Sahara Desert in Africa. For this reason, many environmental policymakers have been attempting to persuade nomads to give up their roaming lifestyle and become sedentary farmers.

Nancy Farm Männikkö

COMMERCIAL AGRICULTURE

Commercial farmers are those who sell substantial portions of their output of crops, livestock, and dairy products for cash. In some regions, commercial agriculture is as old as recorded history, but only in the twentieth century did the majority of farmers come to participate in it. For individual farmers, this has offered the prospect of larger income and the opportunity to buy a wider range of products. For society, commercial agriculture has been associated with specialization and increased productivity.

Commercial agriculture has enabled world food production to increase more rapidly than world population, improving nutrition levels for millions of people.

Steps in Commercial Agriculture

In order for commercial agriculture to exist, products must move from farmer to ultimate consumer, usually through six stages:

1. Processing, packaging, and preserving to protect the products and reduce their bulk to facilitate shipping.

2. Transport to specialized processing facilities and to final consumers.

3. Networks of merchant middlemen who buy products in bulk from farmers and processors and sell them to final consumers.

4. Specialized suppliers of inputs to farmers, such as seed, livestock feed, chemical inputs (fertilizers, insecticides, pesticides, soil conditioners), and equipment.

5. A market for land, so that farmers can buy or lease the land they need.

6. Specialized financial services, especially loans to enable farmers to buy land and other inputs before they receive sales revenues.

Improvements in agricultural science and technology have resulted from extensive research programs by government, business firms, and universities.

International Trade

Products such as grain, olive oil, and wine moved by ship across the Mediterranean Sea in ancient times. Trade in spices, tea, coffee, and cocoa provided powerful stimulus for exploration and colonization around 1500 CE. The coming of steam locomotives and steamships in the nineteenth century greatly aided in the shipment of farm products and spurred the spread of population into potentially productive farmland all over the world. Beginning with Great Britain in the 1840s, countries were willing to relinquish agricultural self-sufficiency to obtain cheap imported food, paid for by exporting manufactured goods.

Most of the leaders in agricultural trade were highly developed countries, which typically had large amounts of both imports and exports. These countries are highly productive both in agriculture and in other commercial activities. Much of their trade is in high-value packaged and processed goods. Although the vast majority of China's labor force works in agriculture, their average productivity is low and the country showed an import surplus in agricultural products. The same was true for Rus-

sia. India, similar to China in size, development, and population, had relatively little agricultural trade. Australia and Argentina are examples of countries with large export surpluses, while Japan and South Korea had large import surpluses. Judged by volume, trade is dominated by grains, sugar, and soybeans. In contrast, meat, tobacco, cotton, and coffee reflect much higher values per unit of weight.

The United States

Blessed with advantageous soil, topography, and climate, the United States has become one of the most productive agricultural countries in the world. Technological advances have enabled the United States to feed its own residents and export substantial quantities with only 3 percent of its labor force engaged directly in farming. In the 2020s there are about 2 million farms cultivating about 1 billion acres. They produced about US$133 billion worth of products. After expenses, this yielded about US$92.5 billion of net farm income—an average of only about US$25,000 per farm. However, most farm families derive substantial income from nonfarm employment.

There is a great deal of agricultural specialization by region. Corn, soybeans, and wheat are grown in many parts of the United States (outside New England). Some other crops have much more limited growing areas. Cotton, rice, and sugarcane require warmer temperatures. Significant production of cotton occurred in seventeen states, rice in six, and sugarcane in four. In 2018, the top 10 agricultural producing states in terms of cash receipts were (in descending order): California, Iowa, Texas, Nebraska, Minnesota, Illinois, Kansas, North Carolina, Wisconsin, and Indiana Typically the top two states in a category account for about 30 percent of sales. Fruits and vegetables are the main exception; the great size, diversity, and mild climate of California gives it a dominant 45 percent.

Socialist Experiments

Under the dictatorship of Joseph Stalin, the communist government of the Soviet Union established a program of compulsory collectivized agriculture in 1929. Private ownership of land, buildings, and other assets was abolished. There were some state farms, "factories in the fields," operated on a large scale with many hired workers. Most, however, were collective farms, theoretically run as cooperative ventures of all residents of a village, but in practice directed by government functionaries. The arrangements had disastrous effects on productivity and kept the rural residents in poverty. Nevertheless, similar arrangements were established in China in 1950 under the rule of Mao Zedong. A restoration of commercial agriculture after Mao's death in 1976 enabled China to achieve greater farm output and farm incomes.

Most Western countries, including the United States, subsidize agriculture and restrict imports of competing farm products. Objectives are to support farm incomes, reduce rural discontent, and slow the downward trend in the number of farmers. Farmers in the European Union will see aid shrink in the 2021–2027 period to 365 billion euros (US$438 billion), down 5 percent from the current Common Agricultural Policy (CAP). Japan's Ministry of Agriculture, Forestry and Fisheries (MAFF) has requested 2.65 trillion yen (roughly US$24 billion) for the Japan Fiscal Year (JFY) 2018 budget, a 15 percent increase over last year. The budget request eliminates the direct payment subsidy for table rice production, but requests significant funding for a new income insurance program,

agricultural export promotion, and underwriting goals to expand domestic potato production. In 2019, trade wars with China and punishing tariffs have led to increased subsidies by the U.S. government, totaling US$10 billion in 2018 and US$14.5 billion in 2019.

Problems for Farmers

Farmers in a system of commercial agriculture are vulnerable to changes in market prices as well as the universal problems of fluctuating weather. Congress tried to reduce farm subsidies through the Freedom to Farm Act of 1996, but serious price declines in 1997-1999 led to backtracking. Efforts to increase productivity by genetic alterations, radiation, and feeding synthetic hormones to livestock have drawn critical responses from some consumer groups. Environmentalists have been concerned about soil depletion and water pollution resulting from chemical inputs.

Productivity and World Hunger

Despite advances in agricultural production, the problem of world hunger persists. Even in countries that store surpluses of farm commodities, there are still people who go hungry. In less-developed countries, the prices of imported food from the West are too low for local producers to compete and too high for the poor to buy them.

Paul B. Trescott

MODERN AGRICULTURAL PROBLEMS

Ever since human societies started to grow their own food, there have been problems to solve. Much of the work of nature was disrupted by the work of agriculture as many as 10,000 years ago. Nature took care of the land and made it productive in its own intricate way, through its own web of interdependent systems. Agriculture disrupts those systems with the hope of making the land even more

productive, growing even more food to feed even more people. Since the first spade of soil was turned over and the first plants domesticated, farmers have been trying to figure out how to care for the land as well as nature did before.

Many modern problems in agriculture are not really modern at all. Erosion and pollution, for example, have been around as long as agriculture.

However, agriculture has changed drastically within those 10,000 years, especially since the dawn of the Industrial Revolution in the seventeenth century. Erosion and pollution are now bigger problems than before and have been joined by a host of others that are equally critical—not all related to physical deterioration. Modern farmers use many more machines than did farmers of old, and modern machines require advanced sources of energy to unleash their power. The machines do more work than could be accomplished before, so fewer farmers are needed, which causes economic problems.

Cities continue to grow bigger as land—usually the best farmland around—is converted to homes and parking lots for shopping centers. The farmers that remain on the land, needing to grow ever more food, turn to the research and engineering industries to improve their seeds. These industries have responded with recombinant technologies that move genes from one species to another; for example, genes cut from peanuts may be spliced into chickens. This creates another set of cultural problems, which are even more difficult to solve because most are still "potential"—their impact is not yet known.

Erosion

Soil loss from erosion continues to be a huge problem all over the world. As agriculture struggles to feed more millions of people, more land is plowed. The newly plowed lands usually are considered more marginal, meaning they are either too steep, too thin, or too sandy; are subject to too much rain; or suffer some other deficiency. Natural vegetative cover blankets these soils and protects them from whatever erosive agents are active in their regions: water, wind, ice, or gravity. Plant cover also increases the amount of rain that seeps downward into the soil rather than running off into rivers. The more marginal land that is turned over for crops, the faster the erosive agents will act and the more erosion will occur.

Expansion of land under cultivation is not the only factor contributing to erosion. Fragile grasslands in dry areas also are being used more inten-

sively. Grazing more livestock than these pastures can handle decreases the amount of grass in the pasture and exposes more of the soil to wind—the primary erosive agent in dry regions.

Overgrazing can affect pastureland in tropical regions too. Thousands of acres of tropical forest have been cleared to establish cattle-grazing ranges in Latin America. Tropical soils, although thick, are not very fertile. Fertility comes from organic waste in the surface layers of the soil. Tropical soils form under constantly high temperatures and receive much more rain than soils in moderate, midlatitude climates; thus, tropical organic waste materials rot so fast they are not worked into the soil at all. After one or two growing seasons, crops grown in these soils will yield substantially less than before.

Tropical fields require fallow periods of about ten years to restore themselves after they are depleted. That is why tropical cultures using slash-and-burn methods of agriculture move to new fields every other year in a cycle that returns them to the same place about every ten years, or however long it takes those particular lands to regenerate. The heavy forest cover protects these soils from exposure to the massive amounts of rainfall and provides enough organic material for crops—as long as the forest remains in place. When the forest is cleared, however, the resulting grassland cannot provide the adequate protection, and erosion accelerates. Grasslands that are heavily grazed provide even less protection from heavy rains, and erosion accelerates even more.

The use of machines also promotes erosion, and modern agriculture relies on machinery: tractors, harvesters, trucks, balers, ditchers, and so on. In the United States, Canada, Europe, Russia, Brazil, South Africa, and other industrialized areas, machinery use is intense. Machinery use is also on the rise in countries such as India, China, Mexico, and Indonesia, where traditional nonmechanized methods are practiced widely. Farming machines, in gaining traction, loosen the topsoil and inhibit vegetative cover growth, especially when they pull behind them any of the various farm implements designed to rid the soil of weeds, that is, all vegetation except the desired crop. This leaves the soil more exposed to erosive weather, so more soil is carried away in the runoff of water to streams.

Eco-fallow farming has become more popular in the United States and Europe as a solution to reducing erosion. This method of agriculture, which leaves the crop residue in place over the fallow (nongrowing) season, does not root the soil in place, however. Dead plants do not "grab" the soil like live plants that need to extract from it the nutrients they need to live, so erosion continues, even though it is at a slower rate. Eco-fallow methods also require heavier use of chemicals, such as herbicides, to "burn down" weed growth at the start of the growing season, which contributes to accelerated erosion and increases pollution.

Pollution

Pollution, besides being a problem in general, continues to grow as an agricultural problem. With the onset of the Green Revolution, the use of herbicides, insecticides, and pesticides has increased dramatically all over the world. These chemicals are not used up completely in the growth of the crop, so the leftovers (residue) wash into, and contaminate, surface and groundwater supplies. These supplies then must be treated to become useful for other purposes, a job nature used to do on its own. Agricultural chemicals reduce nature's ability to act as a filter by inhibiting the growth of the kinds of plant life that perform that function in aquatic environments. The chemical residues that are not washed into surface supplies contaminate wells.

As chemical use increases, contamination accumulates in the soil and fertility decreases. The microorganisms and animal life in the soil, which had facilitated the breakdown of soil minerals into usable plant products, are no longer nourished because the crop residue on which they feed is depleted, or they are killed by the active ingredients in the chemical. As a result, soil fertility must be restored to maintain yield. Chemical replacement is usually the method of choice, and increased applications of chemical fertilizers intensify the toxicity of this cyclical chemical dependency.

Chemicals, although problematic, are not as difficult to contend with as the increasingly heavy silt load choking the life out of streams and rivers. Accelerated erosion from water runoff carries silt particles into streams, where they remain suspended and inhibit the growth of many beneficial forms of plant and animal life. The silt load in U.S. streams has become so heavy that the Mississippi River Delta is growing faster than it used to. The heavy silt load, combined with the increased load of chemical residues, is seriously taxing the capabilities of the ecosystems around the delta that filter out sediments, absorb nutrients, and stabilize salinity levels for ocean life, creating an expanding dead zone.

This general phenomenon is not limited to the Mississippi Delta—it is widespread. Its impact on people is high, because most of the world's population lives in coastal zones and comes in direct contact with the sea. Additionally, eighty percent of the world's fish catch comes from the coastal waters over continental shelves that are most susceptible to this form of pollution.

Monoculture

Modern agriculture emphasizes crop specialization. Farmers, especially in industrialized regions, often grow a single crop on most of their land, perhaps rotating it with a second crop in successive years: corn one year, for example, then soybeans, then back to corn. Such a strategy allows the farmer to reduce costs, but it also makes the crop, and, thus, the farmer and community, susceptible to widespread crop failure. When the crop is infested by any of an ever-changing number and variety of pests—worms, molds, bacteria, fungi, insects, or other diseases—the whole crop is likely to die quickly, unless an appropriate antidote is immediately applied. Chemical antidotes can do the job but increase pollution. Maintaining species diversity—growing several different crops instead of one or two—allows for crop failures without jeopardizing the entire income for a farm or region that specializes in a particular monoculture, such as tobacco, coffee, or bananas.

Chemicals are not the only modern methods of preventing crop loss. Genetically engineered seeds are one attempt at replacing post-infestation chemical treatments. For example, splicing genes into varieties of rice or potatoes from wholly unrelated species—say, hypothetically, a grasshopper—to prevent common forms of blight is occurring more often. Even if the new genes make the crop more resistant, however, they could trigger unknown side effects that have more serious long-term environmental and economic consequences than the problem they were used to solve. Genetically altered crops are essentially new life-forms being introduced into nature with no observable precedents to watch beforehand for clues as to what might happen.

Urban Sprawl

As more farms become mechanized, the need for farmers is being drastically reduced. There were more farmers in the United States in 1860 than there were in the year 2000. From a peak in 1935 of about 6.8 million farmers farming 1.1 billion acres, the United States at the end of the twentieth century counted fewer than 2.1 million farmers farming 950 million acres. As fewer people care for land, the potential for erosion and pollution to accelerate is likely to increase, causing land quality to decline.

As farmers are displaced and move into towns, the cities take up more space. The resulting urban sprawl converts a tremendous amount of cropland into parking lots, malls, industrial parks, or suburban neighborhoods. If cities were located in marginal areas, then the concern over the loss of farmland to commercial development would be nominal. However, the cities attracting the greatest numbers of people have too often replaced the best cropland. Taking the best cropland out of primary production imposes a severe economic penalty.

James Knotwell and Denise Knotwell

WORLD FOOD SUPPLIES

All living things need food to begin the life process and to live, grow, work, and survive. Almost all foods that humans consume come from plants and animals. Not all of Earth's people eat the same foods, however, nor do they require the same caloric intakes. The types, combinations, and amounts of food consumed by different peoples depend upon historic, socioeconomic, and environmental factors.

The History of Food Consumption

Early in human history, people ate what they could gather or scavenge. Later, people ate what they could plant and harvest and what animals they could domesticate and raise. Modern people eat what they can grow, raise, or purchase. Their diets or food composition are determined by income, local customs, religion or food biases, and advertising. There is a global food market, and many people can select what they want to eat and when they eat it according to the prices they can pay and what is available.

Historically, in places where food was plentiful, accessible, and inexpensive, humans devoted less time to basic survival needs and more time to activities that led to human progress and enjoyment of leisure. Despite a modern global food system, instant telecommunications, the United Nations, and food surpluses at places, however, the problem of providing food for everyone on Earth has not been solved.

According to the United Nations Sustainable Development Goals that were adopted by all Member States in 2015, an estimated 821 million people were undernourished in 2017. In developing countries, 12.9 percent of the population is undernourished. Sub-Saharan Africa has the highest prevalence of hunger; the number of undernourished people increased from 195 million in 2014 to 237 million in 2017. Poor nutrition causes nearly half (45 percent) of deaths in children under five—3.1 million children each year. As of 2018, 22 percent of the global under-5 population were still chronically undernourished in 2018. To meet challenge of Goal 2: Zero Hunger, significant changes both in terms of agriculture and conservation as well as in financing and social equality will be required to nourish the 821 million people who are hungry today and the additional 2 billion people expected to be undernourished by 2050.

World Food Source Regions

Agriculture and related primary food production activities, such as fishing, hunting, and gathering, continue to employ more than one-third of the world's labor force. Agriculture's relative importance in the world economic system has declined with urbanization and industrialization, but it still plays a vital role in human survival and general economic growth. Agriculture in the third millennium must supply food to an increasing world population of nonfood producers. It must also produce food and nonfood crude materials for industry, accumulate capital needed for further economic growth, and allow workers from rural areas to industrial, construction, and expanding intraurban service functions.

Soil types, topography, weather, climate, socioeconomic history, location, population pressures, dietary preferences, stages in modern agricultural development, and governmental policies combine to give a distinctive personality to regional agricultural characteristics. Two of the most productive food-producing regions of the world are North America and Asia. Countries in these regions export large amounts of food to other parts of the world.

Foods from Plants

Most basic staple foods come from a small number of plants and animals. Ranked by tonnage produced, the most important food plants throughout the world are corn (maize), wheats, rice, potatoes, cassava (manioc), barley, soybeans, sorghums and millets, beans, peas and chickpeas, and peanuts (groundnuts).

More than one-third of the world's cultivated land is planted with wheat and rice. Wheat is the dominant food staple in North America, Western and Eastern Europe, northern China, and the Middle East and North Africa. Rice is the dominant food staple in southern and eastern Asia. Corn, used primarily as animal food in developed nations, is a staple food in Latin America and Southeast Africa. Potatoes are a basic food in the highlands of South America and in Central and Eastern Europe. Cassava (manioc) is a tropical starch-producing root crop of special dietary importance in portions of lowland South America, the west coast countries of Africa, and sections of South Asia. Barley is an important component of diets in North African, Middle Eastern, and Eastern European countries. Soybeans are an integral part of the diets of those who live in eastern, southeastern, and southern Asia. Sorghums and millets are staple subsistence foods in the savanna regions of Africa and south Asia, while peanuts are a facet of dietary mixes in tropical Africa, Southeast Asia, and South America.

Food from Animals

Animals have been used as food by humans from the time the earliest people learned to hunt, trap, and fish. However, humans have domesticated only a few varieties of animals. Ranked by tonnage of meat produced, the most commonly eaten animals are cattle, pigs, chickens and turkeys, sheep, goats, water buffalo, camels, rabbits and guinea pigs, yaks, and llamas and alpacas.

Cattle, which produce milk and meat, are important food sources in North America, Western Europe, Eastern Europe, Australia and New Zealand, Argentina, and Uruguay. Pigs are bred and reared for food on a massive scale in southern and eastern Asia, North America, Western Europe, and Eastern Europe. Chickens are the most important domesticated fowl used as a human food source and are a part of the diets of most of the world's people.

Sheep and goats, as a source of meat and milk, are especially important to the diets of those who live in the Middle East and North Africa, Eastern Europe, Western Europe, and Australia and New Zealand.

Water buffalo, camels, rabbits, guinea pigs, yaks, llamas, and alpacas are food sources in regions of the world where there is low consumption of meat for religious, cultural, or socioeconomic reasons. Fish is an inexpensive and wholesome source of food. Seafood is an important component to the diets of those who live in southern and eastern Asia, Western Europe, and North America.

The World's Growing Population
The problem of feeding the world is compounded by the fact that population was increasing at a rate of nearly 82 million persons per year at the end of second decade of the twenty-first century. That rate of increase is roughly equivalent to adding a country the size of Germany to the world every single year.

Also compounding the problem of feeding the world are population redistribution patterns and changing food consumption standards. In the year 2050, the world population is projected to reach approximately 10 billion—4 billion people more than were on the earth in 2000. Most of the increase in world population is expected to occur within the developing nations.

Urbanization
Along with an increase in population in developing nations is massive urbanization. City dwellers are food consumers, not food producers. The exodus of young men and women from rural areas has given rise to a new series of megacities, most of which are in developing countries. By the year 2030, there could be as many as forty-one megacities (cities with populations of 10 million people or more).

When rural dwellers move to cities, they tend to change their dietary composition and food-consumption patterns. Qualitative changes in dietary consumption standards are positive, for the most part, and are a result of copying the diets of what is considered a more prestigious group or positive educational activities of modern nutritional scientists

working in developing countries. During the last four decades of the twentieth century, a tremendous shift took place in overall dietary habits as Western foods became increasingly available and popular throughout the world. While improved nutrition has contributed to a decrease in child mortality, an increase in longevity, and a greater resistance to disease, it is also true that conditions including morbid obesity, Type II diabetes, and hypertension are on the rise.

Strategies for Increasing Food Production
To meet the food demands and the food distribution needs of the world's people in the future, several strategies have been proposed. One such strategy calls for the intensification of agriculture—improving biological, mechanical, and chemical technology and applying proven agricultural innovations to regions of the world where the physical and cultural environments are most suitable for rapid food production increases.

The second step is to expand the areas where food is produced so that areas that are empty or underused will be made productive. Reclaiming areas damaged by human mismanagement, expanding irrigation in carefully selected areas, and introducing extensive agrotechniques to areas not under cultivation could increase the production of inexpensive grains and meats.

Finally, interregional, international, and global commerce should be expanded, in most instances, increasing regional specializations and production of high-quality, high-demand agricultural products for export and importing low-cost basic foods. A disequilibrium of supply and demand for certain commodities will persist, but food producers, regional and national agricultural planners, and those who strive for regional economic integration must take advantage of local conditions and location or create the new products needed by the food-consuming public in a one-world economy.

Perspectives
Humanity is entering a time of volatility in food production and distribution. The world will produce enough food to meet the demands of those

who can afford to buy food. In many developing countries, however, food production is unlikely to keep pace with increases in the demand for food by growing populations.

Factors that could lead to larger fluctuations in food availability include weather variations such as those induced by El Niño and climate change, the growing scarcity of water, civil strife and political in-stability, and declining food aid. In developing countries, decision makers need to ensure that poli-cies promote broad-based economic growth—and in particular agricultural growth—so that their countries can produce enough food to feed them-selves or enough income to buy the necessary food on the world market.

William A. Dando

AGRICULTURE OF NORTH AMERICA

"Agriculture" originally meant simply the cultivation of fields. In modern usage, however, the word has been generalized to mean the entire process of producing food and fiber by the raising of domesticated plants and animals. In addition, a discussion of modern agriculture in North America cannot be complete without some attention to agribusiness, the system of businesses associated with and supporting agricultural production in an industrialized society.

In continental North America, agriculture generally has become mechanized and heavily dependent upon an integrated system of supporting agribusinesses, although traditional practices continue in Mexico on numerous small plots worked by families and small communities. In the United States and Canada, most farmers and ranchers practice monoculture, relying upon a single crop or livestock species for their primary income, and have expanded to very large acreages or herds in order to take advantage of economies of scale. Such farms generally are referred to in terms of the primary crop or species, for example, a dairy farm, a cattle

Corn field near Starbuck, Manitoba, Canada. (Comessy)

Selected Agricultural Products of North America

GREENLAND

Fish

Fish

Timber

CANADA

Vancouver

Spring
Wheat

Seattle

Apples

Pears

Quebec

Timber
Potatoes

Montreal

Timber

Minneapolis

Ottawa★

Dairy
Cattle

Toronto

Boston

Apples

Fruits
and
Vege-
tables

San Francisco

Denver

Winter
Wheat

Chicago

Detroit

New York
Philadelphia

Corn
Oil Seeds

Hogs

Pears

Washington D.C.

Los Angeles

UNITED STATES

San Diego

Soybeans

Cotton
Tobacco
Citrus

Sheep

Oklahoma City

Flax

Beef Cattle

Rice

Waco

Sugar Cane

Citrus

New Orleans

Atlantic
Ocean

Corn

Catfish

Oranges

MEXICO

Gulf

Miami

Monterrey

of

Mexico

Pacific
Ocean

Guadalajara

Mexico
City ★

Oaxaca

Coffee

Caribbean Sea

ranch, or a wheat farm. Some small farms are run by part-time farmers who supplement their income with a second job.

In 2019, there were just over 2 million farms and ranches in the United States. Although farmers represented only 1.3 percent of the employed US population, they successfully fed the country at a high standard of nutritional value and produced grain and other products for export, with agricultural exports accounting for some 20 percent of the total production. In Canada, the world's fifth-largest agricultural exporter, the number of farms in the 2010s dropped to below 200,000 but their average size increased compared to the turn of the century. In 2019, farmers represented 1.7 percent of the Canadian population. In Mexico, there were millions of individual units of agricultural production but over 40 percent of them were less than 4.9 acres (2 hectares) in size; almost 13 percent of the Mexican workforce were employed in agriculture. Many farm laborers, especially in areas with large indigenous populations, still survive on subsistence agriculture. However, the rural workforce has been consistently shrinking.

Regional Patterns of Cultivation

Modern farming techniques in the United States and Canada require specialization in a single cash crop. Such specialized farms tend to cluster by region, where the climate is appropriate to a given crop. Such factors as average temperature and rainfall, soil quality, and the likelihood of frost determine what crops can be grown successfully in a given region. The supporting agribusinesses—such as implement suppliers, and sellers of chemical fertilizers, pesticides, and grain elevators, tend to specialize in the products and activities that support the primary crops of their given area.

Wheat, the most important cereal grain in Western diets, grows in the broad, open lands of the Great Plains, in Kansas, Nebraska, North and South Dakota, and the Canadian provinces of Alberta and Saskatchewan. In the southern part of this region, the primary crop is winter wheat, which is planted in the fall, is dormant during the winter, completes its growth in spring, and is harvested in midsum-

LAND GRANT COLLEGES

In traditional cultures, farming was taught through practical experience. However, as scientific and technological developments evolved, there was a need to apply modern techniques systematically to agriculture. In 1862 a federal law was passed establish agricultural colleges in each state. Because the colleges would be funded by the sale of public lands, they were called "land grant colleges." Although land grant colleges were originally founded to research and to teach agriculture and mechanic arts, many of them soon diversified. Some, such as the University of Illinois at Urbana-Champaign, became leading general universities with world-class colleges of engineering, business, education, and law.

In addition to teaching new agricultural techniques in the classroom, the land grant colleges founded research stations to study practical aspects of farming. These university farms applied the scientific method of investigation to develop new methods, hybrids, and so on. Now protected as a National Historic Landmark, the Morrow Plots at the University of Illinois at Urbana-Champaign are the oldest continually operated experimental crop field in the United States.

Land grant colleges also were mandated to teach their new methods to farmers. The Cooperative Extension System established offices in individual counties, where local extension agents were accessible to farmers. County agents brought in speakers and other educational presentations from the parent university to educate farmers about new developments in agricultural technique. Home advisors educated women about more effective household techniques to improve the standard of living on American farms. To educate young people, the Extension System nationalized 4-H, a youth development organization focused on addressing the nation's top issues through innovation and practical learning.

mer. In many of these areas, a farmer then can plant a crop of soybeans, a practice known as double-cropping. The soybeans often can be harvested in time to plant the following year's wheat crop in the fall. Farther to the north, where the weather is too harsh for wheat to survive the winter, farmers plant spring wheat, which completes its entire growth during the spring and summer and is har-

An orange grove in Florida. (Mmacbeth)

vested in the fall. Wheat is used to make bread, pasta, and many breakfast cereals, and is an ingredient in numerous other products.

Corn (maize), which originally was domesticated by Native Americans in what is now Mexico, is the best producer per acre. It requires a longer growing season than wheat, so the area where it can be grown economically is limited. The Midwestern states—Iowa, Illinois, Indiana, and Ohio—are the principal areas for cultivation of corn and frequently are referred to as the Corn Belt states. Much corn is used as livestock feed, although a considerable amount is processed into human foods as well, often in the form of cornstarch and corn-syrup sweeteners.

In Mexico, the birthplace of corn, traditional corn farmers have struggled to compete with US corn, which has been exported to Mexico at aggressively low prices since the signing of the North American Free Trade Agreement (NAFTA) in 1994.

Many had to downsize or went out of business. At the same time, Mexico significantly increased the export of vegetables and winter fruits to the United States during the NAFTA years. These exports—as well as low-priced, high-quality US farm equipment, seeds, fertilizers, and insecticides imported under NAFTA—have been driving the fast growth of Mexico's agriculture.

Rice requires flooded fields for successful cultivation, so it can be grown only in areas such as Louisiana, where large amounts of water are readily available. Because labor costs are the primary limiter in US agriculture, American rice growers use highly mechanized single-field growing techniques rather than the labor-intensive transplantation technique used in Asian countries, where labor is cheap and the land is at a premium. Laser levels and computerized controls tied into the Global Positioning System (GPS) enable farmers to prepare perfectly smooth fields with the slight slope necessary for efficient flooding and drainage. Because the ground is usually wet during tilling and harvesting, the machinery typically used in growing rice is fitted with tracks instead of wheels to reduce soil compaction. Nearly half of the US rice crop is exported to more than 120 countries worldwide.

Rye, oats, and barley are other major grain crops, although none form the backbone of an area's economy to the extent that wheat, corn, and rice do. Oats, once a staple feed grain for horses, now is used mainly for breakfast cereals, while most barley is malted for the brewing of beer. Rye typically is used in the production of whiskey and the increasingly popular specialty breads.

Legumes such as soybeans and alfalfa form the next major group of crops produced in North America. In addition to being an important source of protein in human and livestock diets, legumes are important in maintaining soil fertility. Nodules on their roots contain bacteria that help to transform nitrogen in the soil into compounds that plants can use. Because of this, soybeans have also become a regular rotation crop with corn in much of the US Midwest.

Both corn and soybeans can be grown using the same machinery and sold to the same markets, although harvesting corn requires a specialized corn-header that pulls down the stalks and breaks loose the cob on which the corn kernels grow, rather than the generalized grain platform used with soybeans and small grains. Soybeans for human consumption generally are processed into cooking oil or become filler in other foods, although there is a market for tofu (bean curd) and other soybean products for Asian cuisines.

Other crops include edible oil seeds, such as rapeseed, sunflower seeds, and safflower seeds; the

Cattle in General Terán, Nuevo León, Mexico. (Jorge.a.izaguirre)

Intensively farmed pigs.

latter two are generally grown as rotation crops with corn or wheat. Like much of corn and soybeans, oil seeds are used in the production of vegetable oils. Canola was bred from rapeseed cultivars at the University of Manitoba, Canada, and the name "canola" is a contraction of the phrase "Canadian oil." It is now a generic term for edible varieties of rapeseed widely grown in Canada's prairie provinces and the northern US states.

Sugarcane is grown in Louisiana and other areas on the coast of the Gulf of Mexico that have the necessary subtropical climate. Many varieties of fruits and vegetables are grown in California's irrigated valleys; Florida grows much of the nation's juice oranges. Other citrus crops are grown in Alabama, Mississippi, and Texas, where these warmth-loving trees will not be damaged by frost. Fruits such as apples and pears, which require a cold period to break dormancy and set fruit, are grown in northern states such as Michigan and Washington, upstate

New York, and the eastern provinces of Canada. The Niagara Peninsula in southern Ontario is known as Canada's Fruit Belt.

Fruits and vegetables—especially avocados, tomatoes, peppers, berries, nuts, citrus fruits, and cucumbers—are among Mexico's most successful commercial crops. Other important crops include sugarcane, sorghum, wheat, beans, barley, blue agave, and coffee. In 2018, the country for the first time entered the list of the ten largest global sellers of agricultural and agro-industrial goods. Mexico is the world's biggest producer and exporter of avocados.

Livestock

Several species of animals are raised for their meat and other products in North America. Dairy cattle, which need to be milked twice daily, generally are raised in hilly areas that are not suitable for growing row crops but produce sufficient grazing yield per acre to support the small spreads that dairying requires. Much of Wisconsin's agricultural income comes from dairy cattle. In the open country of Texas, by contrast, there is low rainfall and each animal must graze over a large area. This region is suitable for the production of beef cattle, which can range freely with little human intervention until they are of marketable size. Texas has been leading the United States in the number of cattle for well over a century. Since the mid-1980s, however, most of the cattle in Texas and elsewhere in the United States have been raised not on grass pastures but in feedlots, or arrays of pens where animals are fed enhanced nutritious forages. Farther west, where the grass cover becomes too scant to support free-range cattle, sheep can find satisfactory grazing. Before the introduction of feedlot operations, there had been frequent and intense conflicts between sheep and cattle ranchers over marginal areas.

Because hogs require a concentrated diet for commercial production, they generally are raised in confinement buildings in areas that produce the foodstuffs of their primary diet. Large hog operations are common in the Corn Belt area of the Midwest, where farmers can feed hogs corn from their own operations. Fowl such as chickens and turkeys generally are raised in factory-style battery cages in large barns with controlled temperature, humidity, food and water delivery, and lighting. Chickens are raised for both their eggs and their meat; turkeys traditionally have been holiday fare, but have now become more commonly used throughout the year. A still small but growing market segment in both the United States and Canada now belongs to locally produced, certified organic poultry and eggs that originate with livestock raised in less-confined settings; these are marketed as "free-range."

Cattle production systems and breeds vary greatly across Mexico and are determined mainly by geography and economic differences between northern, southern, and central Mexico. Most large operations are located in northern Mexico, and small-scale backyard livestock raising is more common in the southeastern states. Mexico and the United States trade in about the same amount of beef with each other, roughly 500 million pounds

Workers harvest catfish from the Delta Pride Catfish farms in Mississippi. (Ken Hammond, USDA OnLine Photography Center)

(225 million kg.). Weaned calves (called feeder cattle) that are born in Mexico get sent to feedlots across the border where it is cheaper to fatten them up on American corn and alfalfa until they are ready for slaughter. After that, many parts of them get sent back to Mexico, particularly parts like heads, stomachs, and tails. These are considered by-products in the United States but have a much higher value in Mexico, where they are staples of national cuisine.

Aquaculture

Although most fish and other seafood are caught in the wild, there has been a growing interest in the United States and Canada in aquaculture—the raising of aquatic organisms in controlled environments to be harvested commercially. In areas where rice is grown, putting catfish or other commercially harvestable fish into the paddies during the flooding period has enabled farmers to make additional income from one parcel of land. In other areas, fish and other seafood are raised in coastal wetlands, ponds, or completely artificial tanks similar to giant aquaria. In Canada, aquaculture was initially used to enhance natural fish stocks but has since become a large-scale commercial industry.

The controlled environment of aquaculture enables predictable harvests, and protection from natural predators and selective breeding for improved varieties allow higher production for a given amount of resources. Fish farming also avoids the problem of high levels of toxic substances in wild fish, the result of industrial pollution in commercial inland fisheries such as those on the Great Lakes.

Fibrous Plants

In addition to food plants, the production of fibers for garments and other textiles has remained a part of American agriculture, although such artificial fibers as nylon and polyester, along with imported natural fabrics, have taken a significant share of the market. Sheep produce wool as well as lamb and mutton. Cotton and flax also are important sources of natural fibers. Cotton requires a long growing season and relatively high rainfall levels; therefore, it generally is grown in an area of the southern United States often referred to as the Cotton Belt as well as in northern and central Mexico. Flax, or linseed, has a shorter growing season and is often planted in rotation with such small grains as wheat and oats. Flax stems are used to produce linen; edible and industrial oils and meal are obtained from its seeds. Canada is the largest producer of industrial flaxseed in the world.

Tree farming involves mostly the growing of trees for timber. During the twentieth century, concerns about the environmental damage done by the clear-cutting of virgin forests for lumber and paper encouraged many companies to start reseeding the cut areas with suitable tree species that could be harvested thirty or forty years later. These trees are either seeded directly on the site or, more commonly, first grown on tree plantations. Canada, the world's

Cotton field, Ware County, Georgia. (Jud McCranie)

second-largest exporter of forestry goods, has become a world leader in sustainable forest management in the twenty-first century. Another form of tree farming, although on a much smaller scale, is the production of evergreens for Christmas trees.

The Business of Farming

Because of the intense specialization of modern mechanized agriculture, farming has become a business interlocked with a number of supporting businesses. Farmers have had to become entrepreneurs and look at their activities as businesses rather than as traditional occupations. Farm management—the control of capital outlay, production costs, and income—has become as vital to a farmer's economic survival as skill in growing the crops themselves. Such organizations as Farm Business/Farm Management help farmers develop the skills and knowledge needed to farm more productively and economically.

Farmers also have had to become actively involved in the marketing of their crops to ensure adequate income. In many areas of the United States, Canada, and Mexico, farmers have banded together in cooperatives to gain economic leverage in buying supplies and selling their products. Some of these cooperatives have taken on preliminary and intermediate steps in transforming the raw farm products into useful goods for consumers, thus increasing the prices farmers receive from buyers.

Modern farmers also are concerned with the management of the resources that support agriculture. In earlier generations, it often was assumed that natural resources were unlimited and could be used and abused without consequence. The result of this ignorance was ecological catastrophes such as the Dust Bowl of the 1930s, in which the topsoil over large areas of Kansas, Oklahoma, and other states dried up and blew away in the wind, ren-

dering the land unfarmable. To prevent more such disasters and the economic dislocation they produced, various soil conservation measures were introduced through government programs that gave farmers financial aid and incentives to change their practices. The use of contour plowing and terracing on steeply sloping hillsides helped to slow the movement of water that could carry away soil, thus preventing the formation of gullies. Reduced tillage techniques allowed more plant residue to remain on the surface of the soil, protecting it from the ravages of both wind and water. A better understanding of an enormous environmental footprint of agriculture has led to the creation of organizations such as the Natural Resources Conservation Service (a federal agency within the US Department of Agriculture), which provides farmers technical and financial assistance in conserving the soil, water, and other natural resources.

Besides agricultural production, the modern agribusiness sector of the economy includes the production of agricultural chemicals and machinery; the many businesses and services involved in food

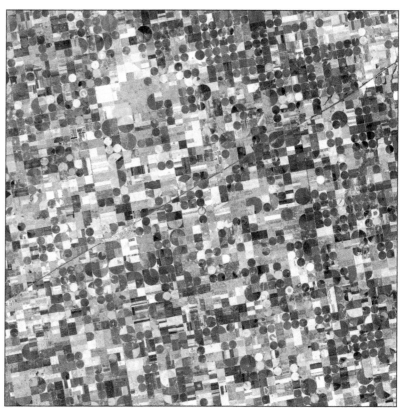

Satellite image of circular crop fields characteristic of center pivot irrigation in Kansas. (NASA)

processing, marketing, and distribution; and research and development institutions working in different areas of agricultural science, food science, and food technology. Large farms (with over $1 million in sales) have especially benefited from modern technology and efficient farm organization. In the mid-2010s, they accounted for only 4 percent of all US farms but for over 65 percent of all agricultural sales. The average size of a US farm, however, has not changed much since the 1970s. The average size of a Canadian farm increased by 7 percent between 2006 and 2018.

Genetically Modified Organisms and Artificial Growth Hormones

Genetically modified organisms (GMOs) have become widely used in the United States and Canada since their initial introduction in the 1990s. The benefits of genetic engineering in agriculture include increased crop yields, reduced production costs, enhanced nutrient composition, resistance to disease, and lower need for pesticides. While genetically modified crops are grown by millions of farmers in dozens of countries, the United States is one of the leading proponents of research into GMOs and surpasses most other countries in their proliferation; it grows some 60 percent of the world's GMO crops.

The predominant modified crops grown in the United States and Canada are soybeans, corn (maize), canola, potatoes, sugar beets, and alfalfa. In the United States, the list also includes cotton. Genetic improvement has been slower for wheat due to this grain's significantly more complex genetics and lower potential returns from research investments. Farmers grow wheat primarily for human food use, and North American food processors have been wary of consumer reaction to products containing genetically modified wheat. No genetically modified wheat is commercially grown in the United States or Canada.

In Mexico—where corn is not just a crop but an essential element of the cultural identity, heritage, and cuisine—a broad coalition of corn farmers, scientists, and consumer advocates has been engaged in a campaign to protect native maize from genetic contamination and modification. Worldwide, opponents of GMOs claim that all the latent risks of genetic modification have not been adequately identified, especially its potential long-term impact on the environment and human health. GMO labeling is required in many countries but not in the United States, Canada, or Mexico.

In the European Union, the use of artificial growth hormones in animal agriculture is banned because their potential long-term effects are considered to be uncertain. In North America, artificial growth hormones are used by most beef cattle producers to increase livestock weight gain and lean meat yield. North American regulators believe that the safety of beef produced with artificial hormones—as well as the safety of GMOs—has been established by scientific consensus and the weight of evidence. In Canada, unlike the United States, artificial hormones are used only in the production of beef cattle and are not administered to dairy cows. Hormones have not been used in poultry production in North America since the mid-twentieth century.

Leigh Husband Kimmel

AGRICULTURE OF THE CARIBBEAN

Agriculture had long been the chief source of livelihood in the Caribbean island region. Until the end of the twentieth century, it was basic to the economy of nearly every island. Since then, tourism has come to dominate the revenue production for most Caribbean islands. Agriculture was relegated to second, and in some cases, third place, after export-oriented manufacturing. As of 2019, Dominica, Guyana, Haiti, and Suriname remained heavily dependent on agriculture, with the sector accounting for between 12 and 17 percent of the national GDP and between 10 and 50 percent of employment. In other nations throughout the region, agriculture accounted for less than 4 percent of the GDP.

Two fundamentally different types of agriculture dominate in the Caribbean island economies: large-scale commercial, or plantation, agriculture and small-scale semi-subsistence farming. Large-scale farming provides most exports, by value, whereas small-scale farming involves far more human labor.

Caribbean agriculture operates under various natural and cultural restraints. First, most of the islands have a rugged terrain, restricting productive agriculture to river valleys and coastal plains. Typically, less than one-third of island land is suitable for crops. Second, the windward (northerly) portions of islands are commonly very wet, whereas their leeward (southerly) areas suffer seasonal or yearlong drought, necessitating irrigation. Third, various hazards affect agriculture, including damaging winds, hurricanes, flooding, accelerated erosion, and landslides. In addition, some crops (notably bananas) have suffered from diseases. Most small-scale farms are restricted to steep, unproductive slopes, while large-scale operations control most of the productive lowland soils. Population pressures and tourism development have led to the loss of some of the best lands in coastal areas, and have caused fragmentation of farmland. Farm labor shortages, climbing wages, and foreign competition have added to the burden.

Commercial Agriculture

Modern commercial large-scale farms, born out of the colonial plantations, differ from their earlier version mainly in that they use hired, often part-time, labor rather than slaves or indentured workers. Otherwise, the system has changed relatively little. Modern large-scale farms own vast tracts of land and still specialize in one crop, most commonly sugarcane, bananas, coconuts, coffee, rice, or tobacco. They are more mechanized and better managed than colonial plantations were, although still largely owned by parties from foreign

Sugarcane flower, Dominica. (Charles J Sharp)

countries, including Britain, France, and the United States. The largest farms are found on the largest islands, especially Hispaniola, Jamaica, and Puerto Rico. Cuba also has large-scale farming, but the operations are government-owned. The farms always have been smaller in the Lesser Antilles archipelago, where relatively little suitable land is available.

Sugar still dominates the agricultural exports of Cuba, the Dominican Republic, Guadeloupe, and Saint Kitts. Among traditional sugar producers in the Caribbean, notably Jamaica, Puerto Rico, Trinidad, and Barbados, sugar exports are exceeded by those of other commodities. Haiti, a leading sugar producer when a French colony, now produces very little. Some islands, including the former Dutch Antilles, the Bahamas, the British Virgin Islands, Barbuda, Nevis, Anguilla, and Montserrat, have never had significant commercial sugar plantations because they lack suitable land. Overall, sugar production in the Caribbean has been on the decline since the 1960s as a result of the variety of problems noted above and the international substitution of cane sugar with corn-based fructose.

Other commercial export crops grown in the Caribbean region include bananas, coffee beans, and tobacco. Bananas, introduced in the sixteenth century by Spanish missionaries, became an important export in the late nineteenth century as markets developed in Europe and North America. Sweet bananas are significant exports for the nations of Saint

Tobacco plantation, Pinar del Río, Cuba. (Henryk Kotowski)

CARIBBEAN AGRICULTURAL PRODUCTS
Commercial Export
sugarcane, bananas, coffee, cacao, cotton, citrus, ganja, rice, coconuts, tobacco
Subsistence Farming
yams, sweet potatoes, plantains, avocados, mangos, sugarcane, breadfruit

Vincent and the Grenadines, the Dominican Republic, Jamaica, Grenada, and Saint Lucia, and the French Caribbean territories of Guadeloupe and Martinique. Until the 1990s, Caribbean nations had preferential trade agreements with the European Union, which allowed them to export local bananas to Europe. If not for these agreements, Caribbean small-scale producers would be unable to compete for exports with large-scale banana operations of Central and South America. As the World Trade Organization started to systematically dismantle such agreements as incompatible with free trade, Caribbean exports of bananas started to decline dramatically.

Coffee beans are raised for export mainly in Haiti, Jamaica, and the Dominican Republic. Jamaica's famous Blue Mountain coffee, from beans grown in the Blue Mountains northeast of Kingston, is among the most prized and expensive coffee varieties in the world. Its production largely depends on the Japanese and US markets. Formerly a state-owned operation, it was acquired by an international investment company in 2013.

Tobacco was important before the sugar era and has seen a recent resurgence in the Greater Antilles, especially in Cuba, Puerto Rico, Jamaica, and the Dominican Republic, mostly for cigars. Marijuana, although illegal throughout the Caribbean region, is of considerable commercial importance. It was brought to the region from British India by East Indian indentured laborers. Used widely for smoking and as a tea, it is known as ganja in Jamaica, its chief producer, and its main foreign destination is the United States. Other significant export crops include cacao (or cocoa beans) and citrus fruits.

Small-Scale Farming

Small-scale farming in the Caribbean began after emancipation in the nineteenth century, when freed slaves sought out the only land available, in the hills and mountains. Unfortunately, this land, with thin and erodible soils, is poorly suited for crop agriculture. Individual small-scale farms average less than 5 acres (2 hectares) in area—often in disconnected plots. A variety of crops are raised, including fruits such as mangoes, plantains, akee, and breadfruit; vegetables such as yams, potatoes, and okra; sugarcane; and coffee beans.

Small-scale farms are semi-subsistence, in which most food is raised for family consumption, and the rest is raised for local markets or for sale to exporters, in the case of coffee. Mechanization is rarely used, and the hoe and machete are the main agricultural implements. Terracing, crop rotation, and fertilization are uncommon. Many small-scale farmers also raise livestock, typically scavengers such as chickens, goats, and swine. Small-scale farming never developed widely in the Lesser Antilles because of the lack of land. Freed slaves here typically remained where they were, but as paid laborers.

Trends in Agriculture

Agriculture's contribution to the economies of the Caribbean region has been declining for decades. In its current state, this sector has only a limited ability to comply with modern food safety and quality standards. Because of this, it has been unable to adequately respond to the population's rapidly growing demand for high-standard food products or to satisfy the demands of the tourism sector. Much of the local food production has been replaced by food imports, especially animal protein (meat, dairy, and fish) and canned prepared foods. On most islands, imported food exceeds 50 percent of consumption, but it is over 80 percent in Puerto Rico, Barbados, Antigua, and Trinidad and Tobago. Agriculture's low earnings and economic uncertainties have made this sector unattractive to young people.

National governments, regional organizations, and international development agencies have started developing agricultural strategies that involve public-private partnerships, regional cooperation, and investing in new acreages and transportation infrastructure. If adapted, agriculture has an important role to play in the region's socioeconomic development, particularly in poverty reduction for households that have been unable to profit from the growth in other economic sectors.

P. Gary White

INDUSTRIES OF NORTH AMERICA

North American mainland countries—Canada, the United States, and Mexico—have some of the most diverse and productive economies in the world. The continent has an abundance of natural resources, including oil and mineral deposits, fertile soils, freshwater, and forests, as well as industrious and skilled human resources. All this has made it possible for North America to become one of the world's most developed regions. The region's economy is built largely on its industrial geography.

National Comparisons

One way to compare two countries is to compare their gross domestic products (GDPs). The GDP is the total value of goods and services produced within a country during a year. When national economies are compared, their GDP is usually expressed in US dollars because the US dollar is the most traded currency in the world. The GDP does not ac-

curately measure how productive the industry is within the country, so economists often divide a country's GDP by its population to determine the average GDP per person. This per-capita figure is a more accurate measure because it reveals how much, on average, of the total economic production value can be attributed to each individual citizen: in other words, how technologically advanced the economy of a country is. Figures for per capita GDP and annual GDP can reveal a great deal about the US, Canadian, and Mexican economies.

The United States has always been the largest of the three economies. In 1960, the US GDP was thirteen times larger than that of Canada and forty-one times larger than that of Mexico. In 2001 its GDP was fourteen times larger than that of Mexico or Canada. At that time, for a brief period, Mexico's GDP was about the same size as Canada's GDP, but this does not mean that the Mexican economy was

Apple Park is the corporate headquarters of Apple Inc., located at One Apple Park Way in Cupertino, California (Arne Müseler)

ever as productive as Canada's because the population of Canada was, and still is, about a third of Mexico's population. In 2018, the US economy was twelve times larger than Canada's and almost seventeen times larger than Mexico's. The per capita GDP in 2018 was $62,900 in the United States, $46,230 in Canada, and $9,670 in Mexico, that is, in the United States it was 1.36 times higher than in Canada and 6.5 times higher than in Mexico. This difference can be explained by looking at the different components of the GDP.

Any detailed comparison of economies must differentiate among types of economic activity. Economists have traditionally presented the industrial components of the GDP by grouping them into three broad categories: agriculture and similar activities such as fisheries and forestry are often combined in one group; another grouping includes activities involved in the production of nonagricultural goods—mining, manufacturing, energy production, and construction; and the third grouping includes services. The service activities are so diverse that some economists further split this sector into three different ones: financial, business, and consumer services; a knowledge sector that includes information and communication technologies as well as media, education, consulting, and research and development; and human services such as hospitality and health care. Other types of categorization are also used but more commonly than not, all industries that create services rather than tangible objects are grouped together as "service industries."

With these categories in mind, one can examine the extent to which a given country depends on each. One measure looks at the proportion that the various sectors contribute to the total market value of the country's GDP. For example, in 2017, the production of nonagricultural goods represented 32 percent of the Mexican GDP, 28 percent of the Canadian GDP, and about 19 percent of the US GDP. Service industries accounted for roughly 80, 70, and 60 percent of the GDP, respectively. Alternatively, one can examine the proportion of the workers in each country employed in a particular sector. The United States and Canada have high proportions of workers in service industries, but almost a quarter of the Mexican population is engaged in low-wage jobs in agriculture, mining, and manufacturing.

Based on a combination of factors related to the GDP—employment, household income, and human development—the United States and Canada belong to the category of developed economies and Mexico, to the category of developing economies; it is one of the most advanced developing countries in the world. Based on predominant types of economic activity, Mexico is considered a newly industrialized country. The United States and Canada, both of which until the late twentieth century belonged to the category of highly industrialized countries, have since moved into the post-industrialized category. In post-industrialized economies, the relative importance of services—especially those related to high technologies, information, and research—is much higher than that of the production of tangible goods.

Analysis of the GDP and the relative share of different economic activities in it shows how the economies of these three countries have been changing over time. However, such data do not show why these changes occurred or how they have manifested themselves at the regional level.

Industrial Geography of North America: A Historical Overview

The US Department of Commerce identifies eight economic regions and tracks the total value of the goods and services produced in each region. The defined regions are the Far West, the Great Lakes, the Middle Atlantic, New England, the Plains, the Rocky Mountains, the Southeast, and the Southwest. (The Great Lakes region of North America is a bi-national Canadian-American region.) These regions have developed at different rates. The original industrial core was located in New England, the Middle Atlantic, and the Great Lakes region of the United States and Canada.

Early centers of commerce developed in these regions first because of technological limitations. Before the advent of steam-driven transportation, commercial centers were located either at natural harbors or where navigable rivers flowed into the Atlantic Ocean. This explains the location of industrial and trading hubs such as Montreal, Boston, and New York. Manufacturing centers grew alongside rivers, where falling water generated energy to operate textile or flour mills. The rail system expanded into the North American interior, prompting the spread of industry into Ohio, Michigan, and Illinois. Steamboats on the Ohio and Mississippi rivers accelerated the growth of factories in St. Louis, Pittsburgh, and Cincinnati. In Canada, Montreal and Toronto grew rapidly as canals and locks enabled travel among the Great Lakes.

Comparable industrial expansion did not occur in the US South. Its slave-based agrarian economy was oriented toward exporting cotton to European markets, not expanding into the US interior. The Civil War also affected the regional structure of the US industry. Much of the industrial base in the South was destroyed. This widened the economic gap between the industrial North and the agrarian South. As a result, the southern economy languished well into the twentieth century.

In the latter half of the nineteenth century, industrial cities such as Cleveland, Cincinnati, St. Louis, Detroit, and Chicago battled for control over the Midwestern economy. Chicago became the United States' "second city" after New York. The US automobile industry first developed in Detroit. It was there that Henry Ford utilized assembly-line mass production to build affordable cars for consumers. Steel, tire, and other factories emerged in Midwestern cities to support the expanding automotive industry.

Canadian economic growth occurred in the shadow of the US doctrine of Manifest Destiny—the idea that the country was destined to control all land between the Atlantic and Pacific oceans. When

Ford Motor Company Headquarters, Dearborn, Michigan. The Glasshouse; built in 1956. (Dave Parker)

Donald Smith, later known as Lord Strathcona, drives the last spike of the Canadian Pacific Railway, at Craigellachie, 7 November 1885. (Ross, Alexander, Best & Co., Winnipeg)

Canada became essentially independent from the United Kingdom in 1867, it was already transitioning from an economy based on primary commodities (such as timber, furs, fish, and wheat) to an economy based on the processing of natural resources and manufacturing of goods. Fear of US domination prompted Canada to build a transcontinental railroad and to impose tariffs on US imports. The Canadian Pacific Railroad was completed in 1885, connecting eastern Canada to British Columbia on the Pacific Ocean. Until then, British Columbia, Canada's westernmost province, had been virtually an economic appendage of the US Pacific Northwest.

Tariffs on US products coming into Canada and the resulting increased demand for Canadian products spurred industrial growth in Canada. US firms realized that the only way to reach Canadian consumers was to build factories in Canada. Compa-

nies such as General Motors built factories in southern Ontario because of its proximity to the US industrial heartland. Canada's industrialization owes much to US firms who located in Canada to avoid paying Canadian tariffs. During these years of breakneck industrialization and railway building, the so-called Golden Horseshoe region of southern Ontario became, and still remains, Canada's most densely populated and industrial-ized region.

The economy of the industrial core was based on the availability of primary commodities. Iron ore from the upper Great Lakes and coal from Appalachia were used to make steel in places such as Pittsburgh, Cleveland, and Detroit. This steel was used in automobiles, skyscrapers, and countless other products. Toronto became a major industrial and commercial center thanks to northern Ontario's timber forests and mines. Of course, there were service sector jobs too, but the rapid growth of North

America's industrial core was based on smokestack industries, extracting and processing raw materials.

The Rise of the South and the West

A clear divide once existed in the United States between the wealthy industrial North and the poorer South and West. Since the 1970s, however, the fastest-growing regions in the United States have been located in the South and West. At the same time, the economy of the traditional industrial core has undergone profound changes. The areas of Northeast and Midwest once known as the Manufacturing Belt, the Factory Belt, or the Steel Belt became known as the Rust Belt. With time, however, the Rust Belt developed new, service-related sectors. Chicago, for instance, went from a hub of meat-packing, lumber, and farm machinery to one of insurance and finance. Cleveland became a major academic medical center. Minnesota headquarters some of America's largest publicly traded companies.

Regional growth in the South and West can be partially understood by looking at natural resources. Texas, California, New Mexico, and Oklahoma, for instance, are rich in oil and have hugely benefited from selling oil products to the rest of North America. Warm waters and excellent beaches have made it possible for Florida, the Carolinas, and other southern states to develop large service sectors centered around tourism, retire- ment, and real estate.

However, other factors are to a greater extent responsible for the dramatic growth in the South and West. Expanded air and auto travel drew economic activities out of the industrial Northeast and Midwest. The groundwork for this industrial diffusion rested in a set of social policies called the New Deal, implemented in the 1930s during the Great Depression. Before then, unions could not represent workers during wage negotiations. The New Deal

allowed workers to form unions without management interference. Corporate and political leaders, however, feared that powerful unions would weaken the US economy with high wage demands. As a result, a federal law was passed in 1948 allowing individual states to determine labor laws. Twenty-one states, primarily in the South, gave workers the "right to work" without joining a union or paying dues. This legislation weakened unions and lowered wages in those states, thus making them more attractive for corporations—a crucial fact for understanding North America's economic geography.

Military expenditures also have affected the South and West. A disproportionate amount of military spending on bases and equipment occurred in the so-called Sun Belt (the southern tier of the United States). Important technology centers began to emerge during this time. Places where radar, aviation, nuclear, and other technologies were developed during and after World War II provided the context for private-sector technological research and development. Many high technology firms

A disused grain elevator in Buffalo, New York. (Fortunate4now)

A maquiladora-factory in Mexico. (Guldhammer)

emerged and are located in the Silicon Valley of California, in Texas, and in the Seattle, Washington, area. The two largest research parks in the United States are located in the Raleigh-Durham area of North Carolina and in Huntsville, Alabama.

During the late 1960s, US manufacturers had to compete against more efficient firms from West Germany and Japan. Inflation led US workers to strike for higher wages, while their employers had to find ways to cut costs in order to compete. As a result, many factories moved from union strongholds in the northern Rust Belt states to southern and western Sun Belt states where unions were weak and wages low. Other US firms moved to northern Mexico, where wages were lower still.

Mexico's industrial sector has always lagged behind those of the United States and Canada. Mexico has relied on its abundant mineral resources, such as gold, silver, copper, and iron. Mexico is also a large producer of oil and oil products. Oil and chemical production spurred economic growth in the traditional industrial core region of central Mexico. While Mexico's mining industry has been dominated by Canadian companies, there is also substantial Mexican capital involved in mines producing silver, gold, and other strategically important metals and non-metal minerals.

In 1965 the Mexican government initiated the Border Industrialization Program along the Mexico-United States border. This region was intended to lure US firms to Mexico with the promise of low-wage, nonunion labor. Firms could ship unfinished products to Mexico without tariffs, and final assembly could occur in Mexican factories called *maquiladoras*. The finished goods could then be shipped back to the United States.

While the *maquiladora* program was expanding growth in northern Mexico, changes were occur-

Mexico City is the financial center of Mexico. (Eneas De Troya)

ring in the United States and Canada too. There was a growing understanding that import tariffs and other trade restrictions were impeding economic growth and preventing the creation of cost-effective supply and distribution chains. This movement culminated in free trade agreements between the United States and Canada in 1989, and the addition of Mexico to this partnership in 1994 under the North American Free Trade Agreement (NAFTA). Most corporations welcomed these developments because NAFTA removed all tariffs on goods traded among the three countries. Workers worried that as free trade made it easier for firms to move to regions with lower wages, they would lose their jobs.

NAFTA was an encompassing and complicated deal. It more than tripled trade and foreign investment in North America; as a result, production costs and consumer prices dropped. NAFTA improved the competitiveness of cars made in North America versus Asian rivals. At the same time, some

manufacturing jobs were lost in the United States while Mexico lost farming jobs in corn production. However, it must be taken into consideration that NAFTA was enacted and implemented at the same time that profound shifts were taking place throughout the industrialized world. The process of industrial automation and agricultural mechanization led to workforce displacement worldwide. In the process of globalization, manufacturing companies were moving across national borders to countries where wages were lower, and other factors determining the cost of doing business were more favorable. In Europe, as in the United States and Canada, manufacturing was being replaced by the service industries.

Leaders of the United States, Canada, and Mexico renegotiated a trilateral agreement in 2018 and the new deal, known in the United States as the United States-Mexico-Canada Agreement, entered into force in 2020. Compared to NAFTA, it includes many new provisions based on twenty-first century realities and trends. Among them are those aimed at securing the automobile industry in North America, the creation of high-paying jobs, and incentivizing greater North American production of textiles and apparel. It also covers trade in pharmaceuticals, medical devices, chemicals, cosmetic products, and information and communication technology.

Twenty-First Century Trends

The de-industrialization of North America and the growing power of China, which has started transforming its essential role in global production and supply chains into political influence, have led to the growing belief that some essential manufacturing sectors should be kept in North America or moved back there. These include such staples of North American economies as the manufacturing of automobiles and aircraft and their parts. The 2020 worldwide COVID-19 pandemic added pharmaceuticals and medical-supply chains to the list.

Economic sectors that have been driving the US economy in the twenty-first century include the energy sector—production of oil, electricity, gasoline, and other fuels; high technologies, especially in information and communication fields; retail trade, including online retail; construction; and health care. Health care, especially, has risen to fill the economic void left by manufacturing. Between 2006 and 2017, there was a 20-percent growth in health-care sector jobs in the United States, while the average growth rate for the economy was 3 percent.

Canada's health-care system is also a major component of the national economy. Employment in this sector, which is about 9 percent of the country's entire workforce, is protected from recessions or other economic calamities. Unlike in the United States, most aspects of the Canadian health-care system are publicly funded, and most health-care costs are covered not by employers (as in the United States) but by provincial and territorial governments. This is a major consideration factor for foreign companies considering opening or keeping production in Canada. Guaranteed health insurance contributes greatly to a more stable workforce in all sectors of the Canadian economy.

The world economy and trade continue to evolve toward the merge of the delivery of goods with the delivery of services. Since the beginning of the digital revolution, services accompanying many manufactured goods have become a critical part of their functionality. Cars with navigation services, personal computers, and smartphones are essentially useless without the Internet. Due to its leading role in many high-technology, education, research, and development fields, the United States has a comparative advantage over other nations in this hybridized economic landscape.

Christopher D. Merrett

Industries of the Caribbean

Caribbean island economies are highly vulnerable to weather events, world market prices, and foreign competition. Tourism and plantation agriculture are the main economic activities of the Caribbean region, followed by export-oriented product assembly. Although these are the largest contributors to the gross domestic product and the biggest sources of jobs, they involve mostly low-skilled and low-paying jobs performed by the older segments of the population. Nearly one in every four young people in the Caribbean is unemployed, compared to two in every twenty-five middle-aged and older working-age people. The tourism industry is controlled by international hotel chains; most higher-paying managerial positions in this sector are held by foreign nationals. The Caribbean region loses an excess number of professionals, technicians, and students to developed regions, especially to continental North America and Europe.

Tourism has become the major industry and source of revenue in the Bahamas, the British Virgin Islands, Saint Lucia, Barbados, Antigua and Barbuda, Dominica, and Jamaica. The Bahamas and the Cayman Islands have prospered not only through tourism but also through international banking and investment management. Together, tourism and financial services sectors comprise up to 85 percent of the Bahamian GDP. The Cayman Islands territory has become a global offshore financial center.

Sugarcane is still the major export crop of Cuba, Barbados, the Dominican Republic, and some of the smaller islands. Bananas are the next leading agricultural export. Many Caribbean countries had preferential trade agreements, including price supports, with European Union members. This means, as an example, that Caribbean bananas were given preference over cheaper Central American bananas by former European colonial powers such as France and Britain. Criticism that these preferential deals were incompatible with the rules of the World Trade Organization led to their replacement by more reciprocal and less exclusive economic partnership agreements; export-oriented agriculture in the Caribbean lost its competitive advantage as a result. Without major restructuring, agriculture's prospects in the Caribbean are bleak, as soils are becoming depleted and eroded, and world markets are increasingly competitive. Since the 1990s, tourism and manufacturing surpassed the sugar industry in economic importance. By the twenty-first century, for example, tourism contributed ten times more to the GDP of Barbados than sugarcane cultivation and refining.

Butterfield Bank in George Town, Cayman Islands. (James Willamor)

Caribbean fishing boats at Aruba. (Dennis Matheson)

Mining

No mining or insignificant amounts of it are done on most islands in the Caribbean region, except in Cuba, Jamaica, Trinidad and Tobago, and the Dominican Republic. Cuba's mining industry produces and exports nickel. Small- and medium-sized mines producing zinc and lead have been mostly closed because of the lack of investment, as also happened to Haitian mines extracting gold and associated metals. Haitian mineral deposits are located in the same mineralized zone of the island of Hispaniola as the Dominican Republic's Pueblo Viejo gold mine, the largest gold mine in the Caribbean. That mine, operated by a Dominican mining group, went bankrupt in 1999. Since 2012, however, operations have been restarted and expanded by a US-Canadian corporation. In recent years, foreign and multinational companies have been exploring a revival of the abandoned mining operations in Cuba and Haiti.

Jamaica is among the world's largest producers of bauxite. About a third of bauxite mined in Jamaica is shipped, unprocessed, to the United States. The rest is processed in the country's four alumina refineries and exported mostly to Europe and North America. Trinidad and Tobago's economy is dominated by the production of oil and natural gas. In the 1990s, natural gas surpassed oil. Today, the country has one of the largest gas-processing facilities in the Western Hemisphere.

Manufacturing

The Caribbean region is dependent on the United States and the European Union for markets, technology, investment, credit, and aid. To break this cycle of dependency, countries in the Caribbean region have tried two strategies to expand the manufacturing sector and free the region from total dependence on imported manufactured products. The first policy, import substitution, was intended to promote manufacturing for local consumption by protecting local industries with tariffs levied on similar imported goods. Import substitution was regulated by the government. This policy encouraged industrialization throughout the region between 1950 and 1975 but failed to produce the desired economic boom. It was replaced by a second development strategy—export-oriented industrialization (EOI).

Export-oriented industrialization—producing manufactured goods that can be sold abroad—has been a major factor in the development strategies of most Caribbean countries. Governments often offer incentives such as tax breaks and duty-free zones to attract foreign investment and manufacturing. Most islands have established export processing zones (EPZs), located near ports or international airports. Foreign parent companies can import materials tax-free, but only if the majority of the final assembled product is re-exported. As a result of EPZs, manufacturing exports from the Caribbean increased dramatically during the 1980s, especially exports to the United States. Many industrial products have been assembled in the Caribbean region, from Barbadian medical products to Trinidadian electronic aircraft equipment. *Maquiladora* export production in the Dominican Republic works much the same as it does on the United States-Mexico border and provides many unskilled, assembly-type jobs.

The main drawback of EOI has been its instability. Foreign companies are mobile and will quickly leave one island and set up on another when local conditions deteriorate or wage rates rise. Islands in the Caribbean have long been in competition with one another and in recent decades, also with Asian and Central American competitors, for export-oriented assembly jobs. Foreign firms have invested, moved profits, and withdrawn operations from various islands with great ease. This has been a destabilizing factor in most Caribbean economies. In countries like Haiti, the United States and other developed nations have attempted to support local manufacturing through grants, but without much success.

In recent years, the Dominican Republic has partially offset the loss of garment manufacturing to Central America and Asia by expanding its production of tobacco, medical equipment, and pharmaceuticals. Jamaica, with Chinese help, has

Tourists at Pigeon Point beach, Tobago. (Kp93)

started developing an integrated transport infrastructure network with the aim of becoming a globally competitive logistics hub.

Manufacturing, mining, and food processing are important contributors to the gross domestic product of the islands; however, they do not contribute more than 20 percent of the total economy of most island nations, with several exceptions. The exceptions include Trinidad and Tobago, where oil and gas account for about 40 percent of the GDP, and Cuba, where the production of pharmaceuticals and vaccines has been expanding in recent years and the industrial production overall accounts for over 35 percent of the economy. The industrialization has failed to provide adequate jobs or sufficient wealth for most of the islands to offset the decline in agricultural production. Tourism and related sectors have shown much more potential for future development. In the twenty-first century, regional and international organizations have been focusing on new strategies of sustainable economic development in the Caribbean, based on public-private partnerships and regional cooperation.

Carol Ann Gillespie

ENGINEERING PROJECTS IN NORTH AMERICA

Adapting natural landscapes through various engineering methods is a recurring theme in North American history. Ingenuity is a characteristic often used to describe both the indigenous peoples of and immigrants to North America, who have been challenged by the continent's geography. Early on, engineering ideas were applied to dwellings and military fortifications; the latter were placed on geographically strategic sites such as bluffs. Then, civil engineering techniques prevailed to develop natural harbors, create transportation routes, and enact flood-control measures.

Throughout the North American continent—from Mexico's diverse arid and tropical areas to Canada's and the United States' contrasting mountains and prairies—North Americans had long used and controlled the environment to benefit humans. The Ancestral Puebloans of what is now the southeastern United States lived in unique apartment-like complexes made from adobe mud and stone or carved into the sides of canyon walls. In parts of what are now Arizona and Sonora, Mexico, the Hohokam developed large-scale irrigation networks. Mineral mining in Mexico dates back to Mesoamerican civilizations. Native American tribes in the north formed timber weirs to dam creeks. Their dwellings—the tipi, wigwam, longhouse, pit house, and igloo—were perfectly suited to the environment, climate, and geography of a specific area.

Later, people in North America developed methods to use water and wind to generate energy to process and transport raw materials. Westward migration resulted in creative engineering techniques that helped settlers to traverse the Rocky Mountains and benefit from topographical features in the new locations. More sophisticated engineering works evolved as scientific comprehension of North American geography advanced.

Mississippi River Improvement

North American demographics emphasize settlement patterns near water, and North American rivers have both helped and hindered movement. Humans require a variety of engineered devices to use waterways efficiently as trade routes. The Mississippi River, which approximately bisects the United States, flows from Minnesota to the Gulf of Mexico. Prior to engineering intervention, it was only erratically useful as a transportation route to Native American tribes, settlers, and traders. The

Pueblo Bonito is the largest and best-known great house in Chaco Culture National Historical Park, northern New Mexico. It was built by the Ancestral Puebloans who occupied the structure between AD 828 and 1126. (Bob Adams)

river's physical characteristics often caused delays or destroyed boats. Currents changed according to the season, and the water flow moved sediment and debris into sandbars that obstructed travel and sometimes snagged even steamboats.

During winter, northern sections of the river became ice-choked; in the spring and summer, the river often flooded because of ice melt and torrential rain. The 1811 and 1812 earthquakes at New Madrid, Missouri, altered the Mississippi River's course and created new channels adjacent to the previous riverbed. The Mississippi River's tributaries poured water and mud into the main river, and, by the 1870s, the mouth at the Gulf of Mexico was clogged with so much sand that vessels could not enter or exit the river.

In an attempt to ease travel and prevent destruction caused by the river, self-educated civil engineer James B. Eads studied hydrodynamic data to develop engineering solutions. Congress had ordered the US Army Corps of Engineers (USACE) to clear the river's mouth and dig a canal upstream from the sandbars. Disagreeing with this approach, Eads suggested instead to build jetties in the Mississippi River's mouth by using the river's currents to move and deposit sediment away from the main channel. The project was accomplished by a mix of USACE and civilian engineers. An underwater grid of willow, rock, and silt formed the foundation of masonry built above the riverbanks to prevent tidal flooding. The jetties forced water into this narrow channel, quickening the current's flow to prevent sandbars from forming. As a result of these jetties, the river became accessible to most ships, even those with deep drafts.

Canals, Shipyards, and Roads

The USACE has erected and managed hydraulic engineering projects on the rivers and other waterways. The locks and dams it has built on the Mississippi River are the most significant of its projects. Built in the 1920s and 1930s, this dam system, which also spans the Illinois River, enables barge traffic to move along the rivers. Without the twenty-nine dams located between St. Louis and St. Paul, the river would not be navigable. Each dam

pools the water to sufficient depths to accommodate large vessels.

The dam system is designed to ensure that the river will behave consistently. Without the dams, the river would be unpredictable, going through cycles of flooding and drought and forming sandbars and other unsafe sections. Corps engineers dredge the river to improve passage; repair the system; and stress the need for improvements, such as increasing the size of the locks, which cannot accommodate large convoys of around fifteen barges tied together. These convoys must be split up to move through each lock. Delays as long as two hours create traffic jams on the river and affect farmers' profits. The dams also harm habitats: millions of dollars are spent annually to repair environmental damages. At the same time, navigation on the river generates about $600 billion in annual economic activity. The river transports 60 percent of all grain products in the United States, the world's largest grain producer.

Before the twentieth century, canals offered travelers and traders the most direct routes to desired destinations and the least expensive means of inland transportation. The post-Revolutionary Patowmack Canal was built because George Washington wanted to expand trade from the Atlantic Ocean to the frontier. The narrow, twisting Potomac River often flooded in the spring and dried up in the summer, impeding navigation. In the seventeen-year period beginning in 1785, workers dredged the riverbed and blasted rock to build a diversion canal with five locks at the Great Falls, where the river dropped 80 feet (24 meters) in a 1-mile (1.6-km.) stretch. The canal took much longer to build than the new federal capital, Washington, D.C. The company was bankrupted by 1828 thanks to excessive costs, but the project was significant for inspiring further canal construction based on lock engineering. Canada's Rideau Canal—the oldest continuously operating canal system in North America—represents how engineers adapted local conditions and materials instead of relying on European traditions. At the Hog's Back, a steep drop on the Rideau River, engineers used stone and timber to dam unpredictable floodwaters in order to

The Rideau Canal in Ottawa, Canada. (Bobak Ha'Eri)

complete construction. Built between 1826 and 1832, the canal connected Canada's capital city of Ottawa to the Great Lakes and the St. Lawrence River, and, ultimately, to the Atlantic Ocean.

The Erie Canal, connecting Lake Erie to the Atlantic Ocean, is perhaps the best-known North American waterway. When it opened in 1825, it was a 4-foot (1.2-meter)-deep by 40-foot (12.2-meter)-wide and 363-mile (584-km.)-long engineering feat accomplished by local farmers, African American workers and recent Irish and German immigrants used shovels, pickaxes, oxen, and horses. Later deepened to about 14 feet (4.27 meters), the canal could hold barges of more than 3,000 tons. It made New York City the financial capital of the world and triggered commercial and agricultural development of—as well as large-scale immigration to—the sparsely populated frontiers of western New York, the Midwest, and points farther west. Railroads and

highways later made the canal obsolete, but the Erie Canal has nevertheless remained one of history's best examples of how engineering can play a vital role in a nation's industrial and economic growth.

Later canals linked major cities and included more-sophisticated technology such as hydraulic locks. For example, the St. Lawrence Seaway, completed in 1959, was a joint engineering effort by Canada and the United States. The Great Lakes-St. Lawrence system has fifteen locks and is 370 miles (600 km.) long. The project relied on cooperation between the Canadian government and the USACE, who addressed such concerns as water-level fluctuations, coastal erosion, and shoreline protection and management. Those engineers used the Geographic Information System to facilitate their work.

Ports and shipyards along the coasts and large inland bodies of water enhanced local economies by encouraging commerce. Engineers developed natural harbors, such as the Chesapeake Bay in the Mid-Atlantic region of the United States, Campeche and Manzanillo bays on the coasts of Mexico, and Hudson Bay in northeastern Canada. Established in 1774, the Portsmouth shipyard in New Hampshire became the first naval shipyard in the United States, implementing technological procedures that were later adopted at other North American ports. Businesses, manufacturers, and industries benefited, producing steamships and naval vessels and providing fuel and other necessities for crews. North American ports benefited from the construction of the Panama Canal, completed in 1914, which shortened voyages between the Atlantic and Pacific oceans by up to 3,500 nautical miles (6,500 km.). The canal itself was built under the supervision of USACE officers by workers from the United States, the Caribbean islands (West Indies), Europe, and Asia.

In the twenty-first century, many North American ports with infrastructures dating back to the second half of the twentieth century face the challenge of keeping up with the ever-increasing size of cargo ships and, consequently, the higher volume of multimodal traffic among sea, land, and air carriers. This is especially important for the United States, which leads all nations of the world in the value of imports and exports. Improvements have been implemented or are planned at many US ports, particularly on the West Coast, where navigation channels generally are deeper than those in other areas. The Port of Tacoma, Washington, a leading container port, already has on-dock train depots, and similar facilities were under development in Los Angeles in 2020. The Ports of Los Angeles and Long Beach have been using landfill to construct artificial islands in San Pedro Bay, to be outfitted with state-of-the-art shipping terminals, new berths, and miles of new, deep navigation channels. The first part of the plan, the Pier 400 Dredging and Landfill Program, was completed in 2000 by the USACE; it is the largest dredging and landfill project ever undertaken in the United States.

Port of Tacoma, where the Puyallup River (left) feeds into Commencement Bay. (D Coetzee)

In Mexico, the nation's largest port, Manzanillo, as well as many other ports, have been operating at full capacity for several years. The Mexican government has instituted plans for a massive upgrade of twenty-five ports. The expansion of the Port of Veracruz, expected to be completed by 2024, is the most important port infrastructure project in Mexico in the last 100 years. Veracruz services both the huge industrial hub of Mexico City and the manufacturing sector of central Mexico. Its expansion in-

volves the building of a terminal capable of handling two mega-ships at once, as well as the construction of four specialized terminals for the handling of grain, minerals, liquids, and bulk cargo.

Like shipyards and ports, road connections encourage domestic and international trade. Interstate highways and railroads cross North America, both conforming to topographical landscapes and transforming areas to create direct routes between metropolitan areas. The Trans-Canada Highway,

The Trans-Canada Highway passing through Glacier National Park in British Columbia. (Jesse Allen - NASA Earth Observatory)

The Verrazzano-Narrows Bridge is a suspension bridge connecting the New York City boroughs of Staten Island and Brooklyn. (Ajay Suresh)

stretching 4,860 miles (7,820 km.), was the world's longest paved highway when completed in 1970. The CanAm Highway, an international route that connects Mexico to Canada through the United States has become North America's vital trade route, especially after 1994, when the newly passed North American Free Trade Agreement (NAFTA) led to a huge increase in the truck traffic among the three countries.

In the United States, the CanAm Highway travels along US Route 85 (US 85) and Interstate 25 (I-25), passing through six US states. In Canada, it continues as Highway 102 and comprises Saskatchewan Highway 35 (SK 35), SK 39, SK 6, SK 3, and SK 2. In Mexico, the route continues as Mexican Federal Highway 45 (Fed. 45). Between 1993, just before the signing of NAFTA, and 2018, when the agreement became subject to renegotiation, total merchandise trade between Canada and the United States tripled and total merchandise trade between Canada and Mexico grew almost tenfold. The North American trade region has become the world's largest free-trade market.

In the twenty-first century, several of Canada's largest engineering projects have been devoted to passenger urban-rail transit, or more specifically,

light rail. Those include the Green Line light-rail project in Calgary, set for completion in 2026; the Réseau Express Métropolitain (Metropolitan Express Network), being built in the Greater Montreal area; and the Line 5 Eglinton in Toronto, which was nearing completion in 2020. Canada is far ahead of the United States and Mexico in the development of modern urban rail transit.

Above and Below Ground

Bridge design and materials differ according to local conditions and needs. Some bridges span narrow streams, others stretch across wide bays. Notable North American bridges include the Verrazzano-Narrows (4,260 feet/1,298 meters), Mackinac Straits (3,800 feet/1,158 meters), George Washington (3,500 feet/1,067 meters), and Pont de Québec (1,800 feet/549 meters).

Bridge construction long perplexed engineers, who had to determine how to pour concrete piers in water and balance loads. Engineers such as James B. Eads (who would later become famous for the Mississippi River jetty system) innovated in bridge construction methods and the use of materials such as steel to connect the eastern and western banks of the Mississippi River. Although that bridge-building project was deemed technologically impossible by many engineers, Eads studied how the river's sands shifted from currents and sunk granite piers 90 feet (27.45 meters) into the riverbed to a foundation of solid bedrock. For building the piers, he used pressurized underwater chambers, or pneumatic caissons, introducing that building technology to North America. The Eads Bridge was the longest arch bridge in the world when it was completed in 1874. Still in use today, it is the only large bridge in the world named after its engineer.

John A. Roebling, who designed the Brooklyn Bridge, created suspension bridges that prevented the structures from being damaged by floating ice and boats. The 4,200-foot (1,280-meter)-long Golden Gate Bridge enables over 110,000 vehicles a day to travel between the city of San Francisco and suburbs within the San Francisco Bay area. Other bridges connect countries, such as the Peace Bridge between Buffalo in upstate New York and Fort Erie in Ontario, Canada. Many bridges, such as the

Ashtabula and Quebec bridges, have collapsed during erection or due to stress, killing workers and motorists. By contrast, the Frankford Avenue Bridge in Philadelphia, Pennsylvania, built in 1697, was still in use in the year 2020. It is the oldest surviving roadway bridge in the United States.

Mexico's Baluarte Bicentennial Bridge, completed in 2012, is one of the world's tallest cable-stayed bridges and the highest in the Americas, with a clearance of 1,322 feet (403 meters) above the valley below. The longest bridge in North America is located in Louisiana. At 23.83 miles (38.35 km.) in length, the Lake Pontchartrain Causeway is actually composed of two parallel bridges. The causeway was the longest bridge over water in the world for over a half-century, between 1956, when it was opened, and 2011, when it was surpassed by a bridge in China's Shandong province.

The Evergreen Point Floating Bridge, carrying Washington State Route 520 across Lake Washington from Seattle to its eastern suburbs, is the longest

The Baluarte Bicentennial Bridge is the third-highest cable-stayed bridge in the world, the seventh-highest bridge overall and the highest bridge in the Americas. (panoramio)

BUILDING THE TRANS-ALASKA PIPELINE

Stretching 800 miles (1,300 km) between Prudhoe Bay and Valdez, the Trans-Alaska pipeline transports around half a million barrels of oil daily, on average. The Prudhoe Bay oil field, located on the North Slope of Alaska, contains an estimated 25 billion barrels of petroleum resources. The Valdez Marine Terminal is an oil port on the Gulf of Alaska and can store over 9 million barrels of oil at the southern end of the pipeline.

In 1970, a consortium of major oil companies formed the Alyeska Pipeline Service Company to build the pipeline, which crosses the Arctic Circle, 3 mountain ranges, 350 rivers, and seismically active areas. More than 70,000 workers were involved and the total cost was $8 billion. Construction began in 1974 and was completed in 1977. Each pipe section is approximately 48 inches (122 centimeters) in diameter and 40 or 60 feet (12 or 18 meters) long. Half (420 miles) of the pipeline is elevated to prevent the permafrost from melting because of the oil's frictional heat. Lifted by 78,000 supports cooled by liquid ammonia, the pipe is coated with teflon so it can move within the supports during earthquakes. Part of the pipeline is positioned underground where caribou migrate, and is insulated to reinforce the frozen soil supporting the pipe. Concrete weighs down sections buried underneath rivers. A pipeline bridge spans the 2,300-foot-wide (701 meters) Yukon River. Valves are located at vulnerable sites to shut off flow if the pipe ruptures.

The Trans-Alaska Pipeline System. (Luca Galuzzi)

Engineers locate some trade connectors underground because of weather conditions or to increase the usability of crowded urban areas. Subways move workers and goods beneath city streets, and tunnels transport cargo through mountains and under bays. The St. Clair Tunnel (1891) between the city of Sarnia in the Canadian province of Ontario and Port Huron in Michigan was the first full-sized underwater tunnel built in North America. Engineers mastered excavation techniques to deal with soft, moist earth and lined the tunnel with cast iron. The twin-tube Holland Tunnel, connecting New York and New Jersey (1927), was the world's first major underwater tunnel for automobiles and the first mechanically ventilated tunnel.

Pipelines above and below ground move petroleum and natural gas from deposits to distribution centers. By the late twentieth century, Canada was connected to the United States by several natural gas pipelines, supplying the United States with more energy than it received from Saudi Arabia, Mexico, and Venezuela combined. The El Paso Natural Gas system of pipelines brings gas from New Mexico, Colorado, and Texas to several US states and northern Mexico. Mexico itself has numerous crude oil pipelines. In 2020, the country had six new natural gas pipelines under construction, and more were planned to be built with US participation.

and widest floating bridge in the world. Its length is 7,710 feet (2,350 meters) and the width at the midpoint is 116 feet (35 meters). Opened in 1997, Canada's Confederation Bridge carries the Trans-Canada Highway across a straight between Prince Edward Island and the province of New Brunswick on the mainland. At a length of 8 miles (12.9 km.), it is Canada's longest bridge and the world's longest bridge over ice-covered water.

Dams and Aqueducts

Engineers design and supervise construction of dams throughout North America for flood control and hydroelectric power generation. In 2020, Mexico's Chicoasén Dam (856 feet/261 meters) was the tallest North American dam, tenth in height globally. British Columbia's Mica Dam stands 794 feet (242 meters) high, the second-tallest on the North American continent.

California's Oroville Dam (770 feet/ 235 meters) is the tallest US earthen dam. Besides producing energy, it prevents an estimated billion dollars in flood damages annually. Quebec's James Bay Project consists of a series of dams and hydroelectric power stations on the La Grande River, covering an area the size of England or the state of New York. It is already one of the largest hydroelectric systems in the world, and if fully expanded to include all originally planned dams, it will be the largest. Its full implementation had been delayed by environmentalist objections and the provincial government's conflict with the Crees and Inuit over land rights. In 2002, the government reached an agreement with the Crees and the construction of new power stations has resumed. However, an associated Great Whale River Project, north of La Grande River, has been indefinitely suspended since 1994.

The Tennessee Valley Authority (TVA) dams economically transformed the southeastern United States during the twentieth century. Established by Congress as a New Deal agency in 1933, the TVA offered employment to laborers during the Great Depression. A series of dams was built along the Tennessee River system to generate energy and control floods. The TVA enhanced the economies of communities in the Tennessee River Valley by offering affordable and consistent renewable energy supplies. Electricity drastically changed people's lives at home and at work. To meet World War II munitions demands, the TVA built sixteen hydroelectric stations and a steam plant to power war industries. By 1945 the TVA had created a 650-mile (1,050-km.)-long navigation channel that stretched along the Tennessee River's entire length. The TVA became the United States' largest public power provider. In the twenty-first century, the TVA remained

BUILDING THE HOOVER DAM

Built to control the Colorado River, the Hoover Dam was erected in Black Canyon during the early 1930's and demonstrates how technology can alter nature. Engineers planned to collect flood water to irrigate the Nevada and Arizona desert and create hydroelectric power for industries and homes. Funded by the federal government, the dam cost $48,890,955 and provided jobs for five thousand workers during the Great Depression. Hundreds of laborers suffered from the region's extreme heat, and 112 people died from illness and accidents while building the dam.

When it was constructed, Hoover Dam was the largest dam ever envisioned. Workers blasted river diversion tunnels in the bedrock before dredging the canyon floor and pouring 4.1 million cubic meters of concrete through a grid of blocks and pipes, which were flushed with chilled water so that the concrete cooled in two years instead of the century it would have taken to cool naturally. The billions of watts of energy produced by the 60-story-tall dam's enormous generators and turbines encouraged settlement of southwestern cities such as Las Vegas and Los Angeles. The dammed water created Lake Mead, Earth's largest artificial lake. Considered the eighth wonder of the world, Hoover Dam inspired innovative engineering construction techniques.

Aerial view of Hoover Dam, Nevada-Arizona. (Mario Roberto Durán Ortiz)

the largest public utility in the nation. In 2020, TVA's power mix included thirty hydroelectric power facilities; twenty-seven coal, natural gas, and nuclear plants; fourteen solar energy sites; and one wind energy site.

Some dams had been poorly placed, built, or maintained. The St. Francis Dam was built on a fault at the southern end of the Los Angeles aqueduct system. Collecting water from the Sierra Nevada, the dam failed within a week of its completion in 1926, killing more than 430 people. The South Fork Dam in Pennsylvania collapsed from the pressure of 20 million tons of water from torrential rains, killing more than 2,200 people in the Johnstown flood in 1889. The Teton Dam in Idaho suffered a catastrophic failure as it was filling for the first time on June 5, 1976, killing 11 people and 13,000 cattle. The 1996 Saguenay flood in Quebec was caused by a storm that overwhelmed poorly maintained dikes and dams; ten people were killed and almost 16,000 were forced to flee their homes.

Aqueducts are essential to provide ample clean water for urban populations, especially in arid regions. Engineers design extensive waterworks and pumping systems to deliver water to consumers. The United States has some of the world's longest systems of artificial channels, pipes, tunnels, and supporting structures built for conveying water to cities. These include the 120-mile (190-km.)-long Catskill Aqueduct delivering water to New York City; the Colorado River Aqueduct, which runs for almost 250 miles (400 km.) supplying water to the Los Angeles area; and the Central Arizona Project,

the largest and most expensive aqueduct constructed in the United States. It is a complex system whereby 336 miles (541 km.) of canals and fourteen pumping plants lift the water over 2,000 feet (600 meters) between the Lake Havasu reservoir on the Colorado River and its final destinations in the cities of Phoenix and Tucson. The canals transverse some of the driest and hottest terrains in North America.

However impressive, North America's engineering achievements of the past pale in comparison with some of the megaprojects contemplated for the future. The most ambitious among them is a Bering Strait Crossing—a proposal to build a tunnel or bridge between Russia's Chukotka Peninsula and the Seward Peninsula in Alaska, thus connecting the United States and Russia and the continent of North America with Eurasia. Technical concerns—including long, dark winters, extremely cold weather, and ice floes—have so far prevented the project from being realized, but with the rapid technological advances of the twenty-first century, it looks more and more realistic.

Elizabeth D. Schafer

TRANSPORTATION IN NORTH AMERICA

In a geographical context, the word "transportation" refers to the movement of people and goods among towns, cities, regions, and countries. Transportation is the foundation by which people and commodities move for exploration, economics, and communication. Transportation allows for areas to specialize. For example, California grows citrus fruit and the Midwest grows corn. Transportation enables each area to develop, exchange, and sell products that best suit its particular physical geography. It was transportation that made it possible for the United States and Canada to spread across the North American continent from ocean to ocean during the era of westward expansion. It also helped connect far-flung parts of the growing nations.

Animals

Before the Industrial Revolution, which brought with it the development of railroad networks, people were only able to cover long distances in a relatively short time by using animals or boats. In North America, horses provided the most versatile and readily available mode of transportation.

Conestoga Wagon, about 1840-1850, displayed at the National Museum of American History. (Smithsonian Institution)

A sled dog team of 11 in Denali National Park and Preserve. (AlbertHerring)

Horses had become extinct in North America between 13,000 and 11,000 years ago, at the end of the Pleistocene Epoch. The reintroduction of the horse to the Americas by the Europeans allowed indigenous peoples of the Great Plains and Canadian prairies to revolutionize their way of life; agrarian economies of European colonists depended on the horse for their survival. For several centuries, people in North America traveled on horseback and moved goods by horse and buggy or horse-drawn carriages. The heavy Conestoga wagon, pulled by teams of horses, mules, or oxen, was widely used in the eastern regions of the young United States for hauling freight. But it was a smaller and lighter descendant of the Conestoga wagon, the prairie schooner, that made possible the large-scale westward migration of settlers in the nineteenth century.

Today, animals are still used in some regions of North America. Anabaptist communities of the Old Order Amish, found throughout the United States and Canada, do not use electricity or gasoline-powered engines and transport themselves and their goods by horse and buggy. In the Great Plains and prairies, ranchers still travel great distances on horseback to watch and herd their livestock. Although the horse is the animal most commonly used for transportation, other animals also provide means of movement. In northern Canada and Alaska, dog sleds are used during the winter months for transportation across the ice and snow. Donkeys, brought to Mexico by Spanish conquistadors, have become a symbol of rural life in that country. Mexican donkeys, or burros, are still used as pack animals due to their strength, endurance, and ability to acclimate to heat and water shortage.

Water Transportation

Boats have long been used for fishing and whaling, exploration, transportation of commodities, and passenger travel. Many indigenous peoples of the Americas built bark canoes that were easy to repair, light enough to be portaged, and strong enough to carry significant loads. Northern indigenous groups made dugout canoes and boats constructed of whalebones, sealskin, and driftwood. After Europeans arrived in North America in the fifteenth and sixteenth centuries, early explorers of the continent adopted the canoe and relied almost entirely on canoe travel. The coastline of North America and its major rivers were accurately mapped hundreds of years before the interior lands were mapped. Most of the early European arrivals who first settled and lived in the New World lived along the coast or near rivers. French missionaries, fur traders, and explorers traveled great distances by canoe and other boats, eventually settling much of the St. Lawrence River Valley, the coast of the Great Lakes, and down the banks of the Mississippi River all the way to the Gulf of Mexico.

Because early European settlements in North America were colonies of European countries and the colonial economies were based on the extraction of natural resources, transportation of goods between the eastern seaboard and Europe was vitally important. Hudson Bay and the St. Lawrence River Valley housed major transportation ports where boats were loaded with exports destined for Europe. Canadian coastal waters provided many commodities that were in high demand in Europe. Whale products provided millions of tons of oil,

while fish provided a good source of protein. Inland, settlers gathered furs, fell trees, and produced agricultural products that were sent to Europe. Ships came back to North America carrying European manufactured goods and new settlers.

In the eighteenth and nineteenth centuries, many canals were built to transport goods in areas that did not have navigable rivers. Boats and barges were used to transport materials. Many canals linked major waterways; for example, the Erie Canal linked the Atlantic Ocean with the Great Lakes, and the Delaware and Hudson Canal carried coal from Pennsylvania's mines to the New York City market. Steamboats opened the era of modern ocean transportation. They also played an important role in the development of the Midwestern and Southern US states drained by the Mississippi River system—the Mississippi River and its many tributaries, the largest of which are the Missouri and Ohio rivers. In Mexico, however, the almost total absence of navigable rivers, lakes, and canals deprived the country of the advantages of inexpensive inland transportation and had presented a major impediment to early industrial development.

Large boats or freight ships still transport heavy goods on inland waterways and along ocean coasts; maritime shipping is the main means of the transportation of goods overseas. Many large ports are located along the Atlantic and Pacific coasts of North

RMS Queen Mary *in Long Beach Harbor. (Mfield)*

America. In the late 1950s the system of canals and channels on the St. Lawrence River was developed into the deep-draft Great Lakes-St. Lawrence Seaway. This enabled large seagoing vessels to transport goods from the Great Lakes to anywhere else in the world. Many of North America's manufactured goods were made in the Midwest; thus the Great Lakes and St. Lawrence Seaway became important assets in North America's economy.

At the turn of the twenty-first century, Los Angeles became North America's largest container port, with almost 5 million TEUs (20-foot equivalent units, a prevailing measure of container cargo volume) handled every year. The Port of South Louisiana handles the largest amount of shipping, in tonnage, of all North American ports. It ships some 60 percent of all grain exports of the United States. In 2018, the list of the twenty-five largest ports in North America included seventeen US ports, four Canadian ports, and four Mexican ports.

Ships also carry passengers. Cruise boats transport millions of North Americans every year. Popular destinations for both Canadian and US passengers include the Caribbean islands, the Mexican Riviera (the western coast of Mexico), and the coast of British Columbia and Alaska. Canadians take cruises to New York and Los Angeles. The Port of Miami (styled as PortMiami) is the busiest cruise port in the world. In some areas, ferries provide a convenient form of transportation for people. Washington, British Columbia, and Alaska have a series of coastal ferry routes that enable people to travel from one place to another. In northern Quebec and Nunavut, ferryboats in the summer months similarly allow people to travel from one island to another.

Railroads

The development of the railroad by the mid-nineteenth century drastically changed transportation patterns throughout North America. Between 1850 and 1920, hundreds of thousands of miles of rail track were laid down in North America. The railroads improved upon what the canals had tried to do: they connected natural resources with manufacturing plants and final products with the market-

place and consumers. The joining of the Union Pacific rail line with the Central Pacific line in May 1869 at Promontory Summit, Utah, created a transcontinental railroad linking the United States from east to west, from the Atlantic to the Pacific. Canada's first transcontinental railroad, the Canadian Pacific Railway, was opened in 1885.

The United States, Canada, and Mexico are all vast countries, and large territories in all of them were still very sparsely settled in the mid-nineteenth century. Railroads provided them much closer ties with the rest of the nations than ever before. Towns began to be settled all along the railroad tracks, growing the population in the western states and provinces. As more people moved to and settled the American West, more natural resources were discovered, which in turn brought more prospectors, manufacturing plants, and eventually more settlers. And with the invention of the refrigerated boxcar, perishable agricultural products could be easily and quickly transported from prime agricultural areas to the eastern seaboard, where most of the population lived.

The railroad affected the geography and society of North America in other ways. For the first time, people could travel the vast continent relatively quickly. Towns that in the past had been secluded and remote could now get news and information more quickly. People started to demand standardized items, so that if something they had purchased in Boston broke while they were in North Dakota, they could find similar parts with which it could be fixed. Mail-order retailing became a big business.

The spread of railways throughout the United States helped force the national standardization of time zones. Clocks in various cities were also synchronized within the four time zones used today—eastern, central, mountain, and Pacific—so that trains could run on schedules that everyone could use. Many people began to travel, and national parks were established in both the United States and Canada to protect and preserve unique geographic wonders and cultural treasures.

The proliferation of railway lines throughout the eastern and southern United States by the mid-nineteenth century played a major role in the

An electric Amtrak train led by an ACS-64 locomotive running through Maryland on the Northeast Corridor. (Ryan Stavely)

Civil War. The speed and ease with which trains moved troops and equipment helped make the Civil War what many have called the first modern mechanized war. Indeed, such was the importance of the railways that trains, tracks, and railway bridges themselves became targets of military assaults.

Railways also played a major role in Mexico's early twentieth-century revolution. Both rebels and government troops used them to move across great distances, thereby spreading the fighting through a large part of the country.

Decline and Resurgence of the Railroads

In the face of new competition from automobiles, trucks, and airplanes, the twentieth century saw a decrease in the total mileage of railroad tracks throughout the United States and Canada. In some instances, these old, deserted lines have been changed into rails-to-trails. These are long pathways that provide a means of recreational transportation where people can hike or ride bikes or horses.

In Mexico, passenger rail service has been suspended indefinitely since 1997, in the process of privatization of the country's state-owned railroad company.

North American railroads have experienced a resurgence in the twenty-first century. Technological progress led to enhanced speed and cargo capacity, and freight rail became much more efficient and cost-effective. The previously built rail lines still in use have heavy traffic loads again. At the same time, significant investments are being made into additional track, modern rail yards, bridge and tunnel improvements, and intermodal (rail, ship, and truck) terminals resembling inland ports.

Short-distance commuter rail lines have never lost their popularity in densely populated areas. Metropolitan areas of New York City, Boston, Baltimore, Chicago, San Francisco, Washington, D.C., Toronto, and Vancouver have busy commuter rail lines that transport people to and from work. Mexico City's suburban rail system began operation in 2008. All three North American mainland countries have plans to develop high-speed passenger rail services.

The United States has the largest and most extensive rail network in the world, totaling some 140,000 miles (225,000 km.) of railroad tracks; railway lines for freight transportation account for about 80 percent of the network. Canada has the world's fifth-largest railroad system, with over 32,000 miles of track (51,500 km.).

Automobiles

In the twentieth century, transportation patterns in North America were affected profoundly by the automobile. Because of this new mode of transport, people and goods were no longer tied to the routes dictated by the railroad but could travel whenever and wherever they pleased. To make it possible, however, many more—and much better—roads were needed than North America had at the time. Overland routes by which settlers had traveled to the American West were called trails for a reason. In the nineteenth century, there were only a few high-

way-quality roads of nationwide significance in North America, such as the National (or Cumberland) Road in the United States. Within years after the beginning of mass production of automobiles, many of the existing wagon roads throughout the United States and Canada were upgraded to automobile roads, and many new ones were being built.

Today, the United States, Canada, and parts of Mexico have well-developed, extensive road systems that allow people and goods to be transported quickly, easily, and relatively cheaply. Americans, especially, place strong importance on their automobiles and their mobility.

Henry Ford developed the relatively inexpensive Model T and the idea of the assembly line, which quickly made the automobile accessible to the majority of the population, not just the wealthy. In less than two decades, the United States became a motorized country. By the Great Depression of the 1930s, the automobile industry was the largest consumer of raw materials and oil in the United States.

The Harbor Freeway is often heavily congested at rush hour in Downtown Los Angeles.

After World War II, North America grew more prosperous, with two-thirds of all the world's goods being produced in the area from western Pennsylvania to St. Louis and north into Ontario around the Great Lakes. As North Americans became wealthier, they bought more cars. In 2017, 24 percent of US households had three or more vehicles. There were 838 vehicles (cars, vans, buses, and trucks) per 1,000 people in the United States, 685 in Canada, and almost 300 in Mexico. In Mexico, the number of cars per 1,000 people doubled between the years 2000 and 2011.

The use of the automobile and the transportation of people and goods in the United States were enhanced by the development of the national Interstate Highway System. The idea was developed in the 1930s, but construction did not begin until the 1950s. A federal gasoline tax provided 90 percent of the funds for the highways, and each state had to provide the remaining 10 percent of the cost. In the following decades, tens of thousands of miles were constructed.

The Interstate Highway System is relatively evenly distributed geographically. Roads were not concentrated where the population lived or the manufacturing plants existed; the goal was to make the highway a national transportation system. The Interstate Highway System accounts for only a small fraction of the total number of road miles in the United States (just over 1 percent), but it carries nearly one-quarter of all roadway traffic. This makes it a major contributor to the nation's economy in general and to the vehicle tourism industry in particular.

Two other highway systems are noteworthy in North America. First is the Alaska (or Alaska-Canadian) Highway. It was built during World War II when the United States was concerned that the Japanese might invade Alaska. The second is the Trans-Canada Highway, which was completed in 1962. It links St. John's, Newfoundland and Labrador, to Victoria, British Columbia, cutting across lower Canada and connecting Canada's major cities.

Mexico has had difficulty in building an integrated highway network because the country is crisscrossed by numerous and high mountain ranges. While all significant economic and population centers are incorporated in the national highway system, thousands of small rural communities do not have paved or surfaced roads. Inadequate roads have prevented large areas of Mexico from participating in the country's economic development. It is mainly because of transportation problems that Mexico has far fewer large cities than the United States or Canada.

Trucking

Trucks also have benefited from the extensive road systems. In cities, trucks have dominated freight transportation since the 1920s. Trucks enable manufacturers and retailers new geographic options for setting up shop. In the past, transporting goods and materials away from a central delivery point such as a factory, rail depot, or a port was expensive. Many manufacturers and retailers tried to locate themselves close to the downtown to minimize transportation costs. Trucks allowed people and retailers to move to the suburbs where more open space and cheaper land were available. Automobiles and trucks have contributed to a major change in residential and commuting patterns—the explosive growth of suburbia and long-distance commuting.

About 70 percent of all Canadian trucking is cross-border, in and out of the United States. The wages in Canada's trucking industry, although higher than the median wage, have not kept up with the rapid increase in the cost of living in Canada, causing driver shortages. In the United States and Mexico, domestic delivery of essential goods and materials is done largely by trucking. The vital role of the trucking community in the national economies is one of the reasons why truck driving is a well-respected profession in both countries. Trucks also carry about 70 percent of goods between Mexico and the United States. Every day, more than 30,000 trucks cross the northern and southern US borders.

Urban Transportation

Urban transportation in North America is generally well-developed and diversified, but in the

United States, urban public transportation has become increasingly outdated and strained in the twenty-first century. With the country's long reliance on the automobile as the main transportation mode, public investment in urban transportation infrastructure in US cities has lagged behind that of Canadian and Mexican cities. Nevertheless, over 20 percent of US urban residents use public transit on a regular basis, and the number is much higher in the large urbanized areas of the Northeast—New York City, Chicago, Boston, Philadelphia, and Washington, D.C.

Buses and subways (also known as rapid transit or heavy rail) are the main forms of urban transportation in North America. The extensive Mexico City Metro system is North America's second-largest rapid transit network after the New York City Subway. In the 2010s, the average daily subway ridership in Mexico City was close to 5 million; in New York, it approached 5.5 million. Canada, Toronto, and Montreal also have extensive subway systems, while cities like Edmonton, Calgary, and Vancouver rely on light-rail transit. In 2017, public buses supplied some 5 billion rides in the United States alone, but that huge number is down 11 percent since 2000.

Aviation

Airplanes became a common form of transportation for both goods and passengers in the second half of the twentieth century. Canada, Mexico, and the United States have large cities spaced far apart. The three countries make up about 6.5 percent of the world's population, but they create over 22 percent of the world's air traffic (2018 data).

In 2018, the United States accounted for eighteen of the twenty busiest (in terms of passenger traffic) airports in North America; the remaining two were in Toronto, Canada, and Mexico City, Mexico. The busiest US airports are located in Atlanta, Chicago, Dallas-Fort Worth, Los Angeles, Denver, New York City, and San Francisco. Since 2000, Hartsfield-Jackson Atlanta International Airport has been the busiest airport in the world; it is also home to the world's largest airline hub, for Delta Air Lines.

Altogether, there are more than 20,000 airports in North America. In Mexico, all cities with more than 500,000 inhabitants have an airport. Due to the popularity of numerous Mexican locations as tourist destinations, Mexico has more international airports than Canada, but they are much smaller than Canadian airports. In sparsely populated northern and western parts of the North American continent—such as Alaska, the Yukon, and Montana, where little towns are spaced far away from larger cities—many individuals own and fly private airplanes, serving their own needs and those of small remote communities.

Pipelines

An important form of transportation of goods is often ignored because it is out of sight and usually buried under the ground: the transportation of crude oil, oil products, and natural gas via pipelines. Most US underground pipelines originate at the source of the oil and gas in the south-central United States, in places like Texas and Louisiana, and travel underground to the Northeast, where most of the consumers are located. The exception to this is in Alaska, where the pipeline is above the surface and travels from Prudhoe Bay southward to the seaport of Valdez.

Most of the Canadian pipelines begin in the oil-rich province of Alberta and go west to British Columbia, north to the Northwest Territories, south to Texas, and east to Quebec. Part of US natural gas exports goes to Mexico from Texas via a pipeline; several more gas pipelines between the two countries were under negotiation or construction in 2019.

Alison Philpotts

TRANSPORTATION IN THE CARIBBEAN

Through most of the history of the Caribbean island region, its islands have been linked more closely with the outside world than with each other. Although this changed somewhat in the late twentieth century, as air transportation began moving tourists among the islands, other kinds of transportation links among the Caribbean Islands have remained weak in the twenty-first century. This is especially true for inter-island sea transportation. The improvement of inter-island connections is the greatest transportation challenge facing the region.

Since World War II, many Caribbean islands have been integrated into international transport systems. As the Caribbean emerged as a major tourist destination, especially for North Americans and Europeans, improved transportation infrastructure was essential to accommodate the millions of tourists who visited it annually. Many islands are served by airlines from North and South America, Europe, and Asia. New and improved airports and other transportation infrastructure were built on most of the islands, and port and harbor facilities were expanded to accommodate the increasing number of cruise ships that called at Caribbean ports. Air transport moves not only tourists but, increasingly, imports and exports. The agricultural sector in the Caribbean has been contracting for decades, and most food products are now imported. They are delivered both by sea and by air, as are manufactured goods produced in the Caribbean export processing zones (EPZs).

Most island residents are more dependent on bus transportation than are North Americans. On most islands where private motor vehicles are scarce, well-developed urban and intercity bus systems move people quickly and efficiently. Bus service in rural areas, however, usually has been inferior to urban and intercity transit. Nevertheless, in terms of cost and reliability, public transportation in many Caribbean countries is superior to that in much of North America. Trinidad and Tobago, Barbados, and the Dominican Republic serve as good examples.

On most Caribbean islands, an extensive road network with good infrastructure is a legacy of the time when plantation agriculture was the main economic activity and source of export earnings. Plantation agriculture had been established in the colonial period; it has been dominated by companies

Oasis of the Seas entering the port at Nassau, Bahamas. (Baldwin040)

from former colonial powers ever since. Roads were built to deliver sugarcane, bananas, coffee, and other cash crops to ports for export.

A few countries, notably Haiti, have inadequate road systems; road infrastructure has been underdeveloped due to political instability. Many of Haiti's roads are impassable during the frequent torrential rains. On many major roads, bridges have not been constructed, forcing vehicles frequently to ford streams. Many small interior towns and villages can be reached only by foot, often over rugged terrain. Such inadequate transportation systems contribute to widespread poverty.

In most Caribbean countries, rail systems have not been well developed and on many of the small islands, never constructed. During the nineteenth century, rail lines were built to serve the mines and sugarcane plantations in several of the larger island countries and territories, including Cuba, Puerto Rico, Jamaica, Haiti, and the Dominican Republic. In some of them, these lines have remained important in the twenty-first century. In the Dominican Republic, trains move half of the sugarcane from the fields to the mills. A rapid transit system, Santo Domingo Metro, opened in the capital in 2008. Since 2007, the government of Cuba has been ordering new rolling stock (locomotives and coaches) from China for the country's old, state-run railway system. Railway service on the island has markedly improved in reliability; the railway network has been expanding.

Like all island regions, the Caribbean region is highly dependent on sea transport for the movement of goods. The major ports in the region are served by a diverse array of ships and have modern port facilities. Most ports handle both passenger traffic and cargo shipments such as containers. Caribbean ports that handle the largest cargo volumes include Freeport (Bahamas), Caucedo (Dominican Republic), Kingston (Jamaica), and San Juan (Puerto Rico). Some of the most popular ports of call for cruise ships are Nassau (Bahamas), Bridgetown (Barbados), Castries (Saint Lucia), Montego Bay (Jamaica), and Havana (Cuba). A new, modern cruise ship terminal opened in Falmouth, Jamaica, in 2019. It can handle the same number of passengers per week as the town's population—about 9,000.

Jamaica Logistics Hub has been planned by the government of Jamaica as the fourth global logistic connection point on a scale similar to Dubai, Singapore, and Rotterdam. The implementation of the Global Logistics Hub Initiative started in 2016 with the upgrading of the Kingston container terminal. The master plan was unveiled in 2018. The ambitious project involves railroads, airports, a special economic zone, warehousing facilities, and more. Strategically located in the center of the Caribbean island region, between the Panama Canal and the huge North American market, the country is well-positioned to accomplish its goal of becoming a major transshipment point for larger cargo ships.

Robert R. McKay

DISCUSSION QUESTIONS: TRANSPORTATION

Q1. What essential roles did the development of railroads play in the westward expansion of settlement in the United States and Canada? What aspects of everyday life were transformed by the railroads? Why did the railroads ultimately decline?

Q2. How has the automobile transformed North American society? How did the development of the Interstate Highway System impact the nation's economy? What is the role of trucking, and how has it enhanced commerce among the United States, Canada, and Mexico?

Q3. What role does mass transit play in North American urban society? What is the difference between subways and light-rail systems? Why does North America have (or need) so many airports?

TRADE IN NORTH AMERICA

Canada, the United States, and Mexico make up the political and economic entities of mainland North America. European explorers and colonists opened up the continent to world trade in the fifteenth and sixteenth centuries. Before the arrival of Europeans, numerous and diverse indigenous Native American tribes both warred and traded with each other. That complex and fragile world was destroyed in the process of colonial conquests and land grabs through warfare, disease epidemics, the loss of food sources, and the resulting sharp decline of indigenous populations.

Canada, the largest of the North American countries, with an area of 3.8 million square miles (9.85 million sq. km.), has the smallest population—approximately 38 million inhabitants in 2020. The first major wave of European settlers in Canada came from France, and later, Britain, which took over control of the French colony in 1763. Subsequently, independent Canada retained close cultural, political, and economic ties with Europe. Historically, Canada's economic development was strongly dependent on the export of fur, fish, grains, and timber, and trade has always played a central role in the country's economic history. By the 1950s, Canada had become a highly industrialized country; the share of raw materials in its exports markedly declined in favor of processed materials and manufactured goods. The trade and economy of Canada have become tightly aligned with those of its closest neighbor, the United States.

The United States, although slightly smaller in area than Canada, possesses a broader spectrum of natural resources and a more temperate climate. Accordingly, during the period of European settlement and westward expansion, it attracted more immigrants from a greater number of countries than did its northern neighbor. The United States developed as a microcosm of much of the rest of the globe, a world in itself. In the early days of the nation's history, its government and business community mostly concentrated on developing the domestic economy and internal commerce. Many times throughout its history, the country has had a strong impulse toward trade protectionism, using tariffs, quotas, and other government regulations to limit imports of foreign goods in order to boost native industries and agricultural production. Only after the Great Depression of the 1930s, and especially after World War II, did it start to approach open trade as an important tool for building peaceful and systematic relations among nations as well as a means of advancing its own economic interests.

Colonial Mexico was part of New Spain, a huge colonial territory of the Spanish Empire. The Spanish rulers envisioned New Spain as a supplier of wealth to the Spanish crown. During the colonial era, Mexican silver accounted for over three-quarters of all colonial exports to Spain. Cochineal, a natural dye, was Mexico's second-largest export. Since the entire Mexican economy was based on these two export products, there was no national market in Mexico at the end of the colonial era, and regional markets were poorly developed. When Mexico became an independent country in 1821, its economy and trade ground to a halt. It would take decades for the non-mining-related economic sectors to develop in the post-colonial period.

As the only three countries in mainland North America, Canada, Mexico, and the United States have had a long and varied history of bilateral and, in recent decades, trilateral trade relations.

Trade and Export Products

For the better part of the twentieth century, oil was Mexico's primary export. Oil production in Mexico is a government monopoly. The state-owned company Petróleos Mexicanos, commonly referred to as

PEMEX-Refinery Tula de Allende, in Hidalgo, Mexico. (Presidencia de la República Mexicana)

PEMEX, handles the exploration, production, and distribution of petroleum products throughout the country, as well as sales to foreign customers. In 1980, oil accounted for almost 62 percent of the country's GDP. By 2000, total Mexican exports had more than doubled and the share of oil dropped to 7.3 percent. Oil's contribution to Mexico's GDP had fallen to 1.72 percent by 2017.

Mexico produces a variety of agricultural products, many for export, such as corn, cotton, wheat, beans, coffee, wine and beer, and snack foods. The country is the world's leading exporter of tropical fruit and tomatoes; the latter go almost exclusively to the United States. Various metals and other minerals are mined and traded as well, including the traditional silver (of which Mexico remains the world's largest exporter), copper, gold, lead, and zinc. Minerals such as fluorspar, graphite, and strontium are critical for US industries.

Besides the United States, Mexico's major trading partners are Canada and Germany. In terms of manufacturing and commercial agriculture, Mexico has a critical competitive advantage: labor costs there are lower than in the United States, Canada,

and Europe. In addition to agricultural products, oil, and mining products, Mexico exports many manufactured goods such as vehicles and vehicle parts, electronics and machinery, and optical and medical instruments.

In 2019, almost half of Mexico's imports originated from the United States and Canada; Asian countries accounted for some 36 percent of import purchases, and about 12 percent were purchased from European countries. The top imports were electrical machinery and equipment, computers, vehicles, plastics, optical and medical apparatus, articles of iron and steel, and organic chemicals.

Canada
In recent decades, Canada's principal export products were oil and related energy products, vehicles and machinery, precious stones and metals, aluminum, plastics, aircraft, assorted wood products, precision instruments, iron and steel articles, and furniture. The global decline in oil prices that began in mid-2014 significantly reduced export earnings from energy products, and annual receipts from automotive exports have been consistently

larger than those from energy products. Other non-energy exports that have also recorded notable gains in recent years include food products, pharmaceuticals, building and packaging materials, and communications equipment.

The United States, China, Japan, the United Kingdom, Germany, Mexico, and South Korea are Canada's major trading partners for both exports and imports. Much of the overall growth in Canadian exports following the global recession of 2008–2009 was due to higher shipments to the United States, the United Kingdom, and non-EU countries, most notably China. Exports to China have risen to about one-half of the value of Canada's exports to the European Union.

United States

The United States is among the world's largest producers of manufactured goods. Having a wide variety of natural resources, it rarely depends on raw materials from the rest of the world in its manufac-

turing sector, except for some rare minerals and oil. In the production of oil, the US domestic output, while being very large, does not meet all of the country's industrial needs. Some grades of US oil cannot be used for specific manufacturing purposes, and these are exported to other countries. Also, the United States has more oil-refining capacity than it needs, so heavy crude oil is imported and refined oil products exported. For these and other reasons related to supply, demand, transportation costs, and world market prices, the United States has been both the exporter and importer of oil and oil products.

Several major trading partners of the United States, most notably Japan and China, have consistently shipped more to the United States than they received from it in terms of exports. This unfavorable balance of trade resulted in Japan and China holding the largest—among all foreign countries—amounts of US public debt, which is held in the form of US Treasury securities.

Ford's Oakville Assembly in the Greater Toronto Area. Central Canada is home to several auto factories of the major American and Japanese automakers. (Whpq)

Boeing plant in Ridley Park, Pennsylvania. (Smallbones)

For much of the twentieth century, Canada, Japan, and Western European nations were the country's top trading partners, but since the beginning of the twenty-first century, China has eclipsed Japan, and the North American Free Trade Agreement has moved Mexico up among the leaders in exporting to the United States. In 2020, Mexico for the first time became the largest trading partner of the United States, Canada was the second-largest, and China slipped from first to third place, but the gap among all three was quite small. Japan and Germany remain in the top five.

The largest US import categories include passenger cars, cell phones, and pharmaceuticals. The top exports of the United States include refined and crude oil, aircraft, automobiles, and instrumentation. The country also exports a wide variety of other industrial and agricultural products—from computers and plastics to soybeans, corn, and rice.

Overall, the United States has had a large and growing deficit in foreign trade for over forty years, exclusively due to an imbalance of trade in goods.

In the other major component of trade, which is services, the country runs a substantial and increasing trade surplus. This component of trade, which includes financial, professional, and technical (including digital) services, entertainment, license fees, royalties and franchises, real estate-related services, transportation, tourism, and related services, has given a major impetus to America's economic growth and job creation.

The United States is the largest single-nation exporter and importer of private-sector (commercial) services in the world. In 2018, for example, the three leading European economies—the United Kingdom, Germany, and France—together captured 17 percent of the world services market while the United States, alone, contributed 14 percent of global services. A bigger share of US trade is tied to services than that of most other industrialized nations. A large proportion of services exports occur through the commercial presence of US companies in foreign markets, in the form of majority-owned foreign affiliates.

The North American Free Trade Agreement and the United States-Mexico-Canada Agreement

Trade among the three North American countries was hampered for many decades by government regulation. Duties of various types were imposed by the respective governments for two purposes: those duties, or tariffs, produced revenue for the governments of the countries that imposed them, and taxing imports on incoming goods provided an economic shelter for domestic producers by making imports less competitive with local products. These and other protectionist measures, such as import quotas, government subsidies to domestic producers, and import licensing, significantly slowed the movement of goods across North American borders.

In 1990 the governments of the three North American continental countries, recognizing the need for a trade-reform program, began to plan a comprehensive free trade agreement. Realizing the challenge that the already well-integrated European Union represented to outsiders and the limitations that the Japanese imposed on their trading partners with regard to incoming goods, this new alliance of North American countries was aimed at creating a competitive trading bloc and promised expanded trade horizons within the continent itself. On December 17, 1992, the leaders of the United States, Canada, and Mexico signed the North American Free Trade Agreement (NAFTA). After it was ratified in each country, NAFTA came into force on January 1, 1994. The agreement stimulated the construction of many new factories and encouraged the formation of business partnerships across the national borders. As a result of NAFTA, average tariff rates among the three countries dropped to only 2 percent on goods shipped.

Some segments in the industries of all three participants had reservations about the pact. Labor in both the United States and Canada feared the loss of jobs to Mexico's lower wage scales. While Canadians' fears of job loss did not materialize, their hopes

President Donald J. Trump is joined by Mexican President Enrique Pena Nieto and Canadian Prime Minister Justin Trudeau at the USMCA signing ceremony Friday, Nov. 30, 2018, in Buenos Aires, Argentina. (Shealah Craighead, The White House)

that NAFTA would help minimize the gap in industrial productivity between Canada and the United States were not realized either. The United States, however, experienced net job loss in manufacturing during the NAFTA years, but its causes were more complicated than just NAFTA. In the late twentieth century, the United States was already transitioning from the industry-based economy to the services-based economy. While some factories did move from the United States to Mexico, much of the manufacturing job loss during the NAFTA years can be attributed to automation and the general de-industrialization of North America, the same processes that also took place in Europe and were a result of globalization and the shift of heavy industries and manufacturing to Asia and other parts of the world where production costs were the lowest.

Implementation of the agreement in Mexico was not without controversy either. The expansion of factory facilities in northern Mexico, which became an economic extension of the United States, had serious repercussions with regard to the standards of living in that area. Increased demand on local water resources, as a result of the influx of waves of new workers, caused northern Mexico's water table to drop. Many of the new workers' communities that sprung up since the upswing in trade among the three countries were without running water or electricity. Adequate new housing construction lagged. The demand for heating fuel strained the region's capability to deliver resources. Industrial pollution in the border area increased. For Mexican farmers and rural residence in general, the increase in export-oriented agricultural production led to poorer working and living conditions, as Mexico did not invest in the infrastructure that is necessary to support large-scale agribusiness.

The movement of contraband goods, already endemic in the border area, increased as legitimate trade grew. The substantial increase in the number of incoming trucks led to the use of some of them as conveyances for illicit cargos, primarily for illegal drugs but also for human trafficking. To preclude the entry of illegal cargo, each truck entering the United States from Mexico would have to be carefully searched. Such close supervision at the border was impossible due to the sheer volume of traffic.

For Canada, one of the adverse consequences of NAFTA resulted from the agreement's energy proportionality provision, under which Canada had to export large volumes of oil, gas, and electricity to the United States instead of using them for domestic needs.

A new trilateral agreement, replacing NAFTA, was negotiated in 2017-2018 and entered into force as of July 1, 2020. In the United States, it is officially called the United States-Mexico-Canada Agreement (USMCA); in Canada, it is referred to as the Canada-United States-Mexico Agreement (CUSMA); and in Mexico, it is known as the Tratado entre México, Estados Unidos y Canadá (T-MEC). The new agreement is more than an "improved NAFTA." It is intended to create more balanced trade, beneficial to broader segments of economies and societies of the participating nations. It contains specific provisions supporting the creation of higher-paying jobs, creating incentives for automobile manufacturers to make cars and car parts in North America, protecting intellectual property, regulating trade in services, and covering digital trade, anticorruption, labor violations, and good regulatory practices. And unlike NAFTA, it contains a sunset clause that requires the participating countries to review the agreement every six years. If any of them decides not to continue with the USMCA, it will expire upon the end of a subsequent sixteen-year term.

Carl Henry Marcoux

TRADE IN THE CARIBBEAN

Caribbean economies are small and linked in many ways to the economies of the industrialized countries that trade with them. In spite of their proximity to the large markets of the United States and Canada, the international competitiveness of Caribbean exports is impeded by their small size, high production costs, and significant transport costs.

During the 1960s, Caribbean sugar and banana exports commanded high prices in the world markets. In the 1970s, world oil prices soared to unheard-of heights and caused deep recessions in most industrialized nations. These recessions in the 1970s and 1980s led to a decrease in demand and falling prices for the Caribbean's chief exports—sugar, bananas, coffee, citrus, bauxite, and aluminum. As a result, the volume of trade between the Caribbean and the industrialized nations dropped. At the same time the Caribbean region's exports dropped, and their imports increased, causing a trade deficit. By the end of the twentieth century, light manufacturing and component assembly for export started generating more export earnings in some Caribbean economies than traditional exports. But most Caribbean island nations have moved directly from the primary-sector-led economies to those based on tourism-related services and—in the case of the Bahamas and the British overseas territory of the Cayman Islands—services related to offshore banking.

Most Caribbean countries import manufactured goods and most of their food and energy supplies. Trinidad and Tobago, Barbados, and Puerto Rico, for example, import more than three-quarters of their food needs. International prices and terms of trade are determined by nations outside the region. Transnational corporations provide the marketing and refining of raw materials; multinational hotel chains dominate the tourism and hospitality sector. The manufactured goods imported into the Caribbean region are relatively expensive because of shipment costs and the low purchasing power of large segments of Caribbean populations.

The United States is the Caribbean region's most significant trade partner. China, Canada, Mexico, the European Union (EU), and neighboring Caribbean nations also play an important trade role. The French dependencies in the Caribbean—Guadeloupe, Martinique, and Saint Martin—are part of the EU. Many Caribbean nations have preferential trade agreements and economic partnership agreements with the EU. Without these guaranteed European markets, many Caribbean countries could not compete internationally.

The United States is not a trading partner with Cuba. After the Bay of Pigs invasion (1961) and the Cuban missile crisis (1962), the United States placed an economic embargo on Cuba and prohibited US citizens from traveling to the island. While the travel ban has been somewhat relaxed in recent years, the trade embargo has remained in place.

Puerto Rico enjoys a special trade relationship as a self-governing commonwealth of the United States. Goods produced in Puerto Rico are not subject to tariffs on entering the United States. Puerto Rico is one of the world's largest producers of rum

Cigar production in Santiago de Cuba. (BluesyPete)

but imports sugar, tobacco, and coffee—items that once were exports. Much of Puerto Rico's food is imported from the United States.

Not all trade in the Caribbean region is legal. Marijuana, called ganja in Jamaica, was introduced by Hindu East Indians in the nineteenth century who used it in religious rituals. Most ganja is grown by small farmers in Jamaica; that country is a significant supplier of marijuana to the United States. The Caribbean region serves as a channel through which South American drugs such as cocaine pass. The Bahamas and some other Caribbean island nations have become transshipment centers for illegal drugs headed for US markets.

Caribbean Community and Other Regional Economic Organizations

The Caribbean Community (CARICOM), an association of Caribbean islands, was formed in 1973 in an effort to create a common market and wield more influence in world markets. Until 1995, CARICOM was made up of thirteen former British colonies: Antigua and Barbuda, the Bahamas, Barbados, Belize, Dominica, Guyana, Jamaica, Grenada, Saint Kitts and Nevis, Saint Vincent and the Grenadines, Montserrat (the only full member that is not an independent country but a British overseas territory), Saint Lucia, and Trinidad and Tobago. In 1995 Suriname became the first non-British island to become part of CARICOM. Haiti joined in 2002; the Dominican Republic, Puerto Rico, Mexico, Colombia, Venezuela, and some of the Dutch dependencies have observer status.

After several decades of incremental success, CARICOM has focused its strategy for achieving economic integration on the implementation of a multiphase initiative called the Caribbean Single Market and Economy (CSME), formally established in 2006. Its ultimate goal is the integration of all member states into a single economic unit. The economic integration that the Caribbean region seeks is impeded by water, distance, the relatively small size of its market, and the political, cultural, and economic diversity among the Caribbean island nations and territories.

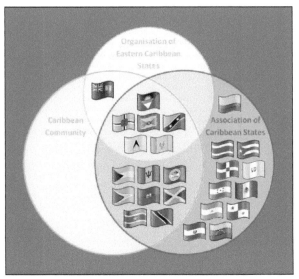

A Venn diagram depicting the relations between the ACS, CARICOM, and OECS countries. (Wdcf)

Smaller groupings of similar countries may have better chances of achieving economic integration sooner than a large and diverse grouping such as CARICOM or an even larger Association of Caribbean States (ACS). Created in 1994, the organization includes both the Caribbean island nations and those mainland countries of the Americas that border the Caribbean Sea. The ACS serves mainly as a forum for political dialogue.

A still smaller regional group, the Organization of Eastern Caribbean States (OECS), consists mostly of former British colonies, which have close historical, cultural, and economic ties and are in close geographic proximity. Its seven full members (Antigua and Barbuda, Dominica, Grenada, Montserrat, Saint Kitts and Nevis, Saint Lucia, and Saint Vincent and the Grenadines), along with the four associate members, form a continuous archipelago across the Leeward and Windward Islands. The organization was created in 1981 by the Treaty of Basseterre, named after the capital of Saint Kitts and Nevis. The OECS created the Eastern Caribbean Central Bank, and the Eastern Caribbean dollar became the common currency of all seven full members and one associate member. In 2010, member states signed the Revised Treaty of Basseterre, which established the OECS as an economic union.

Carol Ann Gillespie

COMMUNICATIONS IN NORTH AMERICA

Since the American Industrial Revolution, continental North America has played a critical role in the development and deployment of innovative communications systems and technology throughout the world. When Europeans began to colonize the New World, messages traveled only as fast as humans could move on horseback or aboard ships. In the twenty-first century, messages are delivered worldwide at essentially the speed of light.

Print Communication

The first book printed in North America was issued in Mexico City in 1539 by a print shop affiliated with one of the leading printing houses of imperial Spain. The first printing press in the British North American colonies was set up a century later, in 1638, at Harvard College. Presses such as the ones at Harvard and Mexico City produced religious materials as well as educational pamphlets and books.

The first newspaper in North America was published by Boston printer Benjamin Harris in 1690. Newspapers became a political force during the fierce debate over American independence, and have remained on the political forefront ever since. In 1865, fifty-four newspapers were published in New York City alone. In the first half of the twentieth century, as many as seventeen offices of various newspapers lined just one section of Boston's Washington Street, which became known as Newspaper Row.

Since the beginning of the digital revolution in the last decades of the twentieth century, the number of individual newspapers published in North America has declined, but those that endured have largely kept their role in the formation of public opinion. In the 2010s, US newspapers and their websites still reached up to 75 percent of opinion leaders—people who consistently vote in local, state, and national elections. In Mexico, newspapers have continued to play an important role in exposing organized crime and political corruption.

In an era when people get news instantaneously, and from multiple sources, newspapers have

Publick Occurrences Both Forreign and Domestick was the first multi-page newspaper published in the Americas.

changed their focus from news reporting to feature stories, analysis of political events, and interpretation of the news. As a result of the adoption and proliferation of digital computing and advanced telecommunication technologies, overall newspaper readership has been in decline. Online editions help maintain the core readership but have failed to increase it. By 2020, website traffic has leveled off after some years of growth.

In 2020, the largest US newspapers were *USA Today*, *The Wall Street Journal*, *The New York Times*, *New York Post*, *Los Angeles Times*, and *The Washington Post*. Ranked first by circulation, *USA Today* was a middle-market newspaper, that is, a publication that caters to a readership that likes entertainment as well as news coverage.

In Canada, three of the ten largest newspapers—*The Globe and Mail*, *The National Post*, and *The Toronto Star*—were all published in Toronto, Ontario, and three others—*Le Journal de Montreal*, *La Presse*, and *Montreal Gazette*—in Quebec City, the capital of Quebec. In Mexico, *Excélsior*, *La Jornada*, *Reforma*, *El Universal*, and *El Financiero* were among the most influential newspapers. Sensationalism characterized the biggest-selling Mexican dailies such as the tabloids *La Prensa* and *El Grafico*.

In 1741, Andrew Bradford published the first magazine in the American colonies, beginning a tradition of providing a compilation of essays and articles for readers. By the end of the American Civil War in 1865, compulsory education had helped to increase the literacy rate in the United States, and railroads had improved the mail system. As a result, the US magazine became a staple of popular culture. By the beginning of the twentieth century, political and social reformers were using magazines as a means of communicating their ideas about social change. Magazine journalists used their forum to expose corruption and scandal in US society. This new approach to journalism was called "muckraking."

In the years that followed, the magazine continued to develop as a form of mass communication. Ranging from comic books to trade newsletters, the magazine has become a critical means of reaching segmented audiences. By the end of the twentieth

In 2016, The Globe and Mail *moved its headquarters to the Globe and Mail Centre on King Street East in Toronto, Ontario, Canada. (Stephen Andrew Smith)*

century, the role of political magazines declined; magazines with the highest circulation were those belonging to such narrowly focused special-interest categories as sports, fitness, fashion, lifestyles, or finance. This was also true for Canada and Mexico.

In the twenty-first century, magazines in North America did not follow the trend of newspapers: the number of magazine titles did not drop. In the United States and Mexico, it was nearly the same in 2017 as it was in 1997; in Canada, it grew by 160 percent as Canadians' preferences changed from US magazine brands to those covering Canada-specific topics. Top Canadian magazines included the news magazine *Maclean's*, *Canadian Living*, and *Chatelaine*, a women's magazine. In Mexico, the Spanish version of the US general-interest family magazine *Reader's Digest* had a higher circulation than the entire circulations of Mexico's eight largest magazines put together.

Now downloadable on Web-enabled cell phones, magazines have continued to draw a young audience. Some of the leading magazine brands—such as *National Geographic* and *Better Homes and Gardens* —still reach the majority of their readers via print editions, others—such as *Forbes*—reach most of their readers online. Still others, such as *ESPN The Magazine,* are available in online versions only.

Postal Systems

Mexico's postal system dates back to the Aztec Empire, where the rulers operated a system of messengers. The system worked well enough that the conqueror of the Aztecs, Hernán Cortés, continued using it after the 1521 conquest and the establishment of the colony of New Spain. An official postal service was established in 1580, mainly for communications between the colony and Spain.

In early colonial times, the delivery of mail in British North America was entirely dependent upon travelers who were willing to carry correspondence. In 1691 the British monarchy established the groundwork for a postal system in the British North American colonies. Benjamin Franklin was appointed joint postmaster general for all these colonies in 1753 and immediately began to improve postal service. Franklin put in place the foundation for many aspects of the modern mail system of the United States and Canada (Canada was ceded by France to Britain in 1763) and went on to become the first postmaster general of the United States.

As technology advanced, so did the mails. Moving from foot to horseback, stagecoach, steamboat, rail, automobile, and airplane, postal systems were able to distribute mail more quickly and efficiently. The success of mail service in North America had allowed business and personal communication to be transacted economically across long distances. By 1998 the United States Postal Service (USPS), an agency of the federal government, was delivering 107 billion pieces of first-class mail each year. That number dropped to some 66 billion in 2020 as the worldwide digital revolution brought with it

View of the Palacio de Correos (Main Post Office) in Mexico City, Mexico from the northwest. (Thomas Ledl)

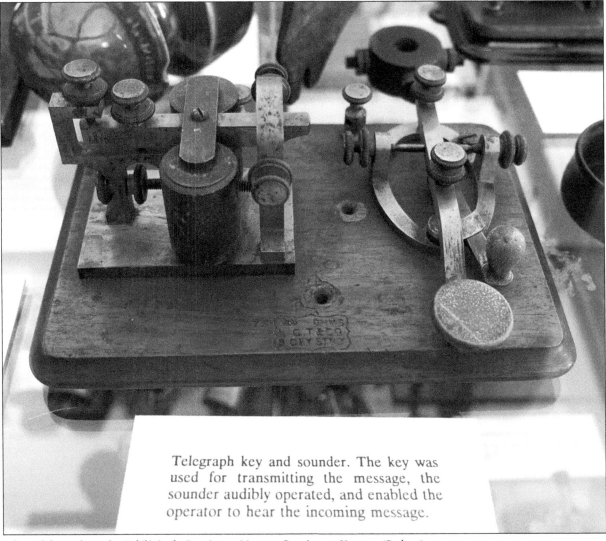

Telegraph key and sounder. The key was used for transmitting the message, the sounder audibly operated, and enabled the operator to hear the incoming message.

Telegraph key and sounder. Exhibit in the Bennington Museum, Bennington, Vermont. (Daderot)

a rapid transition of written communications to electronic media. The number of delivery points, however, grew from 130 million in 1998 to 160 million in 2020. The USPS was also handling more parcels than in 1998, thanks to another phenomenon caused by the digital revolution: online shopping.

Canada has the second-largest postal system in North America. Unlike the USPS, Canada Post functions independently from the government. With Canada's vast territory and a population density of just 11 persons per 1 square mile (4 persons per 1 sq. km.), Canada Post prides itself with the fact that 88 percent of its consumers have a postal outlet within 3.1 miles (5 km.) of their location.

Twice restructured (in 1986 and 2008) and once renamed (in 2008), Mexico's postal service, Correos de México, is government-run, but it has a separate entity, Mexpost, working as a private company. Mexpost is more expensive than the regular postal service but also more efficient, competing with multinational delivery services like the United Parcel Service (UPS) and Federal Express (FedEx).

From Telegraphy to Cellular Phones
In 1844, Samuel F. B. Morse and his partners designed a means of sending sound waves over wires. By 1861 a transcontinental line was completed, linking the east and west coasts of the United States. The telegraph was limited in that it could only

transmit in coded signals, the dots and dashes of Morse code (the code was actually invented by Morse's colleague, Alfred Vail).

Italian inventor Antonio Meucci developed a transducer that allowed voices to be transmitted over telegraph wires. In 1876, Scottish-born Canadian-American engineer Alexander Graham Bell patented a workable system implementing Meucci's ideas, which became known as the telephone. By 1877 the first permanent outdoor telephone wires had been strung. As early as 1882, American Bell (a company built on Bell's telephone patent) took in more than $1 million per year in revenue.

The next logical step was to find a means of transmitting signals wirelessly, through the air. The first step in that direction came in 1901 when Italian electrical engineer Guglielmo Marconi successfully transmitted Morse code signals across the Atlantic. In 1906, American inventor Lee De Forest perfected the Audion vacuum tube, which enabled radio receivers to pick up complex sounds such as voices and music. De Forest's creation led to radio-telephones and paved the way for long-distance service. On January 25, 1915, the first transcontinental telephone service was established between New York City and San Francisco. By 1927 the first transcontinental line had been established between New York and London. This was quickly followed by the establishment of service to other countries.

The next big evolutionary development for telephone service and all communications came on July 1, 1948, when Bell System (a corporate group that evolved from American Bell and other early telephone companies) unveiled the transistor, which allowed for the amplification of the signal. This amplification allowed a new form of transmission of telephone messages: microwaves. On August 17, 1951, Bell Laboratories opened the first transcontinental microwave system. Comprising 107 relay stations spaced about 30 miles (48 km.) apart, the microwave system spanned the United States.

In 1945, Arthur C. Clarke proposed that artificial satellites containing radio repeaters be placed in geostationary orbit around Earth's equator. The first artificial satellite, *Sputnik 1*, was launched by

the Soviet Union in 1957. It had four radio antennas broadcasting radio pulses. The US satellite *SCORE* was the first to relay a human voice from space back to Earth, sending President Dwight D. Eisenhower's Christmas message to Earth-bound receivers in December 1958. By 1965 the commercial communication satellite *Intelsat 1* (also known as Early Bird) was successfully placed into orbit. Intelsat provided 240 high-quality relays for television signals and allowed the exchange of programs between Western Europe and North America.

Cellular technology combines microwave and satellite technology to allow mobile, wireless telephones. The modern cellular phone was invented by American engineer Martin Cooper of Motorola, a company that was producing television and radio equipment in the United States and Canada. Cooper first demonstrated the hand-held portable telephone in 1973. Cooper's phone was dubbed "the brick" because its battery weighed 2.5 pounds (1.1 kg.). In 1984, the Motorola DynaTAC 8000X be-

Dr. Martin Cooper, the inventor of the cell phone, with DynaTAC prototype from 1973. (Rico Shen)

A restored 1941 Canadian General Electric KL-500 4-band radio, with AM and 3 SW bands. (Dghutt)

came the first commercially available cellular mobile phone. Digital cellular networks and Wi-Fi (wireless fidelity) technology appeared at the same time, in the 1990s. After that, cellular phones became much more than just mobile telephones. They became multi-purpose mobile computing and telecommunication devices.

The traditional landline phone has not become obsolete. Its advantage is that it provides secure and reliable communication, especially in cases of emergency or in places where there is no Internet connection. North America has developed a complex telephony network. A combination of fiber-optic cable and coaxial cable, microwave relay stations, satellites, and Internet routers and modems provides telephone service to nearly everyone on the continent. In 2018, there were thirty-four landline phones and 129 cell phones per 1,000 people in the United States; thirty-seven and ninety, respectively, in Canada; and seventeen and ninety-five in Mexico.

Radio

In 1912, the US Congress ordered the Department of Commerce to begin issuing licenses for commercial radio operators. This was done in an effort to keep public signals from interfering with government transmissions. In 1920, the first commercial station, KDKA in Pittsburgh, Pennsylvania, went on the air. In 1927, Congress established the Federal Radio Commission (which later became the Federal Communications Commission, or FCC) to regulate the airwaves in the interests of the public. By the 1930s, radio was firmly established as a part of US popular culture. The advent of radio revolutionized the reporting of the news, because coverage of events was much more rapid than was possible with newspapers. Live coverage of events in the 1940s, especially the work of World War II correspondents, allowed listeners to vicariously experience events that would have seemed more remote when reported only through newspapers.

Prior to World War II, frequency modulation (FM) was introduced. FM operates on a higher frequency than amplitude modulation (AM), and it sends its signals in a straight line, whereas AM bounces its signals off the ionosphere. As a result, FM produces a higher-quality broadcast signal. In the late 1970s and early 1980s, FM overtook AM as the most popular radio format in the United States and became widespread in Canada . In Mexico , the process of transition from AM to FM took place largely around 2010 and was driven by the government.

Radio is a massive medium, providing North Americans with entertainment and information. Despite the vast size of the audience and the prevalence of syndicated programming, radio remains a primarily local medium in that the information and entertainment that each station provides are tailored to serve a local population. Because of this and thanks to being portable and largely free, radio has survived competition from both television and Internet technologies.

In the United States by 2018, the FCC had licensed more than 1,500 radio stations. US radio also broadcasts to foreign countries under the auspices of the Voice of America, funded by Congress. National Public Radio (NPR) had a network of some 900 member stations.

Canada had about 1,200 public and commercial radio stations, broadcasting in English and French; several stations in the north offered broadcasting in eight indigenous languages. The public radio broadcaster, Canadian Broadcasting Corporation, operated four national radio networks and Radio Canada International.

In 2018, Mexico had almost 2,000 radio stations; over 70 percent of them were privately owned. Public radio in Mexico has been traditionally widespread; it is mostly financed by state governments. The state-owned System of Indigenous Cultural Broadcasters offers programming in more than thirty indigenous languages.

Television

In 1928, General Electric began experimenting with broadcasting television signals. The National Broadcasting Corporation (NBC) began experi-

> ### The Canadian Radio-Television and Telecommunications Commission
>
> The Canadian Radio-Television and Telecommunications Commission (CRTC) is the governmental agency that regulates all electronic communication systems in Canada. The CRTC is a licensing agency for radio and television stations, but its focus is to ensure that a certain percentage of programs aired in Canada have Canadian content. These content quotas are aimed at maintaining Canadian cultural identity despite the overwhelming influence of U.S. programming. The CRTC also encourages production of Canadian radio and television programs.

mental telecasting in 1930; by 1939, the NBC station in New York was broadcasting a regular schedule of programs. Television remained a technology for the wealthy until 1948, when both NBC and the Columbia Broadcasting System (CBS) turned their full attention to developing television as a medium of mass communication. By 1952, there were more than 17 million television receivers in the United States.

By 2000, television sets were in use in about 97 percent of US households; by 2020, most households had multiple television sets. One reason for this expansion was the development of cable television. Cable, first known as Community Antenna Television (CATV), originated in 1949 as a means to bring television service to remote geographical areas of the United States that otherwise faced inadequate reception of television signals. By the 1970s, specialty programming was being created just for cable systems. This programming later included channels devoted to 24-hour news coverage (such as CNN), sports reporting (ESPN), and music videos (MTV).

Cable's influence expanded, thanks to the satellite-based distribution of programming. These systems allow individuals to use a satellite receiving dish as small as 12 inches (30.5 centimeters) across as a means of receiving television and radio signals directly from geostationary orbiting satellites. Subscribers can receive hundreds of programming channels. In the twenty-first century, however, the popularity of cable and satellite channels with a paid monthly subscription has been diminishing. In 2019, the penetration rate of pay television, or premium television, was at 75 percent, a drop of almost 10 percent from 2014. There has been a growing trend among consumers (especially younger generations) to abandon cable and satellite TV altogether and watch streaming services broadcasting online; many of those have been offering free viewing or much lower subscription rates compared to traditional providers.

The Broadcasting Act of Canada requires that all Canadian television and radio broadcasters, including cable and satellite channels, must air a certain percentage of content produced in Canada. In 2019, over 93 percent of Canadian adults were reached by television each week, primarily through digital cable. The Canadian Broadcasting Corporation (CBC) is a public broadcaster. It is financed by the Canadian government but allowed to operate independently. CBC was set up in the 1930s in response to the growing influence of American radio

in Canada. Since the launch of television broadcasting in Canada in 1952, CBC has operated both television and radio networks. CBC and the privately owned Canadian Television Network (CTV) dominate the Canadian television market.

Mexico's television services include both commercial broadcasters and networks run by state governments. Most services are provided by three companies operating nationwide: TV Azteca, Imagen Televisión, and Televisa. In 2016, Mexico became the first country in the Western Hemisphere south of the United States to complete the transition from analog to digital TV transmissions. In 2019, 92.5 percent of all Mexican households were equipped with television sets. Free TV reached 78 percent of viewers weekly, and subscription TV had a weekly reach of 52 percent. For many Mexicans, watching "telenovelas" (soap operas) has been the favorite pastime. In 2019, Mexicans aged four or older spent an average of 5 hours and 16 minutes watching television daily. In the United States, the number was 3 hours and 35 minutes, and in Canada, 3 hours and 50 minutes.

The RCA Model 630-TS was the first mass produced television set and was sold in 1946 and 1947. (Fletcher6)

The heavy usage of smartphones among young people relates to the significant percentage of social media users who are from this demographic. (Tomwsulcer)

Statistics show that in the twenty-first century television has remained the mass medium of choice in North America. However, people have more options among which to choose for television programming. In a 2018 survey conducted in the United States, 33 percent of consumers between the ages of 55 and 64 years and 37 percent of those aged 65 or above described traditional TV viewing as their favorite media activity. Only 7 percent of those between the ages of 25 and 34 and 1 percent of those under the age of 25 said the same; the rest responded that they preferred to view movies or TV programs online via streaming platforms.

The Internet and Social Media

With the development of computing networks and digital electronics, a revolutionary communications system has arisen that has united all existing forms of communication and molded them into a powerful new means of distributing information. Many of these developments occurred in North America.

In 1957, the US Department of Defense shepherded the design and planning of a network of computers linked by telephone lines. The network, ARPANET, had its first node attached at the University of California, Los Angeles (UCLA), on September 2, 1969. ARPANET usage continued to grow at a steady rate until 1974, when the first public network, Telenet, was made available. In 1984, Canada began a one-year push to link all Canadian universities to the growing network. By 1986 the first Freenet, encouraging public use of network services, went online. At the same time, the National Science Foundation created NSFNET, which increased connection speed and provided the opportunity for most US universities to connect to it.

The year 1990 saw two milestones in the development of the Internet: ARPANET went off-line, and the first commercial dial-up access service, The World, made the net available to home computers through telephone lines. In 1990, English computer scientist Tim Berners-Lee working at CERN (the European Organization for Nuclear Research) developed the World Wide Web, an information system allowing data to be accessed from multiple computers. The World Wide Web remained a domain primarily for physicists until the Mosaic browser was released in 1993. As a result of Mosaic, the Web's traffic grew several thousand-fold over the previous year.

The development of Web browsers and Internet communication protocols made it possible for everybody with a computer and Internet connection to view, contribute, and share information online. In 1989, about 15 percent of all households in the United States owned a personal computer; by 2000, the number was up to 51 percent. Meanwhile, smartphones (essentially, handheld minicomputers) and other mobile devices with Internet connectivity emerged. By 2011, over 80 percent of all Americans owned a smartphone, 75 percent had a desktop or laptop computer, and about half owned a tablet computer. In 2020, Internet users included 87 percent of the US population, 66 percent of the Mexican population, and 96 percent of the Canadian population. Canada ranked among the countries with the fastest average mobile Internet speeds worldwide.

Throughout North America, social-media services such as Facebook and Twitter have become not only popular communication platforms but also an important source of news. In Mexico, almost 70 percent of news consumers in 2019 resorted to Facebook to find, read, watch, share, or comment on news, regardless of their age. One of the reasons for this might be the convenience of obtaining news in the palm of one's hand; another reason may be related to the fact that traditional news reporting was not considered comprehensive enough or fast enough to be used as the sole source of news. In the United States and Canada, about 60 percent of adults used social networking sites to keep up with the news in 2019. The overwhelming majority of the population in Canada still considered the traditional mass media as reliable sources. In the United States, that share was 41 percent.

According to the authoritative Pew Research Center, the geographic distribution and demographic composition of the majority of the US mass media have not been representative of that of the US population. As of 2018, one in five newsroom employees lived in New York City, Los Angeles, or Washington, D.C.; about three-quarters were non-Hispanic whites; and over 60 percent were male. These people worked for newspapers, broadcast television and radio, and news media companies originally founded on the Internet.

B. Keith Murphy

COMMUNICATIONS IN THE CARIBBEAN

The thirteen sovereign states and seventeen non-sovereign territories of the Caribbean enjoy the same types of communications systems as continental North America. There are, however, several distinct characteristics in the prevalence of specific communication media and information sources in the Caribbean. One is that radio communication is the primary choice of most island residents and has been the main source of communications for decades. It is affordable, readily available, and also more dependable than other media, especially during and after the severe weather the islands experience quite often.

Caribbean residents get most news, information, and entertainment content from US sources as well as from the former colonial powers. In the local broadcast media, sensationalism often eclipses matters of greater relevance and significance to the people of the region; sources like CNN and the BBC offer Caribbean residents both regional and global news and content. Non-news television content also comes primarily from the United States and is disseminated via local cable operators. Caribbean viewers have had access to US television programming since the late 1970s thanks to the signal overspill from the first US domestic satellites, easily accessed in the region via parabolic dish receivers. It gave rise to numerous cable operators as the majority of the population did not have immediate access to satellite dish antennas. The preeminence of the US media stems not only from geographic proximity but also from the fact that most Caribbean islands share not one, but two common languages with the United States: English and Spanish.

A Jamaican newspaper, *The Gleaner*, is the oldest continuously published newspaper in the Americas. However, newspaper circulations in the Caribbean are generally small and are impeded by high prices. The regional newspaper and magazine industry has been also dominated by US publications. The circulation of American magazines is many times higher than that of local magazines. A Spanish-language edition of Florida's leading daily newspaper, *The Miami Herald*, is printed and circulated in the Caribbean by local newspapers, in places like the Bahamas, Curaçao, the Dominican Republic, and Jamaica.

With the exception of Cuba (where the government owns and controls all broadcast media) and Barbados (where the government controls televi-

RADIO STATIONS IN THE CARIBBEAN REGION

Country	AM	FM	Shortwave
Cuba	263	338	3
Dominican Republic	136	268	0
Puerto Rico	70	71	0
Republic of Trinidad & Tobago	0	37	0
Bahamas	3	28	0
US Virgin Islands	5	24	0
Cayman Islands	0	24	0
Antigua & Barbuda	4	20	0
Barbados	1	20	0
Jamaica	0	16	0
St Christopher-Nevis-Anguilla	3	14	0
St Vincent	0	13	0
British Virgin Islands	1	11	0
St Lucia	0	10	0
Aruba	0	16	0
Guadeloupe	1	8	0

sion broadcasting, including cable service), Caribbean islands are served mostly by commercial, privately owned media. Government broadcasters, including public broadcasters, operate on the margins of the media marketplace.

In the twenty-first century, the telecommunication sector across the Caribbean region has been one of the key growth areas and contributors to local GDPs. The number of companies, businesses, and jobs involved in providing telephone, cable and satellite television, and Internet services has increased dramatically. Cuba, with its centralized, government-controlled economy, has experienced the slowest growth in this sector and has had the lowest number of mobile phones and one of the lowest Internet penetration rates in the region. Cubans

were allowed to buy and use cell phones only in 2008. In 2019, there were about fifty cell phones per 100 people in Cuba; the Internet penetration was at 58 percent of the population. The percentage of Internet users was even lower in the region's poorest country, Haiti, at 32 percent. However, Haitians had more cell phones, about sixty per 100 people.

Among the sovereign Caribbean nations, the Bahamas had the highest Internet penetration in the region, at 90 percent of the population. Among the dependent island territories, Aruba and Saint Barthelemy (Saint Barts) had the highest percentage, over 98 percent. Puerto Rico, a US territory, was close, at over 95 percent.

Earl P. Andresen

DISCUSSION QUESTIONS: COMMUNICATIONS

Q1. How did the cell phone come to combine aspects of the telephone, television, and computer? What technology exists that makes cell phones so portable and multifunctional? Are landline telephones destined for extinction? Why or why not?

Q2. What factors have led to the decline of print newspapers and magazines in the United States? Why have Canada and Mexico been less impacted by this phenomenon? From what sources are North Americans most likely to get their news?

Q3. How has the Internet transformed North American society? What is social media, and what roles does it play in everyday life? To what extent have the Caribbean islands embraced Internet technology?

GAZETTEER

Places whose names are printed in SMALL CAPS *are subjects of their own entries in this gazetteer.*

Acapulco. Old Spanish colonial port on Mexico's Pacific coast. For centuries, Acapulco was the eastern end of the Philippines-Mexico trade route. In the 1950s, a highway linking MEXICO CITY and Acapulco was built. In the 1960s, new high-rise hotels and sightings of show-business personalities boosted tourism. Death-defying cliff divers of La Quebrada provide a spectacular tourist attraction as they climb barefoot to a 60-foot (18-meter)-high cliff to dive into the surf. Population was 687,608 in 2012.

Aguascalientes. State in central Mexico; capital city has the same name. Population in 2015 was 1,312,544, with more than half residing in its capital city. Situated on the edge of the high plain (*altiplano*) in a transition zone between the arid north and the more temperate regions of central Mexico, Aguascalientes is named for its hot springs. The economy is strong and growing steadily, with manufacturing and assembly plants providing many jobs.

Alabama. Became twenty-second state on December 14, 1819. Located in US South. Total area is 52,419 square miles (135,765 sq. km.), with 2019 population of 4.9 million. Named after tribe of the Creek confederacy. First permanent European settlement was in 1702. Alabama Ter-

ritory was created in 1817. In 1847 Montgomery (2018 population, 198,218) was made the capital; also was capital of Confederate States of America in 1861. Largest city is Birmingham (2018 population, 209,880).

Alaska. Became forty-ninth state on January 3, 1959. Located in far northwest NORTH AMERICA. The largest US state in size at 663,267 square miles (1,717,854 sq. km.), its 2019 population was 731,545. Name may have come from Aleut word meaning "mainland." Purchased from Russia for $7.2 million in 1867. Capital is Juneau (2019 population, 31,974) and largest city is Anchorage (2019 population, 288,000).

Alaska Highway. Cross-country artery linking ALASKA and Alaska Highway with the rest of the United States. Formerly known as the ALCAN HIGHWAY, it is 1,523 miles (2,452 km.) long and runs through the YUKON connecting DAWSON CREEK, BRITISH COLUMBIA, with Fairbanks, Alaska. US Army engineers constructed it during World War II at a cost of $110 million to provide an overland military supply route to Alaska.

Alberta. Most western of Canada's three PRAIRIE PROVINCES. Bounded by BRITISH COLUMBIA on the west, SASKATCHEWAN on the east, the NORTHWEST TERRITORIES on the north, and MONTANA in the south. In 2016, 4,067,175 people lived here in an area of 255,541 square miles (661,848 sq. km.), close to the size of the US state of TEXAS. The majority of people live in CALGARY and EDMONTON. Alberta's resources include oil and gas, ranching in the foothills, and farming. Ethnic groups include Germans, Ukrainians, and Scandinavians.

Albuquerque. Largest city in NEW MEXICO. Located on the RIO GRANDE at 5,314 feet (1,619 meters) elevation, with 2019 population of 560,513. Founded in 1706 and named for Duke of Albuquerque, the Spanish viceroy of NEW SPAIN.

Alcan Highway. See ALASKA HIGHWAY.

American Falls. See NIAGARA FALLS.

Americas. Collective term for the lands of the WESTERN HEMISPHERE, including NORTH AMERICA, Central America, South America, and the islands of the CARIBBEAN.

Anegada. One of the British VIRGIN ISLANDS in the CARIBBEAN SEA. Located about 15 miles (24 km.) north of VIRGIN GORDA. Unlike the other islands in this chain, which are hilly and volcanic, Anegada is a flat coral and limestone atoll. It is 9 miles (14 km.) long and 2 miles (3 km.) wide, with a 2010 population of 285. Snorkeling and scuba diving in the beautiful reefs attract many tourists.

Anguilla. Most northerly of the LEEWARD ISLANDS. Located between the CARIBBEAN SEA and the Atlantic Ocean. It stretches from northeast to southwest for about 16 miles (26 km.) and is only 3 miles (5 km.) across at its widest point. The dry, limestone island has no streams or rivers, only saline ponds used for salt production. Named Anguille ("eel") by the French, it has been a British colony since 1650, with a brief period of independence in the 1960s. Population in 2018 was 14,731.

Antigua and Barbuda. Two of the British LEEWARD ISLANDS (171 square miles/443 sq. km.). Includes the island of BARBUDA. Made of limestone coral and covered in dry, grassy flatlands. Sugarcane plantations were the mainstay of the economy two centuries ago, but it is now a popular tourist haven. The population of 98,179 (2020) is almost entirely of African descent. The capital—Saint John's—is a busy harbor town. In

2017, Hurricane Irma struck the country and devastated much of Barbuda.

Appalachian Mountains. Mountain system of eastern North America stretching from southern Canada to ALABAMA. It is an eroded, fold-and-thrust belt, 250 million to 300 million years old. Highest peak is Mount Mitchell, NORTH CAROLINA, at 6,684 feet (2,037 meters).

Mount Mitchell is the highest peak of the Appalachian Mountains and the highest peak in mainland eastern North America. (Brian Stansberry)

Arctic Territories. Often referred to as the NORTHWEST TERRITORIES. A region in northern Canada which, prior to the establishment of NUNAVUT, encompassed the territorial administrative regions of Baffin, Keewatin, Kitikmeot, Fort Smith, and Inuvik. Now made up only of Fort Smith, Inuvik, and one hamlet in Kitikmeot.

Arizona. Became forty-eighth state on February 14, 1912. Located in the US Southwest. Total area of 113,998 square miles (295,254 sq. km.) with 2019 population of 7.28 million. The name is thought to come from the Pima word for "place of the small spring." Explored by Spanish conquistador Francisco Vásquez de Coronado in 1540. The first permanent European settlement was at Tubac in 1752. North of Gila River became part of United States when the Treaty of Guadalupe Hidalgo was signed in 1848 at the end of the Mexican War. In 1850 it was organized as Territory of New Mexico. In 1853 the United States purchased the remainder of the present-day state south of Gila River via the Gadsden Purchase. In 1863 the Arizona Territory was created. Capital and largest city is PHOENIX (2019 population, 1.68 million).

Arkansas. Became twenty-fifth state on June 15, 1836. Located in south-central area of the United States. Total area of 53,179 square miles (137,732 sq. km.) with 2019 population of 3 million. Named after the local Arkansas people. Explored by Hernando de Soto in 1541 and added to the United States in 1803 as part of the Louisiana Purchase. Little Rock is the capital and largest city (2019 population, 197,312).

Aruba. Low, barren, arid island located off the Venezuelan coast, outside the hurricane belt. It is 19.5 miles (31.5 km.) long and 6 miles (9.5 km.) across at its widest point, with an area of 75 square miles (194 sq. km.). Population in 2019 was 112,309. Once a member of the NETHERLANDS ANTILLES, it became independent in 1996. It is popular with tourists because it does not levy sales tax or customs duties and it has a pleasant, sunny climate. The capital is Oranjestad—a charming Dutch colonial town.

Atlanta. Capital and largest city of GEORGIA. Located at 1,050 feet (320 meters) elevation, with a 2019 population of 506,811. Founded in 1837 as Terminus and renamed Atlanta (derived from the Western and Atlantic Railroad) in 1845; became state capital in 1877. Burned during the Civil War, in November, 1864, by federal troops during Sherman's March to the Sea. Site of 1996 Summer Olympic Games.

Atlantic Provinces. Collective term for the three MARITIME PROVINCES (NEW BRUNSWICK, NOVA SCOTIA, PRINCE EDWARD ISLAND) plus NEWFOUNDLAND AND LABRADOR; also referred to as Atlantic Canada. In 2020, the combined population was approximately 2.45 million.

Baffin Island. Largest island in Canada. Located between GREENLAND and the Canadian mainland in the Canadian Arctic. IQALUIT, the capital of NUNAVUT, is located on the southeastern part of the island. Named after William Baffin, the English navigator who discovered it in 1616. The island is 183,810 square miles (476,070 sq. km.) in area, larger than the US state of CALIFORNIA.

Bahamas. Independent commonwealth occupying an archipelago of low coral islands located in the Atlantic Ocean 60 miles (97 km.) off the eastern coast of FLORIDA and southward along the Cuban coast. Capital is NASSAU. Comprises nearly 700 islands and cays (small islands) and almost 2,400 low, barren rock formations. Only twenty-nine of these islands are inhabited, with New Providence and Grand Bahama having four-fifths of the population. The total area is 5,359 square miles (13,880 sq. km.). Population was 337,721 in 2020. Christopher Columbus is believed to have made his first landfall on San Salvador island in 1492. Independent since 1973. Tourism and banking are the main commercial activities. Illegal drug smuggling also plays a role in the economy. In contrast to other Caribbean islands, plantation agriculture was never an important activity because of the poor soil. In 2019, category 5 Hurricane Dorian destroyed much of Grand Bahama and Abaco islands.

Baja California. Mexican peninsula, 800 miles (1,300 km.) long. Thin and mountainous with dry, sandy shores, spectacular desert scenery, and varied wildlife. The name means "Lower California."

Baja California Sur. State in northwestern Mexico that occupies the southern half of BAJA CALIFORNIA peninsula. Volcanoes dominate the central and eastern parts of its 28,447 square miles (73,677 sq. km.) of territory. Some cotton is grown. Industry has been confined to the processing of cotton by-products, fish-packing plants, and saltworks. Its population was 763,929 in 2015.

Baltimore. Largest city in MARYLAND. Located on CHESAPEAKE BAY at 20 feet (6 meters) altitude, with a 2019 population of 593,490. Founded in 1729 and named for the barons Baltimore, the British founders of the Maryland Colony. During the War of 1812, the British bombardment of Fort McHenry inspired Francis Scott Key to write "The Star-Spangled Banner" which in 1931 was adopted as the US national anthem.

Banff National Park. First national park in Canada; founded in 1885. It is a world-class destination for millions of tourists each summer. The city of Banff, located in the center of the park, had a population of 7,847 in 2016.

Barbados. Caribbean island located 100 miles (160 km.) east of the arc of the LESSER ANTILLES; capital is Bridgetown. A coral island, it is actually part of the South American continental shelf. It was colonized by the English in the 1620s. It has one of highest population densities in the world (1,775 persons per square mile). Sugar and rum exports provided early prosperity; tourism has become a major source of employment. Population in 2020 was 294,560.

Barbuda. See ANTIGUA AND BARBUDA.

Bays. See under individual names.

Belle Isle, Strait of. See NEWFOUNDLAND AND LABRADOR.

Bequia. One of the GRENADINE ISLANDS in the CARIBBEAN SEA; located 8.5 miles (13.7 km.) south of ST. VINCENT. The population (about 5,300) depends on fishing, whaling, shipbuilding, and tourism for revenue.

Bering Strait. Sea channel connecting the Arctic Ocean and the Bering Sea and separating North America from Asia by a distance of slightly more than 50 miles (80 km.). During the Ice Age, it is believed, the sea level dropped enough to expose a land bridge—now known as Beringia—over which crossed the first humans to enter the Western Hemisphere.

Bonaire. Low, arid, barren island off coast of Venezuela. Part of the LEEWARD ISLANDS and a special municipality of the Netherlands. Capital is Kralendijk. Some of the finest corals and fish in the CARIBBEAN are found here. It has one of the most unspoiled coral reef systems in the world, which makes it a favorite destination for scuba divers. Salt flats provide salt for export, but tourism is the main economic activity. Population in 2019 was 20,104.

Bonaventure Island. See GASPE.

Boston. Capital and largest city of MASSACHUSETTS. Located at the mouth of the Charles River on Boston Bay at 21 feet (6 meters) elevation, with a 2019 population of 692,600. Named for Boston, England. First permanent settlement by Europeans was in 1630. Site of the Boston Massacre of 1770 and the Boston Tea Party of 1773. Home of Old North Church, Beacon Hill, Paul Revere House, and Bunker Hill—site of first major engagement of the American Revolution on June 17, 1775.

British Columbia. Westernmost Canadian province. Bounded by the Pacific Ocean on the west, ALASKA and the YUKON TERRITORY on the north, ALBERTA on the east, and the US states of WASHINGTON and IDAHO on the south. In 2016, 4.65 million people lived here, the majority in the southern part of Fraser Valley. Logging is a major business. The 357,073 square miles (924,815 sq. km.) of land are equal to the combined size of CALIFORNIA, OREGON, and WASHINGTON.

British Leeward Islands. See LEEWARD ISLANDS.

British Virgin Islands. See VIRGIN ISLANDS, BRITISH

Buffalo. Second-largest city in NEW YORK. Located on the eastern end of Lake Erie at the mouths of the Niagara and Buffalo rivers at 585 feet (178 meters) elevation, with a 2019 population of 255,284. Settled in 1790 by the Dutch.

Cabo San Lucas. Mexican resort city located at the southern tip of BAJA CALIFORNIA, where the Gulf of CALIFORNIA meets the Pacific Ocean. Its jagged, rocky beach is accessible only by boat. At this southernmost tip of the peninsula, called Finisterra (Land's End), granite batholiths (rock formations) mark the end of a spine of rock that stretches south from Alaska's ALEUTIAN ISLANDS. This is one of the world's best game-fishing locations and the only Mexican city to have a marine preserve within its city limits. Population in 2015 was 81,111.

Caicos. British crown colony comprising several small, low-lying coral islands 22 miles (35.5 km.) west of Grand Turk Island in the Atlantic Ocean. The only inhabited islands are South, East, North, Middle, and West Caicos and Providenciales; Pine Cay and Parrot Cay are the only inhabited cays. The name "Caicos" is derived from the Spanish word meaning "string of islands." Salt and cotton were early sources of revenue but have been displaced by banking and insurance.

Calgary. Largest city in ALBERTA, Canada. Located along the Bow River in the southern part of the province, it has been a major city in the oil industry. In 2016 the population was 1,239,220. Calgary hosted the 1988 Winter Olympics and is home of the Calgary Stampede Rodeo.

California. Became thirty-first state on September 9, 1850. Located in the United States' Far West on the Pacific Coast. The third-largest state in area, it covers 163,696 square miles (423,970 sq. km.); with 2019 population of 39.5 million, it is the largest state in terms of population. Name comes from a Spanish novel with a fictional island paradise named "California." In 1542 Spanish explorer Juan Rodriguez Cabrillo claimed the territory for Spain. English explorer Sir Francis Drake visited later in the sixteenth century. In the mid-eighteenth century, Russian traders settled California's northern coast. To prevent Russian claims, the Spanish settled in 1769. Between 1769 and 1823, Franciscan missionaries built twenty-one missions along the coast. California became a territory of the Republic of Mexico in 1825, but in 1848 was ceded to the United States by the Treaty of Guadalupe Hidalgo. In 1848 settlers discovered gold in northern California, sparking the greatest gold rush in U.S. history. SACRAMENTO (2019 population, 513,621) is the state capital; LOS ANGELES (2019 population, 3.97 million) is the largest city.

California, Gulf of. Body of water situated between the west coast of Mexico and the east coast of BAJA CALIFORNIA. Called the Sea of Cortés in Mexico.

Campeche. Mexican state on the southwest coast of the YUCATÁN PENINSULA; capital city has the same name. The state covers 20,013 square miles (51,833 sq. km.) and had a population of nearly 900,000 in 2015. It is arid in the north, but just a few miles south of the border, lush tropical rain forests receive as much as 60 inches (1,500 millimeters) of rainfall each year. Maya archaeological sites draw tourists, and offshore

oil, a thriving agricultural sector, and a commercial fishing industry have boosted Campeche's economy.

Campeche, Bay of. Body of water between the east coast of Mexico and the west coast of Mexico's YUCATÁN PENINSULA.

Canadian Falls. See NIAGARA FALLS.

Canadian Shield. Immense block of ancient bedrock that spans roughly half of Canada. Found from NEWFOUNDLAND AND LABRADOR in the east to the Beaufort Sea in northwestern Canada. Most of QUEBEC and ONTARIO, parts of the PRAIRIE PROVINCES, and the ARCTIC TERRITORIES are included in this area. It is the result of thousands of years of continental glaciation across the continent. Boreal forest, irregular drainage patterns, and thousands of lakes characterize the region, as the glaciers scoured all land under it to the level of the bedrock. Striations and glacial erratics (scratch marks on the bedrock and huge boulders) seen across the Canadian Shield are evidence of the power of glaciation. The region is rich in mineral deposits, which have brought mining to many areas. Settlement overall is sparse.

Cancún. Mexican island on the eastern, Caribbean, side of the YUCATÁN PENINSULA; a city of the same name is located on the mainland. Powdered limestone sand beaches and comfortable temperatures make this a popular Mexican resort. Inhabited by 743,626 (2015) residents, Cancún's main source of revenue is tourism.

Cancún, Mexico. (Dronepicr)

Cape Breton Island. Northeastern portion of NOVA SCOTIA, Canada. The 2-mile(3 km.)-wide Strait of Canso separates it from the rest of the province and the Canadian mainland. The island is 110 miles (177 km.) long and up to 80 miles (129 km.) wide and has an area of 3,970 square miles (10,282 sq. km.). Economic activities include coal mining, forestry, fishing, and summer tourism. Its population was 132,010 in 2016.

Cape Canaveral. Cape in eastern FLORIDA. The cape area and part of nearby Merritt Island house the John F. Kennedy Space Center, operated by the National Aeronautics and Space Administration (NASA). From 1963 to 1973, the cape was known as Cape Kennedy in honor of President John F. Kennedy.

Cape Farewell. Southernmost point of GREENLAND. Located at 59°46′ north latitude, 550 miles (885 km.) south of the Arctic Circle. It was an important geographic point of reference for early explorers.

Cape Morris Jesup. Northernmost tip of GREENLAND. At 83°37′ north latitude—442 miles (711 km.) from the North Pole—it was long thought to be the northernmost piece of land on Earth; however, OODAAQ ISLAND now holds this distinction.

Cape Spear. Most easterly point in North America. Located at 47.31 degrees north latitude, longitude 52.37 degrees west, on the island of NEWFOUNDLAND, Canada, 20 miles (32 km.) from ST. JOHN'S. Cape Spear, and the province of Newfoundlandand Labrador, are closer to Ireland than to WINNIPEG, MANITOBA, in the middle of Canada.

Caribbean Sea. Portion of the western Atlantic Ocean bounded by Central and South America to the west and south, and the islands of the ANTILLES chain on the north and east. Its islands, including PUERTO RICO, the CAYMAN ISLANDS, and the VIRGIN ISLANDS, are popular tourist destinations.

Cascade Range. Mountain range in northwestern United States (CALIFORNIA, OREGON, and WASHINGTON) and adjacent Canada. Approximately 700 miles (1,130 km.) long. Its highest point is Mount Rainier at 14,410 feet (4,392 meters) in WASHINGTON. Mount St. Helens, which

erupted in 1980 and 1982, is also part of this range.

Cayman Brac. One of the CAYMAN ISLANDS. Located 89 miles (143 km.) northeast of GRAND CAYMAN. "Braca" is the Gaelic word for "bluff," and this island's most distinctive feature is a rugged limestone cliff that runs down the center of the island. Cayman Brac is 12 miles (19 km.) long; it had a population of 2,547 in 2018.

Cayman Islands. British colony in the CARIBBEAN SEA. Consists of GRAND CAYMAN, CAYMAN BRAC, and LITTLE CAYMAN islands. Columbus passed the islands in 1503 and, noting that the waters were full of turtles, named them Las Tortugas (later changed to Cayman). The Caymans were uninhabited until the late seventeenth century, when England took over from Spain under the Treaty of Madrid. Caves and coves served as hideaways for pirates, such as Blackbeard and Sir Henry Morgan, who plundered Spain's rich treasure ships. A low crime rate and economic and political stability have made these islands pleasant and prosperous. The Cayman Islands are one of the most important offshore financial centers in the world. Population in 2019 was 65,813.

Charleston. Largest city in SOUTH CAROLINA. Located between the mouths of the Ashley and Cooper rivers at 9 feet (3 meters) elevation, with a 2019 population of 137,566. Founded in 1670 as Charles Town in honor of Charles II, king of England; incorporated in 1783 with its present name. State capital until 1790. The US Civil War began with the firing on Fort Sumter in 1861, after the South Carolina Ordinance of Secession was passed in Charleston in 1860.

Charlotte. Largest city in NORTH CAROLINA. Located at 720 feet (219 meters) elevation, with a 2019 population of 885,708. Settled around 1750. Named for Charlotte Sophia of Mecklenburg-Strelitz, wife of King George III of England.

Charlottetown. Capital of PRINCE EDWARD ISLAND, Canada, located nearly in the center of the island. The site of Canadian confederation, which established the country of Canada in 1867 when the provinces of NOVA SCOTIA, NEW BRUNSWICK, PRINCE EDWARD ISLAND, ONTARIO, and QUEBEC were joined together to form the country. Total population was 36,094 in 2016.

Chesapeake Bay. Inlet of the Atlantic Ocean separating Delmarva Peninsula from the mainlands of VIRGINIA and MARYLAND. Primary feeder is the Delaware River. The Chesapeake Bay Bridge-Tunnel, completed in 1964, crosses the bay at its opening. An important part of the Intracoastal Waterway, with many major ports.

Chiapas. State in southern Mexico; capital is TUXTLA GUTIERREZ. Home to many of Mexico's indigenous communities, who still retain their traditional languages, customs, and dress. Mayan ruins in PALENQUE, the old city of SAN CRISTÓBAL DE LAS CASAS, and numerous Mayan villages make Chiapas rich in Mexican culture. Population was 5,217,908 in 2015.

Chicago. Largest city in ILLINOIS and third-largest in the United States. Located at the southern end of Lake Michigan at 595 feet (181 meters) elevation, with a 2019 population of 2,693,976. After 130 years of exploration in the area, Fort Dearborn was built in 1803. The Potawatomi called the area "Checagou," after a wild onion that grew in the area. Destroyed in 1871 by a great fire. The cultural, transportation, and economic capital of the Midwest and central United States.

Chichén Itzá. One of the finest Mayan archaeological sites in the northern part of the YUCATÁN PENINSULA of Mexico. The site occupies 4 square miles (10.4 sq. km.) and consists of the northern (new) zone, which shows distinct Toltec influence, and the southern (old) zone, which is mostly Puuc architecture.

Chihuahua. Largest and one of the richest states in Mexico; state capital has the same name. Wealth comes from mining, timber, cattle raising, *maquiladoras* (US assembly plants), and tourism. Terrain varies greatly from the towering SIERRA MADRE OCCIDENTAL in the southwest to the desert scrublands and salt lakes of the CHIHUAHUAN DESERT—North America's largest

desert—in the northeast. In between lie grassy basins called *llanos* and wooded rolling hills.

Chihuahuan Desert. The largest of the North American deserts; centered between the SIERRA MADRE OCCIDENTAL and SIERRA MADRE ORIENTAL. The Chihuahuan is a high desert. Many of its peaks rise to 5,000 feet (1,500 meters), and its basins are around 2,300 feet (700 meters) above sea level. Three-fourths of the desert lies in Mexico, the rest in New Mexico and TEXAS.

Christianshåb. See QASIGIANNGUIT.

Churchill. Northernmost seaport in Canada, located in the northeastern corner of MANITOBA. Settlement began with the building of Fort Churchill in 1688. Early settlers to the Canadian Prairies, such as the Scandinavians and Germans, entered the country through this area, led by Lord Selkirk, who founded the Selkirk Settlement in southern Manitoba in 1815. A major grain-exporting port during the summer months, Churchill has the advantage of a short sea route to Europe. The population was estimated at 899 in 2016.

Cincinnati. Third-largest city in OHIO. Located on the OHIO RIVER at 550 feet (168 meters) elevation, with a 2019 population of 303,940. First settled in 1788. In 1789 Fort Washington was built by the US Army; in 1790, the village was renamed Cincinnati in honor of the Society of Cincinnati, an organization of officers of the American Revolution.

Ciudad Juárez. Mexican city across the border from El Paso, Texas. Named El Paso del Norte in 1659, it was renamed in 1888 in honor of Mexicn President Benito Juarez. Located 3,117 feet (950 meters) above sea level on the Rio Grande, at the edge of the Chihuahuan Desert. Maquiladora-based manufacturing changed the size and feel of this border town of 1,321,004 in 2010.

Cleveland. Second-largest city in OHIO. Located at the mouth of the Cuyahoga River on Lake Erie at 660 feet (210 meters) elevation, with a 2019 population of 381,009. Surveyed in 1796 by Moses Cleaveland.

Coahuila. Northeastern Mexican state, located along the United States-Mexico border; capital is Saltillo. Coahuila is less busy than other border crossings, but excellent highways make it a popular crossing area for visitors from the north. Cattle ranching is a major economic activity here, and olives, cotton, and other farm products are cultivated year-round. Population in 2015 was 2,954,915.

Colima. City and capital of the state of Colima in west-central Mexico, standing on the Colima River in the Sierra Madre foothills, 1,667 feet (508 meters) above sea level. It was founded in 1523. Industries center upon the processing of local agricultural products including cotton, rice, and corn. There are also salt refining, alcohol distilling, and the manufacture of shoes and leather goods. Its population was 150,673 in 2010.

Colorado. Became thirty-eighth state on August 1, 1876. Located at the convergence of the ROCKY MOUNTAINS and the GREAT PLAINS. Total area of 104,094 square miles (269,601 sq. km.) with a 2018 population of 5,695,564. Name comes from Spanish word for "red color." It was purchased by the United States from France in 1803 as part of Louisiana Purchase. The first permanent European settlement was in 1858, after gold was discovered. The Colorado Territory was established in 1861. DENVER (2019 population, 727,211) is the capital and largest city.

Colorado River. Longest river west of the ROCKY MOUNTAINS—1,400 miles (2,330 km.). It begins in COLORADO and flows through UTAH and ARIZONA before forming borders between Arizona and NEVADA and Arizona and CALIFORNIA. It then crosses into Mexico (BAJA CALIFORNIA NORTE) and empties into the GULF OF CALIFORNIA (SEA OF CORTÉS). It carved the GRAND CANYON in Arizona and was dammed by HOOVER DAM and other dams. First explored by Spaniard Hernando de Alarcón in 1540; explored more thoroughly by American explorer John Wesley Powell in 1869.

Columbia River. Flows from Canada into WASHINGTON STATE. After confluence with the Snake River, it turns west to form the border between OREGON and Washington to the Pacific Ocean. Length is 1,240 miles (1,996 km.). The Grand Coulee Dam in Washington is one of its large power projects. In 1811 David Thompson, a Canadian explorer, followed the river from source to mouth. Accounts for more than 40 percent of US hydroelectricity.

Columbus. Capital and largest city of OHIO. Located at the junction of the Scioto and Olentangy rivers at 780 feet (238 meters) elevation, with a 2019 population of 898,553. First settled in 1797. In 1812 state legislature chose the central location for state capital. Named in honor of Christopher Columbus.

Connecticut. Fifth of the original thirteen US states; ratified the Constitution on January 9, 1788. Located in New England. Total area of 5,543 square miles (14,357 sq. km.), with a 2018 population of 3,572,665. First European settlement in 1634. Name is thought to come from native term meaning "beside the long tidal river." Capital is Hartford (2019 population, 122,105); largest city is Bridgeport (2019 population, 144,399).

Copper Canyon. See SIERRA TARAHUMARA.

Cordillera Central. Mountain range in the western part of the DOMINICAN REPUBLIC. The core of its highlands rises just west of SANTO DOMINGO, the national capital, and extends northwestward to the border with HAITI. The structurally complex range has a crestline between 5,000 and 8,000 feet (1,500 and 2,400 meters) high, with several isolated higher peaks.

Cozumel. Mexican island situated off the east coast of the YUCATÁN PENINSULA. It was an important center for the Mayan cult of Ixchel, goddess of fertility and childbirth. The Maya called the island Cuzamil, "place of the swallows." Cozumel has become a popular tourist resort, with one of the world's best diving locations. It is 9 miles (14 km.) wide by 31 miles (50 km.) long. Population in 2011 was 100,000.

Crater Lake. National park established in 1902 in the CASCADE RANGE in OREGON. The second-deepest lake in North America, it is in the crater of a prehistoric volcano, Mount Mazama. Wizard Island, in the lake, is the top of the now-extinct volcano. National park established in 1902 in the Cascade Range in Oregon. Lake has an area of 20 square miles (52 sq. km.); park has an area of 286 square miles (742 sq. km.).

Cuba. Largest of the Caribbean islands in the GREATER ANTILLES; capital is HAVANA. Located 90 miles (145 km.) south of FLORIDA. Much of this fertile island is devoted to sugarcane, the most important crop. Cuba is ringed with 4,000 miles (6,400 km.) of sandy beaches, coral cays, islands, and large, secluded bays. The population of 11,193,470 (2019) is mainly Hispanic and Spanish-speaking, with a sizable African minority. Its three mountain ranges are the Sierra de los Organos, whose red earth is covered with tobacco plants; the Sierra de Escambray, halfway along the island; and the Sierra Maestra, in the eastern part of the island. In 1959 a communist government under Fidel Castro came into power. Decades of tensions and a trade embargo with the US followed, although some normalization of the Cuba-US relationship has occurred in recent years.

Cuernavaca. Capital of the Mexican state of MORELOS. Located south of MEXICO CITY, it is a popular weekend retreat for Mexico City's residents. It is called the "City of Eternal Spring." Its mild climate encourages luxurious growth of flowers and trees. Hernán Cortés introduced sugarcane cultivation to the area, and Caribbean slaves were brought in to work in the cane fields. Filled with luxury villas, this city has seen an influx of wealthy foreigners and industrial capital. It hosts several Spanish-language schools and many foreign exchange students. Population in 2015 was 366,321.

Curaçao. Largest of the NETHERLANDS ANTILLES in the CARIBBEAN SEA. Capital is Willemstad. Located 60 miles (97 km.) north of Venezuela and 42 miles (68 km.) east of ARUBA. It is a coral island 38 miles (61 km.) long. The Dutch came to the island in 1634 via the Netherlands West India Company. Curaçao shows the most Dutch influence of these islands. It sits below the hurri-cane zone, and trade winds cool the tropical temperatures. Rocky stretches of washed-up coral have broken down into smooth white or pink sandy beaches. Slave trading once was the major source of wealth. More recently, the economic base has changed from oil refining to tourism. The population (158,655 in 2019) is derived from more than fifty nationalities of Latin, African, and European stock.

Dallas. Third-largest city in TEXAS. Located on Trinity River in north Texas at 435 feet (133 meters) altitude, with 2019 population of 1,343,573. Dallas was settled in 1841 by the French. President John F. Kennedy was assassinated in Dallas on Nov. 23, 1963.

Dawson. Former capital of the YUKON TERRITORY, Canada. Located at the confluence of the Yukon and Klondike rivers. After gold was discovered in Bonanza Creek in 1896, Dawson's population soared to more than 20,000 in two years. Dawson is also famous for sending the Dawson City Nuggets to OTTAWA to play for hockey's Stanley Cup in 1905—a twenty-three-day journey by dogsled, boat, and train—although they did not win. The population of the city was 1,375 in 2016.

Dawson Creek. City in northeastern BRITISH COLUMBIA, Canada, where the ALASKA HIGHWAY begins. Completed in 1942, the highway spans 1,523 miles (2,452 km.) and is the only overland connection to ALASKA. Had a population of 12,178 in 2016.

Death Valley. Located in southern CALIFORNIA, extending into southeastern NEVADA. Established as national monument in 1933; designated a national park in 1994. The largest park in the United States, with an area of 3.4 million acres (1.37 million hectares) of desert. Contains the lowest surface point in the WESTERN HEMISPHERE—282 feet (86 meters) below sea level. In 1913 a temperature of 134 degrees Fahrenheit (56 degrees Celsius) was recorded, the highest ever in the United States.

Delaware. First of the original thirteen U.S. states; ratified Constitution on December 7, 1787. Located midway between NEW YORK CITY and

WASHINGTON, D.C. Total area is 2,489 square miles (6,447 sq. km.), with a 2019 population of 973,764. First European settlement in 1638. Name comes from the name of Thomas West, Third Baron De la Warr, the first colonial governor of VIRGINIA. Capital is Dover (2019 population, 38,166); largest city is Wilmington (2019 population, 70,166).

Denver. Capital and largest city of COLORADO. Located on the South Platte River at 5,280 feet (1,609 meters) altitude. Its 2019 population was 727,211. Settled by gold prospectors in 1858, it became the territorial capital in 1867.

Detroit. Largest city in MICHIGAN. Located on the Detroit River at 585 feet (178 meters) elevation, with a 2019 population of 670,031. Established in 1701; was the state capital from 1837 to 1847. A leading automobile manufacturing center. Name is Anglicized version of French word for "strait," which comes from its location along the strait of the Detroit River between Lakes Erie and Huron.

District of Columbia. See WASHINGTON, D.C.

Dominica. Caribbean island wedged between two French islands, GUADELOUPE to the north and MARTINIQUE to the south. An independent commonwealth; capital is Roseau. It was discovered by Christopher Columbus on his second voyage to the NEW WORLD in 1493. Its name comes from the fact that he sighted it on the Christian Sabbath, Sunday (*Domingo* in Spanish). The chief export is bananas. The interior is covered in luxuriant rain forest and is inaccessible by road. The rugged northwest is last homeland of the Caribs, the region's original inhabitants. This small, poor country has been described as the most ruggedly beautiful island in the CARIBBEAN and called the "Switzerland of the Caribbean." The majority of the 74,243 inhabitants speak a French dialect, and most are Roman Catholic. Dominica is 29 miles (47 km.) long and 16 miles (26 km.) wide. Ecotourism has become a source of income, as a result of the spectacular hiking trails through the dense rain forests. Hurricane Maria, a category 5 storm, devastated much of the island in 2017.

Dominican Republic. Second-largest country in the CARIBBEAN SEA, after CUBA, occupying the eastern two-thirds (18,792 square miles/48,670 sq.km.) of the island of HISPANIOLA. Capital is SANTO DOMINGO, the oldest European city in the AMERICAS. Christopher Columbus discovered the island in 1492 and named it "La Isla Española" ("The Spanish Island"). CORDILLERA CENTRAL runs northwest to southeast, with Pico Duarte rising to 10,417 feet (3,175 meters)—the highest point in the Caribbean. Lake Enriquillo, in the southwest, is a salt lake set in the lowest point of land in the Caribbean—151 feet (46 meters) below sea level. In the north are mountains and swift rivers; in the south, on the Barahona peninsula, is desert. The population of 10,499,707 (2020) is of African and European background with a strong indigenous influence; many live in poverty.

Durango. Mexican city 195 miles (314 km.) northeast of MAZATLÁN. Located on a flat plateau in the center of a rich mining area, it is encircled by heavy industrial plants. Most of the wealth comes from iron ore deposits in the nearby hills. Many Western films have been filmed here. Population in 2015 was 654,876.

Dutch Antilles. See NETHERLANDS ANTILLES.

Edmonton. Provincial capital and largest city in ALBERTA. Located on the North Saskatchewan River, it is the most northerly large city in Canada, with a population of 932,546 (2016). It is a major transportation center and serves areas to the north, including the NORTHWEST TERRITORIES.

Ellesmere Island. Large island in the Arctic Ocean and the most northern landholding of Canada. Located northwest of GREENLAND near the North Pole in NUNAVUT. Mountainous and mostly ice-covered, it has only a few small settlements. The best-known is Alert, a weather station and military outpost at the northern tip of the island that is the northernmost community in North America. The English navigator William Baffin discovered the island in 1616.

English Harbour. Small town on the southern shore of the Caribbean island of ANTIGUA. It was an important stop-off point for English warships in the CARIBBEAN; they were refitted here rather than returning to a shipyard in England.

Erie, Lake. See GREAT LAKES.

Everglades National Park. The largest subtropical wilderness area in the United States, with more than 1.5 million acres (609,000 hectares). Located in FLORIDA; established in 1947. It has suffered tremendous environmental destruction from urbanization, agriculture, introduction of exotic vegetation, and the channelization of the Kissimmee River. Programs now in progress aim to restore the Everglades and conserve its unique ecosystems.

Sunset over the River of Grass. (Everglades NPS)

Federal District (Mexico). District in central Mexico. It is the seat of the national government and is similar in function to the DISTRICT OF COLUMBIA in the United States. It includes a large portion of the MEXICO CITY metropolitan area. The district averages more than 8,000 feet (2,400 meters) in elevation and occupies the southeastern corner of the Valley of Mexico. Its area is 580 square miles (1,500 sq. km.). Its population was 8,918,653 in 2015.

Finisterra. See CABO SAN LUCAS.

Florida. Became twenty-seventh state on March 3, 1845. Located in the US South. Total area is 65,755 square miles (170,304 sq. km.) with a 2019 population of 21.4 million. Spanish explorer Juan Ponce de León was the first European to visit it in 1513; name comes from "flowers," which is part of the Spanish name for Easter (the time of his arrival). Settled by Span-

ish in 1565; ceded to United States in 1819. Tallahassee (2019 population, 194,500) is capital; Jacksonville is largest city (2019 population, 911,507).

Fraser River. River running through BRITISH COLUMBIA, Canada. Named after the explorer Simon Fraser, who canoed down through the ROCKY MOUNTAINS to the coast. It is important both for transporting goods and to the salmon fishing industry. Begins in the north-central part of the province and reaches more than 850 miles (1,360 km.) inland.

Fredericton. Provincial capital of NEW BRUNSWICK, Canada. Located along the St. John River. Founded by British Loyalists who settled 70 miles upriver from ST. JOHN in 1783, it had a population of 58,220 in 2016.

French Antilles. Two Caribbean islands, MARTINIQUE and GUADELOUPE, and Guadeloupe's five dependencies. They are French overseas departments and maintain little contact with the other islands. The inhabitants are French citizens, distinctively French in culture, language, and tradition. Because of their integration with France, the inhabitants have one of the highest standards of living in the developing world. They receive many social welfare benefits, such as health care, social security payments, and unemployment benefits. The administrative region of Guadeloupe is an archipelago, consisting of the tiny islands of Marie-Galante, Les Saintes, La Désirade, and SAINT-BARTHÉLEMY, and the northern part of ST. MARTIN.

French West Indies. See FRENCH ANTILLES.

Fundy, Bay of. Arm of the Atlantic Ocean between the Canadian provinces of NOVA SCOTIA on the east and NEW BRUNSWICK on the west. Located eastward of MAINE, near 45 degrees north latitude, longitude 66 degrees west. Has the largest tides in the world, with 53 feet (16 meters) recorded at the MINAS BASIN, one of the farthest reaches of the bay. The high tidal range causes streams to reverse direction in flow, best exemplified by the REVERSING FALLS in ST. JOHN, New Brunswick. During low tides, water flows through rapids into the Bay of FUNDY, but at high tides, some rapids carry water upstream. Navigation by boats is difficult and can be achieved only by carefully timing the sailing times.

Gaspe. Scenic peninsula in southeastern QUEBEC, Canada. An extension of the APPALACHIAN MOUNTAIN chain, with many small fishing villages along the shores. Major attractions include Percé Rock, a tiny island at the eastern tip of the peninsula, named for a large sea arch visible from the mainland. Nearby Bonaventure Island has the largest breeding colony of gannets and other seabirds along the east coast of North America. Located at approximately 48 degrees north latitude, longitude 65 degrees west.

Georgia. Fourth of original thirteen U.S. states; ratified Constitution on January 2, 1788. Located in the US South. Total area is 59,425 square miles (153,909 sq. km.), with a 2019 population of 10.6 million. Hernando de Soto was the first European to visit the area, in 1540. Spanish controlled it for more than 100 years, until ousted by British. Founded as a colony in 1733 by a group led by British philanthropist James Oglethorpe; named for King George II. ATLANTA is capital and largest city (2019 population, 506,811).

Godthåb. See NUUK.

Godthåbsfjord. Second-largest fjord complex in the world. Located on the southwest coast of GREENLAND at 64°30′ north latitude, longitude 51°23′ west. The early Viking western settlement of Vesterbygden, settled just before the twelfth century, was centered here. The Viking settlement had about ninety farms but disappeared by the fourteenth century for unknown reasons.

Grand Canyon. Established as a US national monument in 1908 and national park in 1918. Covers 1,217,158 acres (492,567 hectares) in ARIZONA. Contains the Grand Canyon of the COLORADO RIVER, a gorge 1 mile (1.6 km.) deep, 18 miles (29 km.) wide, and 217 miles (349 km.) long on the Colorado Plateau in northern Arizona.

Grand Cayman. One of the CAYMAN ISLANDS; capital is George Town. Banking and scuba diving are the main businesses. It has no sales tax and the merchandise for sale is duty-free.

Great Basin. Desert region in the western United States. It has a total area of 186,000 square miles (489,500 sq. km.), comprising most of NEVADA and parts of UTAH, CALIFORNIA, IDAHO, WYOMING, and OREGON. John C. Frémont explored and named the area in 1843-1845.

Great Bear Lake. The eighth-largest lake in the world and the largest lake entirely within Canadian territory. Located in the NORTHWEST TERRITORIES, one of its many inlets extends north of the Arctic Circle. The lake flows into the MACKENZIE RIVER. Europeans working for the Hudson's Bay Company discovered it about 1800. It covers 12,275 square miles (31,791 sq. km.).

Great Lakes. Five large freshwater lakes between north-central United States and south-central Canada formed by glaciation during Pleistocene Epoch. From largest to smallest, they are Superior (the world's largest freshwater lake), Huron, Michigan (the only one completely within the United States), Erie, and Ontario. Primary outlet is the ST. LAWRENCE RIVER to the east. System forms backbone of one of the world's busiest waterways.

Great Plains. Vast high-plateau grassland of central North America stretching from TEXAS northward into the PRAIRIE PROVINCES of Canada. The principal commercial activities in this region are ranching and wheat farming.

Great Salt Lake. Largest body of water between the GREAT LAKES and the Pacific Ocean. Located in northern UTAH, near SALT LAKE CITY, in the GREAT BASIN. It is the largest salt lake in the WESTERN HEMISPHERE and is eight times as salty as ocean water. It changes size drastically based upon precipitation. Although several rivers feed the lake, there is no natural outlet. It is a remnant of Lake Bonneville, an ancient, large freshwater lake.

Great Slave Lake. The tenth-largest lake in the world and the second-largest lake within Canadian territory. Located in the southern part of the NORTHWEST TERRITORIES, its waters flow to the MACKENZIE RIVER and the Arctic Ocean. The city of YELLOWKNIFE is located on its shore. The lake was discovered by Samuel Hearne in 1771 and named after the Slave Indians of the area. It covers 11,170 square miles (28,930 sq. km.).

Great Smoky Mountains National Park. US national park authorized in 1926 and established in 1930. Located in Southeast, in western NORTH CAROLINA and eastern TENNESSEE. Covers area of 520,269 acres (210,546 hectares). The Great Smoky Mountains extend the entire length of the park. Home to some of highest peaks of eastern North America. Clingmans Dome, at 6,643 feet (2,025 meters), located in Tennessee, is the park's highest peak.

Greater Antilles. The four largest islands of the ANTILLES: CUBA, HISPANIOLA (DOMINICAN REPUBLIC and HAITI), JAMAICA, and PUERTO RICO. The islands constitute almost 90 percent of the total land area of the entire WEST INDIES.

Greenland. World's largest island, extending from CAPE MORRIS JESUP in the north to CAPE FAREWELL in the south, with an area of 836,000 square miles (2.16 million sq. km.). A self-governing administrative division of Denmark, Greenland is politically part of Europe but geographically part of North America, with an Arctic-type climate. Approximately 79 percent of the island is covered with the GREENLAND ICE CAP. Small tracts of open land along the southern and central western coasts have sparse arctic vegetation and are home to 57,616 (2020) people, who are mostly Inuit. Capital is NUUK. Other settlements include MANIITSOQ, NARSAQ, QAANAAQ, QAQORTOQ, and QASIGIANNGUIT.

Greenland Ice Cap. Approximately 79 percent of GREENLAND is covered with a massive ice cap that has a maximum thickness of about 10,000 feet (3,000 meters). Extends 656,000 square miles (1.7 million sq. km.), three times the size of the US state of TEXAS. Climate change has reduced the mass of ice here. Should the ice cap melt, the global sea level would increase 19 to 23 feet (6 to 7 meters). Shaped like an irregular, double-headed dome, the cap covers the whole

of the interior of Greenland, acting as a barrier to east-west travel. It is so heavy that it has pushed the surface of land below it below sea level.

Grenada. Most southerly in the chain of WINDWARD ISLANDS in the eastern CARIBBEAN SEA. Located approximately 12 degrees north of the equator, it is 21 miles (34 km.) long and 12 miles (19 km.) wide. Its capital, Saint George's, is one of the most charming capital cities in the Caribbean. A mild climate and fertile soil have made it a major exporter of nutmeg, cinnamon, cocoa, cloves, pimento, and bay leaves. Historically, conflict and violence have marred the tranquility of this island; US intervention in 1983 stabilized the island after a violent military coup. Population in 2020 was 113,094.

Grenadine Islands. See ST. VINCENT AND THE GRENADINES

Guadalajara. Capital of the Mexican state of JALISCO. Founded in 1548, it began as a supply point for the northern mining camps of Mexico after silver was discovered in ZACATECAS in 1546. An industrial boom in the late twentieth century transformed it into a major metropolitan area, second only to MEXICO CITY. Population in 2015 was 1,460,148.

Guadeloupe. Caribbean island in the FRENCH ANTILLES, which is an overseas department of France. Shaped like a butterfly, it consists of two islets of very different landscapes, joined by a small land bridge. The "left wing"—Basse-Terre (Low Land)—is wild, mountainous, and rain-forested. The "right wing"—Grande-Terre (Large Island)—is a flatter, coral-based outcropping from a much older volcanic chain. Most of the beaches and tourist resorts are located on Grande-Terre. Northeast trade winds moderate the temperature. The 395,700 (2016) residents are mostly of African descent or mixed heritage. Sugar has remained a main source of revenue. The island covers 687 square miles (1,780 sq. km.).

Guanajuato. Mexican state in the *bajio*, or heartland of Mexico. Located at an altitude of 5,500 to 7,000 feet (1,675 to 2,135 meters), the state's fertile plains and valleys are full of productive farms. Rich in colonial architecture and history, Guanajuato once supplied one-quarter of NEW SPAIN's silver output. Its capital is a beautiful colonial city of the same name. Population in 2015 was 5,853,677.

Guerrero. Mexican state bordered on the south and west by the Pacific Ocean; capital is Chilpancingo. Steep, rugged mountains called the SIERRA MADRE DEL SUR fall dramatically into the sea. Mixed forests of red cedar, torch pines, and holm oaks cover the high mountains. Tropical forests of fig and copal mahogany take over at lower elevations. Mangroves, grasses, and reeds provide rich habitats for tropical birds near river mouths and lagoons. ACAPULCO, Ixtapa, and Zihuatanejo are popular tourist resorts on Guerrero's coast. Population in 2015 was 3,533,251.

Haida Gwaii. Archipelago of the western coast of British Columbia near the Alaska Panhandle. Discovered by Spaniard Juan Perez in 1774. The Haida Native Americans live on these islands. Known as the Queen Charlotte Islands until 2010. Population in 2008 was 4,761.

Haiti. Country occupying the western third (10,714 square miles/27,750 sq.km.) of the island of HISPANIOLA; capital is PORT-AU-PRINCE. Consists of two mountainous peninsulas that enclose the Gulf of Gonave, as well as the two inhabited islands of Gonâve in the gulf and Tortuga off the north coast. Two-thirds of Haiti is mountainous. Highest peak is the Pico de la Selle (8,793 feet./2,680 meters) in the southeast. Deforestation and soil erosion are serious problems. Originally discovered by the Spanish, Haiti was ceded in 1697 to the French, who named it St. Domingue. Haiti has a history of turmoil and political instability. One of the poorest countries in the WESTERN HEMISPHERE. Its population of 11 million residents has the highest infant mortality rate, lowest life expectancy, and one of the highest illiteracy rates (60 percent) in the CARIBBEAN region. Voodoo, a mixture of African animist beliefs and Roman Catholic doctrine, is widely practiced there. A 2010 earthquake

devastated the island, and it is still struggling to recover.

Halifax. Capital of NOVA SCOTIA, Canada, located on the shores of a natural sheltered harbor. Major seaport and commercial center and the eastern terminus of Canada's two national railway lines. Founded in 1749 by the British, many of its residents are of English, Irish, or Scottish background. Its population was 403,131 in 2016.

Hamilton. Industrial city in ONTARIO, Canada. Located at the western end of Lake Ontario, with a population of 747,545 in 2016. Home to many large steel mills and refineries, it is often compared to PITTSBURGH in the United States. Its settlement began with the arrival of Loyalists, following the American Revolution in 1778. Named for George Hamilton, who laid out the original town in 1813.

Havana. City and capital of the Republic of CUBA and of the province of Havana, with which it is coterminous. Havana is Cuba's industrial, importing, and distributing center. Dominant industries include food processing (mostly sugar), shipbuilding, fishing, and automotive production. The city covers an area of 302 square miles (782 sq. km.). Its population of 2,131,480 in 2018 represented more than a fifth of Cuba's entire population. The city attracts more than 1 million tourists annually.

Hawaii. Became fiftieth state on August 21, 1959. Located in the Pacific Ocean, southwest of the mainland United States. It has a total area of 10,931 square miles (28,311 sq. km.) with 2019 population of 1,415,872. The name may be derived from the native word for "homeland." Hawaii comprises eight primary islands and numerous islets, reefs, and shoals. In 1778 British explorer Captain James Cook sailed to the islands and named them Sandwich Islands in honor of his patron, the Earl of Sandwich. They became a US territory in 1898. Honolulu (2019 population, 345,064) is the capital and largest city.

Hermosillo. Capital of the Mexican state of SONORA. This rapidly growing city in northern Mexico is situated in the middle of a rich agricultural area, where the summer temperatures can reach 111 degrees Fahrenheit (44 degrees Celsius). Population in 2015 was 812,229.

Hidalgo. Mexican state located northeast of Mexico City; capital is Pachuca. It contains El Chico National Park—the premier rock climbing area in Mexico. It was named to honor independence hero Miguel Hidalgo y Costilla. Population in 2015 was 2,858,359.

Hispaniola. Large island in the Greater Antilles in the Caribbean Sea. Christopher Columbus landed here in 1492, and it was the initial focus of Spanish settlement in the New World. Divided into the Spanish-speaking Dominican Republic and French-speaking Haiti. Population in 2014 was 21,396,000.

Hoover Dam. One of highest dams in the world. Located on the Colorado River between Arizona and Nevada. Constructed 1931-1936 to create Lake Mead, one of world's largest artificial lakes. Originally named Boulder Dam, it was renamed in 1947 to honor President Herbert Hoover.

Horseshoe Falls. See Niagara Falls.

Houston. Largest city in Texas. Located on the Houston Ship Channel, which leads to the Gulf of Mexico, at 40 feet (12 meters) elevation. Its 2019 population was 2,320,268. Founded in 1836 and named for Sam Houston, first president of the Republic of Texas.

Hudson Bay. One of the largest ocean inlets in the world, located in east-central Canada. Bounded by Nunavut in the north and west, Manitoba and Ontario in the south, and Quebec in the east, and connected to the Atlantic and Arctic oceans. Discovered by explorer Henry Hudson in 1610. It covers an area of about 320,630 square miles (830,111 sq. km.).

Hudson River. Important commercial waterway originating in eastern New York and flowing into Upper New York Bay at New York City. The first European to explore it was Giovanni da Verrazzano, in 1524. Also explored in 1609 by English navigator Henry Hudson, for whom it is named. Connected to Lake Erie by the Erie Canal in 1825.

Huntsville. City in Alabama that is home to the US Army Aviation and Missile Command at Redstone Arsenal and to the George C. Marshall Space Flight Center, where the Saturn rocket, which took humans to the Moon, was developed. First settled in 1805. Its 2019 population was 200,574.

Huron, Lake. See Great Lakes.

Idaho. Became forty-third state on July 3, 1890. Located in the Northwest of the United States. Its total area is 83,570 square miles (216,446 sq. km.) with 2019 population of 1,415,872. Its name is thought to be a Native American word meaning "gem of the mountains." Visited by Meriwether Lewis and William Clark in 1805. First permanent European settlement was in 1842. Boise (2019 population, 228,959) is the capital and largest city.

Illinois. Became twenty-first state on December 3, 1818. Located in U.S. Midwest. Total area is 57,914 square miles (149,998 sq. km.), with a 2019 population of 12,671,821. Named for the Illinois, or Illiniwek, a confederation of Native Americans who inhabited the region at time of French exploration. Explored by the French in the late seventeenth century; first settlement in 1720. Ceded to the British in 1763 at end of French and Indian War, then ceded to the United States in 1783. Part of Northwest Territory 1787-1800; Illinois Territory was created in 1809. Springfield (2019 population, 114,230) is capital; Chicago (2019 population, 2,693,976) is largest city.

Ilulissat Icefjord (Jakobshavn Glacier). Extending from the Greenland Ice Cap, it moves into the sea with an average speed of 131 feet (40 meters) per day. It produces 16.8 cubic miles (70 cubic km.) of ice daily, discharging it into the sea as icebergs. Located at 69°10′ north latitude, longitude 49°55′ west. Climate change has dramatically impacted the icefjord.

Imperial Valley. Located in southern California along the US border with Mexico. One of the richest agricultural districts in the world, it relies heavily on irrigation. It is a major producer of lettuce, cotton, and other crops.

Indiana. Became nineteenth state on December 11, 1816. Located in US Midwest. Total area is 36,418 square miles (94,321 sq. km.), with a 2019 population of 6,732,219. Name means "land of the Indians." French explorer René-Robert Cavelier, sieur de La Salle, explored Indiana in 1679, and the French controlled the area until the end of the French and Indian War (1754-1763), after which the British assumed control until 1783. Part of NORTHWEST TERRITORY 1787-1800; Indiana Territory was created in 1800. Capital and largest city is INDIANAPOLIS (2019 population, 876,384).

Indianapolis. Largest city and capital of INDIANA. Located on the White River at 710 feet (216 meters) elevation, with a 2019 population of 876,384. Settled in 1820; became capital in 1825. Home of the Indianapolis Motor Speedway, site of annual Indianapolis 500 automobile race.

Iowa. Became twenty-ninth state on December 28, 1846. Located in the Midwest of the United States. Total area of 56,271 square miles (145,743 sq. km.) with 2019 population of 3.155,070. Named for the Ioway people. First explored by the French in the early seventeenth century, the first permanent European settlement was in 1788. Became part of the United States in 1803 via the Louisiana Purchase. Des Moines (2019 population, 214,237) is the capital and largest city.

Iqaluit. Capital of NUNAVUT, Canada, located on the southeastern shore of BAFFIN ISLAND. Established as a trading post in 1914, it became an air base during World War II. It had a population of 7,740 in 2016.

Isla Mujeres. Small, finger-shaped Mexican island 8 miles (13 km.) east of CANCÚN off the YUCATÁN PENINSULA's Caribbean shore. Tourism provides jobs and revenue. The name means "Island of Women." Population in 2010 was 12,642.

Isla Tiburon. See TIBURON ISLAND

Jakobshavn Glacier. See ILULISSAT ICEFJORD

Jalisco. State on Mexico's Pacific coast; capital is GUADALAJARA, Mexico's third-largest city. PUERTO VALLARTA, a popular resort town, is located on Jalisco's coast. Jalisco's residents, nicknamed *tapatios*, have popularized mariachi music, the Mexican hat dance, the *charreada* (Mexican-style rodeo), and tequila. Population in 2018 was 8,256,100.

Jamaica. Third-largest island in the CARIBBEAN SEA (after CUBA and HISPANIOLA), covering 4,244 square miles (10,991 sq. km.). The capital is KINGSTON, the largest English-speaking city south of MIAMI, Florida. It has a huge natural harbor on its southern shore. It gained independence from Great Britain in 1962, but remains a member of the British Commonwealth. Once the most important sugar center in the region, it is now one of the world's leading producers of bauxite. Four-fifths of the island is mountainous; the heavily forested Blue Mountains in the interior grow some of the finest coffees in the world. It is the cultural leader of the Caribbean and has one of the largest tourist industries in the region. Population in 2018 was 2,726,667.

Jost Van Dyke. One of the British VIRGIN ISLANDS in the CARIBBEAN SEA. Situated northwest of TORTOLA; named for an early Dutch settler. It is mountainous with lush vegetation. The island is 4 miles (6.5 km.) long, with a 2010 population of 298.

Juárez. See CIUDAD JUAREZ.

Julianehåb. See QAQORTOQ.

Kansas. Became thirty-fourth state on January 29, 1861. Located in the Midwest of the United States. Total area of 82,277 square miles (213,096 sq. km.) with a 2019 population of 2,913,314. The name is Sioux for "south wind people." In 1541 Spanish explorer Francisco Vásquez de Coronado visited the area. Claimed by France in 1682 and became part of the United States in 1803 via the Louisiana Purchase. Topeka (2019 population, 125,310) is the capital and Wichita (2019 population, 398,938) is the largest city.

Kansas City, Kansas. Located at confluence of Kansas and Missouri rivers at 750 feet (229 meters) elevation. The third-largest city in the state (2019 population, 152,960), it is adjacent to KANSAS CITY, MISSOURI.

Kansas City, Missouri. Largest city in Missouri (2019 population, 495,327). It is located at the confluence of Kansas and Missouri rivers at 750 feet (229 meters) elevation. Located adjacent to KANSAS CITY, KANSAS.

Kentucky. Became fifteenth state on June 1, 1792. Located in the eastern United States between the South and the GREAT LAKES regions. Total area is 40,409 square miles (104,659 sq. km.) with a 2019 population of 3,467,673. Name comes from Cherokee word thought to mean "meadow." Daniel Boone was one of the first to explore it, coming through the Cumberland Gap of the APPALACHIAN MOUNTAINS in 1767. First permanent European settlement in 1774. Although a slave state, it remained neutral during Civil War. Capital is Frankfort (2019 population, 27,755); largest city is Louisville (2019 population, 766,757).

Kingston. Capital and chief port city of JAMAICA on the southeastern coast of the island. Kingston was founded in 1692 after an earthquake destroyed Port Royal, which was located at the mouth of the harbor. Its population was 1,243,072 in 2019.

La Malinche. Mexico's sixth-highest mountain at 14,501 feet (4,420 meters). Located in TLAXCALA state.

Labrador. See NEWFOUNDLAND AND LABRADOR

L'Anse aux Meadows. Canadian national historic site located on the northeastern shore on the island of NEWFOUNDLAND near the Strait of Belle Isle. Remains of Viking houses dated to about 1000 CE were discovered here in 1963 by Norwegian archaeologist Helge Ingstad. Discovery of the site confirms the presence of the Norse Vikings well before the arrival of Columbus in 1492.

Las Vegas. Largest city in NEVADA (2019 population, 651,319). Located in southern Nevada near the COLORADO RIVER, HOOVER DAM, and ARIZONA, at 2,033 feet (620 meters) elevation. The name is Spanish for "the meadows." It was a stop on the Old Spanish Trail between SANTA FE, NEW MEXICO, and Southern CALIFORNIA.

Leeward Islands. A group of small Caribbean islands lying in two lines. In the west, MONTSERRAT, NEVIS, and ST. KITTS are rain-forested peaks of a fairly young volcano chain (about 15 million years old). To the east lie much older, flat, limestone island peaks—ANTIGUA, BARBUDA, and ANGUILLA. Originally agricultural colonies that produced sugarcane, cotton, and coconuts, their chief economic activities are now tourism and manufacturing.

Lesser Antilles. Archipelago of small, volcanically formed islands in the CARIBBEAN SEA, grouped into the LEEWARD ISLANDS and the WINDWARD ISLANDS. Trade winds from the east keep temperatures in this tropical region comfortable. Rich volcanic soils make agriculture productive. Sugar, bananas, and tourism are the leading industries. Population in 2009 was 3,949,250.

Little Cayman. The smallest of three islands in the British colony of the CAYMAN ISLANDS. Little Cayman lies 5 miles (8 km.) west of CAYMAN BRAC. It is 10 miles (16 km.) long, has a maximum width of 2 miles (3 km.), and has a total area of 10 square miles (26 sq. km.). Population in 2018 was 270.

Logan, Mount. Highest mountain peak in Canada. Located in the Saint Elias Mountains of southwestern YUKON TERRITORY at 60.54 degrees north latitude, longitude 140.33 degrees west. It is 19,550 feet (5,959 meters) high, second in North America only to DENALI of nearby ALASKA.

Mount Logan from the southeast. (Gerald Holdsworth)

Los Alamos National Laboratory. Located in Los Alamos, in the central part of NEW MEXICO. The U.S. government used the site for nuclear weapons research during World War II; the first atomic bomb and hydrogen bomb were developed here. Today, Los Alamos works with the US Department of Energy on various national security, science, energy, and environmental issues.

Los Angeles. Largest city in CALIFORNIA, and second-largest in the United States (2019 population, 3,979,576). Located in Southern California at 340 feet (103 meters) elevation. The name is Spanish for "the angels." Settled in 1781 by the Spanish and served as the colonial capital.

Louisbourg. Canadian town on the northeastern shore of NOVA SCOTIA on CAPE BRETON ISLAND. Founded in 1713 as a fishing and shipbuilding center, it later was fortified by the French. The French and English fought many battles for control of this major fort, which was destroyed by the English in 1760. The fort was extensively restored and is now a national historic park. Population was 1,023 in 2011.

Louisiana. Became eighteenth state on April 30, 1812. The southernmost state in the continental United States. Total area of 51,840 square miles (134,264 sq. km.) with 2019 population of 4,648,794. The French explorer René-Robert Cavelier, sieur de La Salle, claimed the entire Mississippi drainage area, including Missouri Valley, for France in 1682, naming it Louisiane for King Louis XIV. NEW ORLEANS was established in 1718. In 1762 transferred to Spain; returned to France in 1800. France sold Louisiana to the United States in the Louisiana Purchase of 1803. The capital is Baton Rouge (2019 population, 220,236); New Orleans (2019 population, 390,144) is the largest city.

Lunenburg. Small, historic fishing village on the southeastern shore of NOVA SCOTIA, south of HALIFAX. Home of the famous Bluenose schooner, pictured on the obverse of the Canadian dime. The first German immigrants to Canada settled here in the early 1750s. The population was 2,263 in 2016.

Mackenzie River. Longest river in Canada (1,060 miles; 1,700 km.), flowing north-northwest from the GREAT SLAVE LAKE in NORTHWEST TERRITORIES to the Arctic Ocean. Including all headwaters, which begin in BRITISH COLUMBIA and drain into the Great Slave Lake, the entire river system runs 2,635 miles (4,240 km.). It is a major transportation route from the south to the Arctic Ocean until the river freezes over. Barges travel along the river during the short summer seasons. Named after Alexander MacKenzie, who discovered and explored the river from Lake Athabasca in northern ALBERTA to its delta in 1789.

McKinley, Mount. See DENALI.

Maine. Became twenty-third US state on March 15, 1820. Located in New England. Total area is 35,385 square miles (91,646 sq. km.), with a 2019 population of 1,344,212. Name comes from ancient French province. First explored in 1520s with first European settlement in 1622. Annexed by MASSACHUSETTS in 1652. Augusta is capital (2019 population, 18,697); Portland is the largest city (2019 population 66,215).

Malinche. See LA MALINCHE.

Mammoth Cave National Park. Largest known cave system in the world, with 336 miles (541 km.) of charted passages. Result of erosion (and deposition) of limestone. Park, established in 1941, covers 52,708 acres (21,330 hectres) in KENTUCKY.

Maniitsoq (SUKKERTOPPEN). Town in GREENLAND. Located at 65 minutes north latitude, longitude 52°40′ west, it had a population of 2,670 in 2013. The earliest settlements were of the Saqqaq culture, which date back approximately 4,000 years, followed by Eskimo and Norse settlements. Danish colonizers founded a trading post in 1755 with the name Sukkertoppen (Sugar Loaf). This area contains some of the oldest bedrock in the world, dating back 3.5 billion years. Greenlandic for "the Uneven."

Manitoba. Most eastern of Canada's three PRAIRIE PROVINCES. Bounded by SASKATCHEWAN on the west, ONTARIO on the east, NUNAVUT and HUDSON BAY on the north, and NORTH DAKOTA and

MINNESOTA on the south. In 2020, 1,379,121 people lived on its 213,271 square miles (552,370 sq. km.), nearly the combined size of NORTH DAKOTA, SOUTH DAKOTA, and NEBRASKA. The majority of the population is concentrated in the city of WINNIPEG. The southern part of the province has rich farmland; the central part is made up of many lakes—the most notable being Lake Winnipeg and Lake Manitoba. The northern part has much boreal forest.

Maritime Provinces. Collective term for the provinces of NEW BRUNSWICK, NOVA SCOTIA, and PRINCE EDWARD ISLAND. During the French period of rule in the 1600s, much of the region was known as Acadia, which was ceded to the British by the Treaty of Utrecht in 1713. The term ATLANTIC PROVINCES describes the Maritime Provinces plus NEWFOUNDLAND AND LABRADOR.

Martinique. Largest of the volcanic WINDWARD ISLANDS in the CARIBBEAN SEA. Settled by the French in the 1630s, by the eighteenth century, it was one of the richest sugar islands in the Caribbean. Its highest peak, the volcano Mount Pelée (4,583 feet/1,397 meters), erupted in 1902, killing 30,333 people and destroying the town of Saint Pierre. Population was 376,480 in 2016.

Maryland. Seventh of thirteen original US states; ratified Constitution on April 28, 1788. Located along mid-Atlantic Coast. Total area is 12,407 square miles (32,133 sq. km.), with a 2019 population of 6,045,680. Named for Queen Henrietta Maria, wife of King Charles I of England. First permanent European settlement in 1634. Capital is Annapolis (2019 population, 39,223); BALTIMORE is largest city (2019 population, 593,490).

Massachusetts. Sixth of the original thirteen U.S. states; ratified Constitution on February 6, 1788. Located in New England. Total area is 10,555 square miles (27,336 sq. km.), with a 2019 population of 6,892,503. Name comes from native word for "large hill place." First European settlement was in 1620, by Pilgrims arriving on the *Mayflower*. Was a leader in movement for independence from Great Britain, and many early and important battles of American Revolution were fought there. BOSTON is capital and largest city (2019 population, 692,600).

Matamoros. One of Mexico's fastest-growing industrial cities. Located on United States-Mexican border not far from the Gulf of Mexico near the mouth of the RIO GRANDE. Matamoras prospered during the US Civil War by supplying Confederate troops in TEXAS through the port of Bagdad, enabling them to avoid the Union blockade of the Texas Gulf Coast. It is one of only a few Mexican cities to have its name preceded by the letter *H*—a designation meaning "Heroic, Loyal, and Unconquered." Population in 2016 was 520,367.

Mazatlán. Mexico's most northern major resort, situated on the central Pacific coast just south of the Tropic of Cancer in the state of SINALOA. It is only two centuries old, young compared to many Mexican cities. It has a mild tropical climate, 502,547 residents in 2017, and about 3.5 million visitors yearly.

Memphis. Largest city in TENNESSEE (2019 population, 651,073). Located on the MISSISSIPPI RIVER at 275 feet (84 meters) elevation. First settled by the Chickasaw; founded as a US city in 1819 by future US president Andrew Jackson.

Mérida. Capital of the Mexican state of YUCATÁN. A significant city during Spanish colonial rule, it enjoyed an economic boom in the early twentieth century. Its wealth came from rope made of agave fiber. It has become an important manufacturing city and a university, business, and cultural center of Mexico. Population in 2015 was 892,363.

Mesa Central. Southern portion of the central MEXICAN PLATEAU. It receives more rainfall and has more moderate temperatures than MESA DEL NORTE. Several of Mexico's major cities—including MEXICO CITY, GUADALAJARA, and PUEBLA—and traditional rural areas are found here. It is the traditional center of Mexico's indigenous population and Mexico's main region of food production.

Mesa del Norte. Arid northern part of the central MEXICAN PLATEAU.

Mexican Plateau. Geologically unstable high central plateau of Mexico. Framed by two mountain ranges—the SIERRA MADRE ORIENTAL on the east and the SIERRA MADRE OCCIDENTAL on the west—it slopes northward from 7,000 to 8,000 feet (2,100 to 2,400 meters) along its southern margin to 3,000 feet (900 meters) or less along the border with TEXAS. Its wide plains average more than 6,000 feet (1,830 meters) above sea level. It is Mexico's largest and most populous region.

Mexico (state). State on the central MEXICAN PLATEAU almost completely surrounding the FEDERAL DISTRICT, which encompasses MEXICO CITY. The state's area is 8,286 square miles (21,461 sq. km.). Its elevation is typically more than 10,000 feet (3,000 meters) above sea level. The state contains many preconquest ruins. Its capital is Toluca. Its agriculture and manufacturing industries include assembly plants, aluminum processing, ironworks, and steelworks. The state has the highest population density of any Mexican state, with a total population of 16,187,608 in 2015.

Mexico City. Capital of Mexico. Located on a high plateau in the mountain-ringed Valley of Mexico in the central highlands of the country, at 7,556 feet (2,303 meters) elevation. The oldest capital of the NEW WORLD and the world's fifth most populated city, it is a hectic, overpopulated, often smog-ridden metropolis. Pollution is a major problem. It is the center of business, government, and culture in Mexico. Hernán Cortés, a Spanish conquistador, led his army into the Aztec capital of TENOCHTITLÁN, which was situated on islands in Lake Texcoco. This later became Mexico City. Because it was built on a soft, dried-up lake bed lacking a rock base, the city is sinking at a rate of 3 feet (1 meter) per year. The area is a highly volcanic region that experiences occasional devastating earthquakes. Called the FEDERAL DISTRICT, or simply "Mexico," by most Mexicans. Population was 8,918,653 in 2015.

Mexico, Gulf of. Mostly enclosed arm of the western Atlantic Ocean, bounded by eastern Mexico and the US states of FLORIDA, ALABAMA, MISSIS-SIPPI, LOUISIANA, and TEXAS. Covers about 619,000 square miles (1,603,000 sq.km.). The island of CUBA closes part of the gap between Mexico's YUCATÁN PENINSULA and the southern tip of Florida that separates the gulf from the CARIBBEAN SEA. Most ocean water enters through the Yucatán passage and exits the gulf around the tip of Florida, becoming the Florida Current.

Mexico, Valley of. High, broad basin of volcanic soil and cool temperatures located at the southern end of the MEXICAN PLATEAU, where MEXICO CITY is situated.

Miami. Second-largest city in FLORIDA. Located at 10 feet (3 meters) elevation, with a 2019 population of 467,9633. Spanish established a mission there in 1567. It has a subtropical environment and a large Cuban population.

Michigan. Became twenty-sixth state on January 26, 1837. Located in GREAT LAKES region of the United States. Total area is 96,716 square miles (250,494 sq. km.), with a 2019 population of 9,983,640. Named after Lake Michigan, which is said to have been derived from Algonquian term "michi gami," meaning "big water" or "great lake." French first explored in 1620; first settlement in 1668. Ceded to the British in 1763 at end of French and Indian War; ceded to United States in 1783. Part of NORTHWEST TERRITORY 1787-1800. Michigan Territory was established in 1805 with DETROIT as capital. Lansing is capital (2019 population, 118,210); Detroit is largest city (2019 population, 713,777).

Michigan, Lake. See GREAT LAKES.

Michoacan. State in central Mexico. The mountains here are the winter nesting grounds of millions of monarch butterflies migrating from Canada. Its terrain varies from coastal lowlands to high peaks, so the climate ranges from tropical to moderate. PARICUTÍN VOLCANO is located here on a high plateau. Population was 4,584,471 in 2015.

Middle Island. See PELEE ISLAND.

Milwaukee. Largest city in WISCONSIN. Located on Lake Michigan at 635 feet (194 meters) elevation, with a 2019 population of 590,157. First settled in 1818 by French, but Germans soon came to dominate the region.

Minas Basin. See Bay of FUNDY.

Minneapolis. Largest city in MINNESOTA. Located on the MISSISSIPPI RIVER across from St. Paul, at 815 feet (248 meters) elevation. Its 2019 population was 429,606. St. Anthony Falls, located here, is the northernmost limit of navigation on the Mississippi River.

Minnesota. Became thirty-second state on May 11, 1858. Located in the Upper Midwest of the United States. Total area of 86,939 square miles (225,171 sq. km.) with 2019 population of 5,639,632. The name is from the Sioux for "sky-tinted water," referring to the Minnesota River. First explored by the French in the seventeenth century; surrendered to Great Britain after the French and Indian War (1754-1763). Controlled by the British from 1763 to the end of the War of 1812 (1812-1815), although Britain gave the United States the portion east of MISSISSIPPI RIVER after the American Revolution (1775-1783). The United States acquired the area of present-day Minnesota west of the Mississippi River as part of the Louisiana Purchase in 1803. St. Paul (2019 population, 308,096) is the capital; MINNEAPOLIS (2019 population, 429,606) is the largest city.

Mississippi. Became twentieth state on December 10, 1817. Located in US South. Total area is 48,430 miles (125,434 sq. km.), with a 2019 population of 2,976,149. Named after the MISSISSIPPI RIVER. Spanish explorer Hernando de Soto is thought to have passed through in 1540. First European settlement was in 1699, by the French. United States took possession in 1783; in 1798, Mississippi Territory was created. Jackson is capital and largest city (2019 population, 160,628).

Mississippi River. The longest river in the United States. Begins in northern MINNESOTA, flowing 2,348 miles (3,779 km.) to the Gulf of Mexico. Name from Native American term meaning "great river." Possibly discovered by Hernando de Soto in 1541, explored by Jacques Marquette and Louis Jolliet in 1673. René-Robert Cavelier, sieur de La Salle, claimed it for France in 1682. Among cities on its banks are MINNEAPOLIS and St. Paul, Minnesota; ST. LOUIS, Missouri; MEMPHIS, Tennessee; and NEW ORLEANS, Louisiana.

Missouri. Became twenty-fourth state on August 10, 1821. Located in the Midwest of the United States. Total area of 69,704 square miles (180,533 sq. km.) with 2019 population of 6,137,428. The name is Algonquin for "river of big canoes." French explorer René-Robert Cavelier, sieur de La Salle, claimed the entire Mississippi drainage area, including the Missouri Valley, for France in 1682, naming it Louisiane. French settlers founded Sainte Genevieve, the first permanent European settlement, about 1750. The United States acquired it in 1803 as part of the Louisiana Purchase. The Missouri Compromise of 1820 ended the debate on statehood and admitted Missouri to the Union in 1821 as a slave state. The capital is Jefferson City (2019 population, 42,708); KANSAS CITY (2019 population, 495,327) is the largest city.

Montana. Became forty-first state on November 8, 1889. Located between the Upper Midwest and Northwest in the United States. Total area of 147,042 square miles (380,838 sq. km.) with 2019 population of 1,068,778. The name is Spanish for "mountainous." Claimed, but largely unexplored, by France in the eighteenth century, acquired by the United States in 1803 via the Louisiana Purchase. Montana Territory was created in 1864. Helena (2019 population, 33,124) is the capital; Billings (2019 population, 109,577) is the largest city.

Monterrey. Northern Mexico's largest city. Located in the foothills of the SIERRA MADRE ORIENTAL 150 miles (240 km.) from the United States-Mexico border. The heart of Mexico's wealthiest state, NUEVO LEÓN, Monterrey is a bustling industrial city with some of the most striking twentieth century architecture in Mexico. Population in 2015 was 1,109,171.

Montreal. Largest city in QUEBEC, Canada, and the second-largest city in Canada. Located at the confluence of the OTTAWA and ST. LAWRENCE RIVERS, the total population of the metropolitan area was 4 million in 2016. Historically, the English population of Quebec has been strongest here. One of the oldest Canadian cities, the site—occupied by a native village—was discovered by explorer Jacques Cartier in 1535. A permanent settlement was begun in 1642 on Isle de Montreal, the easily defended island site in the middle of the river.

Montserrat. Small, mountainous, volcanic island in the Caribbean LEEWARD ISLANDS. Located 25 miles (40 km.) southwest of ANTIGUA. Volcanic activity that began in 1995 and continues into the twenty-first century devastated the southern and eastern parts of the island, including the capital, Plymouth, and caused a massive relocation of residents to the northern part of the island. Called "The Emerald Isle of the CARIBBEAN" because Irish Catholics came here in the 1630s to escape religious persecution by English colonists on ST. KITTS AND NEVIS. Population was 4,649 in 2019.

Morelia. Capital of the Mexican state of MICHOACÁN. This lovely colonial city is located 195 miles (315 km.) northwest of MEXICO CITY. First inhabited by the indigenous Tarascans, the city was founded by the Spanish in 1540 and christened Valladolid. The name was changed to honor the revolutionary hero José María Morelos y Pavon, who once lived there. Population was 784,776 in 2010.

Morelos. State in central Mexico that is bordered by the FEDERAL DISTRICT on the north. Its 1,908 square miles (4,941 sq. km.) territory lies on the southern slope of the central MEXICAN PLATEAU. The state is named after Jose Maria Morelos y Pavon, one of the heroes of Mexico's war for independence. Morelos was also the birthplace of Emiliano Zapata, a hero of one of the revolutionary factions of the Mexican Revolution of the 1910s. Morelos is one of the flourishing agricultural states of Mexico. The population, which includes a large percentage of Native Americans and mestizos, was 1,903,811 in 2015.

Mountains. See under individual names.

Narsaq. Town in GREENLAND. Located at 60 minutes north latitude, longitude 46°0′ west, this area has had a long history of sporadic settlement, dating back 4,000 years to the Saqqaq culture. The Danish trading post of Nordpröven (the "Northern Test") was established here in 1830. The town of Narsaq had a population of 1,346 in 2020, with a local economy based on sheep farming and the preprocessing of fish products. Greenlandic for "the Plain."

Nashville. Capital city of TENNESSEE. Located at 450 feet (137 meters) elevation, with a 2019 population of 694,144. Founded in 1779 as Fort Nashborough; renamed in 1784 and became capital in 1843. Home of the Grand Ole Opry, a musical and tourist institution.

Nassau. Capital of the BAHAMAS. It is a port on the northeastern coast of New Providence Island and one of the world's chief vacation destinations. It is also an international banking center. Population was 274,400 in 2016.

Nayarit. Mexican state on the Pacific coast; capital is Tepic. Nayarit rises from the coast to a fertile, rich plateau of farmland. To the east and south lie high rugged mountains, deep gorges, and green valleys where sugarcane, corn, beans, chiles, bananas, and coconuts are grown. Coastal lagoons and marshes attract hundreds of species of birds. Sea turtles nest on the warm beaches and whales play in the warm waters offshore. Population was 1,181,050 in 2015.

Nebraska. Became thirty-seventh state on March 1, 1867. Located in the Midwest of the United States. Total area of 77,354 square miles (200,345 sq. km.) with 2019 population of 1,934,408. The name is from a Native American word for "flat water," describing the Platte River. The Spanish claimed it in 1541 but did not settle there. In 1682 France claimed all territory drained by the MISSISSIPPI RIVER, including the lower MISSOURI RIVER to the Platte River. Acquired by the United States in 1803 via the Louisiana Purchase. The first permanent settlement was in 1823. By 1854 most of eastern Nebraska was ceded to the United States, and the territories of KANSAS and Nebraska were created. Lincoln (2019 population, 289,102) is the capital; Omaha (2019 population, 478,192) is the largest city.

Netherlands Antilles. Group of Dutch islands in the CARIBBEAN SEA. They lie in two groups separated by 500 miles (800 km.) of sea: The volcanic Dutch WINDWARDS—SINT MAARTEN, SABA, and ST. EUSTATIUS—are near the northern end of the LESSER ANTILLES; the Dutch Leewards—BONAIRE and CURAÇAO—are off the coast of Venezuela. ARUBA was politically part of the latter group until it became an autonomous region in 1986, but many still refer to these as the ABC islands.

Nevada. Became thirty-sixth state on October 31, 1864. Located in the GREAT BASIN of the United States. Total area of 110,561 square miles (286,351 sq. km.) with 2019 population of 3,080,156. Name is Spanish for "snow-capped." Nevada Territory was created in 1861 from part of Utah Territory. Prior to the 1859 discovery of the Comstock Lode (gold), most travel was through Nevada to CALIFORNIA. Carson City (2019 population, 55,274) is the capital; LAS VEGAS (2019 population, 651,319) is the largest city.

Nevis. Lush, fertile island in the CARIBBEAN which, along with its sister island, ST. KITTS, forms an independent nation; capital is Charlestown. Separated from St. Kitts by a narrow channel. Columbus named this island Nieves—Spanish for "snows"—because its clouded volcanic peaks reminded him of the snowcapped Pyrenees. Tourism is the main source of revenue. Population in 2011 was 11,108.

New Brunswick. One of the MARITIME PROVINCES of Canada; FREDERICTON is the provincial capital. Bounded on the east by NOVA SCOTIA, on the north by the NORTHUMBERLAND STRAIT, Gulf of ST. LAWRENCE, and QUEBEC, on the west by the state of MAINE, and on the south by the Bay of FUNDY. Originally part of Nova Scotia, it became a province with the arrival of thousands of Loyalists after the American Revolution in 1785. The only officially bilingual province in Canada. Many French Acadians live along the northern shores; persons of Irish and English background dominate elsewhere. Logging and fishing are major industries. The province covers 27,550 square miles (71,355 sq. km.), about the size of Maine, and had a population of 780,890 in 2020.

New Hampshire. Ninth of the thirteen original US states; ratified Constitution on June 21, 1788. Located in New England. Total area of 9,350 square miles (24,216 sq. km.), with a 2019 population of 1,359,711. Name comes from Hampshire County, England. First explored in 1603 and settled by colonists from MASSACHUSETTS in 1620s and 1630s. First colony to declare independence from Great Britain. Concord (2019 population, 43,627) is capital; largest city is Manchester (2019 population, 112,673).

New Jersey. Third of the original thirteen U.S. states; ratified Constitution on December 18, 1787. Located in Northeast. Total area of 8,721 square miles (22,588 sq. km.) with 2019 population of 8,882,190. Named for island of Jersey in English Channel. English explorer Henry Hudson visited in 1609. First European settlement was in 1660. Trenton is capital (2019 population, 83,203); Newark is largest city (2019 population, 282,011).

New Mexico. Became forty-seventh state on January 6, 1912. Located in the Southwest of the United States. Total area of 121,589 square miles (314,915 sq. km.) with 2019 population of 2,096,829. The name was applied by Spaniards in Mexico to lands north of the RIO GRANDE. In 1540 Francisco Vásquez de Coronado came looking for riches. After 1595, Don Juan de Oñate took possession of New Mexico for Spain. In 1821 Mexico won independence from Spain. The Mexican War ended in 1848 with the Treaty of Guadalupe Hidalgo, which ceded New Mexico to the United States. In 1850 the Territory of New Mexico was created. In 1853 the United States added land south of the Gila River to New Mexico via the Gadsden Purchase. The United States organized ARIZONA Territory in 1863 in the western part of Territory of New Mexico, giving New Mexico its present boundaries. SANTA

Fe (2019 population, 84,683) is the capital; AL-BUQUERQUE (2019 population, 560,513) is the largest city.

New Orleans. Largest city (2019 population, 390,144) in LOUISIANA. Located on the MISSISSIPPI RIVER, at 5 feet (2 meters) elevation. Much of the city lies below sea level. Settled by the French in 1718 and named for Orléans, France. It was the state capital from 1812 to 1849. Flooding from Hurricane Katrina in 2005 devastated the city.

New Spain. Colonial-era name for Mexico.

New World. Term first applied to the WESTERN HEMISPHERE by early sixteenth century geographers. "New World" differentiates the AMERICAS from "Old World," which was seen as consisting of Europe, Africa, and Asia.

New York (State). Eleventh of the thirteen original US states; ratified Constitution on July 26, 1788. Located in Northeast. Total area of 54,566 square miles (141,299 sq. km.) with 2019 population of 19,453,561—the fourth-largest state in population. Named for Duke of York. First European settlement was in 1614. First explored by Samuel de Champlain and Henry Hudson; claimed by the Dutch (who called it New Netherland) in 1624. In 1664 the British took control. Albany is capital (2019 population, 96,460); NEW YORK is largest city (2019 population, 8,336,817).

New York City. Largest city in the United States. Located on New York Bay on the HUDSON RIVER at 55 feet (17 meters) elevation, with a 2019 population of 8,336,817. Founded by the Dutch as New Amsterdam; renamed by the British in honor of the Duke of York. Was US capital 1785-1790. Originally consisted only of Manhattan Island, but was rechartered in 1898 to include the boroughs of Brooklyn, the Bronx, Queens, and Staten Island.

Newfoundland. See NEWFOUNDLAND AND LABRADOR

Newfoundland and Labrador. Most easterly province in Canada. A total area of 143,223 square miles (370,945 sq. km.) is broken into two sections. The island of Newfoundland is bounded on the east by the Atlantic Ocean, on the north by the Strait of BELLE ISLE, and on the west and south by the Gulf of ST. LAWRENCE. The capital is ST. JOHN'S. The island was named for the discovery of "newly found land" by John Cabot, an Italian explorer who sailed for the English in 1497. It became the final province of Canada in 1949, incorporating a second region—LABRADOR. Fishing is the main industry of the province, and rich natural resources abound in the interior of Labrador. A total of 519,716 people lived here in 2016, most having English ancestry.

Niagara Falls. One of the major natural attractions of North America. Located along the Niagara River, which flows from Lake Erie in the west to Lake Ontario in the east, forming part of the border between Canada and the United States. Two sets of waterfalls exist: Horseshoe Falls, in ONTARIO, Canada, reaches a width of 2,400 feet (732 meters) and a crest of 167 feet (51 meters); American Falls, in NEW YORK STATE, reaches a width of 1,200 feet (366 meters) and has a crest of 176 feet (54 meters). The first European to describe the falls was Belgian friar Louis Hennepin, in 1678.

Niagara Falls, Ontario, Canada. (Saffron Blaze)

North Carolina. Twelfth of the thirteen original US states; ratified Constitution on November 21, 1789. Located in the South. Total area of 53,389 square miles (139,389 sq. km.), with a 2019 population of 10,488,084. Named in honor of King Charles I and King Charles II of England. First European settlement was in 1660. Part of province of Carolina until 1691; became separate colony in 1712 and royal colony in 1729. Raleigh is capital (2019 population, 474,069); CHARLOTTE is largest city (2019 population, 885,708).

North Dakota. Became thirty-ninth state on November 2, 1889. Located in the Upper Midwest of the United States. Total area of 70,700 square miles (183,112 sq. km.) with 2019 population of 762,062. "Dakota" is Sioux for "friend." The French first visited the area in 1738. Acquired by the United States in 1803 via the Louisiana Purchase. The first permanent European settlement was in 1812. Dakota Territory was established in 1861 and split into northern and southern in 1867. Bismarck (2019 population, 73,529) is the capital; Fargo (2019 population, 124,662) is the largest city.

Northumberland Strait. Body of water separating PRINCE EDWARD ISLAND from northwestern NOVA SCOTIA at the eastern tip of NEW BRUNSWICK. The 8-mile (12.9-km.)-long Confederation Bridge, opened in 1997, spans the strait at its narrowest point.

Northwest Territories. Second-largest territory in Canada; YELLOWKNIFE is the territorial capital. Bounded on the north by the Arctic Ocean; on the east by NUNAVUT; on the south by the provinces of MANITOBA, SASKATCHEWAN, and ALBERTA; and on the west by the YUKON TERRITORY. The Magnetic North Pole is located here. Barren islands, a frigid Arctic climate, sea ice, and snow are common features. Total area of 440,479 square miles (1,140,835 sq. km.). Home to the Inuit and other native tribes, it was inhabited by a sparse population of 44,982 in 2020.

Northwest Territory. Territory created by the US Continental Congress in 1787 encompassing the region later consisting of the states of OHIO, INDIANA, ILLINOIS, MICHIGAN, WISCONSIN, and the portion of MINNESOTA east of the MISSISSIPPI RIVER. VIRGINIA, NEW YORK, CONNECTICUT, and MASSACHUSETTS had claims to this area, which they ceded to the central government between 1780 and 1800. Land policy and territorial government were established by the Northwest Ordinances of 1785 and 1787.

Nova Scotia. One of the MARITIME PROVINCES of Canada; the capital is HALIFAX. The province was settled by the French in 1605, when Samuel de Champlain founded Port Royal in the Annapolis Valley. The French were expelled in 1755 after the British overtook the region. Many of the French Acadians moved to LOUISIANA, establishing the Cajun population in the United States. Maintains a strong lobster and fishing industry and mining operations, and high-technology industries are developing in its large cities. It covers 20,431 square miles (52,917 sq. km.), comparable in size to the state of WEST VIRGINIA, and had a population of 978,274 in 2020.

Nuevo Laredo. Mexico's quintessential border town, closely linked in culture and economics to Laredo, TEXAS, across the border. A relatively new city by Mexican standards, it compensates for its lack of historical sites with a well-developed downtown sector. Population in 2010 was 373,725.

Nuevo León. The wealthiest state in Mexico; capital is MONTERREY. It is the smallest of the five Mexican states that border the United States, sharing only a 12-mile (19-km.) border with TEXAS. It was a colonial backwater under Spanish rule, with cattle ranching as the chief occupation. Twentieth century industrialization brought prosperity to Monterrey. With a 95 percent literacy rate, Nuevo León has one of Mexico's most well-educated populations. Population was 5,119,504 in 2015.

Nunavut. Territory in northern Canada, established in 1999 by the Canadian government. It is bounded by Baffin Strait and GREENLAND on the east, HUDSON BAY and the provinces of QUEBEC and MANITOBA on the south, the NORTHWEST TERRITORIES on the west, and the Arctic Ocean on the north. The territorial capital, IQALUIT, is located in the southeastern corner of the territory, on the southern shores of BAFFIN ISLAND. The largest land area in Canada, it covers 808,185 square miles (2,093,190 sq. km.), almost as large as ALASKA and MONTANA combined. The population—39,486 in 2020—is made up largely of the native Inuit.

Nuuk (Godthåb). Administrative capital of GREENLAND, located on the southwest coast at the entrance to GODTHÅBSFJORD. The cultural and educational center of Greenland, Nuuk was established in 1721 by the Danes as Godthåb, meaning "good hope." It is one of the world's smallest capital cities, having a population of 18,326 (2020) three times the size of the next largest town in Greenland. It has an extensive fishing industry facilitated by an ice-free harbor. Located at 64°15′ north latitude, longitude 51°35′ west. Greenlandic for "the Headland."

Oak Ridge. Community in TENNESSEE founded in 1942 by US government as part of the Manhattan Project to develop the atomic bomb. Incorporated as a city in 1959. At the end of World War II, the population was more than 75,000, but it had fallen to 29,156 by 2019. It has become a leading research center in environmental management and nuclear energy at Oak Ridge National Laboratory.

Oaxaca. Mexican state on the Pacific Coast; capital city, also called Oaxaca, is a well-preserved colonial city that sits in a fertile valley 5,102 feet (1,555 meters) high in the mountains of the SIERRA MADRE DEL SUR. The city is a major industrial and commercial center in the south of Mexico. Population of the state was 3,967,889 in 2015; city, 300,050 in 2014.

Ohio. Became seventeenth state on March 1, 1803. Located in GREAT LAKES area of the United States. Total area of 44,825 square miles (116,096 sq. km.), with a 2019 population of 11,689,100. Name comes from Iroquois word meaning "great river" or "beautiful river." First explored by René-Robert Cavelier, sieur de La Salle, in 1669. The French controlled the area (first permanent settlement in 1788) until the French and Indian War, after which the British assumed control until 1783. Part of NORTHWEST TERRITORY 1787-1800. Capital and largest city is COLUMBUS (2019 population, 898,553).

Ohio River. Formed by the confluence of the Allegheny and Monongahela rivers in PITTSBURGH,

Pennsylvania. Runs about 981 miles (1,579 km.) to the MISSISSIPPI RIVER at Cairo in southern ILLINOIS. First explored by René-Robert Cavelier, sieur de La Salle, in 1669. CINCINNATI, Ohio, and Louisville, KENTUCKY, are other major cities on this navigable waterway.

Oklahoma. Became forty-sixth state on November 16, 1907. Located in southwest-central United States. Total area of 69,898 square miles (181,035 sq. km.) with 2019 population of 3,956,971. The name is Choctaw for "red people." The Spanish were the first Europeans to visit. In 1682 the French claimed all land drained by the MISSISSIPPI RIVER (including present-day Oklahoma) and named it Louisiana. At the close of the French and Indian War, France ceded Louisiana to Spain, but regained it in 1800. Acquired by the United States in 1803 via the Louisiana Purchase. Following the War of 1812, the US government negotiated removal treaties with the Choctaw, Chickasaw, Creek, Seminole, and Cherokee. In 1834 the Indian Territory—including present-day Oklahoma—was created west of the Mississippi River, where Wichita, Kiowa, and Comanche peoples already lived. Forced removal of Native Americans to Indian Territory took place until 1842, along what was called "The Trail of Tears." OKLAHOMA CITY (2019 population, 655,057) is the capital and largest city.

Oklahoma City. Largest city (2019 population, 655,057) of OKLAHOMA and state capital since 1910. Located on the Canadian and North Canadian rivers at 1,195 feet (364 meters) elevation. Site of the April 19, 1995, terrorist bombing of the Alfred P. Murrah Federal Building, which killed 168 people.

Ontario. Third-largest province in Canada. Bounded on the west by MANITOBA, on the east by QUEBEC, on the north by HUDSON BAY, and on the south by the GREAT LAKES and five states of the United States. A total land area of 350,416 square miles (907,574 sq. km.) makes this comparable to the combined areas of TEXAS and KANSAS. The province is the most populous in Canada (14,745,040 people in 2020). The provincial capital, TORONTO, represents the cultural, economic, and political heartland of the country to many residents.

Ontario, Lake. See GREAT LAKES.

Oodaaq Island. An island northeast of GREENLAND that is considered by some to be the northernmost landmass on Earth. Located 438 miles (705 km.) from the North Pole, it is a small island of gravel about 328 feet (100 meters) across. CAPE MORRIS JESUP formerly was thought to be the northernmost point of land.

Oregon. Became thirty-third state on February 14, 1859. Located in the Pacific Northwest of the United States. Total area of 98,381 square miles (254,805 sq. km.) with 2019 population of 4,217,737. Origin of the name is uncertain. The first European visitor was US Captain Robert Gray, who traded with Native Americans in 1788 and named the COLUMBIA RIVER in 1792. After the War of 1812, Oregon was returned to the United States; nevertheless, the British remained in control. In 1846, US claims to Oregon country south of the forty-ninth parallel were recognized by treaty. Oregon Territory was created in 1848. Salem (2019 population, 174,365) is the capital; PORTLAND (2019 population, 654,741) is the largest city.

Oregon Trail. Pioneer route to the Northwest from Independence, MISSOURI, to the COLUMBIA RIVER in OREGON through present-day NEBRASKA, WYOMING, and IDAHO. The trip usually lasted four to six months; many who attempted it did not finish. The first wagon train reached Oregon in 1842.

Orizaba, Mount. Mexico's highest mountain. This towering, snow-capped volcano, 18,701 feet (5,700 meters) high, is located south of MEXICO CITY.

Ottawa. Capital city of Canada and second-largest city in the province of ONTARIO. In 2016 it had 934,243 residents. Situated on the shores of the Ottawa River, which flows east to the ST. LAWRENCE RIVER and the Atlantic Ocean. Home to many national attractions, including the Canadian parliament buildings and the homes of the prime minister and other officials. Established and chosen in 1855 by Queen Victoria of England, after disputes about which city would become the national capital in a unified Canada.

Palenque. Former Mayan temple ruins located in CHIAPAS, Mexico. Dense topical forests have been cut down to give visitors access to this site. Most of its 1,500 structures have not yet been excavated.

Temple of the Inscriptions at Palenque. (Lousanroj)

Paricutín Volcano. Volcano that erupted without warning in a Mexican farmer's cornfield in 1943. In less than a year, its cone was 1,390 feet (424 meters) high. The volcano was active for nine years, spewing ash and rocks erratically until March 6, 1952, when it ceased as suddenly as it had begun.

Pelée, Mount. Volano on MARTINIQUE whose violent eruption in 1902 destroyed the town of Saint Pierre and killed 30,000 people.

Pelee Island. Island located in Lake Erie a few miles south from POINT PELEE on the southern part of ONTARIO. The island covers 18 square miles (47 sq. km.). The southernmost point in Canada is on a rock named MIDDLE ISLAND, just south of Pelee Island.

Pennsylvania. Second of the original thirteen US states; ratified Constitution on December 12, 1787. Located in Northeast. Total area of 46,055 square miles (119,283 sq. km.), with a 2019 population of 12,801,989. Settled by Swedes in 1634. Royal charter granted to William Penn (after whom it is named) in 1681 as a haven for Society of Friends (Quakers). Most was originally part of Virginia Colony, then part of NEW YORK.

Harrisburg (2019 population, 49,271) is capital; PHILADELPHIA (2019 population, 1,584,064) is largest city.

Percé Rock. See GASPE.

Peter Island. One of the British VIRGIN ISLANDS in the CARIBBEAN SEA. Located about 5 miles (8 km.) directly south across the Sir Francis Drake Channel from Road Town, TORTOLA.

Philadelphia. Largest city of PENNSYLVANIA. Located on Delaware River at 100 feet (30 meters) elevation, with a 2019 population of 1,584,064. Founded by William Penn in 1681 as Quaker (Society of Friends) settlement on a previous Swedish site. First and Second Continental Congresses (1774 and 1775-1776) and Constitutional Convention (1787) were held here. Was the capital of the United States from 1790 to 1800.

Phoenix. Capital, since 1889, and largest city (2019 population, 1,680,992) of ARIZONA. Located on the Salt River at 1,090 feet (332 meters) elevation. The first permanent European settlement was in 1870.

Pittsburgh. Second-largest city in PENNSYLVANIA. Located where confluence of Allegheny and Monongahela rivers form the OHIO RIVER at 745 feet (227 meters) elevation. A 2019 population of 300,286. Fort Duquesne was built on the site by the French around 1750; fell to the British in 1758 and renamed Fort Pitt.

Plateau of Mexico. See MEXICAN PLATEAU.

Point Pelee. Southernmost point of land on the Canadian mainland. Located at 41.55 degrees north latitude, longitude 82.30 degrees west, in the northwestern end of Lake Erie, in southern ONTARIO. Point Pelee is a natural wedge-shaped sandspit jutting into the lake. It lies astride a major flyway of wild ducks, Canadian geese, swans, and other migratory birds, which find sanctuary in Point Pelee National Park.

Popocatepetl. Second-highest mountain in Mexico. Rising to 17,887 feet (5,462 meters), it forms the southeastern lip of the Valley of Mexico. It was a dormant volcano popular with climbers until December 1994, when it began belching clouds of smoke and ash.

Port-au-Prince. Capital, chief port, and commercial center of the Caribbean republic of HAITI. It is located on a bay at the apex of the Gulf of Gonaves, which is protected from the open sea by the island of Gonâve. Its population was 1,234,742 in 2019.

Portland. OREGON's largest city (2019 population, 654,741). Located on the Willamette River near confluence with the COLUMBIA RIVER, at 77 feet (23 meters) elevation. Portland metro area nicknamed "Silicon Forest" for its many tech companies.

Prairie Provinces. Collective term for the Canadian provinces of MANITOBA, SASKATCHEWAN, and ALBERTA in the northern GREAT PLAINS region of North America. These provinces constitute the great wheat-producing region of Canada and have large petroleum and natural gas reserves.

Prince Edward Island. The smallest province in Canada; capital is CHARLOTTETOWN. Bounded on the north by the Gulf of ST. LAWRENCE and on the south by the NORTHUMBERLAND STRAIT. Located at approximately 46 degrees north latitude, longitude 63 degrees west. Known for its fine farmland, potatoes, red soils, and beautiful beaches. It covers 2,195 square miles (5,684 sq. km.), slightly smaller than the state of DELAWARE, and had a population of 142,907 in 2016. Most people are descendants of English, Irish, or Scottish background, many dating to the pre-Loyalist and Loyalist period of the American Revolution.

Puebla. State in east-central Mexico. Its capital is also called Puebla. The state is bounded on the east by VERACRUZ and by the FEDERAL DISTRICT on the west. Its 13,096-square-mile (33,919 sq.-km.) territory occupies the southeastern corner of the Anahuac Plateau. It varies in elevation from 5,000 to 8,000 feet (1,500 to 2,400 meters). Numerous fertile valleys are formed by the SIERRA MADRE ORIENTAL. Coffee, sugarcane, fibers, and corn are the principal crops. Its population was 6,168,883 in 2015. The city of Puebla was founded in 1532. Because of its strategic location on the route between MEXICO CITY and

Veracruz, Puebla has been considered a military key to Mexico. Population of the city was 1,576,259 in 2010.

Puerto Rico. Smallest and most easterly of the GREATER ANTILLES islands in the CARIBBEAN SEA. Its capital, SAN JUAN, is one of the best-preserved Spanish colonial cities in the Caribbean. The island is 119 miles (192 km.) long and 39 miles (63 km.) wide. It is dominated by CORDILLERA CENTRAL, a central mountain range that rises to more than 4,000 feet (1,220 meters). North of this range is rain forest; south of it, the land is arid and covered with cactus. Three-quarters of the island is mountainous with fertile coastal plains. Puerto Rico includes two offshore islands—Vieques and Culebra. The population of 3,193,694 (2019) is predominantly Hispanic, Spanish-speaking, and Roman Catholic. Puerto Rico is a commonwealth territory of the United States; its residents are US citizens, eligible for federal aid and military draft but not required to pay federal income tax. Several public referenda have been held regarding statehood, with support growing to become the fifty-first state. Tourism is the main economic activity.

Puerto Vallarta. Popular Mexican resort located on the Pacific coast. The city rings Bahia de Banderas, the largest natural bay in Mexico and the third-largest in the world. It was a small port and waystation throughout most of the colonial period. After the American film *Night of the Iguana* was made there in the 1960s, tourism accelerated. Now, despite many condos and high-rise buildings, the town retains its quaint Mexican charm. The population in 2015 was 221,200.

Puget Sound. Arm of the Pacific Ocean, extending about 80 miles (130 km.) inland to Olympia, WASHINGTON. SEATTLE and Tacoma, Washington, also are located on the sound. Named for Peter Puget, explorer of the Pacific Northwest in the 1790s.

Qaanaaq (THULE). Administrative capital of northern GREENLAND and the world's northernmost inhabited area. It is the main town (population 646 in 2020) of the municipality of Avanersuaq (population 10,726 in 2020). Located at 77°30′ north latitude, longitude 69°12′ west. Greenlanders were moved here from their original location 60 miles (100 km.) south, now the site of the THULE AIR FORCE BASE. Qaanaaq was the base and the destination for several explorers, including Robert Peary, Elisha Kane, and Knud Johan Victor Rasmussen.

Qaqortoq (JULIANEHÅB). One of the largest towns in southern GREENLAND, with a 2020 population of 3,050. Located at 60°40′ north latitude, longitude 46°0′ west, it has a slightly warmer climate than more northerly areas, and sheep farming is an important aspect of the economy. The largest of the Viking settlements, the Eastern Settlement called "Österbygden"—which was established by the twelfth century and had disappeared by the fifteenth century—was found here. The remains of many Norse farmsteads, including Brattahlid, the home of Erik the Red, have been identified. Greenlandic for "the White."

Qaqortoq, Greenland. (Alankomaat)

Qasigiannguit (CHRISTIANSHÅB). Oldest town on Disko Bay, founded as the first Danish colony in northern West GREENLAND. Located at 68°50′ north latitude, longitude 51°10′ west. Its population in 2013 was 1,734; its economy is dependent on shrimping. Greenlandic for "Small Variegated Seals," because of the plentiful seal population.

Quebec. A French-speaking province of Canada. Bounded on the east by NEWFOUNDLAND AND

LABRADOR, NEW BRUNSWICK, and the Gulf of ST. LAWRENCE; on the north by HUDSON BAY and UNGAVA BAY; on the west by ONTARIO; and on the south by NEW YORK, VERMONT, NEW HAMPSHIRE, and MAINE. Formerly called Lower Canada, it joined the Canadian Confederation in 1867 as one of its founding provinces. It has an area of 523,696 square miles (1,356,367 sq. km.), slightly larger than ALASKA, with a population of 8,552,362 in 2020. Most people are concentrated in the southern portion of the province, where MONTREAL and QUEBEC CITY dominate. Rich natural resources are found in the CANADIAN SHIELD, where logging, mining, and hydroelectric facilities abound. The French ethnicity and language are dominant throughout the province.

Quebec City. Capital city of QUEBEC, Canada. Located on the north shore of the ST. LAWRENCE RIVER at the terminus of the Gulf of ST. LAWRENCE. In 2016 it had a population of 705,103. Founded by the French explorer Samuel de Champlain in 1608, it is the oldest continuous urban settlement in Canada. The city's old section has retained its old European style of narrow winding streets and quaint buildings.

Queen Charlotte Islands. See HAIDA GWAII.

Querétaro. Mexican state north of MEXICO CITY, in the geographic heart of Mexico; capital city has the same name. This mountainous state was the center for the conspiracy that eventually led to the overthrow of the Spanish colony and the establishment of the Republic of Mexico. Querétaro city was the temporary capital of Mexico when the US army took over Mexico City in 1848. Here, the Treaty of Guadalupe Hidalgo was signed, relinquishing much of Mexico's land to the United States. Emperor Maximilian was executed here in 1867, and the present Mexican Constitution was drafted here in 1917. Population was 2,038,372 in 2015.

Quintana Roo. State on the east coast of Mexico's YUCATÁN PENINSULA; its capital is Chetumal, located on the border with Belize. The east coast of the state is lined with Maya ruins marking the heyday of the Putun Maya traders. It was an un-

developed and ignored for many years, until it was named for an army general, Andres Quintana Roo, who had never seen it. When the value of this tropical paradise as a tourist area was recognized, roads were built and resorts such as CANCÚN now flourish. Population was 1,501,562 in 2015.

Red River of the North. River beginning at the confluence of the Bois de Sioux and Otter Tail rivers at the twin cities of Wahpeton, NORTH DAKOTA, and Breckenridge, MINNESOTA. The river flows north for 545 miles (877 km.), and for 440 miles (710 km.) forms the boundary between North Dakota and Minnesota, before entering MANITOBA, Canada, and emptying into Lake WINNIPEG. The Red River Valley region is one of North America's best farming regions.

Regina. Provincial capital of SASKATCHEWAN, Canada. Originally settled along the Canadian Transcontinental Railway line, on flat prairie land with a small creek running through the area. In 2019 the population of Regina was estimated at 226,368. The Royal Canadian Mounted Police is headquartered here.

Research Triangle Park. Largest planned research center in the United States, formed by Duke University, the University of North Carolina, and North Carolina State University. Located in the triangle between Durham, Chapel Hill, and Raleigh, sites of the three universities. Over 55,000 employees in 200 companies conduct research in biotechnology, environmental science, telecommunications, medicine and health, and other areas. Home of the North Carolina Supercomputing Center; has one of the highest concentrations of doctoral degrees in the United States.

Reversing Falls. See Bay of FUNDY.

Reynosa. Modern, industrial city on the United States—Mexico border. Reynosa lacks historic sites because a flood in 1800 destroyed the town and forced its residents to move to its current site on top of a small *meseta* (rise). It is located directly across the border from the small town of Hidalgo, TEXAS. Population was 612,183 in 2012.

Rhode Island. Last of the thirteen original US states; ratified Constitution on May 29, 1790. Located in New England. Total area of 1,545 square miles (4,002 sq. km.) with 2019 population of 1,059,361. Name derived from Dutch term meaning "red island." Smallest state in total area. Settled by religious exiles from MASSACHUSETTS, including Roger Williams, who established Providence in 1636. Granted royal charter in 1663. Official name is "State of Rhode Island and Providence Plantations." Capital and largest city is Providence (2019 population, 179,883).

Rio Bravo del Norte. Mexican name for the border river known to Americans as the RIO GRANDE.

Rio Grande. River that begins in the San Juan Mountains of COLORADO and flows through central NEW MEXICO to the Mexican border at El Paso, where it forms the border between TEXAS and Mexico to the east. Known as RIO BRAVO DEL NORTE in Mexico. Total length is 1,880 miles (3,025 km.), making it one of the longest rivers in North America. It is generally too shallow for navigation and is used primarily for drinking water and agriculture.

Rocky Mountains. One of the greatest chains of mountains in North America, stretching from northeastern Canada into central NEW MEXICO. The highest point is Mount Elbert at 14,440 feet (4,401 meters), in COLORADO.

Rushmore, Mount. Peak in the Black Hills of SOUTH DAKOTA. Memorial busts of US presidents George Washington, Thomas Jefferson, Abraham Lincoln, and Theodore Roosevelt are carved into granite face of the mountain. Established in 1925 and carved from 1927 to 1941 by Gutzon Borglum.

Saba. One of the NETHERLANDS ANTILLES WINDWARD ISLANDS in the CARIBBEAN SEA; capital is The Bottom. The island, 5 square miles (13 sq. km.) in area, is a volcano that has been extinct for 5,000 years; its peak, Mount Scenery, soars to 2,910 feet (887 meters). The population in 2019 was 1,915. There are no beaches because steep cliffs rise sharply from the sea. Well's Bay on the northwestern coast has a black sand beach for a few months during the summer, but it disappears back into the sea the rest of the year. The island's only flat point—named Flat Point—is the site of Saba's airstrip. Columbus saw Saba in 1493, but it remained uninhabited except for a few Carib Indians until the Dutch colonized it in 1640. Saba changed hands twelve times as colonial powers vied for the island.

Sacramento. Capital of CALIFORNIA. Its 2019 population was 513,624. Located at the confluence of the American and Sacramento rivers in the Sacramento Valley, at 25 feet (8 meters) elevation. The first European resident of the area was John A. Sutter, a Swiss immigrant who received the land grant from Mexican government in 1839. In 1848 gold was discovered at Sutter's lumber mill. Sacramento became the state capital in 1854; in 1860, it became the western terminus of the Pony Express.

St. Augustine. Oldest permanent European settlement in the United States. Located in FLORIDA on the Matanzas and San Sebastian rivers. Home of Castillo de San Marcos, begun in 1672, and Cathedral of St. Augustine, erected in 1790s. Established in 1565 by Spanish explorer Don Pedro Menéndez de Avilés. In 1821 the Spanish ceded St. Augustine to the United States. Population in 2019 was 15,415.

St. Croix. Largest of the US-owned VIRGIN ISLANDS. Located 40 miles (65 km.) south of ST. THOMAS, with an area of 82 square miles (210 sq. km.). The island has a lush rain forest in the northwest and dry conditions on the eastern end. Beaches line the coast, attracting many tourists. The capital, Christiansted, is a restored Danish port on a coral-bound bay on the northeastern shore. Population was 50,601 in 2010.

St. Eustatius (STATIA). One of the NETHERLAND ANTILLES islands of the CARIBBEAN SEA; capital and only city is Oranjestad. Located 175 miles (282 km.) east of PUERTO RICO and 35 miles (56 km.) south of ST. MARTIN. Anchored at the north and south ends by extinct volcanoes separated by an arid central plain. Its higher elevations are covered with rich vegetation. Colonized in 1636 by the Dutch, it was once the most powerful trading center in the Caribbean. This island was the first to officially acknowledge the United States as an independent nation. Population in 2019 was 3,138.

St. John. One of the US-owned VIRGIN ISLANDS. Located 3 miles (5 km.) east of ST. THOMAS across the Pillsbury Sound. Administrative capital is Cruz Bay. In 1956 Laurence Rockefeller donated two-thirds of St. John's 20 square miles (53 sq. km.) to the United States as a national park. Steep, volcanic hills are a tropical paradise of lush forests and jungles. Population was 4,170 in 2010.

St. John. Largest city in NEW BRUNSWICK, Canada. Located at the mouth of the St. John River on the eastern shores of the Bay of FUNDY. The first incorporated city in Canada (1784), after thousands of British Loyalists arrived from the United States in 1783, including Benedict Arnold. A major shipping and oil-refining center, the city claims to have the highest concentration of Irish in Canada. Its population was 67,575 in 2016.

St. John's. Capital of NEWFOUNDLAND AND LABRADOR, located at the eastern edge of Canada, and one of the oldest settlements in North America. Its exceptional natural harbor brought the earliest European fishermen to this site. It is the easternmost terminus of the TRANS-CANADA HIGHWAY, which spans the country. Its population was 178,427 in 2016.

St. Kitts. Lush, fertile island in the CARIBBEAN SEA, which, along with the island of NEVIS, forms an independent nation; capital is Basseterre. It has an area of 67 square miles (174 sq. km.). It is known as the mother colony of the WEST INDIES, because it was from here that English and French settlers spread out to settle other islands. Population was 34,983 in 2011.

St. Lawrence, Gulf of. Portion of the Atlantic Ocean north of PRINCE EDWARD ISLAND and east of NEW BRUNSWICK.

St. Lawrence River. River that runs 760 miles (1,224 km.) from Lake Ontario in the west, bordering the United States and Canada between the state of NEW YORK and the province of ON-

TARIO, and drains into the Gulf of ST. LAWRENCE and the Atlantic Ocean near QUEBEC CITY. It has been an important trade route to the interior, and completion of the St. Lawrence Seaway in 1959 opened shipping traffic from the Atlantic Ocean to the GREAT LAKES, benefiting both the United States and Canada.

St. Louis. Second-largest city in MISSOURI. Its 2019 population was 300,576. Located on the MISSISSIPPI RIVER just south of confluence with Missouri River, at 455 feet (139 meters) elevation. Settled by the French in 1764.

St. Lucia. One of the Caribbean WINDWARD ISLANDS; capital is Castries. Located between MARTINIQUE and ST. VINCENT AND THE GRENADINES, 100 miles (160 km.) west of BARBADOS, with an area of 238 square miles (616 sq. km.). The old French capital, Soufriere, was the first settlement on the island; however, it was replaced by Castries, which has a better harbor. A volcano of the same name located nearby is actually a sulfur-spewing volcanic vent that steams continuously. Marigot Bay, on the west coast, once sheltered pirates but is now a scenic port for cruise ships. Off the coast of Soufriere, the Pitons, two pyramid-shaped volcanic mountains, are covered in rain forests that soar from the CARIBBEAN SEA to heights of 2,439 feet (743 meters) and 2,620 feet (799 meters). Population was 166,487 in 2020.

St. Martin. Small Caribbean island. The northern, French half is called Saint-Martin; the southern half is called SINT MAARTEN. This is the smallest island in the world to fly two flags. Capital is Marigot. The coastline has many coves and white sand beaches. The main landmass has many lagoons and salt ponds and one of the largest natural lakes in the Antilles; the Dutch and French share this lake. Population was 77,741 in 2009.

St. Thomas. The capital island of the US-owned VIRGIN ISLANDS in the CARIBBEAN SEA. A mountainous island with a vibrant tourist industry, it has an area of 32 square miles (83 sq. km.). The city of Charlotte Amalie is a bustling port with one of the best waterfronts in the Caribbean. Its excellent harbor is sheltered by hills and protected by islands. Pirate ships and trading vessels have been replaced by cruise ships and yachts. Population was 51,634 in 2010.

St. Vincent and the Grenadines. String of thirty-three islands that make up a single nation; part of the WINDWARD ISLANDS chain in the southern CARIBBEAN SEA. Located 13 degrees north of the equator, the islands extend in a 45-mile (73-km.) arc southwest to GRENADA. La Soufrière is an active volcano that last erupted in 1979 on St. Vincent. Population of 101,390 (2020) is mainly of African descent. St. Vincent is a fertile island that exports bananas and arrowroot.

Saint-Barthélemy. Island dependency of GUADELOUPE and part of an overseas department of France. Commonly referred to as St. Barts. The capital, Gustavia, is named for King Gustav III, the Swedish king who gave the island to France in exchange for trading rights in the Baltic Sea in 1785. It consists of low hills and sheltered inlets with more than twenty small islets in nearby waters; Fourchue is the largest. It has an area of 8 square miles (21 sq. km.). It is located 12 miles (19 km.) southeast of ST. MARTIN, on the same coral geological shelf, which gives it similar beautiful white sand beaches. Known as a retreat for the wealthy, it is a duty-free port. Population was 9,793 in 2016.

Salt Lake City. Capital and largest city (2019 population, 200,567) of UTAH. Located on the Jordan River near the GREAT SALT LAKE, at 4,400 feet (1,340 meters) elevation. Mormons, under the direction of Brigham Young, settled here in 1847.

San Antonio. Second-largest city in TEXAS. Its 2019 population was 1,532,233. Located on the San Antonio River at 650 feet (198 meters) elevation. It was captured by Texans from Mexico in 1835. In 1836 Mexican soldiers stormed the Alamo and killed 187 Texans.

San Cristóbal de las Casas. Colonial capital and market center of the southern Mexican state of CHIAPAS. In 1994 the indigenous people of the area led a revolt—the Zapatista rebellion—against the non-indigenous Mexican towns and the Mexican government over health care, education, and land distribution issues, resulting in the establishment of a strong military presence. Commerce, tourism, and the service industries make up most of the economy. Population in 2010 was 185,917.

San Diego. Second-largest city in CALIFORNIA. Its 2019 population was 1,423,851. Located on San Diego Bay, at 20 feet (6 meters) elevation. Spaniards established California's first settlement and first mission there in 1769.

San Francisco. Located in CALIFORNIA on a peninsula between the Pacific Ocean and San Francisco Bay, at 65 feet (20 meters) elevation. Spanish built a settlement there in 1776. Earthquakes in 1906 and 1989 caused great destruction. Population in 2019 was 881,549.

San Juan. Capital and largest city of PUERTO RICO, located on the northern coast of the island on the Atlantic Ocean. It is the oldest city under US jurisdiction. The Spanish explorer Juan Ponce de León founded the original settlement, Caparra, which became San Juan in 1508. De León is buried in the San Juan cathedral. In the early sixteenth century San Juan was the point of departure for Spanish expeditions to unknown parts of the NEW WORLD. San Juan, along with all of PUERTO RICO, passed to the United States with Spain's defeat in the Spanish-American War of 1898. San Juan is the largest industrial and processing center of the island. Its population was estimated at 395,326 in 2015.

San Luis Potosí. Vast Mexican state; capital city of the same name. Founded in 1592 when gold and silver were discovered in the area, the city is Mexico's gateway to the north and a major transportation hub at the center of Mexico. Highways crisscross here, linking MEXICO CITY and GUADALAJARA with CIUDAD JUÁREZ and NUEVO LAREDO on the United States-Mexico border. Population was 2,717,820 in 2015.

San Miguel de Allende. Mexican city founded in 1542 by a Franciscan friar. During colonial times, it was a busy crossroads for the great mining towns of the *bajio* region, GUANAJUATO and

ZACATECAS. It was declared a Mexican National Monument in 1926 to protect its architectural and cultural integrity. Population in 2005 was 139,297.

Santa Fe. Capital of NEW MEXICO. Its 2019 population was 84,683. Located on the Santa Fe River, at nearly 7,000 feet (2,000 meters) elevation. Established as the capital of New Mexico in 1610 by the Spanish (making it the oldest capital city in the United States); in 1851, as a US territorial capital; and in 1912, as a state capital.

Santo Domingo. Capital of the DOMINICAN REPUBLIC and the oldest permanent city established by Europeans in the WESTERN HEMISPHERE. It is located on the southeast coast of the island of HISPANIOLA at the mouth of the Ozama River. Bartolomeo Columbus, brother of Christopher Columbus, founded the city in 1496. Santo Domingo is the industrial, commercial, and financial center of the Dominican Republic and claims the oldest university in the Western Hemisphere. The colonial era cathedral contains the reputed remains of Christopher Columbus. Its population was 965,040 in 2010.

Saskatchewan. Central province of the Canada's PRAIRIE PROVINCES. Bounded by ALBERTA on the west, MANITOBA on the east, the NORTHWEST TERRITORIES on the north, and the states of MONTANA and NORTH DAKOTA in the south. The 227,135-square-mile (588,276-sq.-km.) area is slightly larger than the states of CALIFORNIA and SOUTH DAKOTA combined. Many residents are of German and Ukrainian heritage. Wheat farming dominates the southern half of the province, with boreal forest in the north. In 2020, 1,181,987 people lived in the province. The provincial capital, REGINA, and SASKATOON are the largest cities.

Saskatoon. Largest city in the province of SASKATCHEWAN, Canada. Located on the banks of the South Saskatchewan River, it is the processing and distribution center for a large agricultural region. In 2016 the population was 246,376. Known as the "City of Bridges."

Savannah. Oldest city in GEORGIA. Located near mouth of Savannah River at 20 feet (6 meters) elevation. Founded by James Oglethorpe in 1733. Was the original capital of Georgia Colony in the eighteenth century. Population was 144,464 in 2019.

Sea of Cortés. See CALIFORNIA, GULF OF.

Seattle. Largest city in WASHINGTON State. Its 2019 population was 753,675. Located between PUGET SOUND and Lake Washington, at 10 feet (3 meters) elevation. Founded in 1851.

Sibley Peninsula. Home to Canada's Sibley Provincial Park and the "Sleeping Giant," a naturally occurring rock formation in the north shore of Lake Superior, which appears to have the shape of a person lying down. When seen from the shores of THUNDER BAY, the "Sleeping Giant" is an anomaly in an otherwise flat, featureless landscape and park. According to local legend, this landscape was formed by the god Nanabijou. Located at 48.5 degrees north latitude, longitude 89 degrees west.

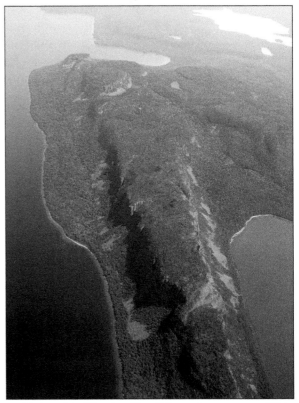

Aerial view of the "Sleeping Giant," the southernmost part of Sibley Peninsula. (P199)

Sierra Madre del Sur. High, rugged mountains in southern Mexico.

Sierra Madre Occidental. Rugged, volcanic mountain range in Mexico. It rises 2,000 to 3,000 feet (600 to 900 meters) above the high arid plains and borders the MEXICAN PLATEAU on the west. The Spanish discovered some of the world's richest silver deposits here in the mid-sixteenth century.

Sierra Madre Oriental. Mountain range in Mexico. It rises abruptly from the Gulf coastal plain to border the MEXICAN PLATEAU on the east. Elevations exceed 10,000 feet (3,050 meters) in many areas. Historically, it has been a rugged barrier to the interior from the east.

Sierra Nevada. Mountain range that extends northwest to southeast, approximating the eastern border of CALIFORNIA with NEVADA, for about 400 miles (640 km.). Location of Mount WHITNEY, the tallest peak in the continental United States.

Sierra Tarahumara. One of the highest and most rugged portions of Mexico's SIERRA MADRE OCCIDENTAL. This is the traditional homeland of Mexico's Tarahumara indigenous people. Also known as Copper Canyon because of the coppery color of the walls of the many deep canyons of this region. Gold, silver, amethyst, and turquoise are mined here.

Copper Canyon in Chihuahua, Mexico. (Jens Uhlenbrock)

Sinaloa. State in northwestern Mexico on the borders of the Gulf of CALIFORNIA and the Pacific Ocean. The 22,429 square miles (58,092 sq. km.) of territory consist of a barren, tropical coastal plain that rises inland to the SIERRA MADRE OCCIDENTAL and is crossed by five main rivers. Its economy is based on agriculture and mining. Its population was 3,216,000 in 2019.

Sint Maarten. Southern, Dutch half of the small Caribbean island of ST. MARTIN; the northern half is SAINT-MARTIN, a French territory. The capital, Philipsburg, is set on a narrow spit of land between Great Bay and the Great Salt Pond, the source of salt that originally attracted the Dutch West Indies Company to the island. The population of 41,486 (2019) relies on tourism for revenue.

Sleeping Giant. See SIBLEY PENINSULA.

Sonora. Mexican state bordered by the United States to the north and the Gulf of CALIFORNIA to the west. Mexico's most important fisheries are located here on the Gulf of California, and citrus fruits, grapes, and vegetables are grown on the YAQUI RIVER and Sonora River plains. Cotton, winter wheat, and oil seed are the principal crops grown here. The SONORAN DESERT is the most diverse desert in the world, with 300 species of cactus adapted to retain water and withstand fierce climatic extremes. Population was 2,850,330 in 2015.

Sonoran Desert. Desert covering 100,000 square miles (260,000 sq. km.) in southwestern ARIZONA and southeastern CALIFORNIA, and including much of the Mexican state of BAJA CALIFORNIA and the western half of the Mexican state of SONORA. Irrigation has produced many fertile agricultural areas such as the IMPERIAL VALLEY in southeastern California.

South Carolina. Eighth of the thirteen original US states; ratified Constitution on May 23, 1788. Located in the South. Total area of 32,020 square miles (82,932 sq. km.), with a 2019 population of 5,148,714. First explored by Spanish in the early sixteenth century. Granted by Charles II of England to eight of his supporters in 1663. Divided into colonies of NORTH CAROLINA and South Carolina in 1729. First state to secede from the Union, in 1860. Columbia is capital and largest city (2019 population, 131,674).

South Dakota. Became fortieth state on November 2, 1889. Located in the Upper Midwest of the United States. Total area of 77,116 square miles (199,731 sq. km.) with a 2019 population of 884,659. "Dakota" is Sioux for "friend." The French first visited in the early 1740s. Acquired by the United States in 1803 via the Louisiana Purchase. Its first permanent European settlement was in 1859. Dakota Territory was established in 1861 and split into northern and southern in 1867. Pierre (2019 population, 13,867) is the capital; Sioux Falls (2019 population, 183,973) is the largest city.

Statia. See ST. EUSTATIUS.

Sukkertoppen. See MANIITSOQ.

Superior, Lake. See GREAT LAKES.

Tabasco. Mexican state on the southern Gulf Coast; capital is VILLAHERMOSA. This green, fertile state was once home to three major pre-Columbian cultures: the Olmecs, the "mother culture" of Mexico; the Totonacs of central VERACRUZ; and the Huastecs. Much of the countryside is covered with thick rain forest, coconut and banana groves, and cacao plantations. A twentieth century oil boom enabled improvement of the state's roads, housing, and employment situation; and archaeological sites, such as Comalcalco, attract many tourists. Population was 2,395,272 in 2015.

Tahoe, Lake. Located on the border of NEVADA and CALIFORNIA in the SIERRA NEVADA at an altitude of about 6,200 feet (1,900 meters). A major US recreational and tourism area, covering 195 square miles (500 sq. km.).

Tamaulipas. Mexican state along the northern coast of the Gulf of Mexico; capital is Ciudad Victoria. It encompasses the eastern escarpment of the SIERRA MADRE ORIENTAL as well as tropical and subtropical coastal plains. Under Spanish influence, Tamaulipas became the cradle of cattle ranching in North America, and the ranchero culture continues today. The *maquiladoras* in the border cities NUEVO LAREDO, REYNOSA, and MATAMOROS are important sources of revenue. The name means high mountains in the

Huastec dialect. Population was 3,441,698 in 2015.

Tampico. Major seaport in Mexico. Located on the Gulf Coast of Mexico, it handles considerable shipping traffic through its Rio Panuco harbor. After oil was discovered here in 1901, Tampico changed from a port ransacked by pirates into an industrial center for the storage, refining, and shipping of oil. Population was 314,418 in 2015.

Tehuantepec, Bay of. Body of water in the Pacific Ocean off the south coast of Mexico's ISTHMUS OF TEHUANTEPEC.

Tehuantepec, Isthmus of. Low, hilly depression west of CHIAPAS, Mexico. The narrowest, and one of the lowest, parts of Mexico. Despite a transcontinental railroad, a highway, and an oil pipeline, it is a relatively isolated and undeveloped region. The Pacific Ocean and Gulf of Mexico are separated here by only about 137 miles (220 km.).

Tennessee. Became sixteenth state on June 1, 1796. Located in the US South. Total area of 42,143 square miles (109,151 sq. km.), with a 2019 population of 6,829,174. Name is thought to have come from "Tanasi," the name of a Cherokee village. First explored by the Spanish in 1540. René-Robert Cavelier, sieur de La Salle, visited MEMPHIS area in 1682; explored by Daniel Boone in 1769. Became part of the United States in 1783. State of Franklin from 1784-1788. NASHVILLE (2019 population, 694,144) is capital; Memphis (2019 population, 651,073) is largest city.

Tennessee River. River in the US South. Flows from confluence of Holston and French Broad rivers in eastern TENNESSEE about 886 miles (1,426 km.) through northern ALABAMA, western Tennessee, and KENTUCKY, to the OHIO RIVER. Extensively dammed beginning in 1933 with creation of the Tennessee Valley Authority for flood control and power production.

Tenochtitlán. Aztec capital city. Built on an island in a lake, it occupied the site that is modern MEXICO CITY. From here, the Aztecs ruled an extensive empire that spread over much of south-central Mexico.

Teotihuacan. One of the largest, most impressive archaeological sites in Mesoamerica. Located near MEXICO CITY, it was built between 100 BCE and CE 900. Its huge ceremonial temple pyramids extended over 8 square miles (20 sq. km.). Little is known about the people who built these structures. When the Aztecs arrived in 1200, they used the abandoned site as a pilgrimage center. According to Aztec legend, the Sun, Moon, and universe were created here.

Texas. Became twenty-eighth state on December 29, 1845. Located in south-central area of the United States. Total area of 268,581 square miles (695,621 sq. km.) with 2019 population of 28,995,881, making it the second-largest state in both area and population. Name is a variant of native word for "friends." The Spanish first visited in the sixteenth century and in 1682 built a mission near present-day El Paso. Mexico gained independence from Spain in 1821. In 1836 Texas declared itself a republic free from Mexico. Nine years later, Texas was admitted to the United States, which led to the Mexican War (1846-1848). US victory established the RIO GRANDE as the southern border of Texas and the United States. The capital is Austin (2019 population, 978,908); HOUSTON (2019 population, 2,320,268) is the largest city.

Thule. See QAANAAQ.

Thule Air Force Base. Giant US air base located at Pituffik on GREENLAND's northwest coast. It was situated here (76°32′ north latitude, longitude 68°51′ west) because of its strategic position between North America and Russia. Built in 1951-1952 during the Cold War, it houses a vast radar installation containing a ballistic missile early warning system.

Thunder Bay. City in ONTARIO, Canada, on the western end of Lake Superior. It was originally two rival cities—Port Arthur and Fort William. The fur-trading Fort William of the Northwest Company was founded in 1812 at the mouth of the Kaministiquia River. Port Arthur began as a public-works settlement in the 1850s. The two cities merged in 1970 to form Thunder Bay. Its many grain elevators handle much of the grain shipped through the GREAT LAKES to the east and abroad. Thunder Bay has the highest concentration of Finnish people in Canada, some 10 percent of the city's population of 107,909 (2016).

Tiburon Island. Continental island, left behind as the San Andreas Fault separated BAJA CALIFORNIA from the rest of Mexico. It is 30 miles (48 km.) long and 20 miles (32 km.) wide. Its desert landscape is similar to that of Mexico's Sonoran coast. It is designated as a special biosphere reserve protecting bighorn sheep and many other SONORAN DESERT species extinct or nearly extinct on the mainland. The name means "Shark Island," for the huge schools of hammerhead sharks that used to congregate in nearby waters.

Tiburon Island from Punta Chueca. (Stevemarlett)

Tijuana. One of Mexico's largest border cities and the world's busiest border crossing. Located less than 20 miles (32 km.) from SAN DIEGO, California. It is one of Mexico's youngest cities, established as a line of defense against *filibusteros* following the Treaty of Guadalupe Hidalgo in 1848. Pollution is a major problem as rapid industrialization progresses. Population was 1,902,385 in 2019.

Tlaxcala. Mexican state; capital city has the same name. The beautiful colonial city was founded jointly by Hernán Cortés and Franciscan friars in 1520. Located 74 miles (120 km.) from MEXICO CITY, it had a population of 1,272,847 in

2015. Its highland location of more than 3,200 feet (2,000 meters) above sea level gives it a cool, sunny, pleasant climate and enables wheat cultivation. Cacaxtla, an important Olmec archaeological site, is nearby.

Tobago. Caribbean island that, along with the island of TRINIDAD, forms a nation in the southernmost LESSER ANTILLES island chain; capital is Port-of-Spain. A peaceful, rural island in contrast to bustling, cosmopolitan Trinidad. Population of the nation was 60,874 in 2011.

Toronto. Capital of ONTARIO, Canada. Located on the northern shore of Lake Ontario, this is the largest city in Canada, with a metropolitan population of 2,731,571 in 2016. Its CN Tower was the tallest freestanding structure in the world until it was surpassed by Burj Khalifa in Dubai, United Arab Emirates, in 2009. Toronto has a large ethnic population of Chinese, Italians, and other groups, and is the heart of Canadian cultural and economic development.

Torreon. City in COAHUILA state, in north-central Mexico. Established in 1887 as a rail center by Mexican, US, British, and French interests, it now is a modern, economically prosperous city in the CHIHUAHUAN DESERT. Population was 679,289 in 2015.

Tortola. Largest and most populated of the British VIRGIN ISLANDS in the CARIBBEAN SEA. It is approximately 10 square miles (26 sq. km.) in area. Population in 2010 was 23,491.

Trans-Canada Highway. World's longest national road, extending, from east to west, across all ten Canadian provinces for more than 4,860 miles (7,825 km.). It runs between VICTORIA, BRITISH COLUMBIA, and ST. JOHN'S, NEWFOUNDLAND AND LABRADOR. From NEW BRUNSWICK, the highway crosses the Confederation Bridge to PRINCE EDWARD ISLAND, and then is linked by car ferries with NOVA SCOTIA, and NEWFOUNDLAND AND LABRADOR. The highway was dedicated in 1962 but was not completed until 1965.

Trinidad. CARIBBEAN island that, along with the island of TOBAGO, forms a nation in the southernmost LESSER ANTILLES island chain; capital is Port-of-Spain. Located 7 miles (11 km.) off the coast of Venezuela, outside the path of most devastating Caribbean hurricanes. Its mountains are thought to be the northernmost Andes Mountains, making it a part of the South American continent. Crude oil production has led to a stable, thriving economy and in 2020, Trinidad and Tobago had the strongest economy in the Caribbean. Population in 2020 was 1,208,789. Steel band music originated here. Its annual pre-Lenten Carnival celebration is the largest and liveliest in the CARIBBEAN region.

Tucson. Second-largest city in ARIZONA (2019 population, 548,073). Located on the Santa Cruz River, at 2,390 feet (728 meters) elevation. Spanish Jesuit Eusebio Kino explored the area in the early 1690s. The Spanish established a settlement in 1776. Tucson was a territorial capital from 1867 to 1877.

Turks. British crown colony comprising two low-lying, coral islands in the Atlantic Ocean. Located 575 miles (927 km.) southeast of MIAMI, Florida, and 90 miles (145 km.) north of HAITI; separated from the CAICOS islands by the Christopher Columbus Passage. Grand Turk Island is the capital and seat of government; Salt Cay has only 186 inhabitants (2006). Banking and insurance institutions are a major part of the economy, as is tourism.

Tuxtla Gutierrez. Capital of the Mexican state of CHIAPAS. A modern, working city, it replaced SAN CRISTÓBAL DE LAS CASAS as capital of Chiapas in 1892. It is at the center of a thriving coffee-growing zone and had a 2015 population of 598,710.

Utah. Became forty-fifth U.S. state on January 4, 1896. Located in the west-central area of the United States. Total area of 84,899 square miles (219,887 sq. km.) with 2019 population of 3,205,958. Name comes from the Navajo word for "higher up." The first European to enter the region was Spaniard Juan Maria Antonio de Rivera in 1765. In 1846 Mormons, persecuted in OHIO, MISSOURI, and ILLINOIS for religious beliefs, moved into what was then Mexico. In 1847 Brigham Young established the region's first permanent European settlement. After the

Mexican War (1846-1848), the region was transferred to the United States. Brigham Young lobbied for creation of new state called Deseret, which included all or part of eight present-day states. In 1850 Congress created new territories, one of which was Utah. Brigham Young was appointed the first territorial governor, but was suspended after reports of disregard of federal authority by Mormon leaders. Utah's petitions for statehood were consistently refused until 1890, when Mormon leadership mandated church members to abstain from polygamy. The capital and largest city is SALT LAKE CITY (2019 population, 200,567).

Vancouver. Largest city in BRITISH COLUMBIA and the third-largest city in Canada. Located near the mouth of the FRASER RIVER in the southwestern part of British Columbia. The city had 631,486 residents in 2016; the metropolitan area, which includes numerous adjacent cities, boosts the population to more than 2.5 million (2016). Originally settled in the 1870s as a sawmill town, Vancouver has grown into a major shipping port and the industrial, financial, and commercial center of the province. A large Asian population —upwards of 45 percent of the city's residents—is found here.

Vancouver Island. Largest island on the Pacific coast of North America. Located off the coast of BRITISH COLUMBIA. Discovered by James Cook in 1778 and named after George Vancouver, who surveyed the island in 1792. Population was 870,297 in 2019.

Veracruz. Mexican state on the lush tropical plains along the Gulf of Mexico; capital is Jalapa. In 1519 Spanish conquistador Hernán Cortés disembarked here, burned his boats, and set off to conquer the Aztecs. For the next three centuries, the vibrant port of Veracruz shipped huge quantities of gold and silver back to Europe. Population was 8,112,505 in 2015.

Vermont. Became fourteenth US state on March 4, 1791. Located in New England. Total area of 9,614 square miles (24,901 sq. km.), with a 2019 population of 623,989. Name comes from French word meaning "green mountain." Explored by Samuel de Champlain in 1609 and permanently settled by British in 1724. NEW HAMPSHIRE, NEW YORK, and MASSACHUSETTS all attempted land claims, leading to the formation of Green Mountain Boys (of which Ethan Allen was a member). Montpelier (2019 population, 7,372) is capital; Burlington (2019 population, 42,819) is largest city.

Victoria. Capital city of BRITISH COLUMBIA. Located at the southern tip of VANCOUVER ISLAND on Canada's west coast, with a population of 85,792 in 2019. Known as the retirement capital of Canada thanks to its mild climate. The city is the western terminus of the TRANS-CANADA HIGHWAY.

Villahermosa. Capital of the Mexican state of TABASCO along the southern Gulf of Mexico coast. Located on the banks of the Grijalva River. Now a friendly, bustling city, it was founded in the late sixteenth century by coastal inhabitants who were forced to move inland by repeated pirate attacks. La Venta, 73 miles (117 km.) to the west of Villahermosa, is the site of the most important ancient Olmec settlement.

Virgin Gorda. Second-largest of the British VIRGIN ISLANDS in the CARIBBEAN SEA. It has an area of 8 square miles (21 sq. km.).

Virgin Islands, British. Approximately fifty islands, inlets, and cays in the CARIBBEAN SEA. TORTOLA, VIRGIN GORDA, JOST VAN DYKE, Great Camanoe, Norman, PETER, Salt, Cooper, Ginger, Dead Chest, and ANEGADA are the largest of these islands. Very hilly and volcanic, except for Anegada. Close to the US-owned VIRGIN ISLANDS, but quieter and more relaxed. Tortola, at 10 square miles (26 sq. km.), is the largest and most populated of the islands; Virgin Gorda is second in population (3,930 in 2010). English is the official language.

Virgin Islands, US. US territory in the CARIBBEAN SEA, comprising the islands of ST. THOMAS, ST. CROIX, and ST. JOHN. Located east of PUERTO RICO, 1,000 miles (1,600 km.) from the southern tip of the US mainland. They were named by Christopher Columbus for their pristine beauty. Collectively, they were called the Danish West

Indies until the United States purchased them from Denmark for strategic reasons in 1917. The islands' 106,977 inhabitants are US citizens. Poor agricultural land and insufficient water supply limit most food production to local markets. Tourism is the main industry, with some industrial development. Charlotte Amalie, on St. Thomas, is the major urban and commercial center of the islands.

Virginia. Tenth of the thirteen original US states; ratified Constitution on June 25, 1788. Located along mid-Atlantic Coast. Total area of 42,774 square miles (110,785 sq. km.) with a 2019 population of 8,535,519. First permanent North American English settlement was Jamestown, Virginia, in 1607. Named for the Virgin Queen, Elizabeth I of England. Richmond (2019 population, 230,436) became capital in 1779 and was the capital of the Confederate States of America 1861-1865. Largest city is Virginia Beach (2019 population, 499,974).

Washington, D.C. City and capital of the United States. The city of Washington, which is coextensive with the DISTRICT OF COLUMBIA, is located at the navigational head of the Potomac River between MARYLAND to the northeast and VIRGINIA in the southwest. Congress chose the District of Columbia in 1790 as the site for a permanent seat of government for the new nation. Its area is 68 square miles (176 sq. km.). Its population was 705,749 in 2019.

Washington State. Became forty-second state on November 11, 1889. Located in the Pacific Northwest of the United States. Total area of 71,300 square miles (184,665 sq. km.) with 2019 population of 7,614,893. Named after George Washington. In 1792 British captain George Vancouver surveyed the coast and inland waters, and US captain Robert Gray discovered the mouth of the COLUMBIA RIVER. Therefore, both countries had claims to the territory. Acquired by the United States in 1803 via the Louisiana Purchase. In 1848 Oregon Territory was created; Washington Territory was created from that in 1853. Olympia (2019 population,

52,882) is the capital; SEATTLE (2019 population, 753,675) is the largest city.

West Indies. Archipelago of more than 7,000 mostly uninhabited islands, cays, coral reefs, and rocks that form the northern and eastern boundaries of the CARIBBEAN SEA. The islands fall into one of three groupings: coral islands such as the BAHAMAS; GREATER ANTILLES islands, such as CUBA, PUERTO RICO, HISPANIOLA, and JAMAICA, which are older, worn-down peaks of sunken mountains; and LESSER ANTILLES islands—an arc of smaller volcanically formed islands. Eastern trade winds moderate the tropical heat, and the rich volcanic or alluvial soils promote agriculture. Sugar and bananas are the major crops. Tourism is a large revenue source. Population was 43,163,817 in 2019.

West Virginia. Became thirty-fifth state on June 20, 1863. Located in US mid-Atlantic area. Total area of 24,229 square miles (62,755 sq. km.), with a 2019 population of 1,792,147. Until 1863, was part of VIRGINIA, but split off on its own owing to Virginia's secession from the United States. Capital and largest city is Charleston (2019 population, 137,566).

Western Hemisphere. Portion of Earth that contains North and South America; generally demarcated as within longitude 20° west and 160° east. Historically known as the NEW WORLD.

Whitehorse. Capital and largest city of the YUKON TERRITORY, Canada. Located in the south-central part of the territory, on the banks of the Yukon River. Became the capital of the territory in 1953. Transportation, communications, mining, and government functions are the main industries. Its population was 25,085 in 2016.

Whitney, Mount. Highest mountain in the contiguous United States at 14,494 feet (4,418 meters). Located in the SIERRA NEVADA, it is named for Josiah Dwight Whitney, the geologist who led the expedition that discovered it in 1864.

Windward Islands. Five volcanic islands—GRENADA, ST. LUCIA, DOMINICA, MARTINIQUE and ST. VINCENT AND THE GRENADINES—in the CARIBBEAN SEA. Still-active volcanoes periodically erupt. The islands were colonized by both the

British and the French; most inhabitants are descendants of African slaves.

Winnipeg. Capital and largest city of MANITOBA, Canada. Located at the confluence of the Red and Assiniboine rivers, it had a population of 705,244 in 2016. Originally a fur-trading fort, it grew into the "Gateway to the West" as thousands of immigrants moving west traveled through it. Its Symington Railway Yard is one of the biggest in the world.

Winnipeg, Lake. The twelfth-largest lake in the world, located in the southern part of MANITOBA, Canada. It is 264 miles (425 km.) long and up to 68 miles (109 km.) wide, with an area of 9,417 square miles (24,389 sq. km.). The name comes from the Cree Indian words for "muddy water." The lake is fairly shallow, with an average depth of about 50 feet (15 meters). It is important for shipping and commercial fishing, and its southern shore is a major resort area for the provincial capital of WINNIPEG, 40 miles (64 km.) south.

Wisconsin. Became thirtieth US state on May 29, 1848. Located in north-central (Upper Midwest) region. Total area is 65,498 square miles (169,639 sq. km.), with a 2019 population of 5,822,434. Name comes from Ojibwa word that may mean "gathering of the waters" or "place of the beaver." First explored by French in 1630s. French controlled area (first permanent settlement in 1766) until French and Indian War, after which the British assumed control until 1783. Part of NORTHWEST TERRITORY 1787-1800. Capital is Madison (2019 population, 259,680) and largest city is MILWAUKEE (2019 population, 590,157).

Wyoming. Became forty-fourth state on July 10, 1890. Located in northern ROCKY MOUNTAINS/ GREAT PLAINS of the United States. Total area of 97,814 square miles (253,336 sq. km.) with 2019 population of 578,759 (making it the smallest state in population). The name is a native word for "at big plains." Acquired by the United States in 1803 via the Louisiana Purchase. The first European settlement was in 1834 at Fort Laramie. The capital and largest city is Cheyenne (2019 population, 64,235).

Yaqui River. River in the SIERRA MADRE OCCIDENTAL of Mexico. It is used to irrigate arid farmlands on its journey to the Gulf of CALIFORNIA.

Yellowknife. Territorial capital and largest city in Canada's NORTHWEST TERRITORIES. Located on the northern shore of GREAT SLAVE LAKE in the south-central part of Northwest Territories. Founded in 1935, one year after gold was discovered in the area, and derived its name from the Yellowknife band of Athabascan Indians who live in the region. Its population was 19,569 in 2016.

Yellowstone National Park. First national park, established in 1872. Located in the ROCKY MOUNTAINS, it covers 3,472 square miles (8,992 sq. km.) in WYOMING, MONTANA, and IDAHO. The park is known for wildlife, geysers, waterfalls, and canyons.

Lower Yellowstone Falls in Yellowstone National Park. (Scott Catron)

York Factory. National historic site on the southwestern shore of HUDSON BAY in north-central Canada, in the northeastern corner of MANITOBA. Originally founded as the Hudson's Bay Company post (Fort Nelson) in 1682, it was later renamed York Factory. It was the chief port, supply depot, and headquarters for the fur-trading centers of northern Canada. The trading post closed in 1957, ending 275 years of nearly continuous operation. It is accessible only by air or canoe.

Yosemite National Park. Established in 1890 in the SIERRA NEVADA, covering 1,189 square miles (3,079 sq. km.) of CALIFORNIA. The park is famous for beautiful scenery, such as waterfalls and cataracts.

Yucatán. Mexican state on the YUCATÁN PENINSULA; capital is MÉRIDA. The state has no rivers or lakes above ground; it depends on subterranean rivers and freshwater wells called *cenotes* for its water supply. Coastal estuaries are sanctuaries for exotic resident birds and many wintering migratory birds. The state's main industries are tourism, fishing, and commerce. CHICHÉN ITZÁ, one of the finest and best-preserved Mayan architectural sites in the northern Yucatán Peninsula, was a commercial, religious, and military center until the thirteenth century. The ruins comprise three distinct zones spread across a plain 75 miles (120 km.) east of Mérida. Population was 2,097,175 in 2015.

Yucatán Peninsula. The flattest part of Mexico, occupying an area of approximately 73,000 square miles (190,000 sq. km.). Includes the states of CAMPECHE, YUCATÁN, and QUINTANA ROO and famous resorts such as CANCÚN and COZUMEL. When rain falls here, it seeps into the earth, gradually dissolving the limestone and creating underground caverns. When the roofs of caverns collapse, sinkholes, called *cenotes*, form; the Maya used them as wells. Some of the finest archaeological sites in the AMERICAS are found here. Mayan ruins at CHICHÉN ITZÁ and Uxmal draw many tourists each year.

Yukon Territory. Region in the northwestern corner of Canada; the capital is WHITEHORSE, through which the ALASKA HIGHWAY passes. Bounded on the east by the NORTHWEST TERRITORIES, on the north by the Arctic Ocean, on the west by the state of ALASKA, and on the south by the province of BRITISH COLUMBIA. The area has many high plateaus and mountains, including the highest peak in Canada. Mining is the main industry and dates to the Klondike gold rushes of the 1890s. The total land area is 183,287 square miles (474,711 sq. km.)—about one-third the size of ALASKA and COLORADO combined; the population was 41,293 in 2020.

Zacatecas. Mexican state; capital city of the same name. The city is Mexico's second-highest, at 8,075 feet (2,461 meters), and sits on a high plateau surrounded by arid hills. Already an old town when the Spaniards arrived in 1546, Zacatecas soon supplied silver to the Spanish crown and became one of Mexico's richest colonial treasures. Population was 1,579,209 in 2015.

Carol Ann Gillespie; Catherine A. Hooey;
James D. Lowry, Jr.; Dana P. McDermott;
Mika Roinila

APPENDIX

THE EARTH IN SPACE

THE SOLAR SYSTEM

Earth's solar system comprises the Sun and its planets, as well as all the natural satellites, asteroids, meteors, and comets that are captive around it. The solar system formed from an interstellar cloud of dust and gas, or nebula, about 4.6 billion years ago. Gravity drew most of the dust and gas together to make the Sun, a medium-size star with an estimated life span of 10 billion years. Its system is located in the Orion arm of the Milky Way galaxy, about two-thirds of the way out from the center.

During the Sun's first 100 million years, the remaining rock and ice smashed together into increasingly larger chunks, or planetesimals, until the planets, moons, asteroids, and comets reached their present state. The resulting disk-shaped solar system can be divided into four regions—terrestrial planets, giant planets, the Kuiper Belt, and the Oort Cloud—each containing its own types of bodies.

Terrestrial Planets
In the first region are the terrestrial (Earth-like) planets Mercury, Venus, Earth, and Mars. Mercury, the nearest to the Sun, orbits at an average distance of 36 million miles (58 million km.) and Mars, the farthest, at 142 million miles (228 million km.). Astronomers call the distance from the Sun to Earth (93 million miles/150 million km.) an astronomical unit (AU) and use it to measure planetary distances.

Terrestrial planets are rocky and warm and have cores of dense metal. All four planets have volcanoes, which long ago spewed out gases that created atmospheres on all but Mercury, which is too close to the Sun to hold onto an atmosphere. Mercury is heavily cratered, like the earth's moon. Venus has a permanent thick cloud cover and a surface temperature hot

enough to melt lead. The air on Mars is very thin and usually cold, made mostly of carbon dioxide. Its dry, rock-strewn surface has many craters. It also has the largest known volcano in the solar system, Olympus Mons, which is 16 miles (25 km.) high.

Average temperatures and air pressures on Earth allow liquid water to collect on the surface, a unique feature among planets within the solar system. Meanwhile, Earth's atmosphere—mostly nitrogen and oxygen—and a strong magnetic field protect the surface from harmful solar radiation. These are the conditions that nurture life, according to scientists. It is widely accepted that Mars had abundant water very early in its history, but all large areas of liquid water have since disappeared. A fraction of this water is retained on modern Mars as both ice and locked into the structure of abundant water-rich materials, including clay minerals (phyllosilicates) and sulfates. Studies of hydrogen isotopic ratios indicate that asteroids and comets from beyond 2.5 AU provide the source of Mars' water. Like Earth, Mars has polar ice caps, although those on Mars are made up mostly of carbon dioxide ice (dry ice), while those on Earth are made up of water ice.

A single natural satellite, the Moon, orbits Earth, probably created by a collision with a huge planetesimal more than 4 billion years ago. Mars has two tiny moons that may have drifted to it from the asteroid belt. A broad ring from 2 to 3.3 AU from the Sun, this belt is composed of space rocks as small as dust grains and as large as 600 miles (1,000 km.) in diameter. Asteroids are made of mineral compounds, especially those containing iron, carbon, and silicon. Although the asteroid belt contains

FORMATION OF THE SOLAR SYSTEM

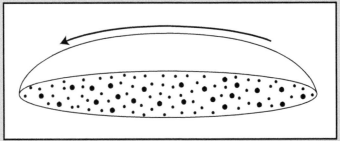

1. The solar system began as a cloud of rotating interstellar gas and dust.

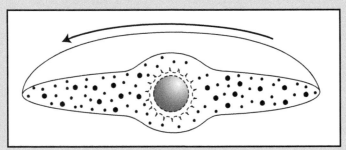

2. Gravity pulled some gases toward the center.

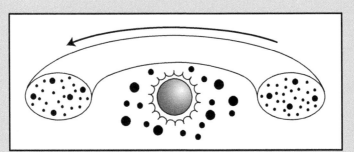

3. Rotation accelerated, and centrifugal force pushed icy, rocky material away from the proto-Sun. Small planetesimals rotate around the Sun in interior orbits.

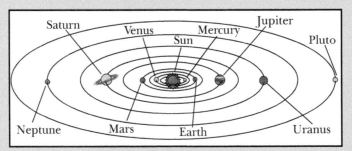

4. The interior, rocky material formed Mercury, Venus, Earth, and Mars. The outer, gaseous material formed Jupiter, Saturn, Uranus, Neptune, and Pluto.

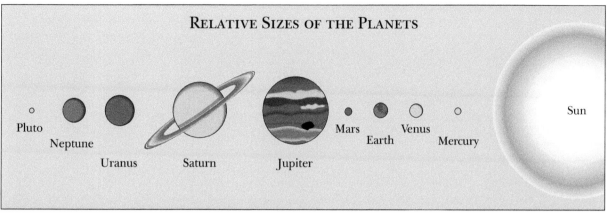

RELATIVE SIZES OF THE PLANETS

Note: The size of the Sun and distances between the planets are not to scale.; Source: Data are from Jet Proulsion Laboratory, California Institute of Technology. The Deep Space Network. Pasadena, Calif.:JPL, 1988, p. 17.

enough material for a planet, one did not form there because Jupiter's gravity prevented the asteroids from crashing together. The belt separates the first region of the solar system from the second.

The Giant Planets

The second region belongs to the gas giants Jupiter, Saturn, Uranus, and Neptune. The closest, Jupiter, is 5.2 AU from the Sun, and the most distant, Neptune, is 30.11 AU. Jupiter is the largest planet in the solar system, its diameter 109 times larger than Earth's. The giant planets have solid cores, but most of their immense size is taken up by hydrogen, helium, and methane gases that grow thicker and thicker until they are like sludge near the core. On Jupiter, Saturn, and Uranus, the gases form wide bands over the surface. The bands sometimes have immense circular storms like hurricanes, but hundreds of times larger. The Great Red Spot of Jupiter is an example. It has winds of up to 250 miles (400 km.) per hour, and is at least a century old.

These planets have such strong gravity that each has attracted many moons to orbit it. In fact, they are like miniature solar systems. Jupiter has the most moons—eighteen—and Neptune has the fewest—eight—but Neptune's moon Triton is the largest of all. Most moons are balls of ice and rock, but Jupiter's Europa and Saturn's Titan may have liquid water below ice-bound surfaces. Several moons appear to have volcanoes, and a wispy atmosphere covers Titan. Additionally, the giant planets have rings of broken rock and ice around them, no more than 330 feet (100 meters) thick. Saturn's hundreds of rings are the brightest and most famous.

The Kuiper Belt

The third region of the solar system, the Kuiper Belt, contains the dwarf planet, Pluto. Pluto has a single moon, Charon. It does not orbit on the same plane, called the ecliptic, as the rest of the planets do. Instead, its orbit diverges more than seventeen degrees above and below the ecliptic. Its orbit's oval shape brings Pluto within the orbit of Neptune for a large percentage of its long year, which is equal to 248 Earth years. Two-thirds the size of the earth's moon, Pluto has a thin, frigid methane atmosphere. Charon is half Pluto's size and orbits less than 32,000 miles (20,000 km.) from Pluto's surface. Some astronomers consider Pluto and Charon to be a double planet.

The Kuiper Belt holds asteroids and the "short-period" comets that pass by Earth in orbits of 20 to 200 years. These bodies are the remains of

OTHER EARTHS

By the year 2000 astronomers had detected twenty-eight planets circling stars in the Sun's neighborhood of the galaxy. Planets, they think, are common. Those found were all gas giants the size of Saturn or larger. Earth-size planets are much too small to spot at such great distances. Where there are gas giants, there also may be terrestrial dwarfs, as in Earth's solar system. Where there are terrestrial planets, there may be liquid water and, possibly, life.

325

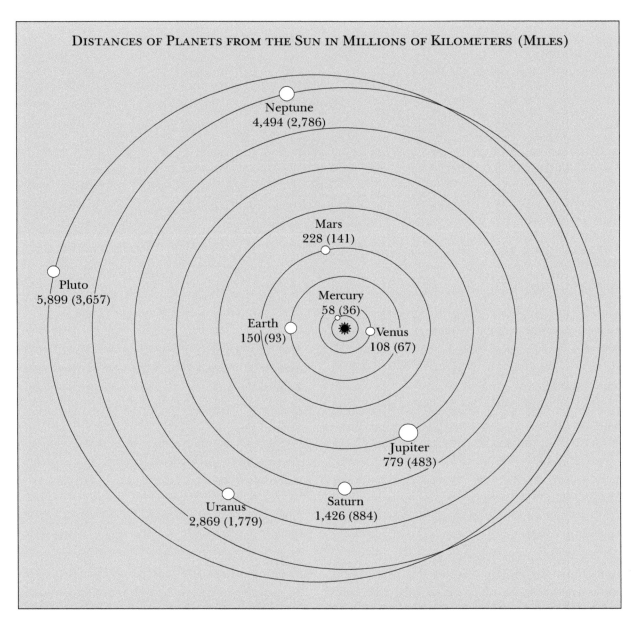

DISTANCES OF PLANETS FROM THE SUN IN MILLIONS OF KILOMETERS (MILES)

Neptune
4,494 (2,786)

Mars
228 (141)

Pluto
5,899 (3,657)

Mercury
58 (36)

Earth
150 (93)

Venus
108 (67)

Jupiter
779 (483)

Uranus
2,869 (1,779)

Saturn
1,426 (884)

planet formation and did not collect into planets because distances between them are too great for many collisions to occur. Most of them are loosely compacted bodies of ice and mineral—"dirty snowballs," as they were termed by the famous astronomer Fred Lawrence Whipple (November 5, 1906–August 30, 2004). An estimated 200 million Kuiper Belt objects orbit within a band of space from 30 to 50 AU from the Sun.

The Oort Cloud

In contrast to the other regions of the solar system, the Oort Cloud is a spherical shell surrounding the entire solar system. It is also a collection of com-

ets—as many as two trillion, scientists calculate. The inner edge of the cloud forms at a distance of about 20,000 AU from the Sun and extends as far out as 100,000 AU. The Oort Cloud thus gives the solar system a theoretical diameter of 200,000 AU—a distance so vast that light needs more than three years to cross it. No astronomer has yet detected an Oort Cloud object, because the cloud is so far away. Occasionally, however, gravity from a nearby star dislodges an object in the cloud, causing it to fall toward the Sun. When observers on Earth see such an object sweep by in a long, cigar-shaped orbit, they call it a long-period comet.

The outer edge of the Oort Cloud marks the farthest reach of the Sun's gravitational power to bind bodies to it. In one respect, the Oort Cloud is part of interstellar space.

In addition to light, the Sun sends out a constant stream of charged particles—atoms and subatomic particles—called the solar wind. The solar wind shields the solar system from the interstellar medium, but it only does so out to about 100 AU, a boundary called the heliopause. That is a small fraction of the distance to the Oort Cloud.

Roger Smith

EARTH'S MOON

The fourth-largest natural satellite in the solar system, Earth's moon has a diameter of 2,159.2 miles (3,475 km.)—less than one-third the diameter of Earth. The Moon's mass is less than one-eightieth that of Earth.

The Moon orbits Earth in an elliptical path. When it is at perigee (when it is closest to Earth), it is 221,473 miles (356,410 km.) distant. When it is at apogee (farthest from Earth), it is 252,722 miles (406,697 km.) distant.

The Moon completes one orbit around Earth every 27.3 Earth days. Because it rotates at about the same rate that it orbits the earth, observers on Earth only see one side of the Moon. The changing angles between Earth, the Sun, and the Moon determine how much of the Moon's illuminated surface can be seen from Earth and cause the Moon's changing phases.

Volcanism

Naked-eye observations of the Moon from Earth reveal dark areas called *maria*, the plural form of the Latin word *mare* for sea. The maria are the remains of ancient lava flows from inside gigantic impact craters; the last eruptions were more than 3 billion years ago. The lava consists of basalt, similar in composition to Earth's oceanic crust and many volcanoes. The maria have names such as Mare Serenitatis (15° to 40°N, 5° to 20°E) and Mare Tranquillitatis (0° to 20°N, 15° to 45°E). Some of the smaller dark areas on the Moon also have names that are water-related: lacus (lake), sinus (bay), and palus (marsh).

Impact Craters

Observing the Moon with an optical aid, such as a telescope or a pair of binoculars, provides a closer view of impact craters. Impact craters of various sizes cover 83 percent of the Moon's surface. More than 33,000 craters have been counted on the Moon.

One of the easiest craters to observe from the Earth is Tycho. Located at 43.3°S, longitude 11.2 degrees west, it is about 50 miles (85 km.) wide. Surrounding Tycho are rays of dusty material, known as

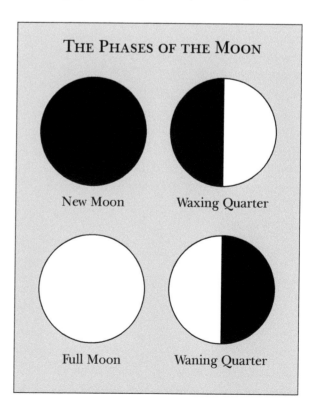

THE PHASES OF THE MOON

New Moon

Waxing Quarter

Full Moon

Waning Quarter

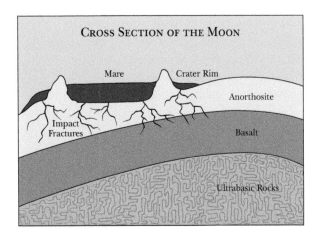

CROSS SECTION OF THE MOON

Mare
Crater Rim
Anorthosite
Impact Fractures
Basalt
Ultrabasic Rocks

ejecta, that appear to radiate from the crater. When an object from space, such as a meteoroid, slams into the Moon's surface, it is vaporized upon impact. The dust and debris from the interior of the crater fall back onto the lunar surface in a pattern of rays. Because the ejecta is disrupted by subsequent impacts, only the youngest craters still have rays. Sometimes, pieces of the ejecta fall back and create smaller craters called secondary craters. The ejecta rays of Tycho extend to almost 1,865 miles (3,000 km.) beyond the crater's edge.

Other Lunar Features

Near the crater called Archimedes is the Apennines mountain range, which has peaks nearly 20,000 feet (60,000 meters) high—altitudes comparable to South America's Andes.

The Moon also has valleys. Two of the most well known are the Alpine Valley, which is about 115 miles (185 km.) long; and the Rheita Valley, located about 155 miles (250 km.) from the Stevinus crater, which is 238 miles (383 km.) long, 15.5 miles (25 km.) wide, and 2,000 feet (609 meters) deep.

Smaller than valleys and resembling cracks in the lunar surface are features called rilles, which are thought to be places of ancient lava flow. Many rilles can be seen near the Aristarchus crater. Rilles are often up to 3 miles (5 km.) wide and can stretch for more than 104 miles (167 km.).

A wrinkle in the lunar surface is called a ridge. Many ridges are found around the boundaries of the maria. The Serpentine Ridge cuts through Mare Serenitatis.

Exploration of the Moon

Robotic spacecraft were the first visitors to explore the Moon. The Russian spacecraft Luna 1 made the first flyby of the Moon in January, 1959. Eight months later, Luna 2 made the first impact on the Moon's surface. In October, 1959, Luna 3 was the first spacecraft to photograph the side of the Moon not visible from Earth. In 1994 the United States' *Clementine* spacecraft was the first probe to map the Moon's composition and topography globally.

The first humans to land on the Moon were the U.S. astronauts Neil Armstrong and Edwin "Buzz" Aldrin. On July 20, 1969, they landed in the *Eagle* lunar module, during the Apollo 11 mission. Armstrong's famous statement, "That's one small step for man, one giant leap for mankind," was heard around the world by millions of people who watched the first humans set foot on the lunar surface, at the Sea of Tranquillity. The last twentieth century human mission to reach the lunar surface, Apollo 17, landed there in December, 1972. Astronauts Gene Cernan and geologist Jack Schmitt landed in the Taurus-Littrow Valley (20°N, 31°E).

Noreen A. Grice

THE SUN AND THE EARTH

Of all the astronomical phenomena that one can consider, few are more important to the survival of life on Earth than the relationship between Earth and the Sun. With the exception of small amounts of residual (endogenic) energy that have remained inside the earth from the time of its formation some 4.5 billion years ago and which sustain some specialized forms of life along some oceanic rift systems, almost all other forms of life, including human, depend on the exogenic light and energy that the earth receives directly from the Sun.

The enormous variety of ecosystems on Earth are highly dependent on the angles at which the Sun's rays strike Earth's spherical surface. These angles, which vary greatly with latitude and time of year, determine many commonly observed phenomena, such as the height of the Sun above the horizon, the changing lengths of day and night throughout the year, and the rhythm of the seasons. Daily and seasonal changes have profound effects on the many climatic regions and life cycles found on earth.

The Sun

The center of Earth's solar system, the Sun is but one ordinary star among some 100 billion stars in an ordinary cluster of stars called the Milky Way galaxy. There are at least 10 billion galaxies in the universe, each with billions of stars. Statistically, the chances are good that many of these stars have their own solar systems. Late twentieth century astronomical observations discovered the presence of what appear to be planets, large ones similar in size to Jupiter, orbiting other stars.

Earth's Sun is an average star in terms of its physical characteristics. It is a large sphere of incandescent gas that has a diameter more than 100 times that of Earth, a mass more than 300,000 times that of Earth, and a volume 1.3 million times that of Earth. The Sun's surface gravity is thirty-four times that of Earth.

The conversion of hydrogen into helium in the Sun's interior, a process known as nuclear fusion, is the source of the Sun's energy. The amount of mass that is lost in the fusion process is miniscule, as evidenced by the fact that it will take perhaps 15 million years for the Sun to lose one-millionth of its total mass. The Sun is expected to continue shining through another several billion years.

Earth Revolution

The earth moves about the Sun in a slightly elliptical orbit called a revolution. It takes one year for the earth

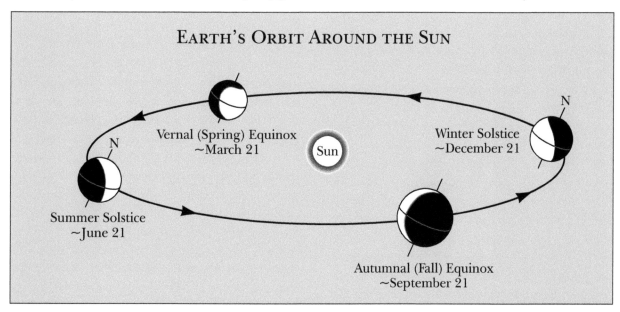

EARTH'S ORBIT AROUND THE SUN

Vernal (Spring) Equinox
~March 21

Winter Solstice
~December 21

Sun

Summer Solstice
~June 21

Autumnal (Fall) Equinox
~September 21

to make one revolution at an average orbital velocity of about 29.6 kilometers per second (18.5 miles per second). Earth-sun relationships are described by a tropical year, which is defined as the period of time (365.25 average solar days) from one vernal equinox to another. To balance the tropical year with the calendar year, a whole day (February 29) is added every fourth year (leap year). Other minor adjustments are necessary so as to balance the system.

Perihelion and Aphelion

The average distance between Earth and the Sun is approximately 93 million miles (150 million km.). At that distance, sunlight, which travels at the speed of light (186,000 miles/300,000 kilometers per second), takes about 8.3 minutes to reach the earth. Since the earth's orbit is an ellipse rather than a circle, the earth is closest to the Sun on about January 3—a distance of 91.5 million miles (147 million km.). This position in space is called perihelion, which comes from the Greek *peri*, meaning "around" or "near," and *helios*, meaning the Sun. Earth is farthest from the Sun on about July 4 at aphelion (Greek *ap*, "away from," and *helios*), with a distance of 152 million kilometers (94.5 million miles).

Axial Inclination

Astronomers call the imaginary surface on which Earth orbits around the Sun the plane of the ecliptic. The earth's axis is inclined 66.5 degrees to the plane of the ecliptic (or 23.5 degrees from the perpendicular to the plane of the ecliptic), and it maintains this orientation with respect to the stars. Thus, the North Pole points in the same direction to Polaris, the North Star, as it revolves about the Sun. Consequently, the Northern Hemisphere tilts away from the Sun during one-half of Earth's orbit and toward the Sun through the other half.

Winter solstice occurs on December 21 or 22, when the tilt of the Northern Hemisphere away from the Sun is at its maximum. The opposite condition occurs during summer solstice on June 21 or 22, when the Northern Hemisphere reaches its maximum tilt toward the Sun. The equinoxes occur midway between the solstices when neither the Southern nor the Northern Hemisphere is tilted toward the Sun. The

---

ECLIPSES

The Sun's diameter is 400 times larger than the moon's; however, the moon is 400 times closer to Earth than the Sun, making the two objects appear nearly the same size in the sky to observers on Earth. As the moon orbits Earth, it crosses the plane of the Earth-Sun orbit twice each month. If one of the orbit-crossing points (called nodes) occurs during a new or full moon phase, a solar or lunar eclipse can occur.

A solar eclipse occurs when the moon and the Sun appear to be in the exact same place in the sky during a new moon phase. When that happens, the moon blocks the light of the Sun for up to seven minutes. Because solar eclipses can be seen only from certain places on Earth, some people travel around the world—sometimes to remote places—to view them.

A lunar eclipse occurs when Earth is positioned between the Sun and the moon and casts its shadow on the moon. In contrast to solar eclipses, lunar eclipses are visible from every place on Earth from which the moon can be seen.

---

vernal and autumnal equinoxes occur on March 20 or 21 and September 22 or 23, respectively.

The axial inclination of 66.6 degrees (or 23.5 degrees from the perpendicular) explains the significance of certain parallels on the earth. The noon sun shines directly overhead on the earth at varying latitudes on different days—between 23.5°S and 23.5°N. The parallels at 23.5°S and 23.5°N are called the Tropics of Capricorn and Cancer, respectively.

During the winter and summer solstices, the area on the earth between the Arctic Circle (at 66.5°N) and the North Pole has twenty-four hours of darkness and daylight, respectively. The same phenomena occurs for the area between the Antarctic Circle (at 66.5°S) and the South Pole, except that the seasons are reversed in the Southern Hemisphere. At the poles, the Sun is below the horizon for six months of the year.

For those living outside the tropics (poleward of 23.5 degrees north and south latitude), the noon sun will never shine directly overhead. Hours of daylight will also vary greatly during the year. For example, daylight will range from approximately nine hours during the winter solstice to fifteen hours during the summer solstice for persons living near 40°N, such as in Philadelphia, Denver, Madrid, and Beijing.

Solar Radiation

Given the size of the earth and its distance from the Sun, it is estimated that this planet receives only about one two-billionth part of the total energy released by the Sun. However, this seemingly small amount is enough to drive the massive oceanic and atmospheric circulation systems and to support all life processes on Earth.

Solar energy is not evenly distributed on Earth. The higher the angle of the Sun in the sky, the greater the duration and intensity of the insolation.

To illustrate this, note how easy it is look at the Sun when it is very low on the horizon—near dawn and sunset. At those times, the Sun's rays have to penetrate much more of the atmosphere, so more of the sunlight is absorbed. When the Sun's rays are coming in at a low angle, the same solar energy is spread over a larger area, thereby leading to less insolation per unit of area. Thus, the equatorial region receives much more solar energy than the polar region. This radiation imbalance would make the earth decidedly less habitable were it not for the atmospheric and oceanic circulation systems (such as the warm Gulf Stream) that move the excess heat from the Tropics to the middle and high latitudes.

Robert M. Hordon

THE SEASONS

Earth's 365-day year is divided into seasons. In most parts of the world, there are four seasons—winter, spring, summer, and fall (also called autumn). In some tropical regions—those close to the equator—there are only two seasons. In areas close to the equator, temperatures change little throughout the year; however, amounts of rainfall vary greatly, resulting in distinct wet and dry seasons. The polar regions of the Arctic and Antarctic also have little variation in temperature, remaining cold throughout the year. Their seasons are light and dark, because the Sun shines almost constantly in the summer and hardly at all in the winter.

The four seasons that occur throughout the northern and southern temperate zones—between the tropics and the polar regions—are climatic seasons, based on temperature and weather changes. Winter is the coldest season; it is the time when days are short and few crops can be grown. It is followed by spring, when the days lengthen and the earth warms; this is the time when planting typically begins, and animals that hibernate (from the French word for winter) during the winter leave their dens.

Summer is the hottest time of the year. In many areas, summer is marked by drought, but other regions experience frequent thunderstorms and humid air. In the fall, the days again become shorter and cooler. This is the time when many crops are harvested. In ancient cultures, the turning of the seasons was marked by festivals, acknowledging the importance of seasonal changes to the community's survival.

Each season is defined as lasting three months. Winter begins at the winter solstice, which is the time when the Sun is farthest from the equator. In the Northern Hemisphere, this occurs on December 21 or 22, when the Sun is directly over the tropic of Capricorn. Summer begins at the other solstice, June 20 or 21 in the Northern Hemisphere, when the Sun is directly over the tropic of Cancer. The winter solstice is the shortest day of the year; the summer solstice is the longest.

Spring and fall begin on the two equinoxes. At an equinox, the Sun is directly above the earth's equator and the lengths of day and night are approximately equal everywhere on Earth. In the Northern Hemisphere, the vernal (spring) equinox occurs on March 21 or 22; in the Southern Hemisphere, it is the autumnal (fall) equinox. The Northern Hemisphere's autumnal equinox (and the Southern Hemisphere's vernal equinox) occurs September 22 or 23.

Seasons and the Hemispheres

The relationship of the seasons to the calendar is opposite in the Northern and Southern Hemispheres. On the day that a summer solstice occurs in the Northern Hemisphere, the winter solstice occurs in the Southern Hemisphere. Thus, when it is summer in the Southern Hemisphere, it is winter in the Northern Hemisphere, and vice versa.

The Sun and the Seasons

The reason why summers and winters differ in the temperate zones is often misunderstood. Many people think that winter happens when the Sun is more distant from Earth than it is in summer. What causes Earth's seasons is not the changing distances between the earth and the Sun, but the tilt of the earth's axis. A line drawn from the North Pole to the South Pole through the center of the earth (the earth's axis) is not perpendicular to the plane of the earth's orbit (the ecliptic). The earth's axis and the perpendicular to the ecliptic make an angle of 23.5 degrees. This tilts the Northern Hemisphere toward the Sun when the earth is on one side of its orbit around the Sun, and tilts the Southern Hemisphere toward the Sun when the earth moves around to the Sun's opposite side. When the Sun appears to be at its highest in the sky, and its rays are most direct, summer occurs. When the Sun appears to be at its lowest, and its rays are indirect, there is winter.

Local Phenomena

Local conditions can have important effects on seasonal weather. At locations near oceans, sea breezes develop during the day, and evenings are characterized by land breezes. Sea breezes bring cooler ocean air in toward land. This results in temperatures at the shore often being 5 to 11 degrees Fahrenheit (3 to 6 degrees Celsius) lower than temperatures a few miles inland.

At night, when land temperatures are lower than ocean temperatures, land breezes move air from the land toward the water. As a result, coastal regions have less seasonal temperature variations than inland areas do. For example, coastal areas seldom become cold enough to have snow in the winter, even though inland areas at the same latitude do.

Hailstorms

Hail usually occurs during the summer, and is associated with towering thunderstorm clouds, called cumulonimbus. Hail is occasionally confused with sleet. Sleet is a wintertime event, and occurs when warmer layers of air sit above freezing layers near the ground. Rain that forms in the warmer, upper layer solidifies into tiny ice pellets in the lower, subfreezing layer before hitting the ground.

Hail is an entirely different phenomenon. When cold air plows into warmer, moist air—called a cold front boundary—powerful updrafts of rising air can be created. The warm, moist air propelled upward by the heavier cold air can reach velocities approaching 100 miles (160 kilometers) per hour. Ice crystals form above the freezing level in the cumulonimbus clouds and fall into lower, warmer parts of the clouds, where they become coated with water. Picked up by an updraft, the coated ice crystals are carried back to a higher, colder levels where their water coatings freeze. This cycle can repeat many times, producing hailstones that have multiple, concentric layers of ice.

Hailstorms can be very damaging. Hail can ruin crops, dent car bodies, crack windshields, and injure people. The Midwest of the United States is particularly susceptible to hailstorms. There, warm, moist air from the Gulf of Mexico often meets much colder, drier air originating in Canada. This combination produces the extreme atmospheric instability necessary for that kind of weather.

Alvin S. Konigsberg

EARTH'S INTERIOR

EARTH'S INTERNAL STRUCTURE

Earth is one of the nine known planets in the Sun's solar system that formed from a giant cloud of cosmic dust called a nebula. This event is thought to have happened between 4.44 billion years ago (based on the age of the oldest-known Moon rock) and 4.56 billion years ago (the age of meteorite bombardment). After Earth's formation, heat released by colliding particles combined with the heat energy released by the decay of radioactive elements to cause some or all of Earth's interior to melt. This melting began the process of differentiation, which allowed the heavier elements, mainly iron and nickel, to sink toward Earth's center while the lighter, rocky components moved upward, as a result of the contrast in density of the earth's forming elements.

This process of differentiation was probably the most important event of Earth's early history. It changed the planet from a homogeneous mixture with neither continents nor oceans to a planet with three layers: a dense core beginning at 1,800 miles (2,900 km.) deep and ending at Earth's center, 3,977 miles (6,400 km.) below the surface; a mantle beginning between 3 and 44 miles (5-70 km.) deep and ending at Earth's core; and a crust going from Earth's surface to about 3-6 miles (5-10 km.) deep for oceanic crust and 22-44 miles (35-70 km.) deep for continental crust.

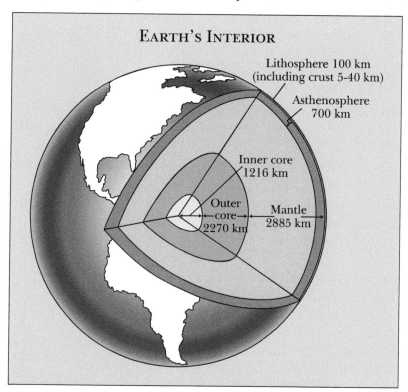

EARTH'S INTERIOR

Lithosphere 100 km (including crust 5-40 km)

Asthenosphere 700 km

Inner core 1216 km

Outer core 2270 km

Mantle 2885 km

Layering of the Earth
Earth's layers can be classified either by their composition (the traditional method) or by their mechanical behavior (strength). Compositional classification identifies several distinct concentric layers, each with its own properties. The outermost layer of Earth is the crust or skin. This is divided into continental and oceanic crusts. The continental crust varies in thickness between 22 and 25 miles (35 and 40 km.) under flat continental regions and up to 44 miles (70 km.) under high mountains. The oceanic crust is made up of igneous rocks rich in iron and magnesium, such as basalt and peridotite. The upper continental crust is composed mainly of alumino-silicates. The old-

PROPERTIES OF SEISMIC WAVES

Seismologists use two types of body waves—primary (P-waves) and secondary (S-waves) waves—to estimate seismic velocities of the different layers within the earth. In most rock types P-waves travel between 1.7 and 1.8 times more quickly than S-waves; therefore, P-waves always arrive first at seismographic stations. P-waves travel by a series of compressions and expansions of the material through which they travel. P-waves can travel through solids, liquids, or gases. When P-waves travel in air, they are called sound waves.

The slower S-waves, also called shear waves, move like a wave in a rope. This movement makes the S-wave more destructive to structures like buildings and highway overpasses during earthquakes. Because S-waves can travel only through solids and cannot travel through Earth's outer core, seismologists concluded that Earth's outer core must be liquid or at least must have the properties of a fluid.

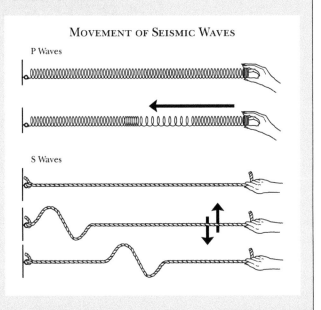

MOVEMENT OF SEISMIC WAVES

P Waves

S Waves

est continental crustal rock exceeds 3.8 billion years, while oceanic crustal rocks are not older than 180 million years. The oceanic crust is heavier than the continental crust.

Earth's next layer is the mantle, which is made up mostly of ferro-magnesium silicates. It is about 1,800 miles (2,900 km.) thick and is separated into the upper and lower mantle. Most of Earth's internal heat is contained within the mantle. Large convective cells in the mantle circulate heat and may drive plate-tectonic processes.

The last layer is the core, which is separated into the liquid outer core and the solid inner core. The outer core is 1,429 miles (2,300 km.) thick, twice as thick as the inner core. The outer core is mainly composed of a nickel-iron alloy, while the inner core is almost entirely composed of iron. Earth's magnetic field is believed to be controlled by the liquid outer core.

In the mechanical layering classification of the earth's interior, the layers are separated based on mechanical properties or strength (resistance to flowing or deformation) in addition to composition. The uppermost layer is the lithosphere (sphere of rock), which comprises the crust and a solid portion of the upper mantle. The lithosphere is divided into many plates that move in relation to each other due to tectonic forces. The solid lithosphere floats atop a semiliquid layer known as the asthenosphere (weak sphere), which enables the lithosphere to move around.

Exploring Earth's Interior

Volcanic activity provides natural samples of the outer 124 miles (200 km.) of Earth's interior. Meteorites—samples of the solar system that have collided with Earth—also provide clues about Earth's composition and early history. The most ambitious human effort to penetrate Earth's interior was made by the former Soviet Union, which drilled a super-deep research well, named the Kola Well, near Murmansk, Russia. This was an attempt to penetrate the crust and reach the upper mantle. The reported depth of the Kola Well is a little more than 7.5 miles (12 km.). Although impressive, the drilled depth represents less than 0.2 percent of the distance from the earth's surface to its center.

A great deal of knowledge about Earth's composition and structure has been obtained through computer modeling, high-pressure laboratory experiments, and meteorites, but most of what is known about Earth's interior has been acquired by studying seismic waves generated by earthquakes and nuclear explosions. As seismic waves are transmitted, reflected, and refracted through the earth, they carry information to the surface about the materials through which they have traveled. Seismic waves are recorded at receiver stations (seismographic stations) and processed to provide a picturelike image of Earth's interior.

Changes in P- and S-wave velocities within Earth reveal the sequence of layers that make up Earth's interior. P-wave velocity depends on the elasticity, rigidity, and density of the material. By contrast, S-wave velocity depends only on the rigidity and density of the material. There are sharp variations in velocity at different depths, which correspond to boundaries between the different layers of Earth. P-wave velocity within crustal rocks ranges from 3.6-4.2 miles (6-7 km.) per second.

The boundary between the crust and the mantle is called the Mohorovičić discontinuity or Moho. At Moho, P-wave velocity increases from 4.2-4.8 miles (7-8 km.) per second. Beyond the crust-mantle boundary, P-wave velocity increases gradually up to about 8.1 miles (13.5 km.) per second at the core-mantle boundary. At this depth, S-waves are not transmitted and P-wave velocity, decreases from 8.1 to 4.8 miles (13.5 to 8 km.) per second, which strongly supports the concept that the outer core is liquid, since S-waves cannot travel through liquids. As P-waves enter the inner core, their velocity again increases, to about 6.8 miles (11.3 km.) per second.

Earth's interior seems to be characterized by a gradual increase with depth in temperature, pressure, and density. Extensive experimental and modeling work indicates that the temperature at 62 miles (100 km.) is between 1,200 and 1,400 degrees Celsius (2,192 to 2,552 degrees Fahrenheit). The temperature at the core-mantle boundary—about 1,802 miles (2,900 km.) deep—is calculated to be about 8,130 degrees Fahrenheit (4,500 degrees Celsius). At Earth's center the temperature may exceed 12,092 degrees Fahrenheit (6,700 degrees Celsius). Although at Earth's surface, heat energy is slowly but continuously lost as a result of outgassing, such as from volcanic eruptions, its interior remains hot.

Seismic Tomography and Future Exploration

Seismic tomography is one of the newest tools that earth scientists are using to develop three-dimensional velocity images of Earth's interior. In seismic tomography, several crossing seismic waves from different sources (earthquakes and nuclear explosions) are analyzed in much the same way that computerized axial tomography (CAT) scanners are used in medicine to obtain images of human organs. Seismic tomography is providing two- and three-dimensional images from the crust to the core-mantle boundary. Fast P-wave velocities have been correlated to cool material—for example, a piece of sinking lithosphere (cool rigid layer) such as in regions underneath the Andes Mountains (subduction zone); slow P-wave velocities have been correlated with hot materials—for example, rising mantle plumes of hot spots such as the one responsible for volcanic activity in the Hawaiian Islands.

Rubén A. Mazariegos-Alfaro

Plate Tectonics

The theory of plate tectonics provides an explanation for the present-day structure of the large landforms that constitute the outer part of the earth. The theory accounts for the global distribution of continents, mountains, hills, valleys, plains, earthquake activity, and volcanism, as well as various associations of igneous, metamorphic, and sedimentary rocks, the formation and location of min-

MAJOR TECTONIC PLATES AND MID-OCEAN RIDGES

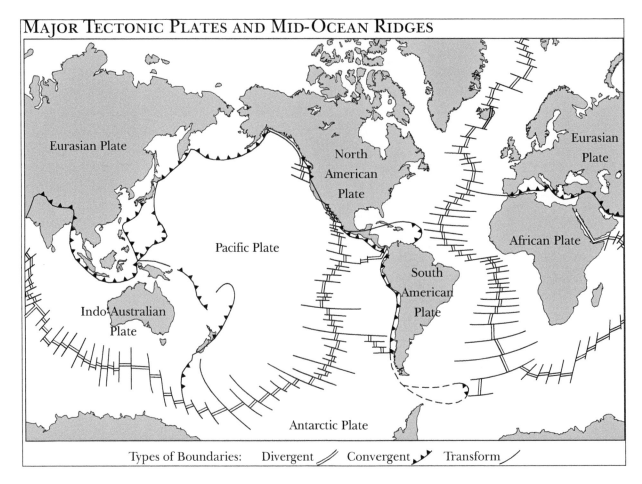

Types of Boundaries: Divergent   Convergent   Transform

eral resources, and the geology of ocean basins. Everything about the earth is related either directly or indirectly to plate tectonics.

Basic Theory

Plate-tectonic theory is based on an Earth model in which a rigid, outer shell—the lithosphere—lies above a hotter, weaker, partially molten part of the mantle called the asthenosphere. The lithosphere varies in thickness between 6 and 90 miles (10 and 150 km.), and comprises the crust and the underlying, upper mantle. The asthenosphere extends from the base of the lithosphere to a depth of about 420 miles (700 km.). The brittle lithosphere is broken into a pattern of internally rigid plates that move horizontally relative to each other across the earth's surface.

More than a dozen plates have been distinguished, some extending more than 2,500 miles (4,000 km.) across. Exhibiting independent motion, the plates grind and scrape against each other,

similar to chunks of ice in water, or like giant rafts cruising slowly on the asthenosphere. Most of the earth's dynamic activity, including earthquakes and volcanism, occurs along plate boundaries. The global distribution of these tectonic phenomena delineates the boundaries of the plates.

Geological observations, geophysical data, and theoretical models support the existence of three types of plate boundaries. Divergent boundaries occur where adjacent plates move away from each other. Convergent boundaries occur where adjacent plates move toward each other. Transform boundaries occur where plates slip past one another in directions parallel to their common boundaries.

The continents were formed by the movement at plate boundaries, and continental landforms were generated by volcanic eruptions and continental plates colliding with each other. The velocity of plate movement varies from plate to plate and even within portions of the same plate, ranging from 0.8 to 8 inches (2 to 20 centimeters) per year. The rates

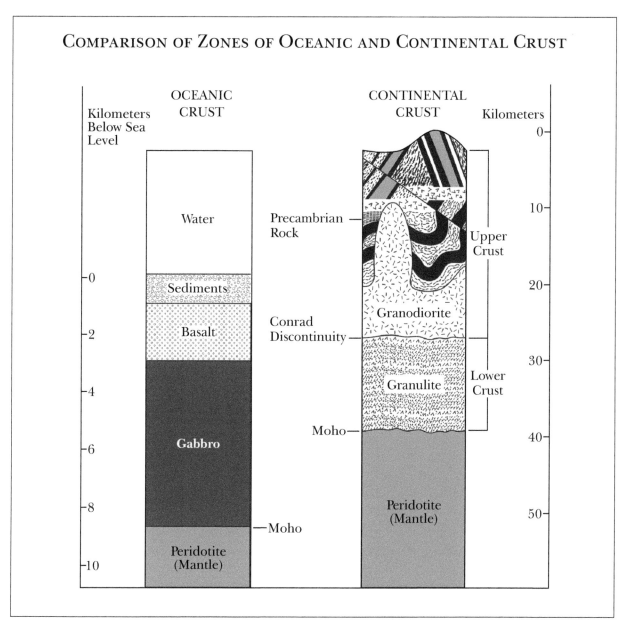

COMPARISON OF ZONES OF OCEANIC AND CONTINENTAL CRUST

are calculated from the distance to the midoceanic ridge crests, along with the age of the sea floor as determined by radioactive dating methods.

Convection currents that are driven by heat from radioactive decay in the mantle are important mechanisms involved in moving the huge plates. Convection currents in the earth's mantle carry magma (molten rock) up from the asthenosphere. Some of this magma escapes to form new lithosphere, but the rest spreads out sideways beneath the lithosphere, slowly cooling in the process. Assisted by gravity, the magma flows outward, dragging the overlying lithosphere with it, thus continu-

ing to open the ridges. When the flowing hot rock cools, it becomes dense enough to sink back into the mantle at convergent boundaries.

A second plate-driving mechanism is the pull of dense, cold, down-flowing lithosphere in a subduction zone on the rest of the trailing plate, further opening up the spreading centers so magma can move upward.

Divergent Plate Boundaries
During the 1950s and 1960s, oceanographic studies revealed that Earth's seafloors were marked by a nearly continuous system of submarine ridges,

THE SUPERCONTINENTS

The theory of plate tectonics explains the present-day distribution of major landforms, seismic and volcanic activity, and physiographic features of ocean basins. Many scientists also use the theory to explain the history of Earth's surface. Evidence indicates that the modern continents once formed a single landmass called Pangaea, meaning "all lands." According to the theory of plate tectonics, approximately 200 million years ago Pangaea began to split into two supercontinents, Laurasia and Gondwanaland. Eventually, as a result of tectonic forces, Laurasia split into North America, Europe, and most of Asia. Gondwanaland broke up into India, South America, Africa, Australia, and Antarctica.

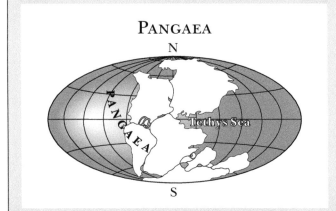

PANGAEA

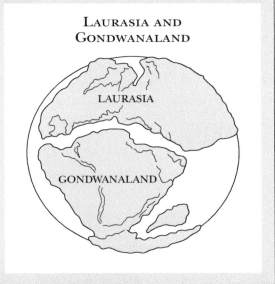

LAURASIA AND GONDWANALAND

more than 40,000 miles (64,000 km.) in length. Detailed investigations revealed that the midoceanic ridge system has a central rift valley that runs along its length and that the ridge system is associated with volcanic and earthquake activity. The earthquakes are frequent, shallow, and mild.

Magnetic studies of the seafloor indicate that the oceanic lithosphere has been segmented into a series of long magnetic strips that run parallel to the axis of the midoceanic ridges. On either side of the ridge, the ocean floor consists of alternating bands of rock, magnetized either parallel to or exactly opposite of the present-day direction of the earth's magnetic field.

Midoceanic ridges, or divergent plate boundaries, are tensional features representing zones of weakness within the earth's crust, where new seafloor is created by the welling up of mantle material from the asthenosphere into cracks along the ridges. As rifting proceeds, magma ascends to fill in the fissures, creating new oceanic crust. Iron minerals within the magma become aligned to the existing Earth polarity as the rock cools and crystallizes. The oceanic floor slowly moves away from the oce-

anic ridge toward deep ocean trenches, where it descends into the mantle to be melted and recycled to the earth's surface to generate new rocks and landforms.

As the seafloor spreads outward from the rift center, about half of the material is carried to either side of the rift, which is later filled by another influx of molten basalt. When the polarity of the earth changes, the subsequent molten basalt is magnetized in the opposite polarity. The continuation of this process over geologic time leads to the young geologic age of the seafloor and the magnetic symmetry around the midoceanic ridges.

Not all spreading centers are underneath the oceans. An example of continental rifting in its embryonic stage can be observed in the Red Sea, where the Arabian plate has separated from the African plate, creating a new oceanic ridge. Another modern-day example of continental divergent activity is East Africa's Great Rift Valley system. If this rifting continues, it will eventually fragment Africa, producing an ocean that will separate the resulting pieces. Through divergence, large plates are made into smaller ones.

Convergent Plate Boundaries

Because Earth's volume is not changing, the increase in lithosphere created along divergent boundaries must be compensated for by the destruction of lithosphere elsewhere. Otherwise, the radius of Earth would change. The compensation occurs at convergent plate boundaries, where plates are moving together. Three scenarios are possible along convergent boundaries, depending on whether the crust involved is oceanic or continental.

If both converging plates are made of oceanic crust, one will inevitably be older, cooler, and denser than the other. The denser plate eventually subducts beneath the less-dense plate and descends into the asthenosphere. The boundary along the two interacting plates, called a subduction zone, forms a trench. Some trenches are more than 620 miles (1,000 km.) long, 62 miles (100 km.) wide, and 6.8 miles (11 km.) deep. Heated by the hot asthenosphere beneath, the subducted plate becomes hot enough to melt.

Because of buoyancy, some of the melted material rises through fissures and cracks to generate volcanoes along the overlying plate. Over time, other parts of the melted material eventually migrate to a divergent boundary and rise again in cyclic fashion to generate new seafloor. The volcanoes generated along the overriding plate often form a string of islands called island arcs. Japan, the Philippines, the Aleutians, and the Mariannas are good examples of island arcs resulting from subduction of two plates consisting of oceanic lithosphere. Intense earthquakes often occur along subduction zones.

If the leading edge of one of the two convergent plates is oceanic crust and the other is continental crust, the oceanic plate is always the one subducted, because it is always denser. A classic example of this case is the western boundary of South America. On the oceanic side of the boundary, a trench was formed where the oceanic plate plunged underneath the continental plate. On the continental side, a fold mountain belt—the Andes—was formed as the oceanic lithosphere pushed against the continental lithosphere.

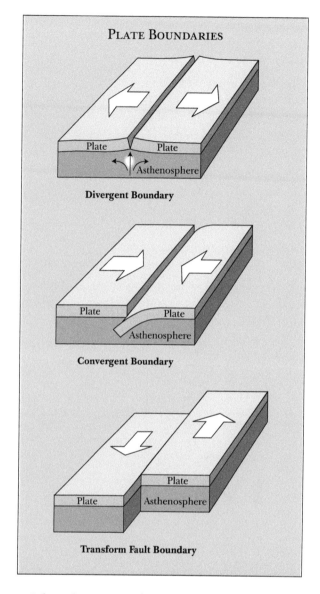

PLATE BOUNDARIES

Plate | Plate
Asthenosphere
Divergent Boundary

Plate | Plate
Asthenosphere
Convergent Boundary

Plate
Plate | Asthenosphere
Transform Fault Boundary

When the oceanic plate descends into the mantle, some of the material melts and works its way up through the mountain belt to produce rather violent volcanoes. The boundary between the plates is a region of earthquake activity. The earthquakes range from shallow to relatively deep, and some are quite severe.

The last type of convergent plate boundary involves the collision of two continental masses of lithosphere, which can result in folding, faulting, metamorphism, and volcanic activity. When the plates collide, neither is dense enough to be forced into the asthenosphere. The collision compresses and thickens the continental edges, twisting and deforming the rocks and uplifting the land to form unusually high fold mountain belts. The prototype example is the collision of India with Asia, resulting in the formation of the Himalayas. In this case, the earthquakes are typically shallow, but frequent and severe.

Transform Plate Boundaries

The actual structure of a seafloor spreading ridge is more complex than a single, straight crack. Instead, ridges comprise many short segments slightly offset from one another. The offsets are a special kind of fault, or break in the lithosphere, known as a transform fault, and their function is to connect segments of a spreading ridge. The opposite sides of a transform fault belong to two different plates that are grinding against each other in opposite directions.

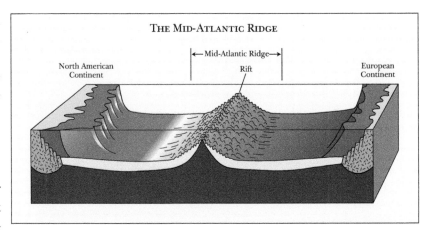

THE MID-ATLANTIC RIDGE

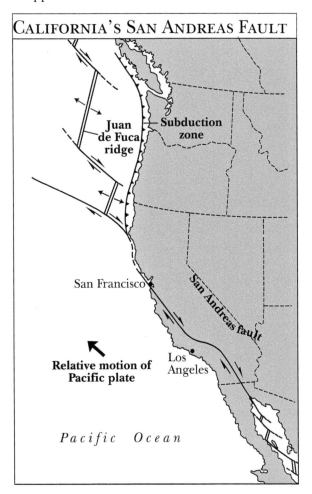

CALIFORNIA'S SAN ANDREAS FAULT

Transform faults form the boundaries that allow the plates to move relative to each another. The classic case of a transform boundary is the San Andreas Fault. It slices off a small piece of western California, which rides on the Pacific plate, from the rest of the state, which resides on the North American plate. As the two plates scrape past each other, stress builds up, eventually being released in earthquakes that can be quite violent.

Mantle Plumes and Hot Spots

Most plate tectonic features are near plate boundaries, but the Hawaiian Islands are not. In the late twentieth century, the only active volcanoes in the Hawaiian Islands were on the island of Hawaii, at the southeast end of the chain. Radiometric dating and examination of states of erosion show that, when proceeding along the chain to the northwest, successive islands are progressively older.

Evidently, the same heat source produced all the volcanoes in the Hawaiian chain. Known as a mantle plume, it has remained stationary while the Pacific plate rides over it, producing a volcanic trail from which absolute motion of the plate can be determined. Since mantle plumes do not move with the plates, the plumes must originate beneath the lithosphere, probably far below it. Resulting volcanoes are called hot spots to distinguish them from subduction-zone volcanoes. Iceland is a good example of a hot spot, as is Yellowstone. At least 100 hot spots are distributed around Earth.

Alvin K. Benson

VOLCANOES

Volcanoes form mountains both on land and in the sea and either do it on a grand scale or merely create minute bumps on the seafloor. Volcanoes do not occur in a random pattern, but are found in distinct zones that are related to plate dynamics. Each of the three types of volcanism on Earth is characterized by specific types of eruptions and magma compositions. Molten magma is the rock material below the earth's crust that forms igneous rock as it cools.

Types of Volcanoes

Geologists generally group volcanoes into four main kinds—cinder cones, composite volcanoes, shield volcanoes, and lava domes.

Cinder cones are built from congealed lava ejected from a single vent. As the gas-charged lava is blown into the air, it breaks into small fragments that solidify and fall as *cinders* around the vent to form a circular or oval cone. Most cinder cones have a bowl-shaped *crater* at the summit and rarely rise more than a thousand feet or so above their surroundings. Cinder cones are numerous in western North America and in other volcanic terrains of the world.

Composite volcanoes —sometimes called stratovolcanoes—include some of the Earth's grandest mountains, including Mount Fuji in Japan, Mount Cotopaxi in Ecuador, Mount Shasta in California, Mount Hood in Oregon, and Mount St. Helens and Mount Rainier in Washington. The essential feature of a composite volcano is a conduit system through which magma deep in the Earth's crust rises to the surface. They are typically steep-sided, symmetrical cones of large dimension built of alternating layers of lava flows, volcanic ash, cinders, blocks, and bombs. They may rise as much as 8,000 feet above their bases. Most have a crater at the summit that contains a central vent or a clustered group of vents. Lavas either flow through breaks in the crater wall or fissures on the flanks of the cone.

Shield volcanoes, the third type of volcano, are built almost entirely of fluid lava flows that pour out in all directions from a central vent, or group of vents, building a broad, gently sloping cone. They are built up slowly as thousands of highly fluid lava flows—basalt lava—spread over great distances, and then cool into thin sheets. Some of the largest volcanoes in the world are shield volcanoes. The Hawaiian Islands are composed of linear chains of these volcanoes including Kilauea and Mauna Loa on the island of Hawaii—two of the world's most active volcanoes. The floor of the ocean is more than 15,000 feet deep at the bases of the islands. As Mauna Loa, the largest of the shield volcanoes (and also the world's largest active volcano), projects 13,679 feet above sea level, its top is over 28,000 feet above the deep ocean floor.

Volcanic Composition

Volcanoes in the midocean ridges and plume environments draw most of their magmas from the earth's mantle and produce mainly dark, magnesium-rich basaltic magmas. When basaltic magmas accumulate in the continental crust (for example, at Yellowstone), the large-scale crustal melting leads to rhyolitic volcanism, the volcanic equivalent of granites. Arc magmas cover a wider range of magmatic compositions, ranging from arc basalt to light-colored, silica-rich rhyolites; the latter are commonly erupted in the form of the silica-rich volcanic rock known as pumice, or the black volcanic glass known as obsidian. Andesites, named after the Andes Mountains, are a common volcanic rock in stratovolcanoes, intermediate in composition between basalt and rhyolite.

Magmas form from several processes that lead to partial melting of a solid rock. The simplest is adding heat—for example, plumes carrying heat from deep levels in the mantle to shallower levels, where melting occurs. Decompressional (lowering the pressure) melting of the mantle occurs where the

SOME VOLCANIC HOT SPOTS AROUND THE WORLD

ocean floor is thinned or carried away by seafloor spreading in midocean ridge environments.

Genesis of Magma

Adding a "flux" to a solid mineral mixture may lower the substance's melting point. The most common theory about arc magma genesis invokes the addition of a low-melting-point substance to the arc mantle, a layer of mantle material at about 60 to 90 miles (100 to 150 km.) below the volcanic arc. The relatively dry arc mantle would usually start to melt at about 2,100 to 2,300 degrees Fahrenheit (1,200 to 1,300 degrees Celsius). However, the addition of water and other gases can lower the melting point of the mixture. The water and its dissolved chemicals are supposedly derived from the subducted slab, the former ocean floor that is pushed back into the earth.

The sequence of events is as follows: New basaltic ocean floor forms at midocean ridge volcanoes. The new hot magma interacts with seawater, leading to vents at the seafloor with their mineralized

deposits. The seafloor becomes hydrated, and sulfur and chlorine from seawater are locked up in newly formed minerals. During subduction, this altered seafloor with slivers of sediment, including limestone, is gradually warmed up and starts to decompose, adding a flux to the surrounding mantle rocks. The mantle rocks then start to melt, and these magmas with minor inherited oceanic materials start to rise and pond at the bottom of the crust. There the magmas sit and wait for an opportunity to erupt, while cooling and crystallizing. Thus, arc magmas bear a chemical signature of subducted oceanic components while their chemical compositions range from basalt to rhyolite.

Volcanic Eruptions

Volcanic eruptions occur as a result of the rise of magma into the volcano (from depths as great as several miles) and then into the throat of the volcano. In basaltic volcanoes, the magmas have relatively little gas, and the magma simply overflows and forms large lava flows, sometimes associated

VOLCANIC ERUPTION AND CALDERA FORMATION

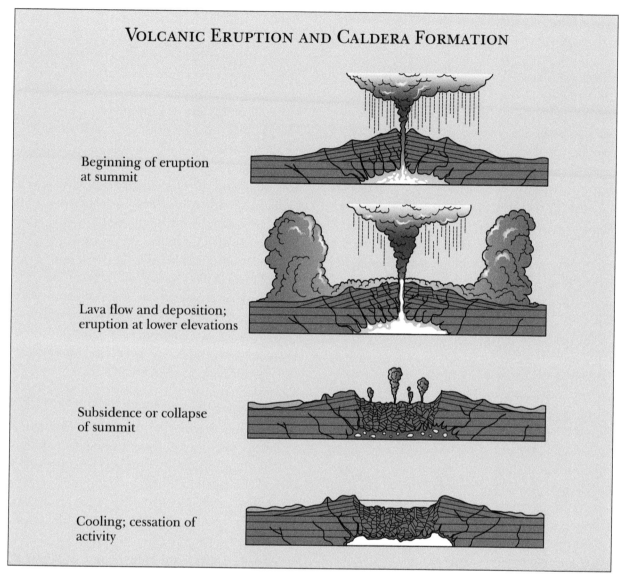

Beginning of eruption
at summit

Lava flow and deposition;
eruption at lower elevations

Subsidence or collapse
of summit

Cooling; cessation of
activity

with fire fountains. Stratovolcanoes can erupt regularly with small explosions or catastrophically after long periods of dormancy. Mount Stromboli, a volcano in Italy, erupts every twenty minutes, with an explosion that creates a column 650 to 980 feet (200 to 300 meters) high. Mount St. Helens in the U.S. state of Washington had a catastrophic eruption in 1980 after about 200 years of dormancy. It emitted an ash plume that reached more than 12 miles (20 km.) into the atmosphere.

After long magma storage periods in the crust, crystallization and melting of crustal material can lead to silica-rich magmas. These are viscous and can have high dissolved water contents—up to 4 to 6 percent by weight. When these magmas break out,

the eruption can be violent and form an eruption column 12 to 35 miles (20 to 55 km.) high. Many cubic miles of magma can be ejected. This leads to so-called plinian ash falls, with showers of pumice and ash over thousands of square miles, with the ash commonly carried around the globe by the high-level winds known as jet streams.

If the volume of ejected magma is large, the volcano empties itself and collapses into the hole, leading to a caldera—a volcanic collapse structure. The caldera at Crater Lake in Oregon is related to a large pumice eruption about 76,000 years ago. Basaltic volcanoes can also form collapse calderas when large volumes of lava have been extruded in a short time. Examples of famous basaltic calderas can be found in Hawaii's Mount Kilauea and the Galapagos Islands.

Volcanic Plumes

The dynamics of volcanic plumes has been studied from eruption photographs, experiments, and theoretical work. The rapidly expanding hot gases force the viscous magma out of the throat of the volcano, where it freezes into pumice. The kinetic energy of the ejected mass carries it 2 to 2.5 miles (3-4 km.) above the volcano. During this phase, air is entrained in the column, diluting the concentration of ash and pumice particles. The hot particles heat the entrained air, the mixture of hot air and solids becomes less dense than the surrounding atmosphere, and a buoyant column rises high into the sky.

The height of an eruption column is not directly proportional to the force of the eruption but is strongly dependent on the rate of heat release of the volcano. If little of the entrained air is heated up, the column will collapse back to the ground and an ash flow forms, which may deposit ash around the volcano. These types of eruptions are among the most devastating, creating glowing ash clouds traveling at speeds up to 60 miles (100 km.) per hour, burning everything in their path. The 1902 eruption of Mount Pelée on Martinique in the Caribbean was such an eruption and killed nearly 30,000 people in a few minutes.

Many volcanoes that are high in elevation are glaciated, and their eruptions lead to large-scale ice melting and possibly mixing of water, magma, and volcanic debris. Massive hot mudflows can race down from the volcano, following river valleys and filling up low areas. The 1980 Mount St. Helens eruption created many mudflows, some of which reached the Pacific Ocean, ninety miles to the west. A catastrophic mudflow event occurred in 1984 at Nevado del Ruiz, a volcano in Colombia, where

20,000 people were buried in mud and perished. When magma intrudes under the ice, meltwater can accumulate and then escape catastrophically, but such meltwater bursts are rare outside Iceland.

Minerals and Gases in Eruptions

The gas-rich character of arc magmas leads to fluid escape at various levels in the volcanoes, and these fluids tend to be rich in chlorine. They can transport metals such as copper, lead, zinc, and gold at high concentrations, and lead to the enrichment of these metals in the fractured volcanic rocks. Many of the world's largest copper ore deposits are associated with older arc volcanism, where erosion has removed most of the volcanic structure and laid the volcano innards bare. Many active volcanoes have modern hydrothermal (hot-water) systems, leading to acid hot springs and crater lakes and the potential to harness geothermal energy. Some areas in Japan, New Zealand, and Central America have an abundance of geothermal energy resources, which are gradually being developed.

Apart from the dangers of eruptions, continuous emissions of large amounts of sulfur dioxide, hydrochloric acid, and hydrofluoric acid present a danger of air pollution and acid rain. Incidences of emphysema and other irritations of the respiratory system are common in people living on the slopes of active volcanoes. The large lava emissions in Iceland in the eighteenth century led to acid fogs all over Europe. Many cattle died in Iceland during this period from the hydrofluoric acid vapors. High levels of fluorine in drinking water can lead to fluorosis, a disease that attacks the bone structure. The discharge of highly acidic fluids from hot springs and crater lakes can cause widespread environmental contamination, which can present a danger for crops gathered from fields irrigated with these waters and for local ecosystems in general.

Johan C. Varekamp

GEOLOGIC TIME SCALE

A major difference between the geosciences (earth sciences) and other sciences is the great enormity of their time scale. One might compare the magnitude of geologic time for geoscientists to the vastness of space for astronomers. Every geological process, such as the movement of crustal plates (plate tectonics), the formation of mountains, and the advance and retreat of glaciers, must be considered within the context of time.

Although certain geologic events, such as floods and earthquakes, seem to occur over short periods of time, the vast majority of observed geological features formed over a great span of time. Consequently, modern geoscientists consider Earth to be exceedingly old. Using radiometric age-dating techniques, they calculate the age of Earth as 4.6 billion years old.

Early miners were probably the first to recognize the need for a scale by which rock and mineral units could be compared over large geographic areas. However, before a time scale—and even geology as a science—could develop, certain principles had to be established. This did not occur until the late eighteenth century when James Hutton, a Scottish naturalist, began his extensive examinations of rock relationships and natural processes at work on the earth. His work was amplified by Charles Lyell in his textbook *Principles of Geology* (1830-1833). After careful observation, Hutton concluded that the natural processes and functions he observed had operated in the same basic manner in the past, and that, in general, natural laws were invariable. That idea became known as the principle of uniformitarianism.

The Birth of Stratigraphy

In 1669 Nicholas Steno, a Danish physician working in Italy, recognized that horizontal rock layers contained a chronological record of Earth history and formulated three important principles for interpreting that history. The principle of superposition states that in a succession of undeformed strata, the oldest stratum lies at the bottom, with successively younger ones above. The principle of original horizontality states that because sedimentary particles settle from fluids under graviational influence, sedimentary rock layers must be horizon-

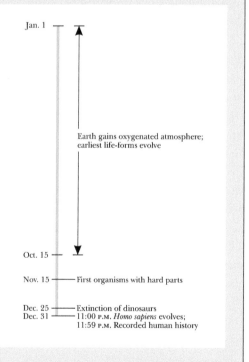

EARTH'S HISTORY COMPRESSED INTO ONE CALENDAR YEAR

One way to visualize events in Earth's history is to compress geologic events into a single calendar year. Earth's birth, 4.6 billion years ago, would occur during the first minute of January 1. The first three-quarters of Earth's history is obscure and would take place from January to mid-October. During this time, Earth gained an oxygenated atmosphere, and the earliest life-forms evolved. The first organisms with hard parts preserved in the fossil record (approximately 570 million years ago) would appear around November 15. The extinction of the dinosaurs (65 million years ago) would occur on Christmas Day. Homo sapiens would first appear at approximately 11 P.M. on December 31, and all of recorded human history would occur in the last few seconds of New Year's Eve.

tal; if not, they have suffered from subsequent disturbance. The principle of original lateral continuity states that strata originally extended in all directions until they thinned to zero or terminated against the edges of the original area of deposition.

In the late eighteenth century, the English surveyor William Smith recognized the wide geographic uniformity of rock layers and discovered the utility of fossils in correlating these layers. By 1815, Smith had completed a geologic map of England and was able to correlate English rock layers with layers exposed across the English Channel in France.

From the need to classify and organize rock layers into an orderly form arose a subdiscipline of modern geology—stratigraphy, the study of rock layers and their age relationships. In 1835 two British geologists, Adam Sedgwick and Roderick Murchison, began organizing rock units into a formal stratigraphic classification. Large divisions, called eras, were based upon well known and characteristic fossils, and included a number of smaller subdivisions, called periods.

The periods are often subdivided into smaller units called epochs. Each period is defined by a representative sequence of rock strata and fossils. For instance, the Devonian period is named for exposures of rock in Devonshire in southern England, while the Jurassic period is defined by strata exposed in the Jura Mountains in northern Switzerland.

Approximately 80 percent of Earth's history is included in the Cryptozoic eon (meaning obscure life). Fossils from the Cryptozoic eon are rare, and the rock record is very incomplete. After the Cryptozoic eon came the Paleozoic (ancient life), Mesozoic (middle life), and Cenozoic (recent life) eras. Most of the life forms that evolved during the Paleozoic and Mesozoic eras are now extinct, whereas 90 percent of the

life-forms that evolved up to the middle Cenozoic era still exist.

The Geologic Time Scale

The geologic time scale is continually in revision as new rock formations are discovered and dated. The ages shown in the table below are in millions of years ago (MYA) before the present and represent the beginning of that particular period. It would be impossible to list all the significant events in Earth's history, but one or two are provided for each period. Note that in the United States, the Carboniferous period has been subdivided into the Mississippian period (older) and the Pennsylvanian period (younger).

The Fossil Record

The word "fossil" comes from the Latin *fossilium*, meaning "dug from beneath the surface of the ground." Fossils are defined as any physical evidence of past life. Fossils can include not only shells, bones, and teeth, but also tracks, trails, and bur-

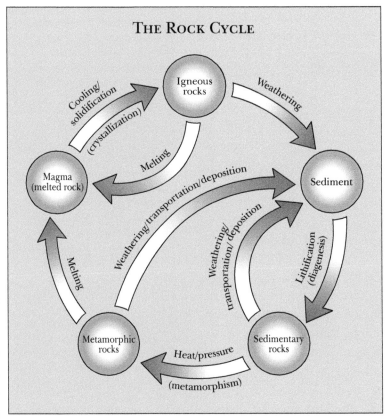

rows. The latter group are referred to as trace fossils. Fossils demonstrate two important truths about life on Earth: First, thousands of species of plants and animals have existed and later became extinct. Second, plants and animals have evolved through time, and the communities of life that have existed on Earth have changed.

Some organisms are slow to evolve and may exist in several geologic time periods, while others evolve quickly and are restricted to small intervals of time within a particular period. The latter, referred to as index fossils, are the most useful to geoscientists for correlating rock layers over wide geographic areas and for recognizing geologic time.

The fossil record is incomplete, because the process of preservation favors organisms with hard parts that are rapidly buried by sediments soon after death. For this reason, the vast majority of fossils are represented by marine invertebrates with exoskeletons, such as clams and snails. Under special circumstances, soft-bodied organism can be preserved, for instance the preservation of insects in amber, made famous by the feature film *Jurassic Park* (1993).

The Rock Cycle

A rock is a naturally formed aggregate of one or more minerals. Three types of rocks exist in the earth's crust, each reflecting a different origin. Igneous rocks have cooled and solidified from molten material either at or beneath Earth's surface. Sedimentary rocks form when preexisting rocks are weathered and broken down into fragments that accumulate and become compacted or cemented together. Fossils are most commonly found in sedimentary rocks. Metamorphic rocks form when heat, pressure, or chemical reactions in Earth's interior change the mineral or chemical composition and structure of any type of preexisting rock.

Over the huge span of geologic time, rocks of any one of these basic types can change into either of the other types or into a different form of the same type. For this reason, older rocks become increasingly more rare. The processes by which the various rock types change over time are illustrated in the rock cycle.

Larry E. Davis

EARTH'S SURFACE

INTERNAL GEOLOGICAL PROCESSES

The earth is layered into a core, a mantle, and a crust. The topmost mantle and the crust make up the lithosphere. Beneath this is a layer called the asthenosphere, which is composed of moldable and partly liquid materials. Heat transference within the asthenosphere sets up convection cells that diverge from hot regions and converge to cold regions. Consequently, the overlying lithosphere is segmented into ridged plates that are moved by the convection process. The hot asthenosphere does not rise along a line. This causes the development of a structure called a transform plate boundary, which is perpendicular to and offsetting the divergent boundary.

The topographic features at Earth's surface, such as mountains, rift valleys, oceans, islands, and ocean trenches, are produced by extension or compression forces that act along divergent, convergent, or transform plate boundaries. The extension and compression forces at Earth's surface are powered by convection within the asthenosphere.

Mountains and Depressions in Zones of Compression

Compression along convergent plate boundaries yields three types of mountain: island arcs that are partly under water; mountains along a continental edge, such as the Andes; and mountains at continental interiors, such as the Alps. At convergent plate boundaries, the denser of the two colliding plates slides down into the asthenosphere and causes volcanic activity to form on the leading edge of the upper plate. Island arcs such as the Aleutians and the Caribbean are formed when an oceanic plate descends beneath another oceanic plate.

Volcanic mountain chains such as the Andes of South America are formed when an oceanic plate descends beneath a continental plate. In both the island arc type and Andean type collisions, a deep depression in the oceans, called a trench, marks the place where neighboring plates are colliding and where the denser plates are pulled downward into the asthensophere. If the colliding plates are of similar density, neither plate will go into the asthenosphere. Instead, the edges of the neighboring plates will be folded and faulted and excess material will be pushed upward to form a block mountain, such as the mountain chain that stretches from the Alps through to the Himalayas. This type of mountain chain is not associated with a trench.

The Appalachians of the eastern United States are an example of the alpine type of mountain belt. When the Appalachians were forming 300 million years ago, rock layers were deformed. The deformation included folding to form ridges and valleys; fracturing along joint sets, with one joint set being parallel to ridges, while the other set is perpendicular; and thrust faulting, in which rock blocks were detached and shoved upward and northwestward.

Millions of years of erosion have reduced the height of the mountains and have produced topographic inversion in the foothills. Topographic inversion occurs because joints create wider fractures at upfolded ridges and narrower fractures at downfolded valleys. Erosion is then accelerated at upfolded ridges, converting ancient ridges into valleys, while ancient valleys stand as ridges. The Valley and Ridge Province of the Appalachians is noted for such topographic inversion.

West of the Valley and Ridge Province of the Appalachians is the Allegheny Plateau, which is bounded by a cliff on its eastern side. In general, plateaus are flat topped because the rock layer that covers the surface is resistant to weathering. The cliff side is formed by erosion along joint or fault surfaces.

The Sierra Nevada range, which formed 70 million years ago, is an example of an Andean type of mountain belt. Millions of years of erosion there has exposed igneous rocks that formed at depth. Over the years, the force of compression that formed the Sierras has evolved to form a zone of extension between the Sierras and the Colorado Plateau.

Mountains and Depressions in Zones of Extension

Extension is a strain that involves an increase in length and causes crustal thinning and faulting. Extension is associated with convergent boundaries, divergent boundaries, and transform boundaries.

Extension Associated with a Convergent Boundary

During the formation of the Sierra Nevada, an oceanic plate that was subducted beneath California declined at a shallow angle eastward toward the Colorado Plateau. Later, the subducted plate peeled off and molten asthenosphere took its place. From the asthenosphere, lava ascended through fractures to form volcanic mountains in Arizona and Utah, and lava flowed and volcanic ash fell as far west as California. The lithosphere has been heated up and has become buoyant, so the Colorado Plateau rises to higher elevations, and rock layers slide westward from it in a zone of extension that characterizes the Basin and Range Province.

In the extension zone, the top rock layers move westward on curved displacement planes that are steep at the surface and nearly horizontal at depth. When rock layers move westward over a curved detachment surface, the trailing edge of the rock layers roll over and are tilted toward the east so they do not leave space in buried rocks. On the other hand, a west-facing slope is left behind on a mountain from which the rock layers were detached. There-

fore, movement along one curved detachment surface creates a valley, and movement along several such detachment surfaces forms a series of valleys separated by ridges, as in the Basin and Range Province. The amount of the displacement along the curved surfaces is not uniform. For example, more displacement has created wide zones of valleys such as the Las Vegas valley in Nevada, and Death Valley in California.

Extension Associated with a Divergent Boundary

The longest mountain chain on Earth lies under the Pacific Ocean. It is about 37,500 miles (60,000 km.) long, 31.3 miles (50 km.) wide, and 2 miles (3 km.) high. The central part of this midoceanic ridge is marked by a depression, about 3,000 feet (1,000 meters) deep, and is called a rift valley. A part of the submarine ridge, called the East Pacific Rise, forms the seafloor sector in the Gulf of California and reappears off the coast of northern California, Oregon, and Washington as the Juan de Fuca Ridge. Another part forms the seafloor sector in the Gulf of Aden and Red Sea seafloor, part of which is exposed in the Afar of Ethiopia. From the Afar southward to the southern part of Mozambique is the longest exposed rift valley on land, the East African Great Rift Valley.

A rift valley is the place where old rocks are pushed aside and new rocks are created. Blocks of rock that are detached from the rift walls slide down by a series of normal fault displacements. The ridge adjacent to the central rift is present because hot rocks are less dense and buoyant. If the process of divergences continues from the rifting stage to a drifting stage, as the rocks move farther away from the central rift, the rocks become older, colder, and denser, and push on the underlying asthenosphere to create basins. These basins will be flooded by oceanic water as neighboring continents drift away. However, not all processes of divergence advance from the rifting to the drifting stage.

Extension Associated with Transform Boundary

The best-known example of a transform boundary is the San Andreas Fault that offsets the East Pacific

Rise from the Juan de Fuca Ridge, and is exposed on land from the Gulf of California to San Francisco. Along transform boundaries, there are pull-apart basins that may be filled to form lakes, such as the Salton Sea in Southern California. An-

other example is the Aqaba transform of the Middle East, along which the Sea of Galilee and the Dead Sea are located.

H. G. Churnet

External Processes

Continuous processes are at work shaping the earth's surface. These include breaking down rocks, moving the pieces, and depositing the pieces in new locations. Weathering breaks down rocks through atmospheric agents. The process of moving weathered pieces of rock by wind, water, ice, or gravity is called erosion. The materials that are deposited by erosion are called sediment.

Mechanical weathering occurs when a rock is broken into smaller pieces but its chemical makeup is not changed. If the rock is broken down by a change in its chemical composition, the process is called chemical weathering.

Mechanical Weathering

Different types of mechanical weathering occur, depending on climatic conditions. In areas with moist climates and fluctuating temperatures, rocks can be broken apart by frost wedging. Water fills in cracks in rocks, then freezes during cold nights. As the ice expands and pushes out on the crack walls, the crack enlarges. During the warm days, the water thaws and flows deeper into the enlarged crack. Over time, the crack grows until the rock is broken apart. This process is active in mountains, producing a pile of rock pieces at the mountain base called talus.

Salt weathering occurs in areas where much salt is available or there is a high evaporation rate, such as along the seashore. Salt crystals form when salty moisture enters rock cracks. Growing crystals settle

in the bottom of the crack and apply pressure on the crack walls, enlarging the crack.

Thermal expansion and contraction occur in climates with fluctuating temperatures, such as deserts. All minerals expand during hot days and contract during cold nights, and some minerals expand and contract more than others. This process continues until the rock loosens up and breaks into pieces.

Mechanical exfoliation can happen to a rock body overlain by a thick rock or sediment layer. If the heavy overlying layer over a portion of the rock body is removed, pressure is relieved and the exposed rock surface will expand in response. This expanding surface will break off into sheets parallel to the surface, but the remaining rock body remains under pressure and unchanged.

When plant roots grow into cracks in rocks, they enlarge the cracks and break up the rocks. Finally, abrasion can occur to rock fragments during transport. Either the fragments collide, breaking apart, or fragments are scraped against rocks, breaking off pieces.

Chemical Weathering

Water and oxygen create two common causes of chemical weathering. For example, dissolution occurs when water or another solution dissolves minerals within a rock and carries them away. Hydrolysis can occur when water flows through earth materials. The hydrogen ions or the hydroxide ions of the water may react with minerals in the

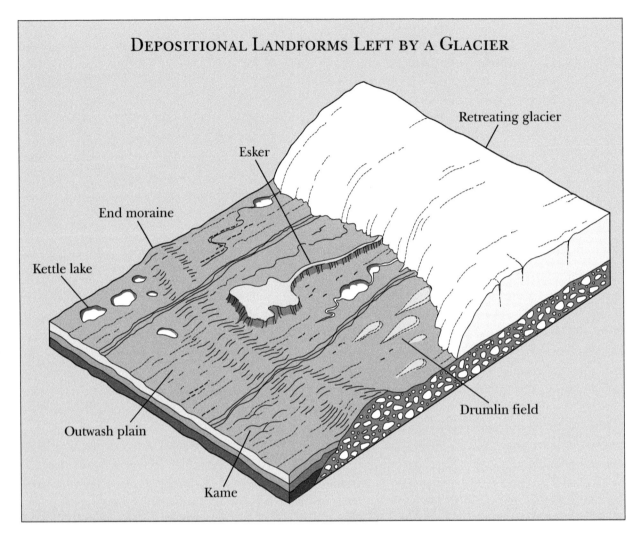

DEPOSITIONAL LANDFORMS LEFT BY A GLACIER

Esker

End moraine

Retreating glacier

Kettle lake

Drumlin field

Outwash plain

Kame

rocks. When this occurs, the chemical composition of the mineral is changed, and a new mineral is formed. Hydrolysis often produces clay minerals.

Some elements in minerals combine with oxygen from the atmosphere, creating a new mineral. This process is called oxidation. Some of these oxidation minerals are commonly referred to as rust.

Mass Movement

Weathered rock pieces (sediments) are transported (eroded) by one or more of four transport processes: water (streams and oceans), wind, ice (glaciers), or gravity. Mass movement transports earth materials down slopes by the pull of gravity. Gravity, constantly working to pull surface materials down, parallel to the slope, is the most important factor affecting mass movement. There is also a force in-

volved perpendicular to the slope that contributes to the effects of friction.

Friction, the second factor, is determined by the earth material type involved. For example, weathering may create cracks in rocks, which form planes of weakness on which the mass movement can occur. Loose sediments always tend to roll downhill.

The third factor is the slope angle. Each earth material has its own angle of repose, which is the steepest slope angle on which the materials remain stable. Beyond this slope angle, earth materials will move downslope.

Water, the fourth factor, affects the stability of the earth material in the slope. Friction is weakened by water between the mineral grains in the rock. For example, water can make clay quite slippery, causing the mass movement.

HOW HYDROLOGY SHAPES GEOGRAPHY

Water and ice sculpt the landscape over time. Fast-flowing rivers erode the soil and rock through which they flow. When rivers slow down in flatter areas, they deposit eroded sediments, creating areas of rich soils and deltas at the mouths of the rivers. Over time this process wears down mountain ranges. The Appalachian Mountain range on the eastern side of the North American continent is hundreds of millions of years older than the Rocky Mountain range on the continent's western side. Although the Appalachians once rivaled the Rockies in size, they have been made smaller by time and erosion.

Canyons are carved by rivers, as the Grand Canyon was carved by the Colorado River, which exposed rocks billions of years old. Ice also changes the landscape. Large ice sheets from past ice ages could have been well over 1 mile (1,600 meters) thick, and they scoured enormous amounts of soil and rock as they slowly moved over the land surface. Terminal moraines are the enormous mounds of soil pushed directly in front of the ice sheets. Long Island, New York, and Cape Cod, Massachusetts, are two examples of enormous terminal moraines that were left behind when the ice sheets retreated.

The rooting system of vegetation, the fifth factor, helps make the surficial materials of the slope stable by binding the loose materials together.

Mass movements can be classified by their speed of movement. Creep and solifluction are the two types of slow mass movement, which are measured in fractions of inches per year. Creep is the slowest mass movement process, where unconsolidated materials at the surface of a slope move slowly downslope. The materials move slightly faster at the surface than below, so evidence of creep appears in the form of slanted telephone poles. During solifluction, the warm sun of the brief summer season in cold regions thaws the upper few feet of the earth. This waterlogged soil flows downslope over the underlying permafrost.

Rapid mass movement processes occur at feet per second or miles per hour. Falls occur when loose rock or sediment is dislodged and drops from a steep slope, such as along sea cliffs where waves erode the cliff base. Topples occur when there is an overturning movement of the mass. A topple can turn into a fall or a slide. A slide is a mass of rock or sediment that becomes dislodged and moves along a plane of weakness, such as a fracture. A slump is a slide that separates along a concave surface. Lateral spreads occur when a fractured earth mass spreads out at the sides.

A flow occurs when a mass of wet or dry rock fragments or sediment moves downslope as a highly viscous fluid. There are several different flow types. A debris flow is a mass of relatively dry, broken pieces of earth material that suddenly has water added. The debris flow occurs on steeper slopes and moves at speeds of 1-25 miles (2-40 km.) per hour. A debris avalanche occurs when an entire area of soil and underlying weathered bedrock becomes detached from the underlying bedrock and moves quickly down the slope. This flow type is often triggered by heavy rains in areas where vegetation has been removed. An earthflow is a dry mass of clayey or silty material that moves relatively slowly down the slope. A mudflow is a mass of earth material mixed with water that moves quickly down the slope.

A quick clay can occur when partially saturated, solid, clayey sediments are subjected to an earthquake, explosion, or loud noise and become liquid instantly.

Sherry L. Eaton

FLUVIAL AND KARST PROCESSES

Earth's landscape has been sculptured into an almost infinite variety of forms. The earth's surface has been modified by various processes for thousands, even hundreds of millions, of years to arrive at the modern configuration of landscapes.

Each process that transforms the surface is classified as either endogenic or exogenic. Endogenic processes are driven by the earth's internal heat and energy and are responsible for major crustal deformation. Endogenic processes are considered constructional, because they build up the earth's surface and create new landforms, such as mountain systems. Conversely, exogenic processes are considered destructional because they result in the wearing away of landforms created by endogenic processes. Exogenic processes are driven by solar energy putting into motion the earth's atmosphere and water, resulting in the lowering of features originally created by endogenic processes.

The most effective exogenic processes for wearing away the landscape are those that involve the action of flowing water, commonly referred to as fluvial processes. Water flows over the surface as runoff, after it evaporates into the atmosphere and infiltrates into the soil. The water that is left over flows down under the influence of gravity and has tremendous energy for sculpturing the earth's surface. Although flowing water is the most effective agent for modifying the landscape, it represents less than 0.01 percent of all the water on Earth's surface. By comparison, nearly 75 percent of the earth's surface water is stored within glaciers.

Drainage Basins
Fluvial processes can be considered from a variety of spatial scales. The largest scale is the drainage basin. A drainage basin is the area defined by topographic divides that diverts all water and material within the basin to a single outlet. Every stream of any size has its own drainage basin, and every portion of the earth's land surfaces are located within a drainage basin. Drainage basins vary tremendously in size, de-

pending on the size of the river considered. For example, the largest drainage basin on earth is the Amazon, which drains about 2.25 million square miles (5.83 million sq. km.) of South America.

The Amazon Basin is so large that it could contain nearly the entire continent of Australia. By comparison, the Mississippi River drainage basin, the largest in North America, drains an area of about 1,235,000 square miles (3,200,000 sq. km.). Smaller rivers have much smaller basins, with many draining only an area roughly the size of a football field. While basins vary tremendously in size, they are spatially organized, with larger basins receiving the drainage from smaller basins, and eventually draining into the ocean. Because drainage basins receive water and material from the landscape within the basin, they are sensitive to environmental change that occurs within the basin. For example, during the twentieth century, the Mississippi River was influenced by many human-imposed changes that occurred either within the basin or directly within the channel, such as agriculture, dams and reservoirs, and levees.

Drainage Networks and Surface Erosion
Drainage basins can be subdivided into drainage networks by the arrangement of their valleys and interfluves. Interfluves are the ridges of higher elevation that separate adjacent valleys. Where an interfluve represents a natural boundary between two or more basins, it is referred to as a drainage divide. Valleys contain the larger rivers and are easily distinguished from interfluves by their relatively low, flat surfaces. Interfluves have relatively steep slopes and, for this reason, are eroded by runoff. The term erosion refers to the transport of material, in this case sediment that is dislodged from the surface.

Runoff starts as a broad sheet of slow-moving water that is not very erosive. As it continues to flow downslope, it speeds up and concentrates into rills, which are narrow, fast-moving lines of water. Because the runoff is concentrated within rills, the water travels faster and has more energy for erosion. Thus, rills are responsible for transporting sediment from higher points of elevation within the basin to the valleys, which are at a lower elevation. Rills can become powerful enough to scour deeply into the surface, developing into permanent channels called gullies.

The presence of many gullies indicates significant erosion on the landscape and represents an expensive and long-lasting problem if it is not remedied after initial development. The formations of gullies is often associated with human manipulation of the earth. For example, gullies can develop after improper land management, particularly intensive agricultural and grazing practices. A change in land use from natural vegetation, such as forests or prairie, can result in a type of land cover that is not suited for preventing erosion. Such land surfaces become susceptible to the formation of gullies during heavy, prolonged rains.

At a smaller scale, fluvial processes can be considered from the perspective of the river channel. River channels are located within the valleys of basins, offering a permanent conduit for drainage. Higher in the basin, river channels and valleys are relatively narrow, but grow larger toward the mouth of the basin as they receive drainage from smaller rivers within the basin. River channels may be categorized by their planform pattern, which refers to their overhead appearance, such as would be viewed from the window of an airplane.

The two major types of rivers are meandering and braided. Meandering rivers have a single channel that is sinuous and winding. These rivers are characterized as having orderly and symmetrical bends, causing the river to alternate directions as it flows across its valley. In contrast, braided rivers contain numerous channels divided by small islands, which results in a disorganized pattern. The islands within a braided river channel are not permanent. Instead, they erode and form over the course of a few years, or even during large flood events. Meandering channels usually have narrow and deep channels, but braided river channels are shallow and wide.

Sediment and Floodplains

Another distinction between braided and meandering river channels is the types of sediment they transport. Braided rivers transport a great amount of sediment that is deposited into midchannel islands within the river. Also, because braided rivers are frequently located higher in the drainage basin, they may have larger sediments from the erosion of adjacent slopes. In contrast, meandering river channels are located closer to the mouth of the basin and transport fine-grained sediment that is easily stored within point bars, which results in symmetrical bends within the river.

The sediments of both meandering and braided rivers are deposited within the valleys onto floodplains. Floodplains are wide, flat surfaces formed from the accumulation of alluvium, which is a term for sediment that is deposited by water. Floodplain sediments are deposited with seasonal flooding. When a river floods, it transports a large amount of sediment from the channel to the adjacent floodplain. After the water escapes the channel, it loses energy and can no longer transport the sediment. As a result, the sediment falls out of suspension and is deposited onto the floodplain. Because flooding occurs seasonally, floodplain deposits are layered and may accumulate into very thick alluvial deposits over thousands of years.

Karst Processes and Landforms

A specialized type of exogenic process that is also related to the presence of water is karst. Karst processes and topography are characterized by the solution of limestone by acidic groundwater into a number of distinctive landforms. While fluvial processes lower the landscape from the surface, karst processes lower the landscape from beneath the surface. Because limestone is a very permeable sedimentary rock, it allows for a large amount of groundwater flow. The primary areas for solution of the limestone occur along bedding planes and joints. This creates a positive feedback by increasing the amount of water flowing through the rock, thereby further increasing solution of the limestone. The result is a complex maze of underground conduits and caverns, and a surface with few rivers because of the high degree of infiltration.

The surface topography of karst regions often is characterized as undulating. A closer inspection reveals numerous depressions that lack surface outlets. Where this is best developed, it is referred to as cockpit karst. It occurs in areas underlain by extensive limestone and receiving high amounts of precipitation, for example, southern Illinois and Indiana in the midwestern United States, and in Puerto Rico and Jamaica.

Sinkholes are also common to karstic regions. Sinkholes are circular depressions having steep-sided vertical walls. Sinkholes can form either from the sudden collapse of the ceiling of an underground cavern or as a result of the gradual solution and lowering of the surface. Sinkholes can fill with sediments washed in from surface runoff. This reduces infiltration and results in the development of small circular lakes, particularly common in central Florida. Over time, erosion causes the vertical walls to retreat, resulting in uvalas, which are much larger flat-floored depressions.

Where there are numerous adjacent sinkholes, the retreat and expansion of the depressions causes them to coalesce, resulting in the formation of poljes. Unlike uvalas, poljes have an irregular shape, and the floor of the basin is not flat because of differences between the coalescing sinkholes.

Caves are among the most characteristic features of karst regions, but can only be seen beneath the surface. Caves can traverse the subsurface for miles, developing into a complex network of interconnected passages. Some caves develop spectacular formations as a result of the high amount of dissolved limestone transported by the groundwater. The evaporation of water results in the accumulation of carbonate deposits, which may grow for thousands of years. Some of the most common deposits are stalactites, which grow downward from the ceiling of the cave, and stalagmites, which grow upward and occasionally connect with stalactites to form large vertical columns.

Paul F. Hudson

GLACIATION

In areas where more snow accumulates each winter than can thaw in summer, glaciers form. Glacier ice, called firn, looks like rock but is not as strong as most rocks and is subject to intermittent thawing and freezing. Glacier ice can be brittle and fracture readily into crevasses, while other ice behaves as a plastic substance. A glacier is thickest in the area receiving the most snow, called the zone of accumulation. As the thickness piles up, it settles down and squeezes the limit of the ice outward in all directions. Eventually, the ice reaches a climate where the ice begins to melt and evaporate. This is called the zone of ablation.

Alpine Glaciation

Varied topographic evidence throughout the alpine environment attests to the sculpturing ability of glacial ice. The world's most spectacular mountain scenery has been produced by alpine glaciation, including the Matterhorn, Yosemite Valley, Glacier National Park, Mount Blanc, the Tetons, and Rocky

A FUTURE ICE AGE

If past history is an indicator, some time in the future conditions again will become favorable for the growth of glaciers. As recently as 1300 to 1600 CE, a cold period known at the Little Ice Age settled over Northern Europe and Eastern North America. Viking colonies perished as agriculture became unfeasible, and previously ice-free rivers in Europe froze over.

Another ice age would probably develop rapidly and be impossible to stop. Active mountain glaciers would bury living forests. Great ice caps would again cover Europe and North America, and move at a rate of 100 feet (30 meters) per day. Major cities and populations would shift to the subtropics and the topics.

Mountain National Park, all of which are visited by large numbers of people annually. Although alpine glaciation is still an active process of land sculpture in the high mountain ranges of the world, it is much less active than it was in the Ice Age of the Pleistocene epoch.

The prerequisites for alpine, or mountain, glaciation to become active are a mountainous terrain with Arctic climatic conditions in the higher elevations, and sufficient moisture to help snow and ice develop into glacial ice. As glaciers move out from their points of origin, they erode into the sides of mountains and increase the local relief in the higher elevations. The erosional features produced by alpine glaciation dominate mountain topography and usually are the most visible features on topographic maps. The eroded material is transported downvalley and deposited in a variety of landforms.

One kind of an erosional feature is a cirque, a hollow bowl-shaped depression. The bowl of the cirque commonly contains a small round lake or tarn. A steep-walled mountain ridge called an arête forms between two cirques. A high pyramidal peak, called horn, is formed by the intersecting walls of three or more cirques.

Erosion is particularly rapid at the head of a glacier. In valleys, moving glaciers press rock fragments against the sides, widening and deepening them by abrasion and forming broad U-shaped valleys. When glaciers recede, tributary streams become higher than the floor of the U-shaped valley and waterfalls occur over these hanging valleys. As the ice continues to melt, residual sediments called moraines may be deposited. Moraines are made up of glacier till, a collection of sediment of all sizes. Bands of sediment along the side of a valley glacier are lateral moraines; those crossing the valley are end or recessional moraines; where two glaciers join, a medial moraine is formed. Meltwater may also sort out the finer materials, transport them downvalley, and deposit them in beds as outwash.

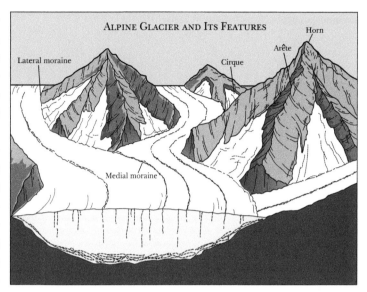

ALPINE GLACIER AND ITS FEATURES

Lateral moraine · Cirque · Arête · Horn · Medial moraine

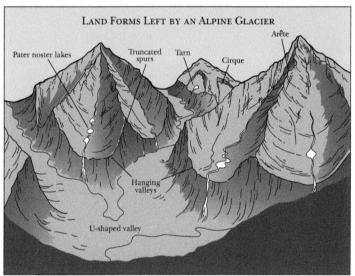

LAND FORMS LEFT BY AN ALPINE GLACIER

Pater noster lakes · Truncated spurs · Tarn · Cirque · Arête · Hanging valleys · U-shaped valley

Continental Glaciation

In the modern world, continental glaciation operates on a large scale only in Greenland and Antarctica. However, its existence in previous geologic ages is evidenced by strata of tillite (a compacted rock formed of glacial deposits) or, more frequently, by surficial deposits of glacial materials.

Much of the geomorphology of the northeastern quadrant of North America and the northwestern portion of Europe was formed during the Ice Age. During that time, great masses of ice accumulated on the continents and moved out from centers near the Hudson Bay and the Fennoscandian Shield, extending over the continents in great advancing and retreating lobes. In North America, the four major

357

FEATURES OF A CONTINENTAL GLACIER

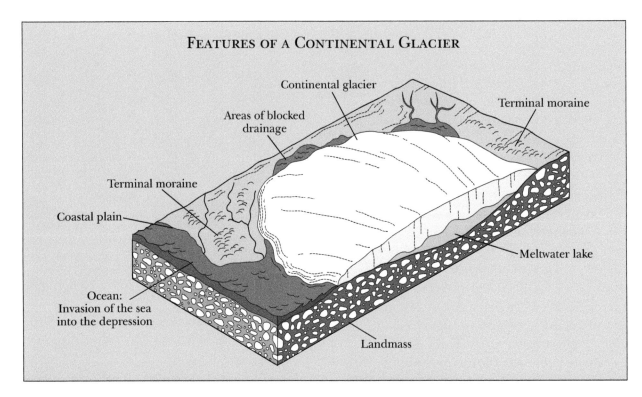

stages of lobe advance were the Wisconsin (the most recent), the Illinoian, the Kansan, and the Nebraskan (the oldest). Between each of these major advances were pluvial periods in which the ice melted and great quantities of water rushed over or stood on the continents, creating distinctive features which can still be detected today.

The two major functions of gradation are accomplished by the processes of scour (degradation) in the areas close to the centers and deposition (aggradation) adjacent to the terminal or peripheral areas of the lobes. Thus, the overall effect of continental glaciation is to reduce relief—to scour high areas and fill in lower regions—unlike the changes caused by alpine glaciation.

Although continental glaciation usually does not result in the spectacular scenery of alpine glaciation, it was responsible for creating most of the Great Lakes and the lakes of Wisconsin, Michigan, Minnesota, Finland, and Canada; for gravel deposits; and for the rich agricultural lands of the Midwest, to mention just a few of its effects.

While glaciers were leveling hilly sections of North America and Europe by scraping them bare

of soil and cutting into the ice itself, they acquired a tremendous load of material. As a glacier warms and melts, there is a tremendous outflow of water, and the streams thus formed carry with them the debris of the glacier. The material deposited by glaciers is called drift or outwash. Glaciofluvial drift can be recognized by its separation into layers of finer sands and coarser gravels.

Kettles and kames are the most common features of the end moraines found at the outermost edges of a glacier. A kettle is a depression left when a block of ice, partially or completely buried in deposits of drift, melts away. Most of the lakes in the upper Great Lakes of the United States are kettle lakes. A kame is a round, cone-shaped hill. Kames are produced by deposition from glacial meltwater. Sometimes, the outwash material poured into a long and deep crevasse, rather than a hole. These tunnels have had their courses choked by debris, revealed today by long, narrow ridges, generally referred to as eskers.

Ron Janke

Desert Landforms

Deserts are often striking in color, form, or both. The underlying lack of water in deserts produces unique features not found in humid regions. Arid lands cover approximately 30 percent of the earth's land surface, an area of about 15.4 million square miles (40 million sq. km.). Arid lands include deserts and surrounding steppes, semiarid regions that act as transition zones between arid and humid lands.

Many of the world's largest and driest deserts are found between 20 and 40 degrees north and south latitude. These include the Mojave and Sonoran Deserts of the United States, the Sahara in northern Africa, and the Great Sandy Desert in Australia. In these deserts, the subtropical high prevents cloud formation and precipitation while increasing rates of surface evaporation.

Some arid lands, like China's Gobi Desert, form because they are far from oceans that are the dominant source for atmospheric water vapor and precipitation. Others, like California's Death Valley, are arid because mountain ranges block moisture from coming from the sea. The combination of mountain barriers and very low elevations makes Death Valley the hottest, driest desert in North America.

Sand Dunes

Many people envision deserts as vast expanses of blowing sand. Although wind plays a more important role in deserts than it does elsewhere, only about 25 percent of arid lands are covered by sand. Broad regions that are covered entirely in sand (such as portions of northwestern Africa, Arabia, and Australia) are referred to as sand seas. Why is wind more effective here than elsewhere?

The lack of soil water and vegetation, both of which act to bind grains together, allows enhanced eolian (wind) erosion. Very small particles are picked up and suspended within the moving air mass, while sand grains bounce along the surface. Removal of material often leaves behind depressions called blowouts or deflation hollows. Moving grains abrade cobbles and boulders at the surface, creating uniquely sculpted and smoothed rocks known as ventifacts. Bedrock outcrops can be streamlined as they are blasted by wind-borne grains to form features called yardangs. As these rocks are ground away, they contribute additional sediment to the wind.

Desert sand dunes are not stationary features—instead, they represent accumulations of moving sand. Wind blows sand along the desert floor. Where it collects, it forms dunes. Typically, dunes have relatively shallow windward faces and steeper slip faces. Sand grains bounce up the windward face and then eventually cascade down the slip face, the movement of individual grains driving movement of the entire dune in a downwind direction.

Four major dune types are found within arid regions. Barchan dunes are crescent-shaped features, with arms that point downwind. They may occur as isolated structures or within fields. They form where winds blow in a single direction and where the supply of sand is limited. With a larger supply of sand, barchan dunes can join with one another to form a transverse dune field.

There, ridges are perpendicular to the predominant wind direction. With quartering winds (that is, winds that vary in direction throughout a range of about 45 degrees) dune ridges form that are parallel to the average wind direction. These so-called longitudinal dunes have no clearly defined windward and slip faces. Where winds blow sand from all directions, star dunes form. Sand collects in the middle of the feature to form a peaked center with arms that spiral outward.

Badlands, Mesas, and Buttes

As scarce as it may be, water is still the dominant force in shaping desert landscapes. Annual precipitation may be low, but the amount of precipitation in a single storm may be a large fraction of the yearly total. An arid landscape that is underlain by poorly cemented rock or sediment, such as that

359

found in western South Dakota, may form badlands as a result of the erosive ability of storm-water runoff. Overall aridity prevents vegetation from establishing the interconnected root system that holds soil particles together in more humid regions.

Cloudbursts cause rapid erosion that forms numerous gullies, deeply incised washes, and hoodoos, which are created when rock or sediment that is more resistant protects underlying material from erosion. Over time, protected sections stand as prominent spires while surrounding material is removed. Landscapes like those found in Badlands National Park in South Dakota are devoid of vegetation and erode rapidly during storms.

Arid regions that are underlain by flat-lying rock units can form mesas and buttes. Water follows fractures and other lines of weakness, forming ever-widening canyons. Over time, these grow into broad valleys. In northern Arizona's Monument Valley, remnants of original bedrock stand as isolated, flat-topped structures. Broad mesas are marked by their flat tops (made of a resistant rock like sandstone or basalt) and steep sides. Buttes are

---

Death Valley Playa

California's Death Valley is the driest desert in the United States, with an average rainfall of only 1.5 inches (38 millimeters) per year at the town of Furnace Creek. It is also consistently one of the hottest places on Earth, with a record high of 134 degrees Fahrenheit (57 degrees Celsius). In the distant past, however, Death Valley held lakes that formed in response to global cooling. Over 120,000 years ago, Death Valley hosted a 295-foot-deep (90 meters) body of water called Lake Manley. Evidence of this lake remains in evaporite deposits that make up the playa in the valley's center, in wave-cut shorelines, and in beach bars.

---

much narrower, with a small resistant cap, but are often as tall and steep as neighboring mesas.

Desert Pavement and Desert Varnish
Much of the desert floor is covered by desert pavement, an accumulation of gravel and cobbles that forms a surface fabric that can interconnect tightly. Fine material has been removed by wind and water, leaving behind larger fragments that inhibit further erosion. In many areas, desert pavements have been stable for long periods of time, as evidenced by their surface patina of desert varnish. Desert varnish is a thin outer coating of wind-deposited clay mixed with iron and manganese oxides. Varying in color from light brown to black, these coatings are thought to adhere to rocks by the action of single-celled microorganisms. Under a microscope, desert varnish can be seen to be made up of very fine layers. A thick, dark patina means that a rock has been exposed for a long time.

Playas
Where neither dunes nor rocky pavements cover the desert floor, one may find an accumulation of saline minerals. A playa is a flat surface that is often blindingly white in color. Playas are usually found in the centers of desert valleys and contain material that mineralized during the evaporation of a lake. Dry lake beds are a common feature of the Great Basin in the western United States. During glacial stages, the last of which occurred about 20,000 years ago, lakes grew in what are now arid, closed valleys. As the climate warmed, these lakes shrank, and many dried completely. As a lake evaporates, minerals that were held in solution crystallize, forming salts, including halite (table salt). These salt deposits frequently are mined for useful household and industrial chemicals.

Richard L. Orndorff

OCEAN MARGINS

Ocean margins are the areas where land borders the sea. Although often referred to as coastlines or beaches, ocean margins cover far greater territory than beaches. An ocean margin extends from the coastal plain—the fertile farming belt of land along the seacoast—to the edge of the gently sloping land submerged in water, called the continental shelf.

Ocean margin constitutes 8 percent of the world's surface. It is rich in minerals, both above and below water, and is home to 25 percent of Earth's people, along with 90 percent of the marine life. This fringe of land at the border of the ocean is ever changing. Tides wash sediment in and leave it behind, just below sea level. This process, called deposition, builds up land in some areas of the coastline. At the same time, ocean waves, winds, and storms wear away or erode parts of the shoreline. As land is worn away or built up, the amount of land above sea level changes. Factors such as climate, erosion, deposition, changes in sea level, and the effects of humans constantly change the shape of the ocean margin on Earth.

Beach Dynamics

The two types of coasts or land formations at the ocean margin are primary coasts and secondary coasts. Primary coasts are formed by systems on land, such as the melting of glaciers, wind or water erosion, and sediment deposited by rivers. Deltas and fjords are examples of primary coasts. Secondary coasts are formed by ocean patterns, such as erosion by waves or currents, sediment deposition by waves or currents, or changes by marine plants or animals. Beaches, coral reefs, salt marshes, and mangrove swamps are examples of secondary coasts.

Sediment carried by rivers to the sea is deposited to form deltas at the mouths of the rivers. Some of the sediment can wash out to sea, causing formations to build up at a distance from the shore. These formations eventually become barrier islands, which are often little more than 10 feet (3 km.) above sea level. As

a consequence, heavy storms, such as hurricanes, can cause great damage to barrier islands. Barrier islands naturally protect the coastline from erosion, however, especially during heavy coastal storms.

Sea level changes also affect the shape of the coastline. As oceans slowly rise, land is slowly consumed by the ocean. Barrier islands, having low sea levels, may slowly be covered with water. The melting of continental glaciers increased the sea level 0.06 inch (0.15 centimeter) per year during the twentieth century. As ocean waters warm, they expand, eating away at sea levels. Global warming caused by carbon dioxide levels in the atmosphere could cause sea levels to rise as much as 0.24 inch (0.6 centimeter) per year as a result of the warming of the water and glacial melting.

Human Influence

The shape of the ocean margin also changes radically as a result of human influence. According to the United States Geological Survey, 39 percent of the people living in the United States live directly on the the coasts. According to UN Atlas of Oceans, about 44 percent of the world's population lives within 93 miles (150 kilometers) of the coast. Pollution from toxins, dredging, recreational boating, and waste disposal kills plants and animals along the ocean margin. This changes the coastal shape, as mangrove forests, coral reefs, and other coastal lifeforms die.

A greater concern along the coastal fringe, however, is human development. Not only are people drawn to the fertile soil along the coastal zone of the continent, but they also develop islands and coves into resort communities. To protect homes and hotels along the coastal zone from coastal erosion, people build breakwalls, jetties, and sand and stone bars called groins.

These human-made barriers disrupt the natural method by which the ocean carries material along the coast. Longshore drift, a zigzag movement, deposits sediment from one area of the beach farther along the shoreline. Breakwalls, jetties, and groins disrupt this flow. As the ocean smashes against a breakwall, the property behind it may be safe for the present, but the coastline neighboring the breakwall takes a greater beating. The silt and sediment from upshore, which would replace that carried downshore, never arrives. Eventually, the breakwall will break down under the impact of the ocean force. Areas with breakwalls and jetties often suffer greater damage in coastal storms than areas that remain naturally open to the changing forces of the ocean.

To compensate for the destructive nature of human-made barriers, many recreational beaches replace lost sand with dredgings or deposit truckloads of sand from inland sources. For example, Virginia Beach in the United States spends between US$2 million and US$3 millon annually to restore beaches for the tourist season in this way.

Despite the changes in the shape of the ocean margin, it continues to provide a stable supply of resources—fish, seafood, minerals, sponges, and other marine plants and animals. Offshore drilling of oil and natural gas often takes place within 200 miles (322 km.) of shorelines.

Lisa A. Wroble

EARTH'S CLIMATES

THE ATMOSPHERE

The thin layer of gases that envelops the earth is the atmosphere. This layer is so thin that if the earth were the size of a desktop globe, more than 99 percent of its atmosphere would be contained within the thickness of an ordinary sheet of paper. Despite its thinness, the atmosphere sustains life on Earth, protecting it from the Sun's searing radiation and regulating the earth's temperature. Storms of the atmosphere carry water to the continents, and weathering by its wind and rain helps shape them.

Composition of the Atmosphere

The earth's atmosphere consists of gases, microscopic particles called aerosol, and clouds consisting of water droplets and ice particles. Its two principal gases are nitrogen and oxygen. In dry air, nitrogen occupies 78 percent, and oxygen 21 percent, of the atmosphere's volume. Argon, neon, xenon, helium, hydrogen, and other trace gases together equal less than 1 percent of the remaining volume.

These gases are distributed homogeneously in a layer called the homosphere, which occurs between the earth's surface and about 50 miles (80 km.) altitude. Above 50 miles altitude, in the heterosphere, the concentration of heavier gases decreases more rapidly than lighter gases.

The atmosphere has no firm top. It simply thins out until the concentration of its gas molecules approaches that of the gases in outer space. The concentration of nitrogen and oxygen remains essentially con-stant in the atmosphere because a balance exists between the production and removal of these gases at the earth's surface. Decaying organic matter adds nitrogen to the atmosphere, while soil bacteria remove nitrogen. Oxygen enters the atmosphere primarily through photosynthesis and is removed through animal respiration, combustion, and decay of organic material, and by chemical reactions involving the creation of oxides.

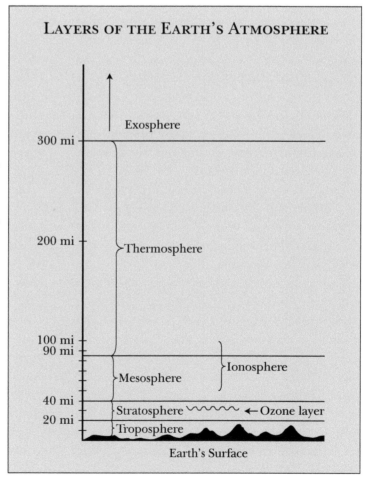

LAYERS OF THE EARTH'S ATMOSPHERE

Exosphere

300 mi

Thermosphere

200 mi

100 mi
90 mi

Ionosphere

Mesosphere

40 mi

Stratosphere — Ozone layer

20 mi

Troposphere

Earth's Surface

THE GREENHOUSE EFFECT

Clouds and atmospheric gases such as water vapor, carbon dioxide, methane, and nitrous oxide absorb part of the infrared radiation emitted by the earth's surface and reradiate part of it back to the earth. This process effectively reduces the amount of energy escaping to space and is popularly called the "greenhouse effect" because of its role in warming the lower atmosphere. The greenhouse effect has drawn worldwide attention because increasing concentrations of carbon dioxide from the burning of fossil fuels result in a global warming of the atmosphere.

Scientists know that the greenhouse analogy is incorrect. A greenhouse traps warm air within a glass building where it cannot mix with cooler air outside. In a real greenhouse, the trapping of air is more important in maintaining the temperature than is the trapping of infrared energy. In the atmosphere, air is free to mix and move about.

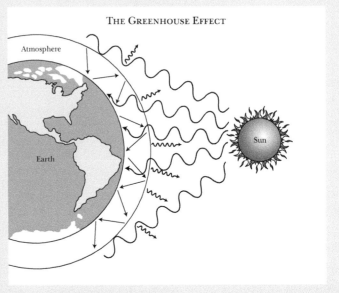

THE GREENHOUSE EFFECT

The atmosphere contains many gases that are present in small, variable concentrations. Three gases—water vapor, carbon dioxide and ozone—are vital to life on Earth. Water vapor enters the atmosphere through evaporation, primarily from the oceans, and through transpiration by plants. It condenses to form clouds, which provide the rain and snow that sustain life outside the oceans. The concentration of water vapor varies from about 4 percent by volume in tropical humid climates to a small fraction of a percent in polar dry climates. Water vapor plays an important role in regulating the temperature of the earth's surface and the atmosphere. Clouds reflect some of the incoming solar radiation, while water vapor and clouds both absorb earth's infrared radiation.

Carbon dioxide also absorbs the earth's infrared radiation. The global average atmospheric carbon dioxide in 2018 was 407.4 parts per million (ppm for short), with a range of uncertainty of plus or minus 0.1 ppm. Carbon dioxide levels today are higher than at any point in at least the past 800,000 years. The annual rate of increase in atmospheric carbon dioxide over the past 60 years is about 100 times faster than previous natural increases, such as those that occurred at the end of the last ice age 11,000–17,000 years ago.

Carbon dioxide enters the atmosphere as the result of decay of organic material, through respiration, during volcanic eruptions, and from the burning of fossil fuels. It is removed during photosynthesis and by dissolving in ocean water, where it is used by organisms and converted to carbonates. The increase in atmospheric carbon dioxide associated with the burning of fossil fuels has raised concerns that the earth's atmosphere may be warming through enhancement of the greenhouse effect.

Ozone, a gas consisting of molecules containing three oxygen atoms, forms in the upper atmosphere when oxygen atoms and oxygen molecules combine. Most ozone exists in the upper atmosphere between 6.2 and 31 miles (10 and 50 km.) in altitude, in concentrations of no more than 0.0015 percent by volume. This small amount of ozone sustains life outside the oceans by absorbing most of the Sun's ultraviolet radiation, thereby shielding the earth's surface from the radiation's harmful effects on living organisms. Paradoxically, ozone is an irritant near the earth's surface and is the major component of photochemical smog. Other gases that contribute to pollution include methane, nitrous oxide, hydrocarbons, and chlorofluorocarbons.

Aerosols represent another component of atmospheric pollution. Aerosols form in the atmosphere during chemical reactions between gases, through mechanical or chemical interactions between the earth, ocean surface, and atmosphere, and during evaporation of droplets containing dissolved or solid material. These microscopic particles are always present in air, with concentrations of about a few hundred per cubic centimeter in clean air to as many as a million per cubic centimeter in polluted air. Aerosols are essential to the formation of rain and snow, because they serve as centers upon which cloud droplets and ice particles form.

Energy Exchange in the Atmosphere

The Sun is the ultimate source of the energy in Earth's atmosphere. Its radiation, called electromagnetic radiation because it propagates as waves with electric and magnetic properties, travels to the surface of the earth's atmosphere at the speed of light. This energy spans many wavelengths, some of which the human eye perceives as colors. Visible wavelengths make up about 44 percent of the Sun's energy. The remainder of the Sun's radiant energy cannot be seen by human eyes. About 7 percent arrives as ultraviolet radiation, and most of the remaining energy is infrared radiation.

The Sun is not the only source of radiation. All objects emit and absorb radiation to some degree. Cooler objects such as the earth emit nearly all their energy at infrared wavelengths. Objects heat when they absorb radiation and cool when they emit radiation. The radiation emitted by the earth and atmosphere is called terrestrial radiation.

The balance between absorption of solar radiation and emission of terrestrial radiation ultimately determines the average temperature of the earth-atmosphere system. The vertical temperature distribution within the atmosphere also depends on the absorption and emission of radiation within the atmosphere, and the transfer of energy by the processes of conduction, convection, and latent heat exchange. Conduction is the direct transfer of heat from molecule to molecule. This process is most important in transferring heat from the earth's surface to the first few centimeters of the at-

---

THE OZONE HOLE

Since the 1970s, balloon-borne and satellite measurements of stratospheric ozone have shown rapidly declining stratospheric ozone concentrations over the continent of Antarctica, termed the "ozone hole." The lowest concentrations occur during the Antarctic spring, in September and October. The decrease in ozone has been associated with an increase in the concentration of chlorine, a gas introduced into the stratosphere through chemical reactions involving sunlight and chlorofluorocarbons—synthetic chemicals used primarily as refrigerants. The ozone hole over Antarctica has raised concern about possible worldwide reduction in the concentration of upper atmospheric ozone.

---

mosphere. Convection, the transfer of heat by rising or sinking air, transports heat energy vertically through the atmosphere.

Latent heat is the energy required to change the state of a substance, for example, from a liquid to a gas. Energy is transferred from the earth's surface to the atmosphere through latent heat exchange when water evaporates from the oceans and condenses to form rain in the atmosphere.

Only 48 percent of the solar energy reaching the top of the earth's atmosphere is absorbed by the earth's surface. The atmosphere absorbs another 23 percent. The remaining 30 percent is scattered back to space by atmospheric gases, clouds and the earth's surface. To understand the importance of terrestrial radiation and the greenhouse effect in the atmosphere's energy balance, consider the solar radiation arriving at the top of the earth to be 100 energy units, with 48 energy units absorbed by the earth's surface and 23 units by the atmosphere.

The earth's surface actually emits 117 units of energy upward as terrestrial radiation, more than twice as much energy as it receives from the Sun. Only 6 of these units are radiated to space—the atmosphere absorbs the remaining energy. Latent heat exchange, conduction, and convection account for another 30 units of energy transferred from the surface to the atmosphere. The atmosphere, in turn, radiates 96 units of energy back to the earth's surface (the greenhouse effect), and 64 units to space. The earth's and atmosphere's energy budget remains in balance, the atmosphere gaining and losing 160 units of energy, and the earth gaining and losing 147 units of energy.

Vertical Structure of the Atmosphere

Temperature decreases rapidly upward away from the earth's surface, to about –60 degrees Farenheit (–51 degrees Celsius) at an altitude of about 7.5 miles (12 km.). Above this altitude, temperature increases with height to about 32 degrees Farenheit (0 degrees Celsius) at an altitude of 31 miles (50 km.). The layer of air in the lower atmosphere where temperature decreases with height is called the troposphere. It contains about 75 percent of the atmosphere's mass. The layer of air above the troposphere, where temperature increases with height, is called the stratosphere. All but 0.1 percent of the remaining mass of the atmosphere resides in the stratosphere.

The stratosphere exists because ozone in the stratosphere absorbs ultraviolet light and converts it to heat. The boundary between the troposphere and stratosphere is called the tropopause. The tropopause is extremely important because it acts as a lid on the earth's weather. Storms can grow vertically in the troposphere, but cannot rise far, if at all, beyond the tropopause. In the polar regions, the tropopause can be as low as 5 miles (8 km.) above the surface, while in the tropics, the tropopause can be as high as 11 miles (18 km.). For this reason, tropical storms can extend to much higher altitudes than storms in cold regions.

The mesosphere extends from the top of the stratosphere, the stratopause, to an altitude of about 56 miles (90 km.). Temperature decreases with height within the mesosphere. The lowest average temperatures in the atmosphere occur at the mesopause, the top of the mesosphere, where the temperature is about –130 degrees Farenheit (–90 degrees Celsius). Only 0.0005 percent of the atmosphere's mass remains above the mesopause. In this uppermost layer, the thermosphere, there are few atoms and molecules. Oxygen molecules in the thermosphere absorb high-energy solar radiation. In this near vacuum, absorption of even small amounts of energy causes a large increase in temperature. As a result, temperature increases rapidly with height in the lower thermosphere, reaching about 1,300 degrees Farenheit (700 degrees Celsius) above 155 miles (250 km.) altitude.

The upper mesosphere and thermosphere also contain ions, electrically charged atoms or molecules. Ions are created in the atmosphere when air molecules collide with high-energy particles arriving from space or absorb high-energy solar radiation. Ions cannot exist very long in the lower atmosphere, because collisions between newly formed ions quickly restore ions to their uncharged state. However, above about 37 miles (60 km.) collisions are less frequent and ions can exist for longer times. This region of the atmosphere, called the ionosphere, is particularly important for amplitude-modulated (AM) radio communication because it reflects standard AM radio waves. At night, the lower ionosphere disappears as ions recombine, allowing AM radio waves to travel longer distances when reflected. For this reason, AM radio station signals can sometimes travel great distances at night.

The top of the atmosphere occurs at about 310 miles (500 km.). At this altitude, the distance between individual molecules is so great that energetic molecules can move into free space without colliding with neighbor molecules. In this uppermost layer, called the exosphere, the earth's atmosphere merges into space.

Robert M. Rauber

GLOBAL CLIMATES

A region's climate is the sum of its long-term weather conditions. Most descriptions of climate emphasize temperature and precipitation characteristics, because these two climatic elements usually exert more impact on environmental conditions and human activities than do other elements, such as wind, humidity, and cloud cover. Climatic descriptions of a region generally cover both mean conditions and extremes. Climatic means are important because they represent average conditions that are frequently experienced; extreme conditions, such as severe storms, excessive heat and cold, and droughts, are important because of their adverse impact.

Important Climate Controls
A region's climate is largely determined by the interaction of six important natural controls: sun angle, elevation, ocean currents, land and water heating and cooling characteristics, air pressure and wind belts, and orographic influence.

Sun angle—the height of the Sun in degrees above the nearest horizon—largely controls the amount of solar heating that a site on Earth receives. It strongly influences the mean temperatures of most of the earth's surface, because the Sun is the ultimate energy source for nearly all the atmosphere's heat. The higher the angle of the Sun in the sky, the greater the concentration of energy, per unit area, on the earth's surface (assuming clear skies). From a global perspective, the Sun's mean angle is highest, on average, at the equator, and becomes progressively lower poleward. This causes a gradual decrease in mean temperatures with increasing latitude.

Sun angles also vary seasonally and daily. Each hemisphere is inclined toward the Sun during spring and summer, and away from the Sun during fall and winter. This changing inclination causes mean sun angles to be higher, and the length of daylight longer, during the spring and summer. Therefore, most locations, especially those outside the tropics, have warmer temperatures during these two seasons. The earth's rotation causes sun angles to be higher during midday than in the early morning and late afternoon, resulting in warmer temperatures at midday. Heating and cooling lags cause both seasonal and daily maximum and minimum temperatures typically to occur somewhat after the periods of maximum and minimum solar energy receipt.

Variations in elevation—the distance above sea level—can cause locations at similar latitudes to vary greatly in temperature. Temperatures decrease an average of about 3.5 degrees Fahrenheit per thousand feet (6.4 degrees Celsius per thousand meters). Therefore, high mountain and plateau stations are much colder than low-elevation stations at the same latitude.

Surface ocean currents can transport masses of warm or cold water great distances from their source regions, affecting both temperature and moisture conditions. Warm currents facilitate the evaporation of copious amounts of water into the atmosphere and add buoyancy to the air by heating it from below. This results in a general increase in precipitation totals. Cold currents evaporate water relatively slowly and chill the overlying air, thus stabilizing it and reducing its potential for precipitation.

The influence of ocean currents on land areas is greatest in coastal regions and decreases inland. The west coasts of continents (except for Europe) generally are paralleled by relatively cold currents, and the east coasts by relatively warm currents. For example, the warm Gulf Stream flows northward off the eastern United States, while the West Coast is cooled by the southward-flowing California Current.

Land can change temperature much more readily than water. As a result, the air over continents typically experiences larger annual temperature ranges (that is, larger temperature differences between summer and winter) and shorter heating and cooling lags than does the air over oceans. This same effect causes continental interiors and the leeward (downwind) coasts of continents typically to have larger temperature ranges than do windward (upwind) coasts. Climates that are dominated by air from landmasses are often described as continental climates. Conversely, climates dominated by air from oceans are described as maritime climates.

The seasonal heating and cooling of continents can also produce a monsoon influence, which has to do with annual shifts of wind patterns. Areas influenced by a monsoon, such as Southeast Asia, tend to have a predominantly onshore flow of moist maritime air during the summer. This often produces heavy rains. An offshore flow of dry air predominates in winter, producing fair weather.

Earth's atmosphere displays a banded, or beltlike, pattern of air pressure and wind systems. High pressure is associated with descending air and dry weather; low pressure is associated with rising air, which produces cloudiness and often precipitation. Wind is produced by differences in air pressure. The air blows outward from high-pressure systems and into low-pressure systems in a constant attempt to equalize air pressures.

The direction and speed of movement of weather systems, such as weather fronts and storms, are controlled by wind patterns, especially those several kilometers above the surface. The seasonal shift of global temperatures caused by the movement of the Sun's vertical rays between the Tropics of Cancer and Capricorn produces a latitudinal migration of both air pressure and wind belts. This shift affects the annual temperature and precipitation patterns of many regions.

Four air-pressure belts exist in each hemisphere. The intertropical convergence zone (ITCZ) is a broad belt of low pressure centered within a few degrees of latitude of the equator. The subtropical highs are high-pressure belts centered between 20 and 40 degrees north and south latitude, which are responsible for many of the world's deserts. The subpolar lows are low-pressure belts centered about 50 or 70 degrees north and south latitude. Finally, the polar highs are high-pressure centers located near the North and South Poles.

The air pressure gradient between these belts produces the earth's major wind belts. The regions between the ITCZ and the subtropical highs are dominated by the trade winds, a broad belt in each hemisphere of easterly (that is, moving east to west) winds. The middle latitudes are mostly situated between the subtropical highs and the subpolar lows and are within the westerly wind belt. This wind belt causes winds, and weather systems, to travel generally from west to east in the United States and Canada. Finally, the high-latitude zones between the subpolar lows and polar highs are situated within the polar easterlies.

The final factor affecting climate— orographic influence—is the lifting effect of mountain peaks or ranges on winds that pass over them. As air approaches a mountain barrier, it rises, typically producing clouds and precipitation on the windward (upwind) side of the mountains. After it crosses the crest, it descends the leeward (downwind) side of the mountains, generally producing dry weather. Most of the world's wettest locations are found on the windward sides of high mountain ranges; some deserts, such as those of the western interior United States, owe their aridity to their location on the leeward sides of orographic barriers.

World Climate Types

The global distribution of the world climate controls is responsible for the development of fourteen widely recognized climate types. In this section, the major characteristics of each of these climates will be briefly described. The climates are discussed in a rough poleward sequence.

Tropical Wet Climate

Sometimes called the tropical rainforest climate, the tropical wet climate exists chiefly in areas lying within 10 degrees of the equator. It is an almost seasonless climate, characterized by year-round warm, humid, rainy conditions that allow land areas to support a dense broadleaf forest cover. The warm temperatures, which for most locations average near 80 degrees Fahrenheit (27 degrees Celsius) throughout the year, result from the constantly high midday sun angles experienced at this low latitude.

WORLD CLIMATE REGIONS

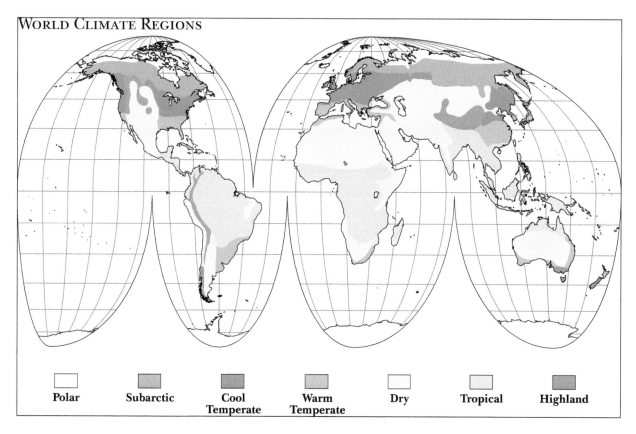

| Polar | Subarctic | Cool Temperate | Warm Temperate | Dry | Tropical | Highland |

The heavy precipitation totals result from the heating and subsequent rising of the warm moist air to form frequent showers and thunderstorms, especially during the afternoon hours. The dominance of the ITCZ enhances precipitation totals, helping make this climate type one of the world's rainiest.

Tropical Monsoonal Climate

The tropical monsoonal climate occurs in low-latitude areas, such as Southeast Asia, that have a warm, rainy climate with a short dry season. Temperatures are similar to those of the tropical wet climate, with the warmest weather often occurring during the drier period, when sunshine is more abundant. The heavy rainfalls result from the nearness of the ITCZ for much of the year, as well as the dominance of warm, moist air masses derived from tropical oceans. During the brief dry season, however, the ITCZ has usually shifted into the opposite hemisphere, and windflow patterns often have changed so as to bring in somewhat drier air derived from continental sources.

Tropical Savanna Climate

The tropical savanna climate, also referred to as the tropical wet-dry climate, occupies a large portion of the tropics between 5 and 20 degrees latitude in both hemispheres. It experiences a distinctive alternation of wet and dry seasons, caused chiefly by the seasonal shift in latitude of the subtropical highs and ITCZ. Summer is typically the rainy season because of the domination of the ITCZ. In many areas, an onshore windflow associated with the summer monsoon increases rainfalls at this time. In winter, however, the ITCZ shifts into the opposite hemisphere and is replaced by drier and more stable air associated with the subtropical high. In addition, the winter monsoon tendency often produces an outflow of continental air. The long dry season inhibits forest growth, so vegetation usually consists of a cover of drought-resistant shrubs or the tall savanna grasses after which the climate is named.

Subtropical Desert Climate

The subtropical desert climate has hot, arid conditions as a result of the year-round dominance of the

369

subtropical highs. Summertime temperatures in this climate soar to the highest readings found anywhere on earth. The world's record high temperature was 134 degrees Fahrenheit (56.7 degrees Celsius), recorded in Furnace Creek Ranch, California (formerly Greenland Ranch) on July 10, 1913. Rainfall totals in this type of climate are generally less than 10 inches (25 centimeters) per year. What rainfall does occur often arrives as brief, sometimes violent, afternoon thunderstorms. Although summer temperatures are extremely hot, the dry air enables rapid cooling during the winter, so that temperatures are cool to mild at this time of year.

Subtropical Steppe Climate

The subtropical steppe climate is a semiarid climate, found mostly on the margins of the subtropical deserts. Precipitation usually ranges from 10 to 30 inches (25 to 75 centimeters), sufficient for a ground cover of shrubs or short steppe grasses. Areas on the equatorward margins of subtropical deserts typically receive their precipitation during a brief showery period in midsummer, associated with the poleward shift of the ITCZ. Areas on the poleward margins of the subtropical highs receive most of their rainfall during the winter, due to the penetration of cyclonic storms associated with the equatorward shift of the westerly wind belt.

Mediterranean Climate

The Mediterranean climate, also sometimes referred to as the dry summer subtropics, has a distinctive pattern of dry summers and more humid, moderately wet winters. This pattern is caused by the seasonal shift in latitude of the subtropical high and the westerlies. During the summer, the subtropical high shifts poleward into the Mediterranean climate regions, blanketing them with dry, warm, stable air. As winter approaches, this pressure center retreats equatorward, allowing the westerlies, with their eastward-traveling weather fronts and cyclonic storms, to overspread this region. The Mediterranean climate is found on the windward sides of continents, particularly the area surrounding the Mediterranean Sea and much of California. This results in the predomi-

nance of maritime air and relatively mild temperatures throughout the year.

Humid Subtropical Climate

The humid subtropical climate is found on the eastern, or leeward, sides of continents in the lower middle latitudes. The most extensive land area with this climate is the southeastern United States, but it is also seen in large areas in South America, Asia, and Australia. Temperature ranges are moderately large, with warm to hot summers and cool to mild winters. Mean temperatures for a given location are dictated largely by latitude, elevation, and proximity to the coast. Precipitation is moderate. Winter precipitation is usually associated with weather fronts and cyclonic storms that travel eastward within the westerly wind belt. During summer, most precipitation is in the form of brief, heavy afternoon and evening thunderstorms. Some coastal areas are subject to destructive hurricanes during the late summer and autumn.

Midlatitude Desert Climate

This type of climate consists of areas within the western United States, southern South America, and Central Asia that have arid conditions resulting from the moisture-blocking influence of mountain barriers. This climate is highly continental, with warm summers and cold winters. When precipitations occurs, it frequently comes in the form of winter snowfalls associated with weather fronts and cyclonic storms. Rainfall in summer typically occurs as afternoon thunderstorms.

Midlatitude Steppe

The midlatitude steppe climate is located in interior portions of continents in the middle latitudes, particularly in Asia and North America. This climate has semiarid conditions caused by a combination of continentality resulting from the large distance from oceanic moisture sources and the presence of mountain barriers. Like the midlatitude desert climate, this climate has large annual temperature ranges, with cold winters and warm summers. It also receives winter rains and snows chiefly from weather fronts and cyclonic storms; summer rains occur largely from afternoon convectional storms. In the Great Plains of the United States, spring can bring very turbulent conditions, with blizzards in early spring and hailstorms and tornadoes in mid to late spring.

Marine West Coast

This type of climate is typically located on the west coasts of continents just poleward of the Mediterranean climate. Its location in the heart of the westerly wind belt on the windward sides of continents produces highly maritime conditions. As a result, cloudy and humid weather is common, along with frequent periods of rainfall from passing weather fronts and cyclonic storms. These storms are often well developed in winter, resulting in extended periods of wet and windy weather. Precipitation amounts are largely controlled by the presence and strength of the orographic effect; mountainous coasts like the northwestern United States and the west coast of Canada are much wetter than are flatter areas like northern Europe. Temperatures are held at moderate levels by the onshore flow of maritime air. As a consequence, winters are relatively mild and summers relatively cool for the latitude.

Humid Continental Climate

The humid continental climate is found in the northern interiors of Eurasia (Europe and Asia) and North America. It does not occur in the Southern Hemisphere because of the absence of large land masses in the upper midlatitudes of that hemisphere. This climate type is characterized by low to moderate precipitation that is largely frontal and cyclonic in nature. Most precipitation occurs in summer, but cold winter temperatures typically cause the surface to be frozen and snow-covered for much of the late fall, winter, and early spring. Temperature ranges in this climate are the largest in the world. A town in Siberia, Verkhoyansk, holds the Guinness world record for the greatest temperature range at a single location is 221 degrees Fahrenheit (105 degrees Celsius), from -90 degrees Fahrenheit (-68 degrees Celsius) to 99 degrees Fahrenheit (37 degrees Celsius). Winter temperatures in parts of both North America and Siberia can fall well below -49 degrees Fahrenheit (-45 degrees Celsius), making these the coldest permanently settled sites in the world.

Tundra Climate

The tundra climate is a severely cold climate that exists mostly on the coastal margins of the Arctic Ocean in extreme northern North America and Eurasia, and along the coast of Greenland. The high-latitude location and proximity to icy water cause every month to have average temperatures below 50 degrees Fahrenheit (10 degrees Celsius), although a few months in summer have means above freezing. As a result of the cold temperatures, tundra areas are not forested, but instead typically have a sparse ground cover of grasses, sedges, flowers, and lichens. Even this vegetation is buried by a layer of snow during most of the year. Cold temperatures lower the water vapor holding capacity of the air, causing precipitation totals to be generally light. Most precipitation is associated with weather fronts and cyclonic storms and occurs during the summer half of the year.

Ice Cap Climate

The most poleward and coldest of the world's climates is called the ice cap climate. It is found on the continent of Antarctica, interior Greenland, and some high mountain peaks and plateaus. Because monthly mean temperatures are subfreezing throughout the year, areas with this climate are glaciated and have no permanent human inhabitants.

The coldest temperatures of all occur in interior Antarctica, where a Russian research station named Vostok recorded the world's coldest temperature of -128.6 degrees Fahrenheit (-89.2 degrees Celsius) on July 21, 1983. This climate receives little precipitation because the atmosphere can hold very little water vapor. A major moisture surplus exists, however, because of the lack of snowmelt and evaporation. This causes the build up of a surface snow cover that eventually compacts to form the icecaps that bury the surface. Snowstorms are often accompanied by high winds, producing blizzard conditions.

Global Warming

Though warming has not been uniform across the planet, the upward trend in the globally averaged temperature shows that more areas are warming than cooling. According to the National Oceanic and Atmospheric Administration (NOAA) 2018 Global Climate Summary, the combined land and ocean temperature has increased at an average rate of 0.13°F (0.07°C) per decade since 1880; however, the average rate of increase since 1981 (0.31°F/ 0.17°C) is more than twice as great. It is strongly suspected that human activities that increase the accumulation of greenhouse gases (heat-trapping gases) in the atmosphere may play a key role in the temperature rise.

Levels of carbon dioxide (CO_2) in the atmosphere are higher now than they have been at any time in the past 400,000 years. This gas is responsible for nearly two-thirds of the global-warming potential of all human-released gases. Levels surpassed 407 ppm in 2018 for the first time in recorded history. By comparison, during ice ages, CO_2 levels were around 200 parts per million (ppm), and during the warmer interglacial periods, they hovered around 280 ppm. The recent rise in CO_2 shows a remarkably constant relationship with fossil-fuel burning, which is understandable when one considers that about 60 percent of fossil-fuel emissions stay in the air. Atmospheric carbon dioxide concentrations are also increased by deforestation, which is occurring at a rapid rate in several tropical countries. Deforestation causes carbon dioxide levels to rise because trees remove large quantities of this gas from the atmosphere during the process of photosynthesis.

Research indicates that if atmospheric concentrations of greenhouse gases continue to increase at the 1990s pace, global temperatures could rise an additional 1.8 to 6.3 degrees Fahrenheit (1 to 3.5 degrees Celsius) during the twenty-first century. That level of temperature increase would produce major changes in global climates and plant and animal habitats and would cause sea levels to rise substantially.

Ralph C. Scott

CLOUD FORMATION

Clouds are visible manifestations of water in the air. Cloud patterns can provide even a casual observer with much information about air movements and the processes occurring in the atmosphere. The shapes and heights of the clouds and the directions from which they have come are valuable clues in understanding weather.

Importance of Cooling

Clouds are formed when water vapor in the air is transformed into either water droplets or ice crystals. Sometimes large amounts of moisture are added to the air, producing clouds, but clouds generally are formed when a large amount of air is cooled. The amount of water vapor that air can hold varies with temperature: Cold air can hold less water vapor than warmer air. If air is cooled to the point at which it can hold no more water vapor, the water vapor will condense into water droplets. The temperature at which condensation begins is called the dew point. At below freezing temperatures, the water vapor will turn or deposit into ice crystals.

Cloud droplets do not necessarily form even if the air is fully saturated, that is, holding as much water vapor as possible at a given temperature. Once formed, cloud droplets can evaporate again very easily. Two factors hasten the production and growth of cloud droplets. One is the presence of

CLOUD FORMATION

The hydrologic cycle is the continuous circulation of the earth's waters through evaporation, condensation, and precipitation. The cycle also moves water through runoff, infiltration, and transpiration.

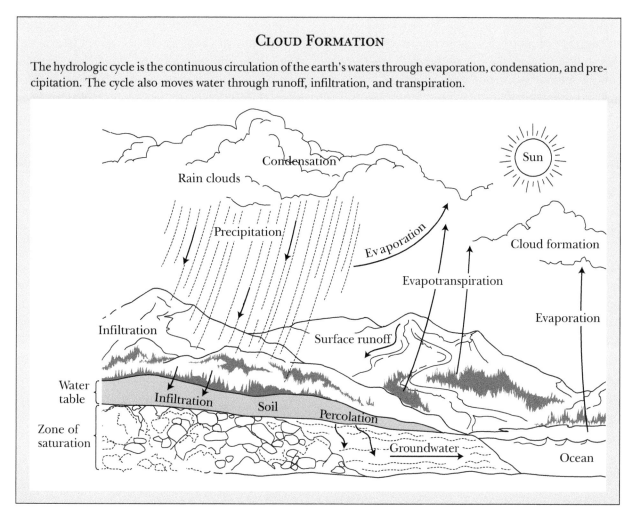

particles in the atmosphere that attract water. These are called hygroscopic particles or condensation nuclei. They include salt, dust, and pollen. Once water vapor condenses on these particles, more condensation can occur. Then the droplets can grow larger and bump into other droplets, growing even larger through this process, called coalescence.

Condensation and cloud droplet growth also is hastened when the air is very cold, at about -40 degrees Farenheit (which is also -40 degrees Celsius). At this temperature ice crystals form, but some water droplets can exist as liquid water. These water droplets are said to be supercooled. The water vapor is more likely to deposit on the ice crystals than on the supercooled water. Thus the ice crystals grow larger and the supercooled water droplets evaporate, resulting in more water vapor to deposit on ice crystals. Whether the cloud droplets start as hygro-

scopic particles or ice crystals, they eventually can grow in size to become a raindrop; around 1 million cloud droplets make one raindrop.

How and Why Rising Air Cools

In order for air to be cooled, it must rise or be lifted. When a volume of air, or an air parcel, is forced to rise through the surrounding air, the parcel expands in size as the pressure of the air around it declines with altitude. Close to the surface, the atmospheric pressure is relatively high because the density of the atmosphere is high. As altitude increases, the atmosphere declines in density, and the still air exerts less pressure. Thus, as an air parcel rises through the atmosphere, the pressure of the surrounding air declines, and the parcel takes up more space as it expands. Since work is done by the parcel as it expands, the parcel cools and its temperature declines.

An alternative explanation of the cooling is that the number of molecules in the air parcel remains the same, but when the volume is larger, the molecules produce less frictional heat because they do not bang into each other as much. The temperature of the air parcel declines, but no heat left the parcel—the change in temperature resulted from internal processes. The process of an air parcel rising, expanding, and cooling is called adiabatic cooling. Adiabatic means that no heat leaves the parcel. If the parcel rises far enough, it will cool sufficiently to reach its dewpoint temperature. With continued cooling, condensation will result—a cloud will be formed. At this height, which is called the lifting condensation level, an invisible parcel of air will turn into a cloud.

Uplift Mechanisms

An initial force is necessary to cause the air parcel to rise and then cool adiabatically. The three major processes are convection, orographic, and frontal or cyclonic.

With certain conditions, convection or vertical movement can cause clouds to form. On a sunny day, usually in the summer, the ground is heated unevenly. Some areas of the ground become warmer and heat the air above, making it warmer and less dense. A stream of air, called a thermal, may rise. As it rises, it cools adiabatically through expansion and may reach its dewpoint temperature. With continued cooling and rising, condensation will occur, forming a cloud. Since the cloud is formed by predominantly vertical motions, the cloud will be cumulus. With continued warming of the surface, the thermals may rise even higher, perhaps producing thunderstorm, or cumulonimbus, clouds. Thus, a sunny summer day can start off without a cloud in the sky, but can be stormy with many thunderstorms by afternoon.

Clouds also can form when air is forced to rise when it meets a mountain or other large vertical barrier. This type of lifting—orographic—is especially prevalent where air moves over the ocean and then is forced to rise up a mountain, as occurs on the west coast of North and South America. As the

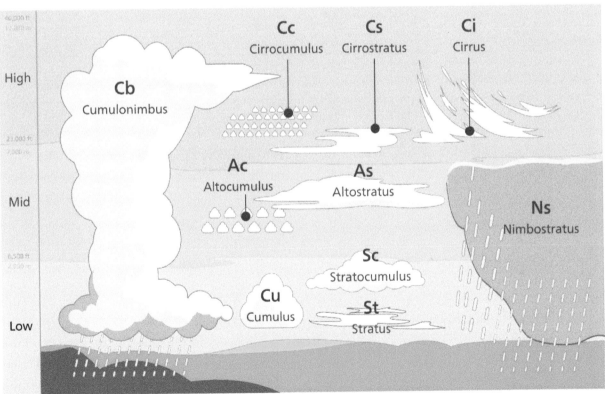

Cloud classification by altitude of occurrence. Multi-level and vertical genus-types not limited to a single altitude level include nimbostratus, cumulonimbus, and some of the larger cumulus species. (Illustration by Valentin de Bruyn)

air rises, it cools adiabatically and eventually becomes so cool that it cannot hold the water vapor. Condensation occurs and clouds form. The air continues to move up the mountain, producing clouds and precipitation on the side of the mountain from which the wind came, the windward side. However, the air eventually must fall down the other side of the mountain, the leeward side. That air is warmed and moisture evaporates, resulting in no clouds.

A third lifting mechanism is frontal, or cyclonic, action. This occurs when a large mass of cold air and a large mass of warm air—often hundreds of miles in area—meet. The warm air mass and the cold air mass will not mix freely, resulting in a border or front between the two air masses. The warm, less dense, air will always rise above the cold, denser, air mass. As the warm air rises, it cools, and when it reaches its dew point, clouds will form. If the warm air displaces the cold air, or a warm front occurs, the warm air will rise gradually, resulting in layered or stratiform clouds. The cloud types will change on an upward diagonal path, with the lowest being stratus, and nimbostratus if rain occurs, followed by altostratus, then cirrostratus, and cirrus.

On the other hand, if the cold air displaces the warm air, the warm air will be forced to rise much more quickly. The clouds formed will be puffy or cumuliform—cumulus at the lowest levels, altocumulus and cirrocumulus at the highest altitudes. Sometimes cumulonimbus clouds will also form.

Sometimes when a cold front meets a warm front, the whole warm air mass is forced off the ground. This forms a cyclone—an area of low pressure—as the warm air rises. As this air rises, it cools. If it reaches its dew point, condensation and clouds will result. In oceanic tropical areas, a cyclone can form within warm, moist air. This air also will cool and, if it reaches its dew point, will condense and form clouds. Sometimes, these tropical cyclones are the precursors of hurricanes. The clouds associated with cyclones are usually cumulus, including cumulonimbus, as they are formed by rapidly rising air.

Margaret F. Boorstein

STORMS

A storm is an atmospheric disturbance that produces wind, is accompanied by some form of precipitation, and sometimes involves thunder and lightning. Storms that meet certain criteria are given specific names, such as hurricanes, blizzards, and tornadoes.

Stormy weather is associated with low atmospheric pressure, while clear, calm, dry weather is associated with high atmospheric pressure. Because of the way atmospheric pressure and wind direction are related, low-pressure areas are characterized by winds moving cyclonically (in a counterclockwise direction in the Northern Hemisphere; clockwise in the Southern Hemisphere) around the center of the low pressure. Storms of all kinds are associated with cyclones, but two classes of cyclones—tropical and extratropical—produce most storms.

Tropical Cyclones
These storms develop during the summer and autumn in every tropical ocean except the South Atlantic and eastern South Pacific Oceans. Tropical cyclones that occur in the North Atlantic and eastern North Pacific Oceans are known as hurricanes; in the western North Pacific Ocean, as typhoons; and in the Indian and South Pacific Oceans, as cyclones.

All tropical cyclones develop in three stages. Arising from the formation of the initial atmospheric disturbance that is characterized by a cluster of thunderstorms, the first stage—tropical depres-

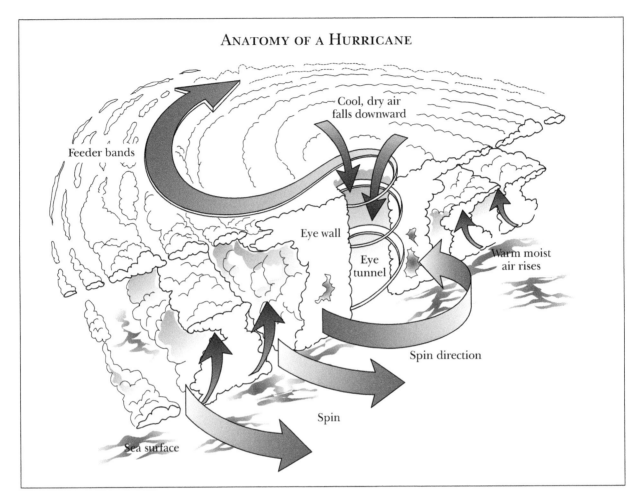

ANATOMY OF A HURRICANE

Cool, dry air
falls downward

Feeder bands

Eye wall

Eye
tunnel

Warm moist
air rises

Spin direction

Spin

Sea surface

sion—occurs when the maximum sustained surface wind speeds (the average speed over one minute) range from 23–39 miles (37–61 km.) per hour. The second stage—tropical storm—occurs when sustained winds range from 40–73 miles (62–119 km.) per hour. At this stage, the storm is given a name. From 80 to 100 tropical storms develop each year across the world, with about half continuing to the final stage—hurricane—at which sustained wind speeds are 74 miles (120 km.) per hour or greater. Moving over land or into colder oceans initiates the end of the hurricane after a week or so by eliminating the hurricane's fuel—warm water.

A mature hurricane is a symmetrical storm, with the "eye" at the center; the eye develops as winds increase and become circular around the central core of low pressure. Within the eye, it is relatively warm, and there are light winds, no precipitation, and few clouds. This is caused by air descending in the center of the storm. Surrounding the eye is the "eye

wall," a ring of intense thunderstorms that can extend high into the atmosphere. Within the eye wall, the strongest winds and heaviest rainfall are found; this is also where warm, moist air, the hurricane's "fuel," flows into the storm. Spiraling bands of clouds, called "rain bands," surround the eye wall. Precipitation and wind speeds decrease from the eye wall out toward the edge of the rain bands, while atmospheric pressure is lowest in the eye and increases outward.

Hurricanes can be the most damaging storms because of their intensity and size. Damage is caused by high winds and the flying debris they carry, flooding from the tremendous amounts of rain a hurricane can produce, and storm surge. A storm surge, which accounts for most of the coastal property loss and 90 percent of hurricane deaths, is a dome of water that is pushed forward as the storm moves. This wall of water is lifted up onto the coast as the eye wall comes in contact with land. For exam-

NAMING HURRICANES

Hurricanes once were identified by their latitudes and longitudes, but this method of naming became confusing when two or more hurricanes developed at the same time in the same ocean. During World War II hurricanes were identified by radio code letters, such as Able and Baker. In 1953 the National Weather Service began using English female names in an alphabetized list. Male names and French and Spanish names were added in 1978. By 2000 six lists of names were used on a rotating basis. When a hurricane causes much death or destruction, as Hurricane Andrew did in August of 1992—its name is retired for at least ten years.

ple, a 25-foot (8-meter) storm surge created by Hurricane Camille in 1969 destroyed the Richelieu Apartments next to the ocean in Pass Christian, Mississippi. Ignoring advice to evacuate, twenty-five people had gathered there for a hurricane party; all but one was killed.

To help predict the damage that an approaching hurricane can cause, the Saffir- Simpson Scale was developed. A hurricane is rated from 1 (weak) to 5 (devastating), according to its central pressure, sustained wind speed, and storm surge height. Michael was a category 5 storm at the time of landfall on October 10, 2018, near Mexico Beach and Tyndall Air Force Base, Florida. Michael was the first hurricane to make landfall in the United States as a category 5 since Hurricane Andrew in 1992, and only the fourth on record. The others are the Labor Day Hurricane in 1935 and Hurricane Camille in 1969.

Extratropical Cyclones

Also known as midlatitude cyclones, these storms are traveling low-pressure systems that are seen on newspaper and television daily weather maps. They are created when a mass of moist, warm air from the south contacts a mass of drier, cool air from the north, causing a front to develop. At the front, the warmer air rides up over the colder air. This causes water vapor to condense and produces clouds and rain during most of the year, and snow in the winter.

Thunderstorms

Thunderstorms also develop in stages. During the cumulus stage, strong updrafts of warm air build the storm clouds. The storm moves into the mature stage when updrafts continue to feed the storm, but cool downdrafts are also occurring in a portion of the cloud where precipitation is falling. When the warm updrafts disappear, the storm's fuel is gone and the dissipating stage begins. Eventually, the cloud rains itself out and evaporates.

Thunderstorms can also form away from a frontal system, usually during summer. This formation is related to a relatively small area of warm, moist air

STORM CLASSIFICATIONS

Tropical Classification	Wind speed
Gale-force winds	>15 meters/second
Tropical depression	20-34 knots and a closed circulation
Tropical storm (named)	35-64 knots
Hurricane	65+ knots (74+ mph)
Saffir-Simpson Scale for Hurricanes	
Category 1	63-83 knots (74-95 mph)
Category 2	83-95 knots (96-110 mph)
Category 3	96-113 knots (111-130 mph)
Category 4	114-135 knots (131-155 mph)
Category 5	>135 knots (>155 mph)

Notes: 1 knot = 1 nautical mile/hour = 1.152 miles/hour = 1.85 kilometers/hour.
Source: National Aeronautics and Space Administration, Office of Space Science, Planetary Data System.
http:/atmos.nmsu.edu/jsdap/encyclopediawork.html

rising and creating a thunderstorm that is usually localized and short lived.

Wind, lightning, hail, and flooding from heavy rain are the main destructive forces of a thunderstorm. Lightning occurs in all mature thunderstorms as the positive and negative electrical charges in a cloud attempt to equal out, creating a giant spark. Most lightning stays within the clouds, but some finds its way to the surface. The lightning heats the air around it to incredible temperatures (54,000 degrees Farenheit/30,000 degrees Celsius), which causes the air to expand explosively, creating the shock wave called thunder. Since lightning travels at the speed of light and thunder at the speed of sound, one can estimate how many miles away the lightning is by counting the seconds between the lightning and thunder and dividing by five. People have been killed by lightning while boating, swimming, biking, golfing, standing under a tree, talking on the telephone, and riding on a lawnmower.

Hail is formed in towering cumulonimbus clouds with strong updrafts. It begins as small ice pellets that grow by collecting water droplets that freeze on contact as the pellets fall through the cloud. The strong updrafts push the pellets back into the cloud, where they continue collecting water droplets until they are too heavy to stay aloft and fall as hailstones. The more an ice pellet is pushed back into the cloud, the larger the hailstone becomes. The largest authenticated hailstone in the United States fell near Vivian, South Dakota, on July 23, 2010. It mea-sured 8.0 inches (20 cm) in diameter, 18 1/2 inches (47 cm.) in circumference, and weighed in at 1.9375 pounds (879 grams).

Tornadoes

For reasons not well understood, less than 1 percent of all thunderstorms spawn tornadoes. Called funnel clouds until they touch earth, tornadoes contain the highest wind speeds known.

Although tornadoes can occur anywhere in the world, the United States has the most, with an average of 1000 per year. Tornadoes have occurred in every state, but the greatest number hit a portion of the Great Plains from central Texas to Nebraska, known as "Tornado Alley." There cold Canadian air and warm Gulf Coast air often collide over the flat land, creating the wall cloud from which most tornadoes are spawned. May is the peak month for tornado activity, but they have been spotted in every month.

Because tornado winds cannot be measured directly, the tornado is ranked according to its damage, using the Fujita Intensity Scale. The scale ranges from an F0, with wind speeds less than 72 miles (116 km.) per hour, causing light damage, to an F5, with winds greater than 260 miles (419 km.) per hour, causing incredible damage. Most tornadoes are small, but the larger ones cause much damage and death.

Kay R. S. Williams

Earth's Biological Systems

Biomes

The major recognizable life zones of the continents, biomes are characterized by their plant communities. Temperature, precipitation, soil, and length of day affect the survival and distribution of biome species. Species diversity within a biome may increase its stability and capability to deliver natural services, including enhancing the quality of the atmosphere, forming and protecting the soil, controlling pests, and providing clean water, fuel, food, and drugs. Land biomes are the temperate, tropical, and boreal forests; tundra; desert; grasslands; and chaparral.

Temperate Forest

The temperate forest biome occupies the so-called temperate zones in the midlatitudes (from about 30 to 60 degrees north and south of the equator). Temperate forests are found mainly in Europe, eastern North America, and eastern China, and in narrow zones on the coasts of Australia, New Zealand, Tasmania, and the Pacific coasts of North and South America. Their climates are characterized by high rainfall and temperatures that vary from cold to mild.

Temperate forests contain primarily deciduous trees—including maple, oak, hickory, and beechwood—and, secondarily, evergreen trees—including pine, spruce, fir, and hemlock. Evergreen forests in some parts of the Southern Hemisphere contain eucalyptus trees.

The root systems of forest trees help keep the soil rich. The soil quality and color is due to the action of earthworms. Where these forests are frequently cut, soil runoff pollutes streams, which reduces fisheries because of the loss of spawning habitat.

Racoons, opposums, bats, and squirrels are found in the trees. Deer and black bear roam forest floors. During winter, small animals such as groundhogs and squirrels burrow in the ground.

Tropical Forest

Tropical forests are in frost-free areas between the Tropic of Cancer and the Tropic of Capricorn. Temperatures range from warm to hot year-round, because the Sun's rays shine nearly straight down around midday. These forests are found in northern Australia, the East Indies, southeastern Asia, equatorial Africa, and parts of Central America and northern South America.

Tropical forests have high biological diversity and contain about 15 percent of the world's plant species. Animal life lives at different layers of tropical forests. Nuts and fruits on the trees provide food for birds, monkeys, squirrels, and bats. Monkeys and sloths feed on tree leaves. Roots, seeds, leaves, and fruit on the forest floor feed deer, hogs, tapirs, antelopes, and rodents. The tropical forests produce rubber trees, mahogany, and rosewood. Large animals in these forests include the Asian tiger, the African bongo, the South American tapir, the Central and South American jaguar, the Asian and African leopard, and the Asian axis deer. Deforestation for agriculture and pastures has caused reduction in plant and animal diversity.

Boreal Forest

The boreal forest is a circumpolar Northern Hemisphere biome spread across Russia, Scandinavia, Canada, and Alaska. The region is very cold. Evergreen trees such as white spruce and black spruce

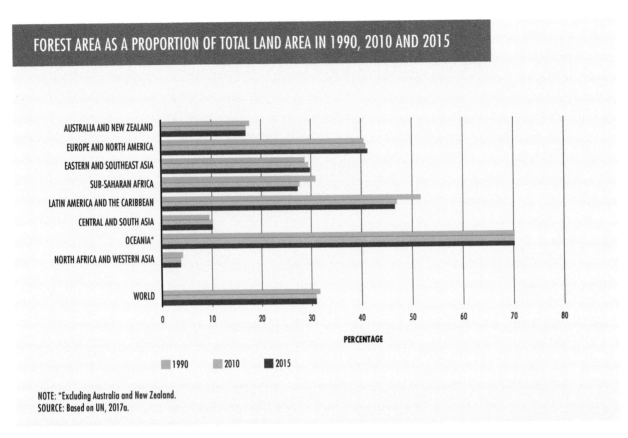

FOREST AREA AS A PROPORTION OF TOTAL LAND AREA IN 1990, 2010 AND 2015

PERCENTAGE

1990   2010   2015

NOTE: *Excluding Australia and New Zealand.
SOURCE: Based on UN, 2017a.

dominate this zone, which also contains larch, balsam, pine, and fir, and some deciduous hardwoods such as birch and aspen. The acidic needles from the evergreens make the leaf litter that is changed into soil humus. The acidic soil limits the plants that develop.

Animals in boreal forests include deer, caribou, bear, and wolves. Birds in this zone include goshawks, red-tailed hawks, sapsuckers, grouse, and nuthatches. Relatively few animals emigrate from this habitat during winter. Conifer seeds are the basic winter food. The disappearing aspen habitat of the beaver has decreased their numbers and has reduced the size of wetlands.

Tundra

About 5 percent of the earth's surface is covered with Arctic tundra, and 3 percent with alpine tundra. The Arctic tundra is the area of Europe, Asia, and North America north of the boreal coniferous forest zone, where the soils remain frozen most of the year. Arctic tundra has a permanent frozen subsoil, called permafrost. Deep snow and low temper-

atures slow the soil-forming process. The area is bounded by a 50 degrees Fahrenheit circumpolar isotherm, known as the summer isotherm. The cold temperature north of this line prevents normal tree growth.

The tundra landscape is covered by mosses, lichens, and low shrubs, which are eaten by caribou, reindeer, and musk oxen. Wolves eat these herbivores. Bear, fox, and lemming also live here. The larger mammals, including marine mammals and the overwintering birds, have large fat layers beneath the skin and long dense fur or dense feathers that provide protection. The small mammals burrow beneath the ground to avoid the harsh winter climate. The most common Arctic bird is the old squaw duck. Ptarmigans and eider ducks are also very common. Geese, falcons, and loons are some of the nesting birds of the area.

The alpine tundra, which exists at high altitude in all latitudes, is acted upon by winds, cold temperatures, and snow. The plant growth is mostly cushion and mat-forming plants.

BIOMES OF THE WORLD

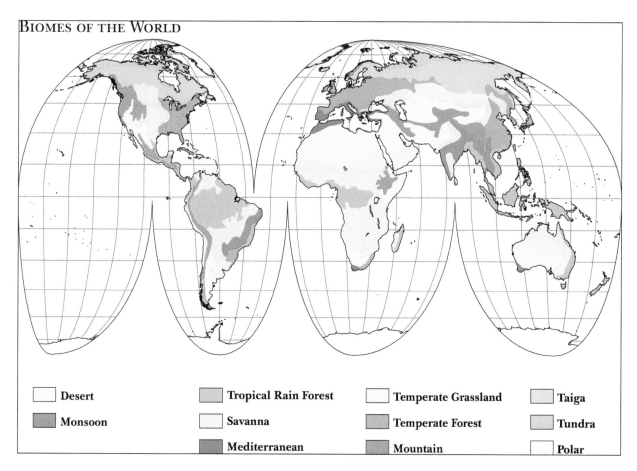

☐ **Desert**	☐ **Tropical Rain Forest**	☐ **Temperate Grassland**	☐ **Taiga**
☐ **Monsoon**	☐ **Savanna**	☐ **Temperate Forest**	☐ **Tundra**
	☐ **Mediterranean**	☐ **Mountain**	☐ **Polar**

Desert

The desert biome covers about one-seventh of the earth's surface. Deserts typically receive no more than 10 inches (25 centimeters) of rainfall a year, but evaporation generally exceeds rainfall. Deserts are found around the Tropic of Cancer and the Tropic of Capricorn. As the warm air rises over the equator, it cools and loses its water content. This dry air descends in the two subtropical zones on each side of the equator; as it warms, it picks up moisture, resulting in drying the land.

Rainfall is a key agent in shaping the desert. The lack of sufficient plant cover removes the natural protection that prevents soil erosion during storms. High winds also cut away the ground.

Some desert plants obtain water from deep below the surface, for example, the mesquite tree, which has roots that are 40 feet (13 meters) deep. Other plants, such as the barrel cactus, store large amounts of water in their leaves, roots, or stems. Other plants slow the loss of water by having tiny leaves or shedding their leaves. Desert plants have

very short growth periods, because they cannot grow during the long drought periods.

Desert animals protect themselves from the Sun's heat by eating at night, staying in the shade during the day, and digging burrows in the ground. Among the world's large desert animals are the camel, coyote, mule deer, Australian dingo, and Asian saiga. The digestive process of some desert animals produces water. A method used by some animals to conserve water is the reabsorption of water from their feces and urine.

Grassland

Grasslands cover about a quarter of the earth's surface, and can be found between forests and deserts. Treeless grasslands grow in parts of central North America, Central America, and eastern South America that have between 10 and 40 inches (250-1,000 millimeters) of erratic rainfall. The climate has a high rate of evaporation and periodic major droughts. The biome is also subject to fire.

Some grassland plants survive droughts by growing deep roots, while others survive by being dormant. Grass seeds feed the lizards and rodents that become the food for hawks and eagles. Large animals include bison, coyotes, mule deer, and wolves. The grasslands produce more food than any other biome. Poor grazing and agricultural practices and mining destroy the natural stability and fertility of these lands. The reduced carrying capacity of these lands causes an increase in water pollution and erosion of the soil. Diverse natural grasslands appear to be more capable of surviving drought than are simplified manipulated grass systems. This may be due to slower soil mineralization and nitrogen turnover of plant residues in the simplified system.

Savannas are open grasslands containing deciduous trees and shrubs. They are near the equator and are associated with deserts. Grasses grow in clumps and do not form a continuous layer. The northern savanna bushlands are inhabited by oryx and gazelles. The southern savanna supports springbuck and eland. Elephants, antelope, giraffe, zebras, and black rhinoceros are found on the savannas. Lions, leopards, cheetah, and hunting dogs are the primary predators here. Kangaroos are found in the savannas of Australia. Savannas cover South America north and south of the Amazon rainforest, where jaguar and deer can be found.

Chaparral

The chaparral or Mediterranean biome is found in the Mediterranean Basin, California, southern Australia, middle Chile, and Cape Province of South America. This region has a climate of wet winters and summer drought. The plants have tough leathery leaves and may contain thorns. Regional fires clear the area of dense and dead vegetation. Fire, heat, and drought shape the region. The vegetation dwarfing is due to the severe drought and extreme climate changes. The seeds from some plants, such as the California manzanita and South African fire lily, are protected by the soil during a fire and later germinate and rapidly grow to form new plants.

Ocean

The ocean biome covers more than 70 percent of the earth's surface and includes 90 percent of its volume. The ocean has four zones. The intertidal zone is shallow and lies at the land's edge. The continental shelf, which begins where the intertidal zone ends, is a plain that slopes gently seaward. The neritic zone (continental slope) begins at a depth of about 600 feet (180 meters), where the gradual slant of the continental shelf becomes a sharp tilt toward the ocean floor, plunging about 12,000 feet (3,660 meters) to the ocean bottom, which is known as the abyss. The abyssal zone is so deep that it does not have light.

Plankton are animals that float in the ocean. They include algae and copepods, which are microscopic crustaceans. Jellyfish and animal larva are also considered plankton. The nekton are animals that move freely through the water by means of their muscles. These include fish, whales, and squid. The benthos are animals that are attached to or crawl along the ocean's floor. Clams are examples of benthos. Bacteria decompose the dead organic materials on the ocean floor.

The circulation of materials from the ocean's floor to the surface is caused by winds and water temperature. Runoff from the land contains polluting chemicals such as pesticides, nitrogen fertilizers, and animal wastes. Rivers carry loose soil to the ocean, where it builds up the bottom areas. Overfishing has caused fisheries to collapse in every world sector. In some parts of the northwestern Altantic Ocean, there has been a shift from bony fish to cartilaginous fish dominating the fisheries.

Human Impact on Biomes

Human interaction with biomes has increased biotic invasions, reduced the numbers of species, changed the quality of land and water resources, and caused the proliferation of toxic compounds. Managed care of biomes may not be capable of undoing these problems.

Ronald J. Raven

Natural Resources

Soils

Soils are the loose masses of broken and chemically weathered rock mixed with organic matter that cover much of the world's land surface, except in polar regions and most deserts. The two major solid components of soil—minerals and organic matter—occupy about half the volume of a soil. Pore spaces filled with air and water account for the other half. A soil's organic material comes from the remains of dead plants and animals, its minerals from weathered fragments of bedrock. Soil is also an active, dynamic, ever-changing environment. Tiny pores in soil fill with air, water, bacteria, algae, and fungi working to alter the soil's chemistry and speed up the decay of organic material, making the soil a better living environment for larger plants and animals.

Soil Formation

The natural process of forming new soil is slow. Exactly how long it takes depends on how fast the bedrock below is weathered. This weathering process is a direct result of a region's climate and topography, because these factors influence the rate at which exposed bedrock erodes and vegetation is distributed. Global variations in these factors account for the worldwide differences in soil types.

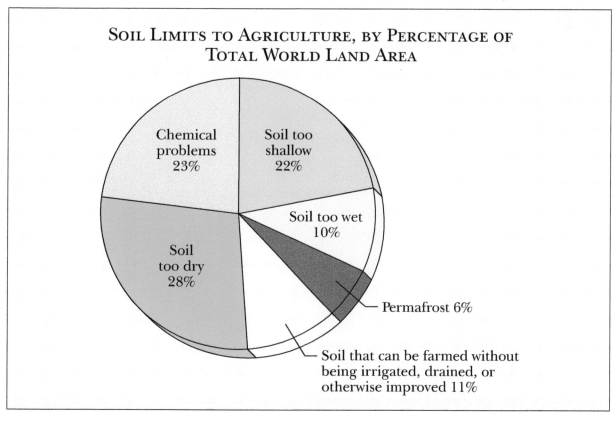

SOIL LIMITS TO AGRICULTURE, BY PERCENTAGE OF TOTAL WORLD LAND AREA

Chemical problems 23%

Soil too shallow 22%

Soil too wet 10%

Soil too dry 28%

Permafrost 6%

Soil that can be farmed without being irrigated, drained, or otherwise improved 11%

Climate is the principal factor in determining the type and rate of soil formation. Temperature and precipitation are the two main climatic factors that influence soil formation, and they vary with elevation and latitude. Water is the main agent of weathering, and the amount of water available depends on how much falls and how much runs off. The amount of precipitation and its distribution during the year influence the kind of soil formed and the rate at which it is formed. Increased precipitation usually results in increased rates of soil formation and deep soils. Temperature and precipitation also determine the kind and amount of vegetation in a region, which determines the amount of available organics.

Topography is a characteristic of the landscape involving slope angle and slope length. Topographic relief governs the amount of water that runs off or enters a soil. On flat or gently sloping land, soil tends to stay in place and may become thick, but as the slope increases so does the potential for erosion. On steep slopes, soil cover may be very thin, possibly only a few inches, because precipitation washes it downhill; on level plains, soil profiles may be several feet thick.

Types of Soil

Typically, bedrock first weathers to form regolith, a protosoil devoid of organic material. Rain, wind, snow, roots growing into cracks, freezing and thawing, uneven heating, abrasion, and shrinking and swelling break large rock particles into smaller ones. Weathered rock particles may range in size from clay to silt, sand, and gravel, with the texture and particle size depending largely on the type of bedrock. For example, shale yields finer-textured soils than sandstone. Soils formed from eroded limestone are rich in base minerals; others tend to be acidic. Generally, rates of soil formation are largely determined by the rates at which silicate minerals in the bedrock weather: the more silicates, the longer the formation time.

In regions where organic materials, such as plant and animal remains, may be deposited on top of regolith, rudimentary soils can begin to form. When waste material is excreted, or a plant or animal dies, the material usually ends up on the earth's surface. Organisms that cause decomposition, such as bacteria and fungi, begin breaking down the remains into a beneficial substance known as humus. Humus restores minerals and nutrients to the soil. It also improves the soil's structure, helping it to retain water. Over time, a skeletal soil of coarse, sandy material with trace amounts of organics gradually forms. Even in a region with good weathering rates and adequate organic material, it can take as long as fifty years to form 12 inches (30 centimeters) of soil. When new soil is formed from weathering bedrock, it can take from 100 to 1,000 years for less than an inch of soil to accumulate.

Water moves continually through most soils, transporting minerals and organics downward by a process called leaching. As these materials travel downward, they are filtered and deposited to form distinct soil horizons. Each soil horizon has its own color, texture, and mineral and humus content. The O-horizon is a thin layer of rotting organics covering the soil. The A-horizon, commonly called topsoil, is rich in humus and minerals. The B-horizon is a subsoil rich in minerals but poor in humus. The C-horizon consists of weathered bedrock; the D-horizon is the bedrock itself.

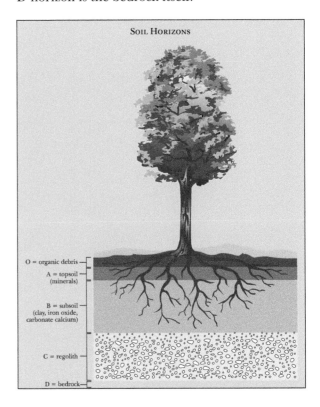

SOIL HORIZONS

O = organic debris
A = topsoil (minerals)
B = subsoil (clay, iron oxide, carbonate calcium)
C = regolith
D = bedrock

Because Earth's surface is made of many different rock types exposed at differing amounts and weathering at different rates at different locations, and because the availability of organic matter varies greatly around the planet due to climatic and seasonal conditions, soil is very diverse and fertile soil is unevenly distributed. Structure and composition are key factors in determining soil fertility. In a fertile soil, plant roots are able to penetrate easily to obtain water and dissolved nutrients. A loam is a naturally fertile soil, consisting of masses of particles from clays (less than 0.002 mm across), through silts (ten times larger) to sands (100 times larger), interspersed with pores, cracks, and crevices.

The Roles of Soil

In any ecosystem, soils play six key roles. First, soil serves as a medium for plant growth by mechanically supporting plant roots and supplying the eighteen nutrients essential for plants to survive. Different types of soil contain differing amounts of these eighteen nutrients; their combination often determines the types of vegetation present in a region, and as a result, influences the number and types of animals the vegetation can support, including humans. Humans rely on soil for crops necessary for food and fiber.

Second, the property of a particular soil is the controlling factor in how the hydrologic system in a region retains and transports water, how contaminants are stored or flushed, and at what rate water is naturally purified. Water enters the soil in the form of precipitation, irrigation, or snowmelt that falls or runs off soil. When it reaches the soil, it will either be surface water, which evaporates or runs into streams, or subsurface water, which soaks into the soil where it is either taken up by plant roots or percolates downward to enter the groundwater system. Passing through soil, organic and inorganic pollutants are filtered out, producing pure groundwater.

Soil also functions as an air-storage facility. Air is pushed into and drawn out of the soil by changes in barometric pressure, high winds, percolating water, and diffusion. Pore spaces within soil provide access to oxygen to organisms living underground as well as to plant roots. Soil pore spaces also contain carbon dioxide, which many bacteria use as a source of carbon.

Soil is nature's recycling system, through which organic waste products and decaying plants and animals are assimilated and their elements made available for reuse. The production and assimilation of humus within soil converts mineral nutrients into forms that can be used by plants and animals, who return carbon to the atmosphere as carbon dioxide. While dead organic matter amounts to only about 1 percent of the soil by weight, it is a vital component as a source of minerals.

Soil provides a habitat for many living things, from insects to burrowing animals, from single microscopic organisms to massive colonies of subterranean fungi. Soils contain much of the earth's genetic diversity, and a handful of soil may contain billions of organisms, belonging to thousands of species. Although living organisms only account for about 0.1 percent of soil by weight, 2.5 acres (one hectare) of good-quality soil can contain at least 300 million small invertebrates—mites, millipedes, insects, and worms. Just 1 ounce (30 grams) of fertile soil can contain 1 million bacteria of a single type, 100 million yeast cells, and 50,000 fungus mycelium. Without these, soil could not convert nitrogen, phosphorus, and sulphur to forms available to plants.

Finally, soil is an important factor in human culture and civilization. Soil is a building material used to make bricks, adobe, plaster, and pottery, and often provides the foundation for roads and buildings. Most important, soil resources are the basis for agriculture, providing people with their dietary needs.

Because the human use of soils has been haphazard and unchecked for millennia, soil resources in many parts of the world have been harmed severely. Human activities, such as overcultivation, inexpert irrigation, overgrazing of livestock, elimination of tree cover, and cultivating steep slopes, have caused natural erosion rates to increase many times over. As a result of mismanaged farm and forest lands, escalated erosional processes wash off or blow away an estimated 75 billion tons of soil annually, eroding away one of civilization's crucial resources.

Randall L. Milstein

WATER

Life on Earth requires water—without it, life on Earth would cease. As human populations grow, the freshwater resources of the world become scarcer and more polluted, while the need for clean water increases. Although nearly three-quarters of Earth's surface is covered with water, only about 0.3 percent of that water is freshwater suitable for consumption and irrigation. This is because more than 97 percent of Earth's water is ocean salt water, and most of the remaining freshwater is frozen in the Antarctic ice cap. Only the small amounts that remain in lakes, rivers, and groundwater is available for human use.

All of earth's water cycles between the ocean, land, atmosphere, plants, and animals over and over. On average, a molecule of surface water cycles from the ocean, to the atmosphere, to the land and back again in less than two weeks. Water consumed by plants or animals takes longer to return to the oceans, but eventually the cycle is completed.

Water's Uses

Water supports the lives of all living creatures. People use for drinking, cooking, cleaning, and bathing. Water also plays a key role in society since humans can travel on it, make electricity with it, fish in it, irrigate crops with it, and use it for recreation. Globally, more than 4 trillion cubic meters of freshwater is used each day. Agriculture accounts for about 70 percent, industry uses 20 percent, and domestic and municipal activities use 10 percent. To produce beef requires between 1,320 and 5,283 gallons (5,000 and 20,000 liters) of water for every 35 ounces (1 kg). A similar amount of wheat requires between 660 and 1056 gallons (2,500 and 4,000 liters) of water. Manufactured goods also require significant amounts of freshwater; a car consumes between 13,000 and 20,000 gallons (49,210 to 75,708 liters), a smartphone has a water footprint of nearly 3,100 gallons (11,734 liters) and a teeshirt uses around 660 gallons (2,498 liters).

The average American family uses more than 300 gallons of drinking quality water per day at home. Roughly 70 percent of this use occurs indoors for drinking, bathing and showering, flushing the toilet, and washing dishes. Outdoor water use for landscape watering, washing cars, and cleaning windows, etc., accounts for 30 percent of household use, although it can be much higher in drier parts of the country. For example, the arid West has some of the highest per capita residential water use because of landscape irrigation.

As the world's population grows, the demand for fresh water will also increase. A study by the World Bank concluded that approximately 80 percent of human illness results from insufficient water supplies and poor water quality caused by lack of sanitation, so careful management of water resources is essential for improving the health of people in the twenty-first century.

Groundwater Supply and Quality

The amount of groundwater in the Earth is seventy times greater than all of the freshwater lakes combined. Groundwater is held within the rocks below the ground surface and is the primary source of water in many parts of the world. In the United States, approximately 50 percent of the population uses some groundwater. However, problems with both groundwater supplies and its quality threaten its future use.

The U.S. Environmental Protection Agency (EPA) found that 45 percent of the large public water systems in the United States that use groundwater were contaminated with synthetic organic chemicals that posed potential health threats. Another major problem occurs when groundwater is used faster than it is replaced by precipitation infiltrating through the ground surface. Many of the arid regions of earth are already suffering from this problem. For example, one-third of the wells in Beijing, China, have gone dry due to overuse. In the United States, the Ogallala Aquifer of the Great Plains, the

The World and North America's Greatest Rivers and Lakes

Longest river	
Nile (North Africa)	4,130 miles (6,600 km.)
Missouri-Mississippi (United States)	3,740 miles ((6,000 km.)
Largest river by average discharge	
Amazon (South America)	6,181,000 cubic feet/second (175,000 cubic meters/second)
Missouri-Mississippi (United States)	600,440 cubic feet/second (17,000 cubic meters/second)
Largest freshwater lake by volume	
Lake Baikal (Russia)	5,280 cubic miles (22,000 cubic km.)
Lake Superior (United States)	3,000 cubic miles (12,500 cubic km.)

largest in North America, is being severely overused. This aquifer irrigates 30 percent of U.S. farmland, but some areas of the aquifer have declined by up to 60 percent. In one part of Texas, so much water has been pumped out that the aquifer has essentially dried up there. Once depleted, the aquifer will take over 6,000 years to replenish naturally through rainfall.

Surface Water Supply and Quality

Surface water is used for transportation, recreation, electrical generation, and consumption. Ships use rivers and lakes as transport routes, people fish and boat on rivers and lakes, and dams on rivers often are used to generate electricity. The largest river on earth is the Amazon in South America, which has an average flow of 212,500 cubic meters per second, more than twelve times greater than North America's Mississippi River. Earth's largest lake by volume—Lake Baikal in Russia—holds 5,521 cubic miles of water (23,013 cubic kilometers), or approximately 20 percent of Earth's fresh surface water. This is a volume of water approximately equivalent to all five of the North American Great Lakes combined.

Although surface water has more uses, it is more prone to pollution than groundwater. Almost every human activity affects surface water quality. For ex-

ample, water is used to create paper for books, and some of the chemicals used in the paper process are discharged into surface water sources. Most foods are grown with agricultural chemicals, which can contaminate water sources. According to 2018 surveys on national water quality from the U.S. Environmental Protection Agency, nearly half of U.S. rivers and streams and more than one-third of lakes are polluted and unfit for swimming, fishing, and drinking.

Earth's Future Water Supply

Inadequate water supplies and water quality problems threaten the lives of more than 1.5 million people worldwide. The World Health Organization estimates that polluted water causes the death of 361,000 children under five years of age each year and affects the health of 20 percent of Earth's population. As the world's population grows, these problems are likely to worsen.

The United Nations estimates that if current consumption patterns continue, 52 percent of the world's people will live in water-stressed conditions by 2050. Since access to clean freshwater is essential to health and a decent standard of living, efforts must be made to clean up and conserve the planet's freshwater.

Mark M. Van Steeter

EXPLORATION AND TRANSPORTATION

EXPLORATION AND HISTORICAL TRADE ROUTES

The world's exploration was shaped and influenced substantially by economic needs. Lacking certain resources and outlets for trade, many societies built ships, organized caravans, and conducted military expeditions to protect their frontiers and obtain new markets.

Over the last 5,000 years, the world evolved from a cluster of isolated communities into a firmly integrated global community and capitalist world system. By the beginning of the twentieth century, explorers had successfully navigated the oceans, seas, and landmasses and gathered many regional economies into the beginnings of a global economy.

Early Trade Systems

Trade and exploration accompanied the rise of civilization in the Middle East. Egyptian pharaohs, looking for timber for shipbuilding, established trade relations with Mediterranean merchants. Phoenicians probed for new markets off the coast of North Africa and built a permanent settlement at Carthage. By 513 BCE, the Persian Empire stretched from the Indus River in India to the Libyan coast, and it controlled the pivotal trade routes in Iran and Anatolia. A regional economy was taking shape, linking Africa, Asia, and Europe into a blended economic system.

Alexander the Great's victory against the Persian Empire in 330 BCE thrust Greece into a dominant position in the Middle Eastern economy. Trade between the Mediterranean and the Middle East increased, new roads and harbors were constructed, and merchants expanded into sub-Saharan Africa, Arabia, and India. The Romans later benefited from the Greek foundation. Through military and political conquest, Rome consolidated its control over such diverse areas as Arabia and Britain and built a system of roads and highways that facilitated the growth of an expanding world economy. At the apex of Roman power in 200 CE, trade routes provided the empire with Greek marble, Egyptian cloth, seafood from Black Sea fisheries, African slaves, and Chinese silk.

The emergence of a profitable Eurasian trade route linked people, customs, and economies from the South China Sea to the Roman Empire. Although some limited activity occurred during the Hellenistic period, East-West trade flourished following the rise of the Han Dynasty in China. With the opening of the Great Silk Road from 139 BCE to 200 CE, goods and services were exchanged between people from three different continents.

The Great Silk Road was an intricate network of middlemen stretching from China to the Mediterranean Sea. Eastern merchants sold their products at markets in Afghanistan, Iran, and even Syria, and exchanged a variety of commodities through the use of camel caravans. Chinese spices, perfumes, metals, and especially silk were in high demand. The Parthians from central Asia added their own sprinkling of merchandise, introducing both the East and the West to various exotic fruits, rare birds, and ostrich eggs.

Romans peddled glassware, statuettes, and acrobatic performing slaves. Since communication lines were virtually nonexistent during this period, trade routes were the only means by which ideas regarding art, religion, and culture could mix. The contacts and exchanges enacted along the Great Silk Road initiated a process of cultural diffusion among a diversity of cultures and increased each culture's knowledge of the vast frontiers of world geography.

The Atlantic Slave Trade

Beginning in the fifteenth century, European navigators explored the West African coastline seeking gold. Supplies were difficult to procure, because most of the gold mines were located in the interior along the Senegal River and in the Ashanti forests. Because mining required costly investments in time, labor, and security, the Europeans quickly shifted their focus toward the slave trade. Although slavery had existed since antiquity, the Atlantic slave trade generated one of the most significant movements of people in world history. It led to the forced migration of more than 10 million Africans to South America, the Caribbean islands, and North America. It ensured the success of several imperial conquests, and it transformed the demographic, cultural, and political landscape on four continents.

Originally driven by their quest to circumnavigate Africa and open a lucrative trade route with India, the Portuguese initiated a systematic exploration of the West African coastline. The architect of this system, Henry the Navigator, pioneered the use of military force and naval superiority to annex African islands and open up new trade routes, and he increased Portugal's southern frontier with every acquisition. In 1415 his ships captured Ceuta, a prosperous trade center located on the Mediterranean coast overlooking North African trade routes. Over the next four decades, Henry laid claim to the Madeira Islands, the Canary Islands, the Azores, and Cape Verde. After his death, other Portuguese explorers continued his pursuit of circumnavigation of Africa.

Diego Cão reached the Congo River in 1483 and sent several excursions up the river before returning to Lisbon. Two explorers completed the Portuguese mission at the end of the fifteenth century. Vasco da Gama, sailing from 1497 to 1499, and Bartholomeu Dias, from 1498 to 1499, sailed past the southern tip of Africa and eventually reached India. Since Muslims had already created a number of trade links between East Africa, Arabia, and India, Portuguese exploration furthered the integration of various regions into an emerging capitalist world system.

When the Portuguese shifted their trading from gold to slaves, the other European powers followed suit. The Netherlands, Spain, France, and England used their expanding naval technology to explore the Atlantic Ocean and ship millions of slaves across the ocean. A highly efficient and organized trade route quickly materialized. Since the Europeans were unwilling to venture beyond the walls of their coastal fortresses, merchants relied on African sources for slaves, supplying local kings and chiefs with the means to conduct profitable slave-raiding parties in the interior. In both the Congo and the Gold Coast region, many Africans became quite wealthy trading slaves.

In 1750 merchants paid King Tegbessou of Dahomey 250,000 pounds for 9,000 slaves, and his income exceeded the earnings of many in England's merchant and landowning class. After purchasing slaves, dealers sold them in the Americas to work in the mines or on plantations. Commodities such as coffee and sugar were exported back to Europe for home consumption. Merchants then sold alcohol, tobacco, textiles, and firearms to Africans in exchange for more slaves. This practice was abolished by the end of the nineteenth century, but not before more than 10 million Africans had been violently removed from their homeland. The Atlantic slave trade, however, joined port cities from the Gold Coast and Guinea in Africa with Rio de Janeiro, Hispaniola, Havana, Virginia, Charleston, and Liverpool, and constituted a pivotal step toward the rise of a unified global economy.

Magellan and Zheng He

The Portuguese explorer Ferdinand Magellan generated considerable interest in the Asian markets when he led an expedition that sailed around the world from 1519 to 1522. Looking for a quick route to Asia and the Spice Islands, he secured financial backing from the king of Spain. Magellan sailed from Spain in 1519, canvassed the eastern coastline of South America, and visited Argentina. He ultimately traversed the narrow straits along the southern tip of the continent and ventured into the uncharted waters of the Pacific Ocean.

Magellan explored the islands of Guam and the Philippines but was killed in a skirmish on Mactan in 1521. Some of his crew managed to return to Spain in 1522, and one member subsequently published a journal of the expedition that dramatically enhanced the world's understanding of the major sea lanes that connected the continents.

China also opened up new avenues of trade and exploration in Southeast Asia during the fifteenth century. Under the direction of Chinese emperor Yongle, explorer Zheng He organized seven overseas trips from 1405 to 1433 and investigated economic opportunities in Korea, Vietnam, the Indian Ocean, and Egypt. His first voyage consisted of more than 28,000 men and 400 ships and represented the largest naval force assembled prior to World War I.

Zheng's armada carried porcelains, silks, lacquerware, and artifacts to Malacca, the vital port city in Indonesia. He purchased an Arab medical text on drug therapy and had it translated into Chinese. He introduced giraffes and mahogany wood into the mainland's economy, and his efforts helped spread Chinese ideas, customs, diet, calendars, scales and measures, and music throughout the global economy. Zheng He's discoveries, coupled with all the material gathered by the European explorers, provided cartographers and geographers with a credible store of knowledge concerning world geography.

Emerging Global Trade Networks

From 1400 to 1900, several regional economic systems facilitated the exchange of goods and services throughout a growing world system. Building on the triangular relationships produced by the slave trade, the Atlantic region helped spread new food-

stuffs around the globe. Plants and plantation crops provided societies with a plentiful supply of sweet potatoes, squash, beans, and maize. This system, often referred to as the Columbian exchange, also assisted development in other regions by supplying the global economy with an ample money supply in gold and silver. Europeans sent textiles and other manufactures to the Americas. In return, they received minerals from Mexico; sugar and molasses from the Caribbean; money, rum, and tobacco from North America; and foodstuffs from South America. Trade routes also closed the distance between the Pacific coastline in the Americas and the Pacific Rim.

Additional thriving trade routes existed in the African-West Asian region. Linking Europe and Africa with Arabia and India, this area experienced a considerable amount of trade over land and through the sea lanes in the Persian Gulf and Red Sea. Europeans received grains, timber, furs, iron, and hemp from Russia in exchange for wool textiles and silver. Central Asians secured stores of cotton textiles, silk, wheat, rice, and tobacco from India and sold silver, horses, camel, and sheep to the Indians. Ivory, blankets, paper, saltpeter, fruits, dates, incense, coffee, and wine were regularly exchanged among merchants situated along the trade route connecting India, Persia, the Ottoman Empire, and Europe.

Finally, a Russian-Asian-Chinese market provided Russia's ruling czars with arms, sugar, tobacco, and grain, and a sufficient supply of drugs, medicines, livestock, paper money, and silver moved eastward. Overall, this system linked the economies of three continents and guaranteed that a nation could acquire essential foodstuffs, resources, and money from a variety of sources.

Several profitable trade routes existed in the Indian Ocean sector. After Malacca emerged as a key trading port in the sixteenth century, this territory served as an international clearinghouse for the global economy. Indians sent tin, elephants, and wood into Burma and Siam. Rice, silk, and sugar were sold to Bengal. Pepper and other spices were shipped westward across the Arabian Sea, while Ceylon furnished India with vast quantities of jewels, cinnamon, pearls, and elephants. The booming interregional trade routes positioned along the Indian coastline ensured that many of the vast commodities produced in the world system could be obtained in India.

The final region of crucial trade routes was between Southeast Asia and China. While the extent of Asian overseas trade prior to the twentieth century is usually downplayed, an abundance of products flowed across the Bay of Bengal and the South China Sea. Japan procured silver, copper, iron, swords, and sulphur from Cantonese merchants, and Japanese-finished textiles, dyes, tea, lead, and manufactures were in high demand on the mainland. The Chinese also purchased silk and ceramics from the Philippines in exchange for silver. Burma and Siam traded pepper, sappan wood, tin, lead, and saltpeter to China for satin, velvet, thread, and labor. As goods increasingly moved from the Malabar coast in India to the northern boundaries of Korea and Japan, the Pacific Rim played a prominent role in the global economy.

Robert D. Ubriaco, Jr.

ROAD TRANSPORTATION

Roads—the most common surfaces on which people and vehicles move—are a key part of human and economic geography. Transportation activities form part of a nation's economic product: They strengthen regional economy, influence land and natural resource use, facilitate communication and commerce, expand choices, support industry, aid agriculture, and increase human mobility. The need for roads closely correlates with the relative location of centers of population, commerce, industry, and other transportation.

History of Road Making

The great highway systems of modern civilization have their origin in the remote past. The earliest travel was by foot on paths and trails. Later, pack animals and crude sleds were used. The development of the wheel opened new options. As various ancient civilizations reached a higher level, many of them realized the importance of improved roads.

The most advanced highway system of the ancient world was that of the Romans. When Roman civilization was at its peak, a great system of military roads reached to the limits of the empire. The typical Roman road was bold in conception and construction, built in a straight line when possible, with a deep multilayer foundation, perfect for wheeled vehicles.

After the decline of the Roman Empire, rural road building in Europe practically ceased, and roads fell into centuries of disrepair. Commerce traveled by water or on pack trains that could negotiate the badly maintained roads. Eventually, a commercial revival set in, and roads and wheeled vehicles increased.

Interest in the art of road building was revived in Europe in the late eighteenth century. P. Trésaguet, a noted French engineer, developed a new method of lightweight road building. The regime of French dictator Napoleon Bonaparte (1800–1814) encouraged road construction, chiefly for military purposes. At about the same time, two Scottish engineers, Thomas Telford and John McAdam, also developed road-building techniques.

Roads in the United States

Toward the end of the eighteenth century, public demand in the United States led to the improvement of some roads by private enterprise. These improvements generally took the form of toll roads, called "turnpikes" because a pike was rotated in each road to allow entry after the fee was paid, and generally were located in areas adjacent to larger cities. In the early nineteenth century, the federal government paid for an 800-mile-long macadam road from Cumberland, Maryland, to Vandalia, Illinois.

With the development of railroads, interest in road building began to wane. By 1900, however, demand for better roads came from farmers, who wanted to move their agricultural products to market more easily. The bicycle craze of the 1890s and the advent of motorized vehicles also added to the demand for more and better roads. Asphalt and concrete technology was well developed by then; now, the problem was financing. Roads had been primarily a local issue, but the growing demand led to greater state and federal involvement in funding.

The Federal-Aid Highway Act of 1956 was a milestone in the development of highway transportation in the United States; it marked the beginning of the largest peacetime public works program in the history of the world, creating a 41,000-mile National System of Interstate and Defense Highways, built to high standards. Later legislation expanded funding, improved planning, addressed environmental concerns, and provided for more balanced transportation. Other developed countries also developed highway programs but were more restrained in construction.

Roads and Development

Transportation presents a severe challenge for sustainable development. The number of motor vehicles at the end of the second decade of the twenty-first century—estimated at more than 1.2 billion worldwide—is growing almost everywhere at higher rates than either population or the gross domestic product. Overall road traffic grows even more quickly. The tiny nation of San Marino has nearly 1.2 cars per person. Americans own one car for every 1.88 residents. In Great Britain, there is one car for every 5.3 people.

Highways around the world have been built to help strengthen national unity. The Trans-Canada Highway, the world's longest national road, for ex-

---

HIGHWAY CLASSIFICATION

Modern roads can be classified by roadway design or traffic function. The basic type of roadway is the conventional, undivided, two-way road. Divided highways have median strips or other physical barriers separating the lanes going in opposite directions.

Another quality of a roadway is its right-of-way control. The least expensive type of system controls most side access and some minor at-grade intersections; the more expensive type has side access fully controlled and no at-grade intersections. The amount of traffic determines the number of lanes. Two or three lanes in each direction is typical, but some roads in Los Angeles have five lanes, while some sections of the Trans-Canada Highway have only one lane. Some highways are paid for entirely from public funds; if users pay directly when they use the road, they are called tollways or turnpikes.

Roads are classified as expressway, arterial, collector, and local in urban areas, with a similar hierarchy in rural areas. The highest level—expressway—is intended for long-distance travel.

---

ample, extends east-west across the breadth of the country. Completed in the 1960s, it had the same goal as the Canadian Pacific Railroad a century before, to improve east-west commerce within Canada.

Sometimes, existing highways need to be upgraded; in less-developed countries, this can simply mean paving a road for all-weather operation. An example of a late-1990s project of this nature was the Brazil-Venezuela Highway project, which had this description: Improve the Brazil-Venezuela highway link by completion of paving along the BR-174, which runs northward from Manaus in the Amazon, through Boa Vista and up to the frontier, so opening a route to the Caribbean. Besides the investment opportunities in building the road itself, the highway would result in investment opportunities in mining, tourism, telecommunications, soy and rice production, trade with Venezuela, manufacturing in the Manaus Free Trade Zone, ecotourism in the Amazon, and energy integration.

Growing road traffic has required increasingly significant national contributions to road construction. Beginning in the 1960s, the World Bank began to finance road construction in several countries. It required that projects be organized to the highest technical and economic standards, with private contracting and international competitive bidding rather than government workers. Still, there were questions as to whether these economic assessments had a road-sector bias and properly incorporated environmental costs. Sustainability was also a question—could the facilities be maintained once they were built?

In the 1990s, the World Bank financed a program to build an asphalt road network in Mozambique. Asphalt makes very smooth roads but is very maintenance-intensive, requiring expensive imported equipment and raw materials. By the end of the decade, the roads required resurfacing but the debt was still outstanding. Alternative materials would have given a rougher road, but it could have been built with local materials and labor.

The European Investment Bank has become a major player in the construction of highways linking Eastern and Western Europe to further European integration. Some of the fastest growth in the world in ownership of autos has been in Eastern Europe. There is a two-way feedback effect between highway construction and auto ownership.

Environment Consequences

Highways and highway vehicles have social, economic, and environmental consequences. Compromise is often necessary to balance transportation needs against these constraints. For example, in Israel, there has been a debate over construction of the Trans-Israel highway, a US$1.3 billon, six-lane highway stretching 180 miles (300 km.) from Galilee to the Negev.

Demand on resources for worldwide road infrastructure far exceeds available funds; governments increasingly are looking to external sources such as tolls. Private toll roads, common in the nineteenth century, are making a comeback. This has spread from the United States to Europe, where private and government-owned highway operators have begun to sell shares on the stock market. Private companies are not only operating and financing roads in Europe, they are also designing and building them. In Eastern Europe, where road construction languished under communism, private financing and toll collecting are seen as the means of supporting badly needed construction.

Industrial development in poor countries is adversely affected by limited transportation. Costs are high-unreliable delivery schedules make it necessary to maintain excessive inventories of raw materials and finished goods. Poor transport limits the radius of trade and makes it difficult for manufacturers to realize the economies of large-scale operations to compete internationally.

In more difficult terrain, roads become more expensive because of a need for cuts and fills, bridges, and tunnels. To save money, such roads often have steeper grades, sharper curves, and reduced width than might be desired. Severe weather changes also damage roads, further increasing maintenance costs.

Stephen B. Dobrow

RAILWAYS

Railroads were the first successful attempts by early industrial societies to develop integrated communication systems. Today, global societies are linked by Internet systems dependent upon communication satellites orbiting around Earth. The speed by which information and ideas can reach remote places breaks down isolation and aids in the developing of a world community. In the nineteenth century, railroads had a similar impact. Railroads were critical for the creation of an urban-industrial society: They linked regions and remote places together, were important contributors in developing nation-states, and revolutionized the way business was conducted through the creation of corporations. Although alternative forms of transportation exist, railroads remain important in the twenty-first century.

The Industrial Revolution and the Railroad

Development of the steam engine gave birth to the railroad. Late in the eighteenth century, James Watt perfected his steam engine in England. Water was superheated by a boiler and vaporized into steam, which was confined to a cylinder behind a piston. Pressure from expanding steam pushes the cylinder forward, causing it to do work if it is attached to wheels. Watt's engine was used in the manufacturing of textiles, thus beginning the Industrial Revolution whereby machine technology mass produced goods for mass consumption. Robert Fulton was the first innovator to commercially apply the steam engine to water transportation. His steamboat *Clermont* made its maiden voyage up the Hudson River in 1807.

Not until the 1820s was a steam engine used for land transportation. Rivers and lakes were natural features where no road needed to be built. Applying steam to land movement required some type of roadbed. In England, George Stephenson ran a locomotive over iron strips attached to wooden rails. Within a short time, England's forges were able to roll rails made completely of iron shaped like an inverted "U."

The amount of profit a manufacturer could make was determined partially by the cost of transportation. The lower the cost of moving cargo and people, the higher the profitability. Several alternatives existed before the emergence of railroads, although there were drawbacks compared to rail transportation. Toll roads were too slow. A loaded wagon pulled by four horses could average 15 miles (25 km.) a day. Canals were more efficient than early railroads, because barges pulled by mules moved faster over waterways. However, canals could not be built everywhere, especially over mountains.

The application of railroad technology, using steam as a power source, made it possible to overcome obstacles in moving goods and people over considerable distances and at profitable costs. Railroads transformed the way goods were purchased by reducing the costs for consumers, thus raising the living standards in industrial societies. Railroads transformed the human landscape by strengthening the link between farm and city, changed commercial cities into industrial centers, and started early forms of suburban growth well before automobiles arrived.

Financing Railroads

Constructing railroads was costly. Tunnels had to be blasted through mountains, and rivers had to be crossed by bridges. Early in the building of U.S. railroads, the nation's iron foundries could not meet the demands for rolled rails. Rails had to be imported from England until local forges developed more efficient technologies. Once a railway was completed, there was a constant need to maintain the right-of-way so that traffic flow would not be disrupted. Accidents were frequent, and it was an early practice to burn damaged cars because salvaging them was too expensive.

In some countries, railroads were built and operated by national governments. In the United States, railroads were privately owned; however, it was impossible for any single individual to finance and operate a rail system with miles of track. Businessmen raised money by selling stocks and bonds. Just as investors buy stocks in modern high-technology companies, investors purchased stocks and bonds in railroads.

Investing in railroads was good as long as they earned profits and returned money to their investors, but not all railroads made sufficient profits to reward their investors. Competition among railroads was heavy in the United States, and some railroads charged artificially low fares to attract as much business as they could. When ambitious investment schemes collapsed, railroads went bankrupt and were taken over by financiers.

Selling shares of common stock and bonds was made possible by creating corporations. Railroads were granted permission from state governments to organize a corporation. Every investor owned a portion of the railroad. Stockholders' interests were served by boards of directors, and all business transactions were opened for public inspection. One important factor of the corporation was that it relieved individuals from the responsibilities associated with accidents. The railroad, as a corporation, was held accountable, and any compensation for claims made against the company came out of corporate funds, not from individual pockets. This had an impact on the law profession, as law schools began specializing in legal matters relevant to railroads and interstate commerce.

The Success of Railroads

Railroads usually began by radiating outward from port cities where merchants engaged in transoceanic trade. A classic example, in the United States, is the country's first regional railroad—the Baltimore and Ohio. Construction commenced from Baltimore in 1828; by 1850, the railroad had crossed the Appalachian Mountains and was on the Ohio River at Wheeling, Virginia.

Once trunk lines were established, rail networks became more intensive as branch lines were built to link smaller cities and towns. Countries with extremely large continental dimensions developed interior articulating cities where railroads from all directions converged. Chicago and Atlanta are two such cities in the United States. Chicago was surrounded by three circular railroads (belts) whose only function was to interchange cars. Railroads from the Pacific Coast converged with lines from the Atlantic Coast as well as routes moving north from the Gulf Coast.

Mechanized farms and heavy industries developed within the network. Railroads made possible the extraction of fossil fuels and metallic ores, the necessary ingredients for industrial growth. Extension of railroads deep into Eastern Europe helped to generate massive waves of immigration into both

North and South America, creating multicultural societies.

Building railroads in Africa and South Asia made it possible for Europe to increase its political control over native populations. The ultimate aim of the colonial railroad was to develop a colony's economy according to the needs of the mother country. Railroads were usually single-line routes transhipping commodities from interior centers to coastal ports for exportation. Nairobi, Kenya, began as a rail hub linking British interests in Uganda with Kenya's port city of Mombasa. Similar examples existed in Malaysia and Indonesia.

Railroads generated conflicts among colonial powers as nations attempted to acquire strategic resources. In 1904–1905 Russia and Japan fought a war in the Chinese province of Manchuria over railroad rights; Imperial Germany attempted to get around British interests in the Middle East by building a railroad linking Berlin with Baghdad to give Germany access to lucrative oil fields. India was a region of loosely connected provinces until British railroads helped establish unification. The resulting sense of national unity led to the termination of British rule in 1947 and independence for India and Pakistan.

In the United States, private railroads discontinued passenger service among cities early in the 1970s and the responsibility was assumed by the federal government (Amtrak). Most Americans riding trains do so as commuters traveling from the suburbs to jobs in the city. The U.S. has no true high-speed trains, aside from sections of Amtrak's Acela line in the Northeast Corridor, where it can reach 150 mph for only 34 miles of its 457-mile span. Passenger service remains popular in Japan and Europe. France, Germany, and Japan operate high-speed luxury trains with speeds averaging above 100 miles (160 km.) per hour.

Railroads are no longer the exclusive means of land transportation as they were early in the twentieth century. Although competition from motor vehicles and air freight provide alternate choices, railroads have remained important. France and England have direct rail linkage beneath the English Channel. In the United States, great railroad mergers and the application of computer technology have reduced operating costs while increasing profits. Transoceanic container traffic has been aided by railroads hauling trailers on flatcars. Railroads began the process of bringing regions within a nation together in the nineteenth century just as the computer and the World Wide Web began uniting nations throughout the world at the end of the twentieth century.

Sherman E. Silverman

AIR TRANSPORTATION

The movement of goods and people among places is an important field of geographic study. Transportation routes form part of an intricate global network through which commodities flow. Speed and cost determine the nature and volume of the materials transported, so air transportation has both advantages and disadvantages when compared with road, rail, or water transport.

Early Flying Machines

The transport of people and freight by air is less than a century old. Although hot-air balloons were used in the late eighteenth century for military purposes, aerial mapping, and even early photography, they were never commercially important as a means of transportation. In the late nineteenth century, the German count Ferdinand von Zeppelin began experimenting with dirigibles, which added self-propulsion to lighter-than-air craft. These aircraft were used for military purposes, such as the bombing of Paris in World War I. However, by the 1920s zeppelins had become a successful means of passenger transportation. They carried thousands of passengers on trips in Europe or across the Atlantic Ocean and also were used for exploration. Nevertheless, they had major problems and were soon superseded by flying machines heavier than air. The early term for such a machine was an aeroplane, which is still the word used for airplane in Great Britain.

Following pioneering advances with the internal combustion engine and in aerodynamic theory using gliders, the development of powered flight in a heavier-than-air machine was achieved by Wilbur and Orville Wright in December, 1903. From that time, the United States moved to the forefront of aviation, with Great Britain and Germany also making significant contributions to air transport. World War I saw the further development of aviation for military purposes, evidenced by the infamous bombing of Guernica.

Early Commercial Service

Two decades after the Wright brothers' brief flight, the world's first commercial air service began, covering the short distance from Tampa to St. Petersburg in Florida. The introduction of airmail service by the U.S. Post Office provided a new, regular source of income for commercial airlines in the United States, and from these beginnings arose the modern Boeing Company, United Airlines, and American Airlines. Europe, however, was the home of the world's first commercial airlines. These include the Deutsche Luftreederie in Germany, which connected Berlin, Leipzig, and Weimar in 1919; Farman in France, which flew from Paris to London; and KLM in the Netherlands (Amsterdam to London), followed by Qantas—the Queensland and Northern Territory Aerial Services, Limited—in Australia. The last two are the world's oldest still operating airlines.

Aircraft played a vital role in World War II, as a means of attacking enemy territory, defending territory, and transporting people and equipment. A humanitarian use of air power was the Berlin Airlift of 1948, when Western nations used airplanes to deliver food and medical supplies to the people of West Berlin, which the Soviet Union briefly blockaded on the ground.

Cargo and Passenger Service

The jet engine was developed and used for fighter aircraft during World War II by the Germans, the British, and the United States. Further research led to civil jet transport, and by the 1970s, jet planes accounted for most of the world's air transportation. Air travel in the early days was extremely expensive, but technological advances enabled longer flights with heavier loads, so commercial air travel became both faster and more economical.

Most air travel is made for business purposes. Of business trips between 750 and 1,500 miles (1,207 and 2,414 km.), air captures almost 85 percent, and of trips more than 1,500 miles (2,414 km.), 90 percent are made by air. The United States had 5,087 public airports in 2018, a slight decrease from the 5,145 public airports operating in 2014. Conversely, the number of private airports increased over this period from 13,863 to 14,549.

The biggest air cargo carriers in 2019 were Federal Express, which carried more than 15.71 billion freight tonne kilometres (FTK), Emirates Skycargo (12.27 billion FTK), and United Parcel Service (11.26 billion FTK).

The first commercial supersonic airliner, the British-French Concorde, which could fly at more than twice the speed of sound, began regular service in early 1976. However, the fleet was grounded after a Concorde crash in France in mid-2000. The first space shuttle flew in 1981. There have been 135 shuttle missions since then, ending with the successful landing of Space Shuttle Orbiter Atlantis on July 21, 2011. The shuttles have transported 600 people and 3 million pounds (1.36 million kilograms) of cargo into space.

Health Problems Transported by Air

The high speed of intercontinental air travel and the increasing numbers of air travelers have increased the risk of exotic diseases being carried into destination countries, thereby globalizing diseases previously restricted to certain parts of the world. Passengers traveling by air might be unaware that they are carrying infections or viruses. The worldwide spread of HIV/AIDS after the 1980s was accelerated by international air travel.

Disease vectors such as flies or mosquitoes can also make air journeys unnoticed inside airplanes. At some airports, both airplane interiors and passengers are subjected to spraying with insecticide upon arrival and before deplaning. The West Nile virus (West Nile encephalitis) was previously found only in Africa, Eastern Europe, and West Asia, but in the 1990s it appeared in the northeastern United States, transported there by birds, mosquitos, or people.

It was feared in the mid-1990s that the highly infectious and deadly Ebola virus, which originated in tropical Africa, might spread to Europe and the United States, by air passengers or through the importing of monkeys. The devastation of native bird communities on the island of Guam has been traced to the emergence there of a large population of brown tree snakes, whose ancestors are thought to have arrived as accidental stowaways on a military airplane in the late 1940s.

Ray Sumner

ENERGY AND ENGINEERING

ENERGY SOURCES

Energy is essential for powering the processes of modern industrial society: refining ores, manufacturing products, moving vehicles, heating buildings, and powering appliances. In 1999 energy costs were half a trillion dollars in the United States alone. All technological progress has been based on harnessing more energy and using it more effectively. Energy use has been shaped by geography and also has shaped economic and political geography.

Ancient to Modern Energy

Energy use in traditional tribal societies illustrates all aspects of energy use that apply in modern human societies. Early Stone Age peoples had only their own muscle power, fueled by meat and raw vegetable matter. Warmth for living came from tropical or subtropical climates. Then a new energy source, fire, came into use. It made cold climates livable. It enabled the cooking of roots, grains, and heavy animal bones, vastly increasing the edible food supply. Its heat also hardened wood tools, cured pottery, and eventually allowed metalworking.

Nearly as important as fire was the domestication of animals, which multiplied available muscle energy. Domestic animals carried and pulled heavy loads. Domesticated horses could move as fast as the game to be hunted or large animals to be herded.

Increased energy efficiency was as important as new energy sources in making tribal societies more successful. Cured animal hides and woven cloth were additional factors enabling people to move to cooler climates. Cooking fires also allowed drying meat into jerky to preserve it against times of limited supply. Fire-cured pottery helped protect food against pests and kept water close by. However, energy benefits had costs. Fire drives for hunting may have caused major animal extinctions. Periodic burning of areas for primitive agriculture caused erosion. Trees became scarce near the best campsites because they had been used for camp fires—the first fuel shortage.

Energy Fundamentals

Human use of energy revolves about four interrelated factors: energy sources, methods of harnessing the sources, means of transporting or storing energy, and methods of using energy. The potential energies and energy flows that might be harnessed are many times greater than present use.

The Sun is the primary source of most energy on Earth. Sunlight warms the planet. Plants use photosynthesis to transform water and carbon dioxide into the sugars that power their growth and indirectly power plant-eating and meat-eating animals. Many other energies come indirectly from the Sun. Remains of plants and animals become fossil fuels. Solar heat evaporates water, which then falls as rain, causing water flow in rivers. Regional differences in the amount of sunlight received and reflected cause temperature differences that generate winds, ocean currents, and temperature differences between different ocean layers. Food for muscle power of humans and animals is the most basic energy system.

Energy Sources

Biomass—wood or other vegetable matter that can be burned—is still the most important energy source in much of the world. Its basic use is to provide heat for cooking and warmth. Biomass fuels are often agricultural or forestry wastes. The advantage of biomass is that it is grown, so it can be replaced. However, it has several limitations. Its low energy content per unit volume and unit mass makes it unprofitable to ship, so its use is limited to the amount nearby. Collecting and processing biomass fuels costs energy, so the net energy is less. Biomass energy production may compete with food production, since both come from the soil. Finally, other fuels can be cheaper.

Greater concentration of biomass energy or more efficient use would enable it to better compete against other energy sources. For example, fermenting sugars into fuel alcohol is one means of concentrating energy, but energy losses in processing make it expensive.

Fossil fuels have more concentrated chemical energy than biomass. Underground heat and pressure compacts trees and swampy brush into the progressively more energy-concentrated peat, lignite coal, bituminous coal, and anthracite or black coal, which is mostly carbon. Industrializing regions turned to coal when they had exhausted their firewood. Like wood, coal could be stored and shoveled into the fire box as needed. Large deposits of coal are still available, but growth in the use of coal slowed by the mid-twentieth century because of two competing fossil fuels, petroleum and natural gas.

Petroleum includes gasoline, diesel fuel, and fuel oil. It forms from remains of one-celled plants and animals in the ocean that decompose from sugars into simpler hydrogen and carbon compounds (hydrocarbons). Petroleum yields more energy per unit than coal, and it is pumped rather than shoveled. These advantages mean that an oil-fired vehicle can be cheaper and have greater range than a coal-fired vehicle.

There are also hydrocarbon gases associated with petroleum and coal. The most common is the natural gas methane. Methane does not have the energy density of hydrocarbon liquids, but it burns cleanly and is a fuel of choice for end uses such as homes and businesses.

Petroleum and natural gas deposits are widely scattered throughout the world, but the greatest known deposits are in an area extending from Saudi Arabia north through the Caucasus Mountains. Deposits extend out to sea in areas such as the Persian Gulf, the North Sea, and the Gulf of Mexico. Other sources, such as oil tar sands and shale oil, are currently seen as a potentially important source of energy, but controversies surrounding the extraction, refining, and delivery processes make these energy sources a matter of significant debate and concern.

Heat engines transform the potential of chemical energies. James Watt's steam engine (1782) takes heat from burning wood or coal (external combustion), boils water to steam, and expands it through pistons to make mechanical motion. In the twentieth century, propeller-like steam turbines were developed to increase efficiency and decrease complexity. Auto and diesel engines burn fuel inside the engine (internal combustion), and the hot gases expand through pistons to make mechanical motion. Expanding them through a gas turbine is a jet engine. Heat engines can create energy from other sources, such as concentrated sunlight, nuclear fission, or nuclear fusion. The electrical generator transforms mechanical motion into electricity that can move by wire to uses far away. Such transportation (or wheeling) of electricity means that one power plant can serve many customers in different locations.

Flowing water and wind are two of the oldest sources of industrial power. The Industrial Revolution began with water power and wind power, but they could only be used in certain locations, and they were not as dependable as steam engines. In the early twentieth century, electricity made river power practical again. Large dams along river valleys with adequate water and steep enough slopes enabled areas like the Tennessee Valley to be industrial centers. In the 1970s wind power began to be used again, this time for generating electricity.

Solar energy can be tapped directly for heat or to make electricity. Although sunlight is free, it is not concentrated energy, so getting usable energy requires more equipment cost. Consequently, fossil-fueled heat is cheaper than solar heat, and power from the conventional utility grid has been much less expensive than solar-generated electricity. However, prices of solar equipment continue to drop as technologies improve.

Future Energy Sources

Possible future energy sources are nuclear fission, nuclear fusion, geothermal heat, and tides. Fission reactors contain a critical mass of radioactive heavy elements that sustains a chain reaction of atoms splitting (fissioning) into lighter elements—releasing heat to run a steam turbine. Tremendous amounts of fission energy are available, but reactor costs and safety issues have kept nuclear prices higher than that of coal.

Nuclear fusion involves the same reaction that powers the Sun: four hydrogen atoms fusing into one helium atom. However, duplicating the Sun's heat in a small area without damaging the surrounding reactor may be too expensive to allow profitable fusion reactors.

Geothermal power plants, tapping heat energy from within the earth, have operated since 1904, but widespread use depends on cheaper drilling to make them practical in more than highly volcanic areas. Tidal power is limited to the few bays that concentrate tidal energy.

Energy and Warfare

Much of ancient energy use revolved about herding animals and conducting warfare. Horse riders moved faster and hit harder than warriors on foot. The bow and arrow did not change appreciably for thousands of years. Herders on the plains rode horses and used the bow and arrow as part of tending their flocks, and the small amounts of metal needed for weapons was easily acquired. Consequently, the herders could invade and plunder much more advanced peoples. From Scythians to Parthians to Mongols, these people consistently destroyed the more advanced civilizations.

The geographical effect was that ancient civilizations generally developed only if they had physical barriers separating them from the flat plains of herding peoples. Egypt had deserts and seas. The Greeks and Romans lived on mountainous peninsulas, safe from easy attack. The Chinese built the Great Wall along their northern frontier to block invasions.

Nomadic riders dominated until the advent of an energy system of gunpowder and steel barrels began delivering lead bullets. With them, the Russians broke the power of the Tartars in Eurasia in the late fifteenth century, and various peoples from Europe conquered most of the world. Energy and industrial might became progressively more important in war with automatic weapons, high explosives, aircraft, rockets, and nuclear weapons.

By World War II, oil had become a reason for war and a crucial input for war. The Germans attempted to seize petroleum fields around Baku on the Caspian. Later in the war, major Allied attacks targeted oil fields in Romania and plants in Germany synthesizing liquid fuels. During the Arab-Israeli War of 1973, Arabs countered Western support of Israel with an oil boycott that rocked Western economies. In 1990 Iraq attempted to solve a border dispute with its oil-rich neighbor, Kuwait, by seizing all of Kuwait. An alliance, led by the United States, ejected the Iraqis.

Other wars occur over petroleum deposits that extend out to sea. European nations bordering on the North Sea negotiated a complete demarcation of economic rights throughout that body. Tensions between China and other Asian countries continue to mount over rights to the resources available in the South China Sea. Current estimates suggest that there may be 90 trillion cubic feet of natural gas and 11 billion barrels of oil in proved and probable reserves, with much more potentially undiscovered. The area is claimed by China, Vietnam, Malaysia, and the Philippines. Turkey and Greece have not resolved ownership division of Aegean waters that might have oil deposits.

Energy, Development, and Energy Efficiency

Ancient civilizations tended to grow and use locally available food and firewood. Soils and wood supplies often were depleted at the same time, which often coincided with declines in those civilizations.

The Industrial Revolution caused development to concentrate in new wooded areas where rivers suitable for power, iron ore, and coal were close together, for example, England, Silesia, and the Pittsburgh area. The iron ore of Alsace in France combined with nearby coal from the Ruhr in Germany fueled tremendous growth, not always peacefully.

By the late nineteenth century, the development of Birmingham, Alabama, demonstrated that railroads enabled a wider spread between coal deposits, iron ore deposits, and existing population centers. By the 1920s, the Soviet Union developed entirely new cities to connect with resources. By the 1970s, unit trains and ore-carrying ships transported coal from the thick coal beds in Montana and Wyoming to the United States' East Coast and to countries in Asia.

The mechanized transport of electrical distribution and distribution of natural gas in pipelines also changed settlement patterns. Trains and subway trains allowed cities to spread along rail corridors in the late nineteenth century and early twentieth century. By the 1940s, cars and trucks enabled cities such as Los Angeles and Phoenix to spread into suburbs. The trend continues with independent solar power that allows houses to be sited anywhere.

Advances in technology have allowed people to get more while using less energy. For example, early peoples stampeded herds of animals over cliffs for food, which was mostly wasted. Horseback hunting was vastly more efficient. Likewise, fireplaces in colonial North America were inefficient, sending most of their heat up the chimney. In the late eighteenth century, inventor and statesman Benjamin Franklin developed a metallic cylinder radiating heat in all directions, which saved firewood.

The ancient Greeks and others pioneered the use of passive solar energy and efficiency after they exhausted available firewood. They sited buildings to absorb as much low winter sun as possible and constructed overhanging roofs to shade buildings from the high summer sun. That siting was augmented by heavy masonry building materials that buffered the buildings from extremes of heat and cold. Later,

metal pipes and glass meant that solar energy could be used for water and space heating.

The first seven decades of the twentieth century saw major declines in energy prices, and cars and appliances became less efficient. That changed abruptly with the energy crises and high prices of the 1970s. Since then, countries such as Japan, with few local energy resources, have worked to increase efficiency so they will be less sensitive to energy shocks and be able to thrive with minimal energy inputs. This trend could lead eventually to economies functioning on only solar and biomass inputs.

Solid-state electronics, use of light emitting diode (LED) or compact fluorescent lamps (CFLs) rather than incandescent bulbs, and fuel cells, which convert fuel directly into electricity more efficiently than combustion engines, all could lead to less energy use. The speed of their adoption depends on the price of competing energies. According to the U.S. Energy Information Administration's (EIA) International Energy Outlook 2019 (IEO2019), the global supply of crude oil, other liquid hydrocarbons, and biofuels is expected to be adequate to meet world demand through 2050. However, many have noted that continuing to burn fossil fuels at our current rate is not sustainable, not because reserves will disappear, but because the damage to the climate would be unacceptable.

Energy and Environment

Energy affects the environment in three major ways. First, firewood gathering in underdeveloped countries contributes to deforestation and resulting erosion. Although more efficient stoves and small solar cookers have been designed, efficiency increases are competing against population increases.

Energy production also frequently causes toxic pollutant by-products. Sulfur dioxide (from sulfur impurities in coal and oil) and nitrogen oxides (from nitrogen being formed during combustion) damage lungs and corrode the surfaces of buildings. Lead additives in gasoline make internal combustion engines run more efficiently, but they cause low-grade lead poisoning. Spent radioactive fuel from nuclear fission reactors is so poisonous that it must be guarded for centuries.

Finally, carbon dioxide from the burning of fossil fuels may be accelerating the greenhouse effect, whereby atmospheric carbon dioxide slows the planetary loss of heat. If the effect is as strong as some research suggests, global temperatures may increase several degrees on average in the twenty-first century, with unknown effects on climate and sea level.

Roger V. Carlson

ALTERNATIVE ENERGIES

The energy that lights homes and powers industry is indispensable in modern societies. This energy usually comes from mechanical energy that is converted into electrical energy by means of generators—complex machines that harness basic energy captured when such sources as coal, oil, or wood are burned under controlled conditions. This energy, in turn, provides the thermal energy used for heating, cooling, and lighting and for powering automobiles, locomotives, steamships, and airplanes. Because such natural resources as coal, oil, and wood are being used up, it is vital that these nonrenewable sources of energy be replaced by sources that are renewable and abundant. It is also desirable that alternative sources of energy be developed in order to cut down on the pollution that results from the combustion of the hydrocarbons that make the nonrenewable fuels burn.

The Sun as an Energy Source

Energy is heat. The Sun provides the heat that makes Earth habitable. As today's commonly used fuel resources are used less, solar energy will be used increasingly to provide the power that societies need in order to function and flourish.

There are two forms of solar energy: passive and active. Humankind has long employed passive solar energy, which requires no special equipment. Ancient cave dwellers soon realized that if they inhabited caves that faced the Sun, those caves would be warmer than those that faced away from the Sun. They also observed that dark surfaces retained heat and that dark rocks heated by the Sun would radiate the heat they contained after the Sun had set. Modern builders often capitalize on this same knowledge by constructing structures that face south in the Northern Hemisphere and north in the Southern Hemisphere. The windows that face the Sun are often large and unobstructed by draperies and curtains. Sunlight beats through the glass and, in passive solar houses, usually heats a dark stone or brick floor that will emit heat during the hours when there is no sunlight. Just as an automobile parked in the sunlight will become hot and retain its heat, so do passive solar buildings become hot and retain their heat.

Active solar energy is derived by placing specially designed panels so that they face the Sun. These panels, called flat plate collectors, have a flat glass top beneath which is a panel, often made of copper with a black overlay of paint, that retains heat. These panels are constructed so that heat cannot escape from them easily. When water circulated through pipes in the panels becomes hot, it is either pumped into tanks where it can be stored or circulated through a central heating system.

Some active solar devices are quite complex and best suited to industrial use. Among these is the focusing collector, a saucer-shaped mirror that centers the Sun's rays on a small area that becomes extremely hot. A power plant at Odeillo in the French Pyrenees Mountains uses such a system to concentrate the Sun's rays on a concave mirror. The mirror directs its incredible heat to an enormous, confined body of water that the heat turns to steam, which is then used to generate electricity.

Another active solar device is the solar or photovoltaic cell, which gathers heat from the Sun and turns it into energy directly. Such cells help to power spacecraft that cannot carry enough conventional fuel to sustain them through long missions in outer space.

Geothermal Heating

The earth's core is incredibly hot. Its heat extends far into the lower surfaces of the planet, at times causing eruptions in the form of geysers or volcanoes. Many places on Earth have springs that are warmed by heat from the earth's core.

In some countries, such as Iceland, warm springs are so abundant that people throughout the country bathe in them through the coldest of winters. In Iceland, geothermal energy is used to heat and light homes, making the use of fossil fuels unnecessary.

Hot areas exist beneath every acre of land on Earth. When such areas are near the surface, it is easy to use them to produce the energy that humans require. As dependence on fossil fuels decreases, means will increasingly be found of drawing on Earth's subterranean heat as a major source of energy.

Wind Power

Anyone who has watched a sailboat move effortlessly through the water has observed how the wind can be used as a source of kinetic energy—the kind of energy that involves motion—whose movement is transferred to objects that it touches. Wind power has been used throughout human history. In its more refined aspects, it has been employed to power windmills that cause turbines to rotate, providing generators with the power they require to produce electricity.

Windmills typically have from two to twenty blades made of wood or of heavy cloth such as canvas. Windmills are most effective when they are located in places where the wind regularly blows with considerable velocity. As their blades turn, they cause the shafts of turbines to rotate, thus powering generators. The electricity created is usually transmitted over metal cables for immediate use or for storage.

Modern vertical-axis wind turbines have two or three strips of curved metal that are attached at both ends to a vertical pole. They can operate efficiently even if they are not turned toward the wind. These windmills are a great improvement over the old horizontal axis windmills that have been in use for many years. From 2000 to 2015, cumulative wind capacity around the world increased from 17,000 megawatts to more than 430,000 megawatts. In 2015, China also surpassed the EU in the number of installed wind turbines and continues to lead installation efforts. Production of wind electricity in 2016 accounted for 16 percent of the electricity generated by renewables.

Oceans as Energy Sources

Seventy percent of the earth's surface is covered by oceans. Their tides, which rise and fall with predictable regularity twice a day, would offer a ready source of energy once it becomes economically feasible to harness them and store the electrical energy they can provide. The most promising spots to build facilities to create electrical energy from the tides are places where the tides are regularly quite dramatic, such as Nova Scotia's Bay of Fundy, where the difference between high and low tides averages about 55 feet (17 meters).

Some tidal power stations that currently exist were created by building dams across estuaries. The sluices of these dams are opened when the tide comes in and closed after the resulting reservoir fills. The water captured in the reservoir is held for several hours until the tide is low enough to create a considerable difference between the level of the wa-

---

OCEAN ENERGY

The oceans have tremendous untapped energy flows in currents and tremendous potential energy in the temperature differences between warmer tropical surface waters and colder deep waters, known as ocean thermal energy conversion. In both cases, the insurmountable cost has been in transporting energy to users on shore.

---

ter in the reservoir and that outside it. Then the sluice gates are opened and, as the water rushes out at a high rate of speed, it turns turbines that generate electricity.

The world's first large-scale tidal power plant was the Rance Tidal Power Station in France, which became operational in 1966. It was the largest tidal power station in terms of output until Sihwa Lake Tidal Power Station opened in South Korea in August 2011.

Future of Renewable Energy

As pollution becomes a huge problem throughout the world, the race to find nonpolluting sources of energy is accelerating rapidly. Scientists are working on unlocking the potential of the electricity generated by microbes as a fuel source, for example. New technologies are making renewable energy sources economically practical. As supplies of fossil fuels have diminished, pressure to become less dependent on them has grown worldwide. Alternative energy sources are the wave of the future.

R. Baird Shuman

ENGINEERING PROJECTS

Human beings attempt to overcome the physical landscape by building forms and structures on the earth. Most structures are small-scale, like houses, telephone poles, and schools. Other structures are great engineering works, such as hydroelectric projects, dams, canals, tunnels, bridges, and buildings.

Hydroelectric Projects

The potential for hydroelectricity generation is greatest in rapidly flowing rivers in mountainous or hilly terrain. The moving water turns turbines that, in turn, generate electricity. Hydroelectric power projects also can be built on escarpments and fall lines, where there is tremendous untapped energy in the falling water.

Most of the potential for hydroelectricity remains untapped. Only about one-sixth of the suitable rivers and falls are used for hydroelectric power. Certain areas of the world have used more of their potential than others. The percent of potential hydropower capacity that has not been developed is 71 percent in Europe, 75 percent in North America, 79 percent in South America, 95 percent in Africa, 95 percent in the Middle East, and 82 percent in Asia-Pacific. China, Brazil, Canada, and the United States currently produce the most hydroelectric power.

In Africa, only Zambia, Zimbabwe, and Ghana produce significant hydroelectricity. The region's total generating capacity needs to increase by 6 percent per year to 2040 from the current total of 125 GW to keep pace with rising electricity demand. In Southeast Asia, countries continue to grapple with the need to build up their hydroelectric plants without causing harm to the rivers that are used to supply food, water, and transportation.

Dams

Dams serve several purposes. One purpose is the generation of hydroelectric power, as discussed above. Dams also provide flood control and irrigation. Rivers in their natural state tend to rise and fall with the seasons. This can cause serious problems for people living in downstream valleys. Flood-control dams also can be used to regulate the flow of water used for irrigation and other projects. A final reason to build dams is to reduce swampland, in order to control insects and the diseases they carry.

Famous dams are found in all regions of the world. In North America, two of the most notable dams are Hoover Dam, completed in 1936, on the Colorado River between Arizona and Nevada; and the Grand Coulee Dam, completed in 1942, on the Columbia River in Washington State.

In South America, the most famous dam is the Itaipu Dam, completed in 1983, on the Paraná River between Brazil and Paraguay. In Africa, the Aswan High Dam was completed in 1970, on the Nile River in Egypt, and the Kariba Dam was completed in 1958, on the Zambezi River between Zambia and Zimbabwe. In Asia, the Three Gorges Dam spans the Yangtze River by the town of Sandouping in Hubei province, China. The Three Gorges Dam has been the world's largest power station in terms of installed capacity (22,500 MW) since 2012.

Bridges

Bridges are built to span low-lying land between two high places. Most commonly, there is a river or other body of water in the way, but other features that might be spanned include ravines, deep valleys and trenches, and swamps. A related engineering

---

ENGINEERING WORKS AND ENVIRONMENTAL PROBLEMS

Although engineering allows humans to overcome natural obstacles, works of engineering often have unintended consequences. Many engineering projects have caused unanticipated environmental problems.

Dams, for instance, create large lakes behind them by trapping water that is released slowly. This water typically contains silt and other material that eventually would have formed soil downstream had the water been allowed to flow naturally. Instead, the silt builds up behind the dam, eventually diminishing the lake's usefulness. As an additional consequence, there is less silt available for soil-building downstream.

Canals also can cause environmental harm by diverting water from its natural course. The river from which water is diverted may dry up, negatively affecting fish, animals, and the people who live downstream.

The benefits of engineering works must be weighed against the damage they do to the environment. They may be worthwhile, but they are neither all good nor all bad: There are benefits and drawbacks in building any engineering project.

---

project is the causeway, in which land in a low-lying area is built up and a road is then constructed on it.

The longest bridge in the world is the Akashi Kaikyo in Japan near Osaka. It was built in 1998 and spans 6,529 feet (1,990 meters), connecting the island of Hōnshū to the small island of Awaji. The Storebælt Bridge in Denmark, also completed in 1998, spans 5,328 feet (1,624 meters), connecting the island of Sjaelland, on which Copenhagen is situated, with the rest of Denmark. Another bridge spanning more than 5,300 feet is the Osman Gazi Bridge in Turkey. The bridge was opened on 1 July 1, 2016, ad to become the longest bridge in Turkey and the fourth-longest suspension bridge in the world by the length of its central span. The length of the bridge is expected to be surpassed by the Çanakkale 1915 Bridge, which is currently under construction across the Dardanelles strait.

Other long bridges can be found across the Humber River in Hull, England; across the Chiang Jiang (Yangtze River) in China; in Hong Kong, Norway, Sweden, and Turkey and elsewhere in Japan.

The longest bridge in the United States is the Lake Pontchartrain Causeway, Louisiana, which spans 24 miles (38.5 km.), the Verrazano-Narrows Bridge in New York City between Staten Island and Brooklyn was once the longest suspension bridge in the world. Completed in 1964, its main span measures 4,260 feet (1,298 meters).

Canals

Moving goods and people by water is generally cheaper and easier, if a bit slower, than moving them by land. Before the twentieth century, that cost savings overwhelmed the advantages of land travel—speed and versatility. Therefore, human beings have wanted to move things by water whenever possible. To do so, they had two choices: locate factories and people near water, such as rivers, lakes, and oceans, or bring water to where the factories and people are, by digging canals.

One of the most famous canals in the world is the Erie Canal, which runs from Albany to Buffalo in New York State. Built in 1825 and running a length of 363 miles (584 km.), the Erie Canal opened up the Great Lakes region of North America to development and led to the rise of New York City as one of the world's dominant cities.

Two other important canals in world history are the Panama Canal and the Suez Canal. The Panama Canal connects the Atlantic and Pacific Oceans over a length of 50.7 miles (81.6 km.) on the isthmus of Panama in Central America. Completed in 1914, the Panama Canal eliminated the long and dangerous sea journey around the tip of South America. The Suez Canal in Egypt, which runs for 100 miles (162 km.) and was completed in 1856, eliminates a similar journey around the Cape of Good Hope in South Africa.

The longest canal in the world is the Grand Canal in China, which was built in the seventh century and stretches a length of 1,085 miles (2,904 km.). It connects Tianjin, near Beijing in the north of China, with Nanjing on the Chang Jiang (Yangtze River) in Central China. The Karakum Canal runs across the Central Asian desert in Turkmenistan from the Amu Darya River westward to Ashkhabad. It was begun in the 1954, and completed in 1988 and is navigable over much of its 854-mile (1,375-km.) length. The Karakum Canal and carries 13 cubic kilometres (3.1 cu mi) of water annually from the Amu-Darya River across the Karakum Desert to irrigate the dry lands of Turkmenistan.

Many canals are found in Europe, particularly in England, France, Belgium, the Netherlands, and Germany, and in the United States and Canada, especially connecting the Great Lakes to each other and to the Ohio and Mississippi Rivers.

Tunnels

Tunnels connect two places separated by physical features that would make it extremely difficult, if not impossible, for them to be connected without cutting directly through them. Tunnels can be used in place of bridges over water bodies so that water traffic is not impeded by a bridge span. Tunnels of this type are often found in port cities, and cities with them include Montreal, Quebec; New York City; Hampton Roads, Virginia; Liverpool, England; or Rio de Janeiro, Brazil.

Tunnels are often used to go through mountains that might be too tall to climb over. Trains especially

are sensitive to changes in slope, and train tunnels are found all over the world. Less common are automobile and truck tunnels, although these are also found in many places. Train and automotive tunnels through mountains are common in the Appalachian Mountains in Pennsylvania, the Rockies in the United States and Canada, Japan, and the Alps in Italy, France, Switzerland, and Austria.

The Chunnel

Arguably the most famous—and one of the most ambitious—tunnels in the world goes by the name Chunnel. Completed in 1994, it connects Dover, England, to Calais, France, and runs 31 miles (50 km.). "Chunnel" is short for the Channel Tunnel, named for the English Channel, the body of water that it goes under. It was built as a train tunnel, but cars and trucks can be carried through it on trains. In the year 2000 plans were underway to cut a second tunnel, to carry automobiles and trucks, that would run parallel to the first Chunnel.

The Seikan Tunnel in Japan, connects the large island of Hōnshū with the northern island of Hōkkaidō. The Seikan Tunnel is nearly 2.4 miles (4 km.) longer than Europe's Chunnel; however, the undersea portion of the tunnel is not as long as that of the Chunnel.

Buildings

Historically, North America has been home to the tallest buildings in the world. Chicago has been called the birthplace of the skyscraper and was at one time home to the world's tallest building. In 1998, however, the two Petronas Towers (each 1,483 feet/452 meters tall) were completed in Kuala Lumpur, Malaysia, surpassing the height of the world's tallest building, Chicago's Sears Tower (1,450 feet/442 meters), which had been completed in 1974. In 2019, the tallest completed building in the world is the 2,717-foot (828-metre) tall Burj Khalifa in Dubai, the tallest building since 2008.

Of the twenty tallest buildings standing in the year 2019, China is home to ten (Shanghai Tower, Ping An Finance Center, Goldin Finance 117, Guangzhou CTF Finance Center, Tianjin CFT Finance Center, China Zun, Shanghai World Financial Center, International Commerce Center, Wuhan Greenland Center, Changsha); Malaysia (the Petronas towers) and the United States (One World Trade Center and Central Park Towers) boast two each; Vietnam has one (Landmark 81 in Ho Chi Minh City), as does Russia (Lakhta Center), Taiwan (Taipei 101), South Korea (Lotte World Trade Center), Saudi Aragia (Abraj Al-Bait Clock Tower in Mecca).

Timothy C. Pitts

INDUSTRY AND TRADE

MANUFACTURING

Manufacturing is the process by which value is added to materials by changing their physical form—shape, function, or composition. For example, an automobile is manufactured by piecing together thousands of different component parts, such as seats, bumpers, and tires. The component parts in unassembled form have little or no utility, but pieced together to produce a fully functional automobile, the resulting product has significant utility. The more utility something has, the greater its value. In other words, the value of the component parts increases when they are combined with the other parts to produce a useful product.

Employment in Manufacturing
On a global scale, 28 percent of the world's working population had jobs in the manufacturing sector in the third decade of the century. The rest worked in agriculture (28 percent) and services (49 percent). The importance of each of these sectors varies from country to country and from time period to time period. High-income countries have a higher percentage of their labor force employed in manufacturing than low-income countries do. For example, in the United States 19 percent of the labor force worked in manufacturing by 2019, whereas the African country of Tanzania had only 7 percent of its labor force employed in the manufacturing sector at that time.

At the end of the twentieth century, the vast majority of the U.S. labor force (74 percent) worked in services, a sector that includes jobs such as computer programmers, lawyers, and teachers. By the end of the second decade of the twenty-first century, the percentage has risen to slightly more than 79 percent. Only 1 percent worked in agriculture and mining. This employment structure is typical for a high-income country. In low-income countries, in contrast, the majority of the labor force have agricultural jobs. In Tanzania, for example, 66 percent of the labor force worked in agriculture, while services accounted for 27 percent of the jobs.

The importance of manufacturing as an employer changes over time. In 1950 manufacturing accounted for 38 percent of all jobs in the United States. The percentage of jobs accounted for by the manufacturing sector in high-income countries has decreased in the post-World War II period. The decreasing share of manufacturing jobs in high-income countries is partly attributable to the fact that many manufacturing companies have replaced people with machines on assembly lines. Because one machine can do the work of many people, manufacturing has become less labor-intensive (uses fewer people to perform a particular task) and more capital-intensive (uses machines to perform tasks formerly done by people). In the future, manufacturing in high-income countries is expected to become increasingly capital-intensive. It is not inconceivable that manufacturing's share of the U.S. labor force could fall below 10 percent over the course of the twenty-first century.

Geography of Manufacturing
Every country produces manufactured goods, but the vast bulk of manufacturing activity is concentrated geographically. Four countries—China, the United States, Japan, and Germany—produce almost 60 percent of the world's manufactured goods. The concentration of manufacturing activity

in a small number of regions means that there are other regions where very little manufacturing occurs. Africa is a prime example of a region with little manufacturing.

Different countries tend to specialize in the production of different products. For example, 50 percent of the automobiles that were produced in that late 1990s were produced in three countries—Germany, Japan, and the United States. In the production of television sets, the top three countries were China, Japan, and South Korea, which together produced 48 percent of the world's television sets. It is important to note that these patterns change over time. For example, in 1960 the top three automobile-producing countries were Germany, the United Kingdom, and the United States, which together produced 76 percent of the world's automobiles.

Multinational Corporations

A multinational corporation is a corporation that is headquartered in one country but owns business facilities, for example, manufacturing plants, in other countries. Some examples of multinational corporations from the manufacturing sector include the automobile maker Ford, whose headquarters are the in the United States, the pharmaceutical company Bayer, whose headquarters are in Germany, and the candy manufacturer Nestlé, whose headquarters are in Switzerland. Since the end of World War II, multinational corporations have become increasingly important in the world economy. Most multinational corporations are headquartered in high-income countries, such as Japan, the United Kingdom, and the United States.

Companies open manufacturing plants in other countries for a variety of reasons. One of the most common reasons is that it allows them to circumvent barriers to trade that are imposed by foreign governments, especially tariffs and quotas. A tariff is an import tax that is imposed upon foreign-manufactured goods as they enter a country. A quota is a limitation imposed on the volume of a particular good that a particular country can export to another country. The net effect of tariffs and quotas is to increase the cost of imported goods for consumers.

Governments impose tariffs and quotas partly to raise revenue and partly to encourage consumers to purchase goods manufactured in their own country. Foreign manufacturers faced with tariffs and quotas often begin manufacturing their product in the country imposing the tariffs and quotas. As tariffs and quotas apply to imported goods only, producing in the country imposing the quotas or tariffs effectively makes these trade barriers obsolete.

Companies also open manufacturing plants in other countries because of differences in labor costs among countries. While most manufacturing takes place in high-income countries, some low-income countries have become increasingly attractive as production locations because their workers can be hired much more cheaply than in high-income countries. For example, in late 2019, the average manufacturing job in the United States paid more than US$22.50 per hour. By comparison, manufacturing employees in the Philippines earned a few cents more than US$2.50 per hour.

This dramatic differences in labor costs have prompted some companies to close down their manufacturing plants in high-income countries and open up new plants in low-income countries. This has resulted in high-income countries purchasing more manufactured goods from low-income countries.

More than half the clothing imported into the United States came from Asian countries, for example, China, Taiwan, and South Korea, where labor costs were much lower than in the United States. Much of this clothing was made in factories where workers were paid by companies headquartered in the United States. For example, most of the Nike sports shoes that were sold in the United States were made in China, Indonesia, Vietnam, and Pakistan.

Transportation and Communications Technology

The ability of companies to have manufacturing plants in other countries stems from the fact that the world has a sophisticated and efficient transportation and communications system. An advanced transportation and communications system makes it relatively easy and relatively cheap to transfer information and goods between geographically distant locations. Thus, Nike can manufacture soccer balls in Pakistan and transport them quickly and cheaply to customers in the United States.

The extent to which transportation and communications systems have improved during the last two centuries can be illustrated by a few simple examples. In 1800, when the stagecoach was the primary method of overland transportation, it took twenty hours to travel the ninety miles from Lansing, Michigan, to Detroit, Michigan. Today, with the automobile, the same journey takes approximately ninety minutes. In 1800 sailing ships traveling at an average speed of ten miles per hour were used to transport people and goods between geographically distant countries. In the year 2019 jet-engine aircraft could traverse the globe at speeds in excess of 600 miles per hour. Communications technology has also improved over time.

In 1930 a three-minute telephone call between New York and London, England, cost more than US$385 in 2018 dollars. In the year 2019 the same telephone call could be made for less than a dime.

In addition to modern telephones, there are fax machines, email, videoconferencing capabilities, and a host of other technologies that make communication with other parts of the world both inexpensive and swift.

Future Prospects

The global economy of the twenty-first century presents a wide variety of opportunities and challenges. Sophisticated communications and transportation networks provide increasing numbers of manufacturing companies with more choices as to where to locate their factories. However, high-income countries like the United States are increasingly in competition with other countries (both high- and low-income) to maintain existing and manufacturing investments and attract new ones. Persuading existing companies to keep their U.S. factories open and not move overseas has been a major challenge. Likewise, making the United States as an attractive place for foreign companies to locate their manufacturing plants is an equally challenging task.

Neil Reid

Globalization of Manufacturing and Trade

Why are most of the patents issued worldwide assigned to Asian corporations? How did a Taiwanese earthquake prevent millions of Americans from purchasing memory upgrades for their computers? Why have personal incomes in Beijing nearly doubled in less than a decade?

Answers to these questions can be found in the geography of globalization. Globalization is an economic, political, and social process characterized by the integration of the world's many systems of manufacturing and trade into a single and increasingly seamless marketplace. The result: a new world geography.

This new geography is associated with the expansion of manufacturing and trade as capitalist principles replace old ideologies and state-controlled economies. With expanded free markets, the process of manufacturing and trading is constantly changing. Globalization delivers economic growth through improved manufacturing processes, newly developed goods, foreign investment in overseas manufacturing, and expanded employment.

The economies of developing countries are slowly transitioning from agricultural to industrial activities. Nevertheless, more than 65 percent of workers in these countries continue to work in agriculture. Meanwhile, developed countries, such as Australia and Germany, are experiencing high-technology service sector growth and reduced manufacturing employment. In the United States, nearly 30 percent of all workers were employed in manufacturing during the 1950s, but by 2019, less than 8.5 percent were.

In between these extremes, former state-controlled economies, like Romania, are adopting more efficient economic development strategies. Other nations and economic models, such as Indonesia and China, are pulled into the global marketplace by the growth and expansion of market economies. Despite the different economic paths of developing, transitioning, and developed nations, manufacturing and trade link all nations together and represent an economic convergence with important implications for political, business, and labor leaders—as well as all the world's citizens.

The geographies of manufacturing and trade can be examined as the distribution and location of economic activities in response to technological change and political and economic change.

Distribution and Location

Questions about where people live, work, and spend their money can be answered by reading product labels in any shopping mall, supermarket, or automobile dealership. They reveal the fact that manufacturing is a multistage process of component fabrication and final product assembly that can occur continents apart. For example, a shirt may be designed in New Jersey, assembled in Costa Rica from North Carolina fabric, and sold in British Columbia. To understand how goods produced in faraway locations are sold at neighborhood stores, geographers investigate the spatial, or geographic, distribution of natural resources, manufacturing plants, trading patterns, and consumption.

Historically, the geography of manufacturing and trade has been closely linked to the distribution of raw materials, workers, and buyers. In earlier times, this meant that manufacturing and trade were highly localized functions. In the eighteenth century, every North American town had cobblers or blacksmiths who produced goods from local resources for sale in local markets. By the start of the Industrial Revolution, improved transportation and manufacturing techniques had significantly enlarged the geography of manufacturing and trade. As distances increased, new manufacturing and trading centers developed. The location of these centers was contingent upon site and situation. Site and situation refer to a physical location, or site, relative to needed materials, transportation networks, and markets. For example, Pittsburgh, Pennsylvania, became the site of a major steel industry because it was near coal and iron resources. Pittsburgh also benefited from its historical role as a port town on a major river system that provided access to both western and eastern markets.

While relative location and transportation costs continue to be important factors, the geographic distribution of production and movement of goods across space is more complex than the simple calculus of site and situation. New global and local geographies of manufacturing and trade have been fueled by two major factors: technology and political change.

Technological Change

The old saying that time is money partially explains where goods are manufactured and traded. By compressing time and space, technology has enabled people, goods, and information to go farther more quickly. In the process, technology has reduced interaction costs, such as telecommunications. Just as steel enabled railroads to push farther westward, new technologies reduce the distance between places and people.

By increasing physical and virtual access to people, places, and things, technology has eliminated many barriers to global trade. However, improved telecommunications and transportation are only part of technology's contribution to globalization. If time is money, new efficient manufacturing processes also have reduced costs and facilitated globalization.

Armed with more efficient production processes, reliable telecommunications infrastructures, and transportation improvements, businesses can increase profits and remain competitive by seeking out lower-cost labor markets thousands of miles from consumers. As trade and manufacturing are increasingly spatially separate activities, the geographic distribution of manufacturing promotes an uneven distribution of income. The global distribution of manufacturing plants is closely related to industry-specific skill and wage requirements. For example, low-wage and low-skill jobs tend to concentrate in the developing regions of Asia, South America, and Africa. Alternately, high-technology and high-wage manufacturing activities concentrate in more developed regions.

In some cases, high wages and global competition force corporations to move their manufacturing plants to save costs and remain competitive. During the early 1990s, this byproduct of globalization was a major issue during the U.S. and Canadian debates to ratify the North American Free Trade Agreement (NAFTA). Focusing on primarily U.S. and Canadian companies that moved jobs to Mexico, the debate contributed to growing anxiety over job security as plants relocate to low-cost labor markets in South America and around the world.

As global competition increases, the geography of manufacturing and trade is increasingly global and rapidly changing. One company that has adapted to the shifting nature of global trade and manufacturing is Nike. Based in Beaverton, Oregon, Nike designs and develops new products at its Oregon world headquarters. However, Nike has internationalized much of its manufacturing capacity to compete in an aggressive athletic apparel industry. Over the last twenty-five years, Nike's strategy has meant shifts in production from high-wage U.S. locations to numerous low-wage labor markets around Pacific Rim.

Political and Economic Change: A New World Order

In order for companies such as Nike to successfully adapt to changing global dynamics, a stable international, or multilateral, trading system must be in place. In 1948 the General Agreement on Tariffs and Trade (GATT) was the first major step toward developing this stable global trading infrastructure. During that same period, the World Bank and International Monetary Fund were created to stabilize and standardize financial markets and practices. However, Cold War politics postponed complete economic integration for nearly half a century. Since the collapse of communism, globalization has accelerated as economies coalesce around the principles of free markets and capitalism. These important changes have become institutionalized through multilateral trade agreements and international trading organizations.

International trading organizations try to minimize or eliminate barriers to free and fair trade between nations. Trade barriers include tariffs (taxes levied on imported goods), product quotas, government subsidies to domestic industry, domestic content rules, and other regulations. Barriers prevent competitive access to domestic markets by artificially raising the prices of imported goods too high or preventing foreign firms from achieving economies of scale. In some cases, tariffs can also be used to promote fair trade by effectively leveling the playing field.

Because tariffs can be used both to promote fair trade and to unfairly protect markets, trading organizations are responsible for distinguishing between the two. For example, the Asian Pacific Economic Cooperation (APEC) forum has established guidelines to promote fair trade and attract foreign investment. APEC initiatives include a public Web-based database of member state tariff schedules and related links. Through programs such as the APEC information-sharing project, trading organizations are streamlining the international business process and promoting the overall stability of international markets.

The Future

As the globalization of manufacturing and trade continues, a new world geography is emerging. Unlike the Cold War's east-west geography and politics of ideology, an economic politics divides the developed and developing world along a north-south axis. While the types of conflicts associated with these new politics and the rules of engagement are unclear, it is evident that a new hierarchy of nations is emerging.

Globalization will raise the economic standard of living in most nations, but it has also widened the gap between richer and poorer countries. A small group of nations generates and controls most of the world's wealth. Conversely, the poorest countries account for roughly two-thirds of the world's population and less than 10 percent of its wealth.

This fundamental question of economic justice was a motive behind globalization's first major political clash. During the 1999 World Trade Organization (WTO) meetings in Seattle, Washington, approximately 50,000 environmentalists, labor unionists, and human and animal rights activists protested against numerous issues, including cultural intolerance, economic injustice, environmental degradation, political repression, and unfair labor practices they attribute to free trade. While the protesters managed to cancel the opening ceremonies, the United Nations secretary-general, Kofi Annan, expressed the general sentiment of most WTO member states. Agreeing that the protesters' concerns were important, Annan also asserted that the globalization of manufacturing and trade should not be used as a scapegoat for domestic failures to protect individual rights. More important, the secretary-general feared that those issues could be little more than a pretext for a return to unilateral trade policies, or protectionism.

Like the Seattle protesters, supporters of multilateral trade advocate political and economic reforms. Proponents emphasize that open markets promote open societies. Free traders earnestly believe economic engagement encourages rogue nations to improve poor human rights, environmental, and labor records. It is argued that economic engagement raises the expectations of citizens, thereby promoting change.

Conclusion
Technological and political change have made global labor and consumer markets more accessible and established an economic world hierarchy. At the top, one-fifth of the world's population consumes the vast majority of produced goods and controls more than 80 percent of the wealth. At the bottom of this hierarchy, poor nations are industrializing but possess less than 10 percent of the world's wealth. In political, social, and cultural terms, this global economic reality defines the contours and cleavages of a changing world geography. Whether geographers calculate the economic and political costs of a widening gap between rich and poor or chart the flow of funds from Tokyo to Toronto, the globalization of manufacturing and trade will remain central to the study of geography well into the twenty-first century.

Jay D. Gatrell

Modern World Trade Patterns

Trade, its routes, and its patterns are an integral part of modern society. Trade is primarily based on need. People trade the goods that they have, including money, to obtain the goods that they don't have. Some nations are very rich in agriculture or natural resources, while others are centers of industrial or technical activity. Because nations' needs change only slowly, trade routes and trading patterns develop that last for long periods of time.

Types of Trade
The movement of goods can occur among neighboring countries, such as the United States and Mexico, or across the globe, as between Japan and

Italy. Some trade routes are well established with regularly scheduled service connecting points. Such service is called liner service. Liners may also serve intermediate points along a trade route to increase their revenue.

Some trade occurs only seasonally, such as the movement of fresh fruits from Chile to California. Some trade occurs only when certain goods are demanded, such as special orders of industrial goods. This type of service is provided by operators called tramps. They go where the business of trade takes them, rather than along fixed liner schedules and routes.

Many people think of international trade as being carried on great ships plying the oceans of the world. Such trade is important; however, a considerable amount of trade is carried by other modes of transportation. Ships and airplanes carry large volumes of freight over large distances, while trucks, trains, barges, and even animal transport are used to move goods over trade routes among neighboring or landlocked countries.

Trade Routes

Through much of human history, trade routes were limited. Shipping trade carried on sailing vessels, for example, was limited by the prevailing winds that powered the ships. Land routes were limited by the location of water, mountain ranges, and the slow development of roads through thick forests and difficult terrain. The mechanization of transportation eventually freed ships and other forms of transport to follow more direct trade routes. Also, the development of canals and transcontinental highway systems allowed trade routes to develop based solely upon economic requirements.

Other changes in trade routes have occurred with industrialization of transport systems. The world began to have a great need for coal. Trade routes ran to the countries in which coal was mined. Ships and trains delivered coal to the power industry worldwide. Later, trade shifted to locations where oil (petroleum) was drilled. Now, oil is delivered to those same powerplants and industrial sites around the world.

Noneconomic Factors

Some trade is not purely economic in nature. Political relationships among countries can play an important part in their trade relations. For example, many national governments try to protect their countries' automobile and electronics industries from outside competition by not allowing foreign goods to be imported easily. Governments control imports by assessing duties, or tariffs, on selected imports.

Some national governments use the concept of cabotage to protect their home transportation industries by requiring that certain percentages of imported and exported trade goods be carried by their own carriers. For example, the U.S. government might require that 50 percent of its trade use American ships, planes, or trucks. The government might also require that all American carriers employ only American citizens.

Nations also can exert pressure on their trading partners by limiting access to port or airport facilities. Stronger nations may force weaker nations into accepting unequal trade agreements. For example, the United States once had an agreement with Germany concerning air passenger service between the two countries. The agreement allowed United States carriers to carry 80 percent of the passengers, while German carriers were permitted to carry only 20 percent of the passengers.

Multilateral Trade

In situations in which pairs of trading nations do not have direct diplomatic contact with each other, they make their trade arrangements through other nations. Such trade is referred to as multilateral. Certain carriers cater to this type of trade. They operate their ships or planes in around-the-world service. They literally travel around the globe picking up and depositing cargo along the way for a variety of nations.

Trade Patterns

For many years, world populations were coast centered. This means that most of the people in the country lived close to the coast. This was due primarily to the availability of water transportation systems to move both goods and people. At this time, major railroad, highway and airline systems did not exist. As railway and highway systems pushed into the interiors of nations, the population followed, and goods were needed as well as produced in these areas. Thus, over the years many inland population centers have developed that require transportation systems to move goods into and away from this area.

In these cases, international trade to these inland centers required the use of a number of different modes of transportation. Each of the different modes required additional paperwork and time for repackaging and securing of the cargo. For example, cargo coming off ships from overseas was unloaded and placed in warehouse storage. At some later time, it was loaded onto trucks that carried it to railyards. There it would be unloaded, stored, and then loaded onto railcars. At the destination, the cargo would once again be shifted to trucks for the final delivery. During the course of the trip, the cargo would have been handled a number of times, with the possibility of damage or loss occurring each time.

Containerization

As more goods began to move in international trade, the systems for packaging and securing of cargo became more standardized. In the 1960s, shipments began to move in containers. These are highway truck trailers which have been removed from the chassis leaving only the box. Container packaging has become the standard for most cargos moving today in both domestic and international trade. With the advent of containerization of cargo in international trade, cargo movements could quickly move intermodally. Intermodal shipping involves the movement of cargo by using more than a single mode of transportation.

Land, water, and air carriers have attempted to make the intermodal movement of cargo in international trade as seamless as possible. They have not only standardized the box for carrying cargo, but they have also standardized the handling equipment, so that containers move quickly from one mode to another. Advances in communications and

---

THE WORLD TRADE ORGANIZATION AND GLOBAL TRADING

In 1998 domestic political pressures and an expected domestic surplus of rice prompted the Japanese government to unilaterally implement a 355-percent tariff on foreign rice, violating the United Nations' General Agreement on Tariffs and Trade (GATT). On April 1, 1999, Japan agreed to return to GATT import levels and imposed new over-quota tariffs. While domestic Japanese politics could have prompted a trade war with rice-exporting countries, the crisis demonstrates how multilateral trading initiatives promote stability. Without an agreement, rice exporters might not have gained access to Japanese markets. By returning to GATT minimum quotas and implementing over-quota taxes, the compromise addressed the interests of both domestic and foreign rice growers.

---

electronic banking allow the paperwork and payments also to be completed and transferred rapidly.

As the demands for products have grown and as the size of industrial plants has grown, the size of movements of raw materials and containerized cargo has also grown. Thus, the sizes of the ships and trains required to move these large volumes of cargo have also increased.

The development of VLCC's (very large crude carriers) has allowed shippers to move large volumes of oil products. The development of large bulk carriers has allowed for the carriage of large volumes of dry raw materials such as grains or iron ore. These large vessels take advantage of what is known as economies of scale. Goods can be moved more cheaply when large volumes of them are moved at the same time. This is because the doubling of the volume of cargo moved does not double the cost to build or operate the vessels in which it is carried. This savings reduces the cost to move large volumes of cargo.

Intermodal Transportation

Intermodal transportation has allowed cargo to move seamlessly across both international boundaries and through different modes of transportation. This seamless movement has changed ocean trade routes over recent years.

The development of the Pacific Rim nations created a demand for trade between East Asia and both the United States and Europe. This trade has usually taken the all-water routes between Asia and Europe. Ships moving from East Asia across the Pacific Ocean pass through the Panama Canal and cross the Atlantic Ocean to reach Western Europe. This journey is in excess of 10,000 miles (16,000 km.) and usually takes about thirty days for most ships to complete. The all-water route from Asia to New York is similar. The distance is almost as great as that to Europe and requires about twenty-one to twenty-four days to complete.

Intermodal transportation has given shippers alternatives to all-water routes. A great volume of Asian goods is now shipped to such western U.S. ports as Seattle, Oakland, and Los Angeles, from which these goods are carried by trains across the United States to New York. The overall lengths of these routes to New York are only about 7,400 miles (12,000 km.) and take between only fifteen and nineteen days to complete. Cargos continuing to Europe are put back on ships in New York and complete their journeys in an additional seven to ten days. Such intermodal shipping can save as much as a week in delivery time.

Airfreight

Another changing trend in trade patterns is the development of airfreight as an international competitor. Modern aircraft have improved dramatically both in their ability to lift large weights of cargo as well as their ability to carry cargos over long distances. Because of the speed at which aircraft travel in comparison to other modes of transportation, goods can be moved quickly over large distances. Thus, high-value cargos or very fragile cargos can move very quickly by aircraft.

The drawback to airfreight movement of cargo is that it is more expensive than other modes of travel. However, for businesses that need to move perishable commodities, such as flowers of the Netherlands, or expensive commodities, such as Paris fashions or Singapore-made computer chips, airfreight has become both economic and essential.

Robert J. Stewart

POLITICAL GEOGRAPHY

FORMS OF GOVERNMENT

Philosophers and political scientists have studied forms of government for many centuries. Ancient Greek philosophers such as Plato and Aristotle wrote about what they believed to be good and bad forms of government. According to Plato's famous work, *The Republic*, the best form of government was one ruled by philosopher-kings. Aristotle wrote that good governments, whether headed by one person (a kingship), a few people (an aristocracy), or many people (a polity), were those that ruled for the benefit of all. Those that were based on narrow, selfish interests were considered bad forms of government, whether ruled by an individual (a tyranny), a few people (an oligarchy), or many people (a democracy). Thus, democracy was not always considered a good form of government.

Constitutions and Political Institutions

All governments have certain things in common: institutions that carry out legislative, executive, and judicial functions. How these institutions are supposed to function is usually spelled out in a country's constitution, which is a guide to organizing a country's political system. Most, but not all, countries have written constitutions. Great Britain, for example, has an unwritten constitution based on documents such as the Magna Carta, the English Bill of Rights, and the Treaty of Rome, and on unwritten codes of behavior expected of politicians and members of the royal family.

The world's oldest written constitution still in use is that of the United States. All countries have written or unwritten constitutions, and most follow them most of the time. Some countries do not follow their constitutions—for example, the Soviet Union did not; other countries, for example France, change their constitutions frequently.

Constitutions usually first specify if the country is to be a monarchy or a republic. Few countries still have monarchies, and those that do usually grant the monarch only ceremonial powers and duties. Countries with monarchies at the beginning of the twenty-first century included Spain, Great Britain, Lesotho, Swaziland, Sweden, Saudi Arabia, and Jordan. Most countries that do not have monarchies are republics.

Constitutions also specify if power is to be concentrated in the hands of a strong national government, which is a unitary system; if it is to be divided between a national and various subnational governments such as states, provinces, or territories, which is a federal system; or if it is to be spread among various subnational governments that might delegate some power to a weak national government, which is a confederate system.

Examples of countries with unitary systems include Great Britain, France, and China; federal systems include the United States, Germany, Russia, Canada, India, and Brazil. There were no confederate systems by the third decade of the twenty-first century, although there are examples from history as well as confederations of various groups and nations. The United States under its eighteenth-century Articles of Confederation and the nineteenth-century Confederate States of America, made up of the rebelling Southern states, were confederate systems. Switzerland was a confederation for much of the nineteenth century. The concept of dividing power between the national and subnational governments is called the vertical axis of power.

419

MONARCHIES OF THE WORLD

Realm/Kingdom	Monarch	Type
Principality of Andorra	Co-Prince Emmanuel Macron; Co-Prince Archbishop Joan Enric Vives Sicília	Constitutional
Antigua and Barbuda	Queen Elizabeth II	Constitutional
Commonwealth of Australia	Queen Elizabeth II	Constitutional
Commonwealth of the Bahamas	Queen Elizabeth II	Constitutional
Barbados	Queen Elizabeth II	Constitutional
Belize	Queen Elizabeth II	Constitutional
Canada	Queen Elizabeth II	Constitutional
Grenada	Queen Elizabeth II	Constitutional
Jamaica	Queen Elizabeth II	Constitutional
New Zealand	Queen Elizabeth II	Constitutional
Independent State of Papua New Guinea	Queen Elizabeth II	Constitutional
Federation of Saint Kitts and Nevis	Queen Elizabeth II	Constitutional
Saint Lucia	Queen Elizabeth II	Constitutional
Saint Vincent and the Grenadines	Queen Elizabeth II	Constitutional
Solomon Islands	Queen Elizabeth II	Constitutional
Tuvalu	Queen Elizabeth II	Constitutional
United Kingdom of Great Britain and Northern Ireland	Queen Elizabeth II	Constitutional
Kingdom of Bahrain	King Hamad bin Isa	Mixed
Kingdom of Belgium	King Philippe	Constitutional
Kingdom of Bhutan	King Jigme Khesar Namgyel	Constitutional
Brunei Darussalam	Sultan Hassanal Bolkiah	Absolute
Kingdom of Cambodia	King Norodom Sihamoni	Constitutional
Kingdom of Denmark	Queen Margrethe II	Constitutional
Kingdom of Eswatini	King Mswati III	Absolute
Japan	Emperor Naruhito	Constitutional
Hashemite Kingdom of Jordan	King Abdullah II	Constitutional
State of Kuwait	Emir Sabah al-Ahmad	Constitutional
Kingdom of Lesotho	King Letsie III	Constitutional
Principality of Liechtenstein	Prince Regnant Hans-Adam II (Regent: The Hereditary Prince Alois)	Constitutional
Grand Duchy of Luxembourg	Grand Duke Henri	Constitutional
Malaysia	Yang di-Pertuan Agong Abdullah	Constitutional
Principality of Monaco	Sovereign Prince Albert II	Constitutional

MONARCHIES OF THE WORLD *(continued)*

Realm/Kingdom	Monarch	Type
Kingdom of Morocco	King Mohammed VI	Constitutional
Kingdom of the Netherlands	King Willem-Alexander	Constitutional
Kingdom of Norway	King Harald V	Constitutional
Sultanate of Oman	Sultan Haitham bin Tariq	Absolute
State of Qatar	Emir Tamim bin Hamad	Mixed
Kingdom of Saudi Arabia	King Salman bin Abdulaziz	Absolute theocracy
Kingdom of Spain	King Felipe VI	Constitutional
Kingdom of Sweden	King Carl XVI Gustaf	Constitutional
Kingdom of Thailand	King Vajiralongkorn	Constitutional
Kingdom of Tonga	King Tupou VI	Constitutional
United Arab Emirates	President Khalifa bin Zayed	Mixed
Vatican City State	Pope Francis	Absolute theocracy

Whether governments share power with subnational governments or not, there must be institutions to make laws, enforce laws, and interpret laws: the legislative, executive, and judicial branches of government. How these branches interact is what determines whether governments are parliamentary, presidential, or mixed parliamentary-presidential. In a presidential system, such as in the United States, the three branches—legislative, executive, and judicial—are separate, independent, and designed to check and balance each other according to a constitution. In a parliamentary system, the three branches are not entirely separate, and the legislative branch is much more powerful than the executive and judicial branches.

Great Britain is a good example of a parliamentary system. Some countries, such as France and Russia, have created a mixed parliamentary-presidential system, wherein the three branches are separate but are not designed to check and balance each other. In a mixed parliamentary-presidential system, the executive (led by a president) is the most powerful branch of government.

Looking at political systems in this way—how the legislative, executive, and judicial branches of government interact—is to examine the horizontal axis of power. All governments are unitary, federal, or confederate, and all are parliamentary, presidential, or mixed parliamentary-presidential. One can find examples of different combinations. Great Britain is unitary and parliamentary. Germany is federal and parliamentary. The United States is federal and presidential. France is unitary and mixed parliamentary-presidential. Russia is federal and mixed parliamentary-presidential. Furthermore, virtually all countries are either republics or monarchies.

Types of Government

Constitutions describe how the country's political institutions are supposed to interact and provide a guide to the relationship between the government and its citizens. Thus, while governments may have similar political institutions—for example, Germany and India are both federal, parliamentary republics—how the leaders treat their citizens can vary widely. However, governments may have political systems that function similarly although they have different forms of constitutions and institutions. For example, Great Britain, a unitary, parliamentary monarchy with an unwritten constitution, treats its citizens very similarly to the United States, which is a federal, presidential republic with a written constitution.

The three most common terms used to describe the relationships between those who govern and those who are governed are democratic, authoritarian, and totalitarian. Characteristics of democracies are free, fair, and meaningfully contested elections; majority rule and respect for minority rights and opinions; a willingness to hand power to the opposition after an election; the rule of law; and civil rights and liberties, including freedom of speech and press, freedom of association, and freedom to travel. The United States, Canada, Japan, and most European countries are democratic.

An authoritarian system is one that curtails some or all of the characteristics of a democratic regime. For example, authoritarian regimes might permit token electoral opposition by allowing other political parties to run in elections, but they do not allow the opposition to win those elections. If the opposition did win, the authoritarian regime would not hand over power. Authoritarian regimes do not respect the rule of law, the rights of minorities to dissent, or freedom of the press, speech, or association. Authoritarian governments use the police, courts, prisons, and the military to intimidate and threaten their citizens, thus preventing people from uniting to challenge the existing political rulers. Afghanistan, Cuba, Iran, Uzbekistan, Saudi Arabia, Chad, Syria, Libya, Sudan, Belarus, and China are examples of countries with authoritarian regimes.

Totalitarian regimes are similar to authoritarian regimes but are even more extreme. Under a totalitarian regime, there is no legal opposition, no freedom of speech, and no rule of law whatsoever. Totalitarian regimes attempt to control totally all members of the society to the point where everyone always must actively demonstrate their loyalty to and support for the regime. Nazi Germany under Adolf Hitler's rule (1933-1945) and the Soviet Union under Joseph Stalin's rule (1928-1953) are examples of totalitarian regimes. As of 2019, only Eritrea and North Korea are the still have governments classified as totalitarian dictatorships.

Forms of Government: Putting it All Together

In *The Republic*, Plato asserts that people have varied dispositions, and, therefore, there are various types of governments. In recent years, regimes have been created that some call mafiacracies (rule by criminal mafias), narcocracies (rule by narcotics gangs), gerontocracies (rule by very old people), theocracies (rule by religious leaders), and so forth. Such variations show the ingenuity of the human mind in devising forms of government.

Whatever labels that are given to a political system, there remain basic questions to be asked about that regime: Is it a monarchy or a republic? Is all power concentrated in the hands of a national government, or is power shared between a national government and the states or provinces? Are its institutions those of a parliamentary, presidential, or mixed parliamentary-presidential system? Is it democratic, authoritarian, or totalitarian? Finally, does it live up to its constitution, both in terms of how power is supposed to be distributed among institutions and in its relationship between the government and the people? To paraphrase Aristotle, how many rulers are there, and in whose interests do they rule?

Nathaniel Richmond

POLITICAL GEOGRAPHY

Students of politics have been aware that there is a significant relationship between physical and political geography since the time of ancient Greece.

The ancient Greek philosopher Plato argued that a *polis* (politically organized society) must be of limited geographical size and limited population or it would lack cohesion. The ideal *polis* would be only as geographically large as required to feed about 5,000 people, its maximum population.

Plato's illustrious pupil, Aristotle, agreed that stable states must be small. "One can build a wall around the Hellespont," the main territory of ancient Greece, he wrote in his treatise *Politics*, "but that will not make it a polis." Today human ideas differ about the maximum area of a successful state or nation-state, but the close influence of physical geography on political geography and their profound mutual effects on politics itself are not in question.

Geographical Influences on Politics

The physical shape and contours of states may be called their physical geography; the political shape and contours of states, starting with their basic structure as unified state, federation, or confederation, are primary features of their political geography. The idea of "political geography" also can refer to variations in a population's political attitudes and behavior that are influenced by geographical features. Thus, the combination of plentiful land and sparse population tend toward an independent spirit, especially where the economy is agriculturally based. This has historically been the case in the western United States; in the Pampas region of Argentina, where cattle are raised by independent-minded gauchos (cowboys); and on the Brazilian frontier, where government regulation is routinely resisted.

Likewise, where physical geography presents significant difficulties for inhabitants in earning a living or associating, as where there is rough terrain and poor soil or inhospitable climate, the populace is likely to exhibit a hardy, self-reliant character that strongly influences political preferences. Thus, physical geography helps to shape national character, including aspects of a nation's politics.

Furthermore, it is well known that where physical geography isolates one part of a country's population from the rest, political radicalism may take root. This tendency is found in coastal cities and remote regions, where labor union radicalism has often been pronounced. Populations in coastal loca-

tions with access to foreign trade often show a more liberal, tolerant, and outgoing spirit, as reflected in their political opinions. In ancient Greece, the coastal access enjoyed by Athens through a nearby port in the fifth century BCE had a strong influence on its liberal and democratic political order. In modern times, China's coastal cities, such as Tientsin, and North American cities such as San Francisco, show similar influences.

The Geographical Imperative

In many instances, political geography is shaped by what may be called the "geographical imperative." Physical geography in these instances demands, or at least strongly suggests, that political geography follow its course. The numerous valleys of mountainous Greece strongly influenced the emergence of the small, often fiercely independent, polis of ancient times. The formation and borders of Asian states such as Bhutan, Nepal, and Tibet have been strongly influenced by the Himalaya Mountains, and the Alps, which shape Switzerland.

As another example, physical geography demands that the land between the Pacific Ocean and the Andes Mountains along the western edge of South America be organized as a separate country—Chile. Island geography often plays a decisive role in its political geography. The qualified political unity of Great Britain can be directly traced to its insular status. Small islands often find themselves combined into larger units, such as the Hawaiian Islands.

The absence of the geographical imperative, however, leaves political geography an open question. For example, Indonesia comprises some 1,300 islands stretching 3,000 miles in bodies of water such as the Indian Ocean and the Celebes Sea. With so many islands, Indonesia lacks a geographical imperative to be a unified state. It also lacks the imperative of ethnic and cultural homogeneity and cohesion, a circumstance mirrored in its political life, since it has remained unified only through military force. As control by the military waned after the fall of the authoritarian General Suharto in 1998, conflicts among the nation's diverse peoples have threatened its breakup. No such threat, however, confronts Australia, an immense island continent where a European majority dominates a fragmented and primitive aboriginal minority. In Australia, the geographical imperative suggests a unity supported by the cultural unity of the majority.

As many examples show, the geographical imperative is not absolute. For example, mountainous Greece is politically united in the twenty-first century. Although long shielded geographically, Tibet lost its political independence after it was successfully invaded by China. The formerly independent Himalayan state Sikkim was taken over by India. Thus, political will trumps physical geography.

The frequency of exceptions to the geographical imperative illustrates that human freedom, while not unlimited, often plays a key role in shaping political geography. As one example, the Baltic Republics—Lithuania, Latvia, and Estonia—historically have been dominated, or largely swallowed up, by neighboring Russia. By the start of the twenty-first century, however, they had regained their independence through the political will to self-rule and the drive for cultural survival.

Strategically Significant Locations

Locations of great economic or military significance become focal points of political attention and, potentially, of military conflict. There are innumerable such places in the world, but several stand out as models of how important physical geography can be for political geography in the context of international politics.

One significant example is the Panama Canal, without which ships must sail around South America. The Suez Canal, which connects European and Asian shipping, is a similar waterway, saving passage around Africa. The canal's significance was reduced after 1956, however, when its blockage after the Arab-Israeli war of that year led to the building of supertankers too large to traverse it. Another example is Gibraltar, whose fortifications command the entrance to the Mediterranean Sea from the Atlantic Ocean. A final example is the Bosporus, the tiny entrance from the Black Sea to waters leading to the Mediterranean Sea. It is the only warm-water route to and from Eastern Russia and therefore is of great military and economic importance for regional and world power politics.

Charles F. Bahmueller

GEOPOLITICS

Geopolitics is a concept pertaining to the role of purely geographical features in the relations among states in international politics. Geopolitics is especially concerned with the geographical locations of the states in relationship to one another. Geopolitical relationships incorporate social, economic, political, and historical features of the states that interact with purely geographical elements to influence the strategic thinking and behavior of nations in the international sphere.

Coined in 1899 by the Swedish theorist Rudolf Kjellen, the term "geopolitics" combines the logic of the search for security and competition for dominance among states with geographical methodology. *Geopolitics* must not, however, be confused with *political geography*, which focuses on individual states' territorial sizes, boundaries, resources, internal political relations, and relations with other states.

Geopolitical is a term frequently used by military and political strategists, politicians and diplomats, political scientists, journalists, statesmen, and a variety of other government officials, such as policy planners and intelligence analysts.

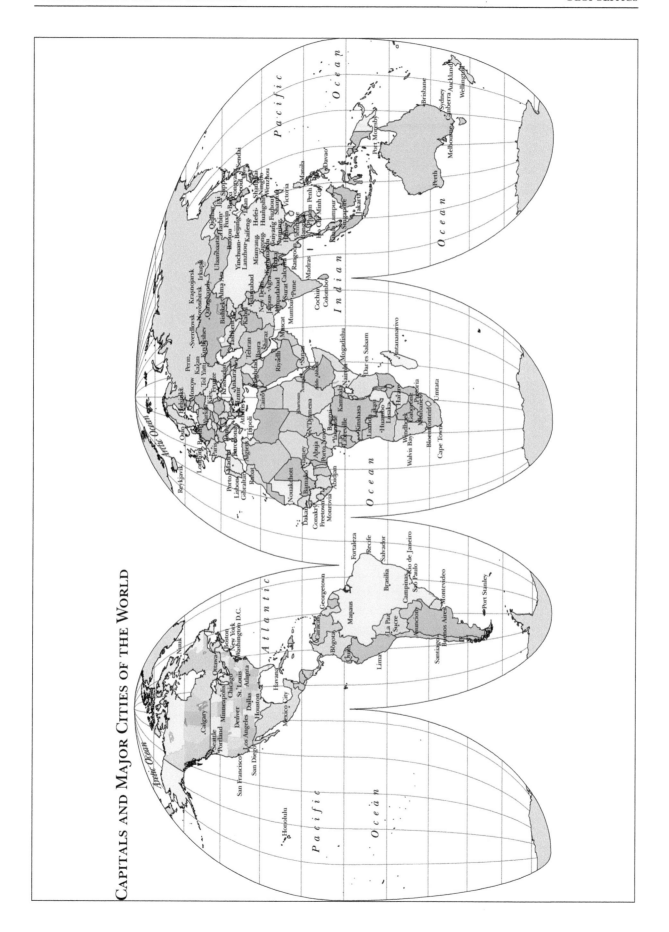

CAPITALS AND MAJOR CITIES OF THE WORLD

Power Struggles Among States

The idea of geopolitics arises in the course of what might be considered the universal struggle for power among the world's most powerful nations, which compete for political and military leadership. How one state can threaten another, for example, is often influenced by geographical factors in combination with technological, social, economic and other factors. The extent to which individual states can threaten each other depends in no small measure on purely geographical considerations.

By the close of twentieth century the Cold War that had dominated world security concerns was over. Nevertheless, the United States still worried about the danger of being attacked by nuclear missiles fired, not by the former Soviet Union, but by irresponsible, fanatical, or suicidal states. American political leaders and military planners were concerned with the geographical position of so-called "rogue states." or "states of concern." In 1994, North Korea, Cuba, Iran, Libya under Muammar Gaddafi, and Ba'athist Iraq were listed as states of concern. By 2019, a list of state sponsors of terrorism included Iran, North Korea, Sudan, and Syria.

Geographical factors play prominent roles in assessments of the different threats that those states presented to American interests. How far those states are located from American territory determines whether their missiles might pose a serious threat. A missile may be able to reach only the periphery of U.S. soil, or it might be able to carry only a small payload. Similar considerations determine the threat such states pose for U.S. forces stationed abroad, as well as for such important U.S. allies as Japan, Western Europe, or Israel. Such questions are thus said to constitute geopolitical, or geostrategic, considerations.

There are many examples of the influence of geopolitical factors on international relations among nations in the past. For example, the Bosporus, the narrow sea lane linking the Black Sea and the Mediterranean where Istanbul is situated, has long been considered of great strategic importance. In the nineteenth century, the Bosporus was the only direct route through which the Russian

---

A PEACEFULLY RESOLVED BORDER DISPUTE

The peaceful resolution of the border dispute between the Southern African states of Botswana and Namibia was hailed by observers of African politics. Instead of resorting to the armed warfare that so often has marked similar disputes on the continent, the two states chose a different course in 1996, when they found negotiations stalemated. They submitted their claims to the International Court of Justice in The Hague and agreed to accept the court's ruling. Late in 1999, by an eleven-to-four vote, the court ruled for Botswana, and Namibia kept its word to embrace the decision. At issue was a tiny island in the Chobe River on Botswana's northern border. An 1890 treaty between colonial rulers Great Britain and Germany had described the border at the disputed point vaguely, as the river's "main channel." The court took the course of the deepest channel to mark the agreed boundary, giving Botswana title to the 1.4-square-mile (3.5-sq. km.) territory.

---

navy could reach southern Europe and the Mediterranean Sea.

Because of Russia's nineteenth century history of expansionism and its integration into the pre-World War I European state system, with its networks of competing military alliances, the Bosporus took on added geopolitical meaning. It was the congested (and therefore vulnerable) space through which Russian naval power had to pass to reach the Mediterranean.

Historical Origins of Geopolitics

Although political geography was a well-established field by the late nineteenth century, geopolitics was just beginning to emerge as a field of study and political analysis at the end of the century. In 1896 the German theorist Friedrich Ratzel published his *Political Geography*, which put forward the idea of the state as territory occupied by a people bound together by an idea of the state. Ratzel's theory embraced Social Darwinist notions that justified the current boundaries of nations. Ratzel viewed the state as a biological organism in competition for land with other states. The ethical implication of his theory seemed to be that "might makes right."

That theme set the stage for later German geopolitical thought, especially the notion of the need for *Lebensraum* (living room)—space into which the people of a nation could expand. German dictator Adolf Hitler justified his attack on Russia during World War II partly upon his claim that the German people needed more *Lebensraum* to the east. To some modern geographers, the use of geopolitical theories to serve German fascism and to justify other instances of military aggression tarnished geopolitics itself as a field of study.

Historical Development of Geopolitics

Modern geopolitics has further origins in the work of the Scottish geographer Sir Halford John Mackinder. In 1904 he published a seminal article, "The Geographical Pivot of History," in which he argued that the world is made up of a Eurasian "heartland" and a secondary hinterland (the remainder of the world), which he called the "marginal crescent." According to his theory, international politics is the struggle to gain control of the heartland. Any state that managed that feat would dominate the world.

A major proposition of Mackinder's theory was that geographical factors are not merely causative factors, but coercive. He tried to describe the physical features of the world that he believed directed human actions. In his view, "Man and not nature initiates, but nature in large measure controls." Geopolitical factors were therefore to a great extent determinants of the behavior of states. If this were true, geopolitics as a science could have deep relevance and corresponding influence among governments.

After Mackinder's time, the concept of geopolitics had a double significance. On the one hand it was a purely descriptive theory of geographic causation in history. On the other hand, its purveyors also believed, as Mackinder argued in 1904, that geopolitics has "a practical value as setting into perspective some of the competing forces in current international politics." Mackinder sought to promote this field of study as a companion to British state-craft, a tool to further Britain's national interest. By extension, geopolitical theory could assist any government in forming its political/military strategy.

As applied to the early twentieth-century world of international politics, however, Mackinder's theory had major weaknesses. Among his most glaring oversights were his failure to appreciate the rise of the United States, which attained considerable naval power after the turn of the century. Also, he failed to foresee the crucial strategic role that air power would play in warfare—and with it the immense change that air power could make in geopolitical considerations. Air power moves continents closer together, revolutionizing their geopolitical relationships.

One of Mackinder's chief critics was Nicolas John Spykman. Spykman argued that Mackinder had overvalued the potential economic, and therefore political, power of the Eurasian heartland, which could never reach its full potential because it could not overcome the obstacles to internal transportation. Moreover, the weaknesses of the remainder of the world—in effect, northern, western and southern Europe—could be overcome through forging alliances.

The dark side of geopolitical thought as handmaiden to political and military strategy became apparent in the Germany of the 1920s. At that time German theorists sought the resurrection of a German state broken by failure in World War I, the harsh terms of the Versailles Treaty that ended the war, and the hyperinflation that followed, wiping out the German middle class. In his 1925 article "Why Geopolitik?" Karl Haushofer urged the practical applications of *Geopolitik*. He urged that this form of analysis had not only "come to stay" but could also form important services for German political leaders, who should use all available tools "to carry on the fight for Germany's existence."

Haushofer ominously suggested that the "struggle" for German existence was becoming increasingly difficult because of the growth of the country's population. A people, he wrote, should study the living spaces of other nations so it could be prepared to "seize any possibility to recover lost ground." This discussion clearly implied that, from geopolitical necessity, Germany should seek additional territory to feed itself—a view carried into effect by Hitler in his quest for *Lebensraum* in attacking the Soviet Union, including its wheat-producing breadbasket, the Ukraine.

After World War II, a chastened Haushofer sought to soft-pedal both the direction and influence of his prewar writings. However, Hitler's morally heinous use of *Geopolitik* left geopolitical theorizing permanently tainted, in some eyes. Nevertheless, there is no necessary connection between geopolitics as a purely analytic description and geopolitics as the basis for a selfish search for power and advantage.

Geopolitics in the Twenty-first Century

Geopolitical considerations were unquestionably of profound relevance to the principal states of the post-World War II Cold War period. After the fall of the Berlin Wall in 1989, however, some theorists thought that the age of geopolitics had passed. In 1990 American strategic theorist Edward N. Luttwak, for example, argued that the importance of military power in international affairs had declined precipitously with the winding down of the Cold War. Military power had been overtaken in significance by economic prowess. Consequently, geopolitics had been eclipsed by what Luttwak called "geoeconomics," the waging of geopolitical struggle by economic means.

The view of Luttwak and various geographers of the declining significance of military power and geopolitical analysis, however, was soon proved to be overdrawn by events. As early as the first months of 1991, before the Soviet Union was officially dismantled, military power asserted itself as a key determinant on the international scene. Led by the United States, a far-flung alliance of nations participated in a war to remove Iraqi dictator Saddam Hussein's forces from neighboring Kuwait, which Iraq had illegally occupied. The decisive and successful use of military power in that war dramatically disproved assertions of its growing irrelevance.

Similarly, in the first three decades of the twenty-first century, military power retained its pre-eminence in the dynamics of international politics, even as economic forces were seen to gather momentum. To states throughout Asia and the West (especially Western Europe and the United States), the relative military capability of potential adversaries, and therefore geopolitics, remained a vital feature of the international order. Central to this view of the world scene is the growing military rivalry of the United States and China in East Asia. As China modernizes and expands its nuclear and conventional forces, it may feel itself capable of challenging America's predominant military power and prestige in East Asia. This possibility heightens the use of geopolitical thinking, giving it currency in analyzing this emerging situation.

Geopolitics as Civilizational Clash

A sometimes controversial expression of geopolitical analysis has been offered by Samuel Huntington of Harvard University. In his *The Clash of Civilizations and the Remaking of World Order* (1996) Huntington constructs a theory to explain certain tendencies of international behavior. He divides the world into a number of cultural groupings, or "civilizations," and argues that the character of various international conflicts can best be explained as conflicts or clashes of civilizations. In his view, Western civilization differs from the civilization of Orthodox Christianity, with a variety of conflicts erupting between the two. An example is the attack by the North Atlantic Treaty Organization (NATO), the bastion of the West, on Serbia, which is part of the Orthodox East.

Huntington's other civilizations include Islamic, Jewish, Eastern Caribbean, Hindu, Sinic (Chinese), and Japanese. The clash between Israel and its neighbors, the struggle between Pakistan and India over Kashmir, the rivalries between the United States and China and between China and India, for example, can be viewed as civilizational conflicts. Huntington has stated, however, that his theory is not intended to explain all of the historical past, and he does not expect it to remain valid long into the future.

Charles F. Bahmueller

NATIONAL PARK SYSTEMS

The world's first national parks were established as a response to the exploitation of natural resources, disappearance of wildlife, and destruction of natural landscapes that took place during the late nineteenth century. Government efforts to preserve natural areas as parks began with the establishment of Yellowstone National Park in the United States in 1872 and were soon adopted in other countries, including Australia, Canada, and New Zealand.

While the preservation of nature continues to be an important benefit provided by national parks, worldwide increases in population and the pressures of urban living have raised public interest in setting aside places that provide opportunities for solitude and interaction with nature.

Because national parks have been established by nations with diverse cultural values, land resources, and management philosophies, there is no single definition of what constitutes a national park. In some countries, areas used principally for recreational purposes are designated as national parks; other countries emphasize preservation of outstanding scenic, geologic, or biological resources. The terminology used for national parks also varies among countries. For example, protected areas that are similar to national parks may be called reserves, preserves, or sanctuaries.

Diverse landscapes are protected within national parks, including swamps, river deltas, dune areas, mountains, prairies, tropical rainforests, temperate forests, arid lands, and marine environments. Individual parks within nations form networks that vary with respect to size, accessibility, function, and the type of natural landscapes preserved. Some national park areas are isolated and sparsely populated, such as Greenland National Park; others, such as Peak District National Park in Great Britain, contain numerous small towns and are easily accessible to urban populations.

The functions of national parks include the preservation of scenic landscapes, geological features, wilderness, and plants and animals within their natural habitats. National parks also serve as outdoor laboratories for education and scientific research and as reservoirs for genetic information. Many are components of the United Nations International Biosphere Reserve Program.

National parks also play important roles in preserving cultures, by protecting archaeological, cultural, and historical sites. The United Nations recognizes several national parks that possess important cultural attributes as World Heritage Sites. Tourism to national parks has become important to the economies of many developing nations, especially in Eastern and Southern Africa, India, Nepal, Ecuador, and Indonesia. Parks are sources of local employment and can stimulate improvements to transportation and other types of infrastructure while encouraging productive use of lands that are of marginal agricultural use.

The International Union for Conservation of Nature has developed a system for classifying the world's protected areas, with Category II areas designated as national parks. Using this definition, there are 3,044 national parks in the world, with a mean average size of 457 square miles (1,183 sq. km.) each. Together, they cover an area of about 1.5 million square miles (4 million sq. km.), accounting for about 2.7 percent of the total land area on Earth.

STEPHEN T. MATHER AND THE U.S. NATIONAL PARK SERVICE

In 1914 businessman and conservationist Stephen T. Mather wrote to Secretary of the Interior Franklin K. Lane about the poor condition of California's Yosemite and Sequoia National Parks. Lane wrote back, "if you don't like the way the national parks are being run, come on down to Washington and run them yourself." Mather accepted the challenge and became an assistant to Lane and later the first director of the U.S. National Park Service, from 1917 to 1929.

North America

In 1916 management of U.S. national parks and monuments was shifted from the U.S. Army to the newly established National Park Service (NPS). The system has since grown in size to protect sixty-one national parks, as well as other natural areas including national monuments, seashores, and preserves.

North America's second-largest system of national parks is Parks Canada, created in 1930. Among the best-known Canadian parks is Banff, established in southern Alberta in 1885. Preserved within this area are glacially carved valleys, evergreen forests, and turquoise lakes. Parks Canada has the goal of protecting representative examples of each of Canada's vegetation and physiographic regions.

Mexico began providing protection for natural areas in the late nineteenth century. Among its system of sixty-seven national parks is Dzibilchaltún, an important Mayan archaeological site on the Yucatán Peninsula. With fewer resources available for park management, the emphasis in Mexico remains the preservation of scenic beauty for public use.

South America

Two of South America's best-known national parks are located within Argentina's park system. Nahuel Huapi National Park preserves two rare deer species of the Andes, while Iguazú National Park, located on the border with Brazil, is home to tapir, ocelot, and jaguar.

Located on a plateau of the western slope of the Andes Mountains in Chile, Lauca National Park is one of the world's highest parks, with an average elevation of more than 14,000 feet (4,267 meters)-an altitude nearly as high as the tallest mountains in the continental United States. Huascarán, another mountain park located in western Peru, boasts twenty peaks that exceed 19,000 feet (5,791 meters) in elevation. The volcanic islands of Galapagos Islands National Park, managed by Ecuador, have been of interest to biologists since British naturalist Charles Darwin studied variation and adaptation in animal species there in 1835.

Australia and New Zealand

Established in 1886, Royal was Australia's first national park. Perhaps better known to tourists, Uluru National Park in Australia's Northern Territory protects two rock domes, Ayer's Rock and Mount Olga, that rise above the plains 15 miles (40 km.) apart.

Along with Australia and other former colonies of Great Britain, New Zealand was a leader in establishing early national parks. The first of these was Tongagiro, created in 1887 to protect sacred lands of the Maori people on the North Island. New Zealand's South Island features several national parks including Fiordland, created in 1904 to preserve high mountains, forests, rivers, waterfalls, and other spectacular features of glacial origin.

Africa

Game poaching continues to be a severe problem in Africa, where animals are slaughtered for ivory, meat, and hides. Many African national parks were established to protect large game. South Africa's national park system began in 1926, when the Sabie Game Preserve of the eastern Transvaal region became Kruger National Park. Among South Africa's greatest attractions to foreign visitors, Kruger is famous for its population of lions and elephants.

East Africa is also known for outstanding game sanctuaries, such as Serengeti National Park, created prior to Tanzania's independence from Great Britain. Another national park in Tanzania, Kilimanjaro, protects Africa's highest and best-known mountain. Other African countries with well-developed park systems include Kenya, the Democratic Republic of the Congo (formerly Zaire), and Zambia. Although there is now a network of national parks in Africa that protects a wide range of habitats in various regions, there remains a need to protect additional areas in the arid northern part of the continent that includes the Sahara Desert.

Europe

In comparison with the United States, the national park concept spread more slowly within Europe. In 1910 Germany set aside Luneburger Heide National Park near the Elbe River, and in 1913, Sweden established Sarek, Stora Sjöfallet, Peljekasje, and Abisko National Parks. Swiss National Park was founded in Switzerland in 1914, in the Lower Engadine region. Great Britain has several national parks, including Lake District, a favorite recreation destination for English poet William Wordsworth. Spain's Doñana National Park, located on its southwestern coast, preserves the largest dune area on the European continent.

Asia

The system of land tenure and rural economy in many Asian countries has made it difficult for national governments to set aside large areas free from human exploitation. Many national parks established by colonial powers prior to World War II were maintained or expanded by countries following independence. For example, Kaziranga National Park is a refuge for the largest heard of rhinocerous in India. Established in 1962, Thailand's Khao Yai National Park protects a sample of the country's wildlife, while Indonesia's Komodo Island National Park preserves the habitat for the large lizards known as Komodo dragons.

In Japan, high population density has made it difficult to limit human activities within large areas. Some Japanese national parks are principally recreation areas rather than wildlife sanctuaries and may contain cultural features such as Shinto shrines. One of the best known national parks in Japan is Fuji-Hakone-Izu, which contains world-famous Mount Fuji, a volcano with a nearly symmetrical shape.

The Future

National parks serve as relatively undisturbed enclaves that protect examples of the world's most outstanding natural and cultural resources. The movement to establish these areas is a relatively recent attempt to achieve an improved balance between human activities and the earth. In recent years, rising incomes and lower costs for international travel have improved the accessibility of national parks to a larger number of persons, meaning that park visitation is likely to continue to rise.

Thomas A. Wikle

BOUNDARIES AND TIME ZONES

INTERNATIONAL BOUNDARIES

International boundaries are the marked or imaginary lines traversing natural terrain of land or water that mark off the territory of one politically organized society—a state or nation-state—from other states. In addition, states claim "air boundaries." While satellites circumnavigate the earth without nations' permission, airplanes and other air vessels that fly much lower must gain the permission of states over whose territory they travel.

The existence of international boundaries is a consequence of the "territoriality" that is a feature of modern human societies. All politically organized societies, except for nomadic tribes, claim to rule some exactly defined geographical territory. International boundaries provide the limits that define this territory.

International boundaries have ancient origins. For example, the oldest sections of the Great Wall of China date back to the Ch'in Dynasty of the second century BCE. The Roman Empire also maintained boundaries to its territories, such as Hadrian's Wall in the north of England, built by the Romans in 122 CE as a defensive barrier against marauders. In these and other ancient instances, however, there was little thought that borders must be exact.

The existence of precisely drawn boundaries among states is relatively recent. The modern state has existed for no more than a few hundred years. In addition, means to determine many boundaries have come into existence only in the nineteenth and twentieth centuries, with the invention of scientific methods and instruments, along with accompanying vocabulary, for determining exact boundaries. The most basic terms of this vocabulary begin with "latitude" and "longitude" and their subdivi-

sions into the "minutes" and "seconds" used in determining boundaries. In modern times, a new attitude toward states' territory was born, especially with the nineteenth century forms of nationalism, which tend to regard every acre of territory as sacred.

Types of Boundaries

There are several types of international boundaries. Some are geographical features, including rivers, lakes, oceans, and seas. Thus boundaries of the United States include the Great Lakes, which border Canada to the north; the Rio Grande, a river that forms part of the U.S. boundary with Mexico to the south; the Atlantic and Pacific Oceans, to the east and west, respectively; and the Gulf of Mexico, to the south. In Africa, Lake Victoria bounds parts of Tanzania, Uganda, and Kenya; and rivers, such as sections of the Congo and the Zambezi, form natural boundaries among many of the continent's states.

Other geographical features, such as mountains, often form international boundaries. The Pyrenees, for example, separate France and Spain and cradle the tiny state of Andorra. In South America, the Andes frequently serve as a boundary, such as between Argentina and Chile. The Himalayas in South Central Asia create a number of borders, such as between India, China, and Tibet and between Nepal, Butan, and their neighbors. When there are no clear geographical barriers between states, boundaries must be decided by mutual consent or the threat of force. In the 2016 presidential campaign, Donald Trump repeatedly called for a wall to be built between the United States and Mexico, claim-

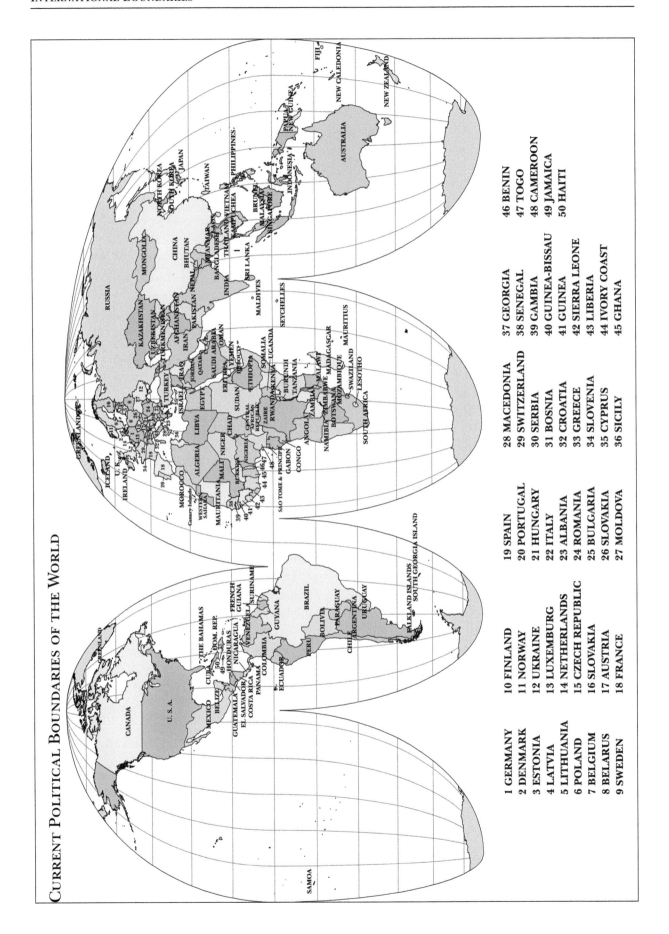

CURRENT POLITICAL BOUNDARIES OF THE WORLD

1 GERMANY
2 DENMARK
3 ESTONIA
4 LATVIA
5 LITHUANIA
6 POLAND
7 BELGIUM
8 BELARUS
9 SWEDEN

10 FINLAND
11 NORWAY
12 UKRAINE
13 LUXEMBURG
14 NETHERLANDS
15 CZECH REPUBLIC
16 SLOVAKIA
17 AUSTRIA
18 FRANCE

19 SPAIN
20 PORTUGAL
21 HUNGARY
22 ITALY
23 ALBANIA
24 ROMANIA
25 BULGARIA
26 SLOVAKIA
27 MOLDOVA

28 MACEDONIA
29 SWITZERLAND
30 SERBIA
31 BOSNIA
32 CROATIA
33 GREECE
34 SLOVENIA
35 CYPRUS
36 SICILY

37 GEORGIA
38 SENEGAL
39 GAMBIA
40 GUINEA-BISSAU
41 GUINEA
42 SIERRA LEONE
43 LIBERIA
44 IVORY COAST
45 GHANA

46 BENIN
47 TOGO
48 CAMEROON
49 JAMAICA
50 HAITI

ing that Mexico would pay for it. As of 2019, the wall has not been completed, however.

Creation and Change of International Boundaries

War and conquest often have been used to determine borders. Such wars, however, historically have created hostility among losers. Political pressures to recover lost lands build up among aggrieved losers, and such irredentist claims provide fuel for future wars. A classic example is the fate of the regions of Alsace and Lorraine between France and Germany. Although natural resources in the form of coal played a substantial role in the dispute over this area, national pride was also a potent element.

Whether boundaries are fixed through compelling geographical imperatives or in their absence, states typically sign treaties agreeing to their location. These may be treaties that conclude wars, or boundary commissions set up by those involved may draw up borders to which states give formal agreement. In 1846, for example, negotiators for Great Britain and the United States settled on the forty-ninth parallel as the boundary between the western United States and Canada, although in the United States, "Fifty-four [degrees latitude] Forty [minutes] or Fight" had been a popular motto in the presidential election campaign of 1844.

Sometimes no accepted borders exist because of chronic hostility between states. Thus, maps of the Kashmir region between India and Pakistan, claimed by both countries, show only a "line of control" or cease-fire line to divide the two warring states. Similarly, only a cease-fire line, drawn at the armistice of the Korean War of 1950-1953, divides North and South Korea; a mutually agreed-upon border remains unfixed.

In rare instances, no true boundary exists to mark where a state's territory begins and ends. Classic cases are found on the Arabian Peninsula, where the land borders of principalities, known as the Gulf Sheikdoms, are vague lines in the sand. Such circumstances usually create no difficulties where nothing is at stake, but when oil is discovered, states must come to agreement or risk coming to blows.

In other instances, negotiations and international arbitration have been effective for determining borders. Perhaps the most important principle for determining the borders of newly created states is found in the Latin phrase, *Uti possidetis iurus*. This principle is used when states become independent after having been colonies or constituent parts of a larger state that has broken up. The principle holds that states shall respect the borders in place when they were colonies. *Uti possidetis* was first extensively used in South America in the nineteenth century, when European colonial powers withdrew, leaving several newly born states to determine their own boundaries. The principle may be used as a basis for border agreements among the fifteen states of the former Soviet Union.

Besides war and negotiation, purchase has sometimes been a means of creating international boundaries. For example, in 1853 the United States purchased territory from Mexico in the southwest; in 1867, it purchased Alaska from Russia.

In rare cases, natural boundaries may change naturally or be changed deliberately by one side, incurring resentment among victims. An example occurred in 1997, when Vietnam complained that China had built an embankment on a border river embankment that caused the river to change its course; China countered that Vietnam had built a dam altering the river's course.

Other border difficulties among states include conflicts over water that flows from one country to another. In the 1990s, for example, Mexico complained of excessive U.S. use of Colorado River waters and demanded adjustment.

Border Disputes

Border disputes among states in the past two centuries have been numerous and lethal. In the twentieth century, numerous such controversies degenerated into violence. In Asia, India and Pakistan fought over Kashmir, beginning in 1947-1949 and recurring in 1965 and 1999. China has been involved in violent border disputes with India, especially in 1962; Vietnam in 1979; and Russia in 1969. In South America, border wars between Ecuador and Peru broke out in 1941, 1981, and 1995. This dispute was settled by negotiation in 1998. In Africa, among numerous recent armed conflicts, the bloody border conflict between Eritrea and Ethiopia in the 1990s was notable.

Other recent disputes have ended peacefully. Eritrea avoided violence with Yemen over several Red Sea islands by accepting arbitration by an international tribunal. In 1995 Saudi Arabia and the United Arab Emirates negotiated a peaceful agreement to their border dispute involving oil rights.

As of 2019, there are four ongoing border conflicts: Israeli-Syrian ceasefire line incidents during the Syrian Civil War, the War in Donbass, India-Pakistan military confrontation, and the 2019 Turkish offensive into north-eastern Syria, code-named by Turkey as Operation Peace Spring. Many unresolved boundary disputes might yet lead to conflicts. Among the most complex is the multinational dispute over the 600 tiny Spratly Islands in the South China Sea. Uninhabited but potentially valuable because of oil, the Spratlys are claimed by China, Brunei, Malaysia, Indonesia, the Philippines, Taiwan, and Vietnam.

Border Policies

Problems with international borders are not limited to territorial disputes. Policies regarding how borders should be operated—including the key questions of who and what should be allowed entrance and exit under what conditions—can be expected to continue as long as independent states exist. While the members of the European Union have agreed to allow free passage of people and goods among themselves, this policy does not extent to nonmembers.

The most important purpose of states is to protect the lives and property of their citizens. One of the principal purposes of international boundaries is to further this purpose. Most states insist on controlling their borders, although borders seem increasingly porous. Given the imperatives of control and the increasing difficulties of maintaining it, issues surrounding international borders are expected to continue indefinitely in the twenty-first century.

Charles F. Bahmueller

GLOBAL TIME AND TIME ZONES

Before the nineteenth century, people kept time by local reckoning of the position of the Sun; consequently, thousands of local times existed. In medieval Europe, "hours" varied in length, depending upon the seasons: Each hour was determined by the Roman Catholic Church. In the sixteenth century, Holy Roman emperor Charles V was the first secular ruler to decree hours to be of equal length. As the industrial and scientific revolutions swept Europe, North America, and other areas, some form of time standardization became necessary as communities and regions increasingly interacted. In 1780 Geneva, Switzerland, was the first locality known to employ a standard time, set by the town-hall clockkeeper, throughout the town and its immediate vicinity.

The growth and expansion of railroads, providing the first relatively fast movement of people and goods from city to city, underscored the need for a standard system in Great Britain. As early as 1828, Sir John Herschel, Astronomer Royal, called for a national standard time system based on instruments at the Royal Observatory at Greenwich. That practice began in 1852, when the British telegraph system had developed sufficiently for the Greenwich time signals to be sent instantly to any point in the country.

As railroads expanded through North America, they exposed a problem of local time variation similar to that in Great Britain but on a far larger scale, since the distances between the East and West Coasts were much greater than in Great Britain. In order for long-distance train schedules to work, different parts of the country had to coordinate their clocks. The first to suggest a standard time framework for the United States was Charles F. Dowd, president of Temple Grove Seminary for Women in Saratoga Springs, New York. Initially, Dowd proposed putting all U.S. railroads on a single standard time, based on the time in Washington, D.C. When he realized that the time in California would be behind such a standard by almost four hours, he produced a revised system, establishing four time zones in the United States. Dowd's plan, published in 1870, included the first known map of a time zone system for the country.

Not everyone was happy with the designation of Washington, D.C., as the administrative center of time in the United States. Northeastern railroad executives urged that New York, the commercial capital of the nation, be used instead: Many cities and towns in the region already had standardized to New York time out of practical necessity. Dowd proposed a compromise: to set the entire national time zone system in the United States using the Greenwich prime meridian, already in use in many parts of the world for maritime and scientific purposes. In 1873 the American Association of Railways (AAR) flatly rejected the proposal.

In the end, Dowd proved to be a visionary. In 1878 Sandford Fleming, chief engineer of the government of Canada, proposed a worldwide system of twenty-four time zones, each fifteen degrees of longitude in width, and each bisected by a meridian, beginning with the prime meridian of Greenwich. William F. Allen, general secretary of the AAR and armed with a deep knowledge of railroad practices and politics, took up the crusade and persuaded the railroads to agree to a system. At noon on Sunday, November 18, 1883, most of the more than six hundred U.S. railroad lines dropped the fifty-three arbitrary times they had been using and adopted Greenwich-indexed meridians that defined the times in each of four times zones: eastern,

central, mountain, and Pacific. Most major cities in the United States and Canada followed suit.

Time System for the World

Almost at the same time that American railroads adopted a standard time zone system, the State Department, authorized by the United States Congress, invited governments from around the world to assemble delegates in Washington, D.C., to adopt a global system. The International Meridian Conference assembled in the autumn of 1884, attended by representatives of twenty-five countries. Led by Great Britain and the United States, most favored adoption of Greenwich as the official prime meridian and Greenwich mean time as universal time.

There were other contenders: The French wanted the prime meridian to be set in Paris, and the Germans wanted it in Berlin; others proposed a mountaintop in the Azores or the tip of the Great Pyramid in Egypt. Greenwich won handily. The conference also agreed officially to start the universal day at midnight, rather than at noon or at sunrise, as practiced in many parts of the world. Each time zone in the world eventually came to have a local name, although technically, each goes by a letter in the alphabet in order eastward from Greenwich.

Once a global system was in place, there was a new issue: Many jurisdictions wanted to adjust their clocks for part of the year to account for differences in the number of hours of daylight between summer and winter months. In 1918 Congress decreed a system of daylight saving time for the United States but almost immediately abolished it, leaving state governments and communities to their local options. Daylight saving time, or a form of it, returned in the United States and many Allied nations during World War II. In the Uniform Time Act of 1966, Congress finally established a national system of daylight saving time, although with an option for states to abstain.

To the extent that it indicates how human communities want to manipulate time for social, political, or economic reasons, the issue of daylight saving time, rather than the establishment of a system of world time zones, is a better clue to the geographical issues involved in time administration. Both the history and the present format of the world time zone system show that the mathematically precise arrangement envisioned by many of the pioneers of time zones is not as important as things on the ground.

In the United States, the railroad time system adopted in 1883 drew the boundary between eastern time and central time more or less between the thirteen original states and the trans-Appalachian West: The entire Midwest, including Ohio, Indiana, and Michigan, fell in the central time zone. As the center of population migrated westward, train speeds increased, highways developed, and New York emerged as the center of mass media in the United States, the boundary between the eastern and central time zones marched steadily westward. In 1918 it ran down the middle of Ohio; by the 1960s, it was at the outskirts of Chicago.

One of the principal reasons for the popularity of Greenwich as the site of the prime meridian (zero degrees longitude), is that it places the international date line (180 degrees longitude)—where, in effect, time has to move forward to the next day rather than the next hour—far out in the Pacific Ocean where few people are affected by what otherwise would be an awkward arrangement. However, even this line is somewhat irregular, to avoid placing a small section of eastern Russia and some of the Aleutian Islands of the United States in different days.

Coordinated Universal Time Coordinated Universal Time (or UTC) is the primary time standard by which the world regulates clocks and time. It is within about 1 second of mean solar time at 0° longitude, and is not adjusted for daylight saving time. In some countries, the term Greenwich Mean Time is used. The co-ordination of time and frequency transmissions around the world began on January 1, 1960. UTC was first officially adopted as CCIR Recommendation 374, Standard-Frequency and Time-Signal Emissions, in 1963, but the official abbreviation of UTC and the official English name of Coordinated Universal Time (along with the French equivalent) were not adopted until 1967. UTC uses a *slightly* different second called the *SI second*. That is based on *atomic clocks*. Atomic clocks are more regular than the slightly variable Earth's rotation period. Hence, the essential difference between GMT and UTC is that they use different definitions of exactly how long one second of time is.

By 1950 most nations had adopted the universal time zone system, although a few followed later: Saudi Arabia in 1962, Liberia in 1972. Despite adhering to the system in principle, many nations take considerable liberties with the zones, especially if their territory spans several. All of Western Europe, despite covering an area equivalent to two zones, remains on a single standard. The People's Republic of China, which stretches across five different time zones, arbitrarily sets the entire country officially on Beijing time, eight hours behind Greenwich. Iran, Afghanistan, India, and Myanmar, each of which straddle time zone boundaries, operate on half-hour compromise systems as their time standards (as does Newfoundland). As late as 1978, Guyana's standard time was three hours, forty-five minutes in advance of Greenwich.

It can be argued that adoption of a worldwide system of time zones in the late nineteenth century was one of the earliest manifestations of the emergence of a global economy and society, and has been a crucial factor in the unfolding of this process throughout the twentieth century and beyond.

Ronald W. Davis

GLOBAL EDUCATION

THEMES AND STANDARDS IN GEOGRAPHY EDUCATION

Many people believe that the study of geography consists of little more than knowing the locations of places. Indeed, in the past, whole generations of students grew up memorizing states, capitals, rivers, seas, mountains, and countries. Most students found that approach boring and irrelevant. During the 1990s, however, geography education in the United States underwent a remarkable transformation.

While it remains important to know the locations of places, geography educators know that place name recognition is just the beginning of geographic understanding. Geography classes now place greater emphasis on understanding the characteristics of and the connections between places. Three things have led to the renewal of geography education: the five themes of geography, the national geography standards, and the establishment of a network of geographic alliances.

The Five Themes of Geography

One of the first efforts to move geography education beyond simple memorization was the National Geographic Society's publication of five themes of geography in 1984: location, place, human-environment interactions, movement, and regions. Not intended to be a checklist or recipe for understanding the world, these themes merely provided a framework for teachers—many of whom did not have a background in the subject—to incorporate geography throughout a social studies curriculum. The five themes were promoted widely by the National Geographic Society and are still used by some teachers to organize their classes.

Location is about knowing where things are. Both the absolute location (where a place is on earth's surface) and relative location (the connections between places) are important. The concept of place involves the physical and human characteristics that distinguish one place from another. The theme of human/environment interaction recognizes that people have relationships within defined places and are influenced by their surroundings. For example, many different types of housing have been created as adaptations to the world's diverse climates. The theme of movement involves the flow of people, goods, and ideas around the world. Finally, regions are human creations to help organize and understand Earth, and geography studies how they form and change.

The National Geography Standards

Geography was one of six subjects identified by President George H. W. Bush and the governors of the U.S. states when they formulated the National Education Goals in 1989. While the goals themselves foundered amid the political debate that followed their adoption, one tangible result of the initiative was the creation of Geography for Life: The National Geography Standards. More than 1,000 teachers, professors, business people, and government officials were involved in the writing of Geography for Life. The project wassupported by four geography organizations: the American Geographical Society, the Association of American Geographers, the National Council for Geographic Education, and the National Geographic Society. The resulting book defines what every U.S. student

GEOGRAPHY STANDARDS

The geographically informed person knows and understands the following:

- how to use maps and other geographic representations, tools, and technologies to acquire, process, and report information from a spatial perspective;
- how to use mental maps to organize information about people, places, and environments in a spatial context;
- how to analyze the spatial organization of people, places, and environments on Earth's surface;
- the physical and human characteristics of places;
- that people create regions to interpret Earth's complexity;
- how culture and experience influence people's perceptions of places and regions;
- the physical processes that shape the patterns of Earth's surface;
- the characteristics and spatial distribution of ecosystems on Earth's surface;
- the characteristics, distribution, and migration of human populations on Earth's surface;
- the characteristics, distribution, and complexity of Earth's cultural mosaics;
- the patterns and networks of economic interdependence on Earth's surface;
- the processes, patterns, and functions of human settlement;
- how the forces of cooperation and conflict among people influence the division and control of Earth's surface;
- how human actions modify the physical environment;
- how physical systems affect human systems;
- the changes that occur in the meaning, use, distribution, and importance of resources;
- how to apply geography to interpret the past;
- how to apply geography to interpret the present and plan for the future.

Source: National Geography Standards Project. Geography for Life: National Geography Standards, Second Edition. Washington, D.C.: National Geographics Research and Exploration, 2012.

should know and be able to accomplish in geography.

Each of the eighteen standards is designed to develop students' geographic skills, including asking geographic questions; acquiring, organizing, and analyzing geographic information; and answering the questions. Each standard features explanations, examples, and specific requirements for students in grades four, eight, and twelve.

Geography Alliances and the Future of Geography Education

To publicize efforts in geography education, a network of geography alliances was established between 1986 and 1993. Today, each U.S. state has a geography alliance that links teachers and organizations such as the National Geographic Society and the National Council for Geographic Education to sponsor workshops, teacher training sessions, field experiences, and other ways of sharing the best in geographic teaching and learning.

A 2013 executive summary prepared by the National Geographic Society for the *Road Map for 21st Century Geography Education Project* continues to champion the goal of better geography education in K–12 schools. The Road Map Project represents the collaborative effort of four national organizations: the American Geographical Society (AGS), the Association of American Geographers (AAG), the National Council for Geographic Education (NCGE), and the National Geographic Society (NGS). The project partners share belief that geography education is essential for student success in all aspects of their adult lives—careers, civic lives, and personal decision making. It also is essential for the education of specialists who can help society addressing critical issues in the areas of social welfare, economic stability, environmental health, and international relations.

Eric J. Fournier

GLOBAL DATA

WORLD GAZETTEER OF OCEANS AND CONTINENTS

Places whose names are printed in SMALL CAPS *are subjects of their own entries in this gazetteer.*

Aden, Gulf of. Deep-water area between the RED and ARABIAN SEAS, bounded by Somalia, Africa, on the south and Yemen on the north. Water is warmer and saltier in the Gulf of Aden than in the Red and Arabian Seas, because little water enters from rain or land runoff.

Africa. Second-largest continent, connected to ASIA by the narrow isthmus of Suez. Bounded on the east by the INDIAN OCEAN and on the west by the ATLANTIC OCEAN. Countries of Africa are Algeria, Angola, Benin, Botswana, Burkina Faso, Burundi, Cameroon, Central African Republic, Chad, Congo, Côte d'Ivoire (Ivory Coast), the Democratic Republic of Congo, Egypt, Ethio-pia, Gabon, Gambia, Ghana, Guinea, Kenya, Liberia, Libya, Madagascar, Malawi, Mali, Mauritania, Morocco, Mozambique, Namibia, Niger, Nigeria, Rio Muni (Mbini), Rwanda, Senegal, Sierra Leone, Somalia, South Africa, Sudan, Tanzania, Togo, Tunisia, Uganda, Western Sahara, Zambia, and Zimbabwe. Climate ranges from hot and rainy near the equator, to hot and dry in the huge Sahara Desert in the north and the Kalahari Desert in the south, to warm and fairly mild at the northern and southern extremes. Paleontological evidence indicates that humans originally evolved in Africa.

Agulhas Current. Warm, swift ocean current moving south along East AFRICA's coast. Part moves between AFRICA and MADAGASCAR to form the Mozambique Current. The warm water of the

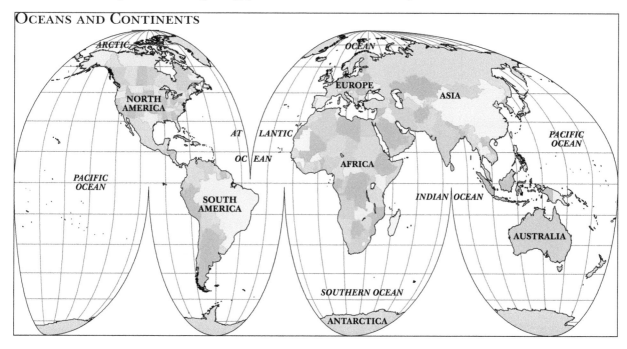

OCEANS AND CONTINENTS

Agulhas Current increases the average temperatures in the eastern part of South Africa.

Agulhas Plateau. Relatively small ocean-bottom plateau that lies south of South AFRICA, at the area where the INDIAN and ATLANTIC OCEANS meet.

Aleutian Islands. Chain of volcanic islands that extends 1,100 miles (1,770 km.) from the tip of the Alaska Peninsula to the Kamchatka Peninsula in Russia and forms the boundary between the North PACIFIC OCEAN and the BERING SEA. The area is hazardous to navigation and has been called the "Home of Storms."

Aleutian Trench. Located on the northern margin of the PACIFIC OCEAN, stretching 3,666 miles (5,900 km.) from the western edge of the Aleutian Island chain to Prince William Sound, Alaska. Depth is 25,263 feet (7,700 meters).

American Highlands. Elevated region on the ANTARCTIC coast between Enderby Land and Wilkes Land, located far south of India. The Lambert and Fisher glaciers originate in the American Highlands and move down to feed the AMERY ICE SHELF.

Amery Ice Shelf. Year-round shelf of relatively flat ice in a bay of ANTARCTICA, located at approximately longitude 70 degrees east, between MAC. ROBERTSON LAND and the AMERICAN HIGHLANDS. The ice shelf is fed by the Lambert and Fisher glaciers.

Amundsen Sea. Portion of the southernmost PACIFIC OCEAN off the Wahlgreen Coast of ANTARCTICA, approximately longitude 100 to 120 degrees west. Named for the Norwegian explorer Roald Amundsen, who became the first person to reach the SOUTH POLE in 1911.

Antarctic Circle. Latitude of 66.3 degrees south. South of this line, the Sun does not set on the day of the summer solstice, about December 22 in the SOUTHERN HEMISPHERE, and does not rise on the day of the winter solstice, about June 21.

Antarctic Circumpolar Current. Eastward-flowing current that circles ANTARCTICA and extends from the surface to the deep ocean floor. The largest-volume current in the oceans. Extends northward to approximately 40 degrees south latitude and is driven by westerly winds.

Antarctic Convergence. Meeting place where cold Antarctic water sinks below the warmer sub-Antarctic water.

Antarctic Ocean. See SOUTHERN OCEAN.

Antarctica. Fifth-largest continent, located at the southernmost part of the world. There are two major regions; western Antarctica, which includes the mountainous Antarctic peninsula, and eastern Antarctica, which is mostly a low continental shield area. An ice cap up to 13,000 feet (4,000 meters) thick covers 95 percent of the continent's surface. Temperatures in the austral summer (December, January, and February) rarely rise above 0 degrees Fahrenheit (-18 degrees Celsius) except on the peninsula. By international treaty, the continent is not owned by any single country, and human access is largely regulated. There has never been a self-supporting human habitation on Antarctica.

Arabian Sea. Portion of the INDIAN OCEAN bounded by India on the east, Pakistan on the north, and Oman and Yemen of the Arabian Peninsula on the west.

Arctic Circle. Latitude of 66.3 degrees north. North of this line, the Sun does not set on the day of the summer solstice, about June 21 in the NORTHERN HEMISPHERE, and does not rise on the day of the winter solstice, about December 22.

Arctic Ocean. World's smallest ocean. It centers on the geographic NORTH POLE and connects to the PACIFIC OCEAN through the BERING SEA, and to the ATLANTIC OCEAN through the GREENLAND SEA. The Arctic Ocean is covered with ice up to 13 feet (4 meters) thick all year, except at its edges. Norwegian explorers on the ship *Fram* stayed locked in the icepack from 1893 to 1896, in order to study the movement of polar ice. They drifted in the ice a total of 1,028 miles (1,658 km.), from the Bering Sea to the Greenland Sea, proving that there was no land mass under the Arctic ice at the top of the world. Also known as Arctic Sea or Arctic Mediterranean Sea.

Argentine Basin. Basin on the floor of the western Atlantic Ocean, off the coast of Argentina in South America. Among ocean basins, this one is unusually circular.

Ascension Island. Isolated volcanic island in the South Atlantic Ocean, about midway between South America and Africa. One of the islands visited by British biologist Charles Darwin during his five-year voyage on the *Beagle*.

Asia. Largest continent; joins with Europe to form the great Eurasian landmass. Asia is bounded by the Arctic Ocean on the north, the western Pacific Ocean on the east, and the Indian Ocean on the south. Its countries include Afghanistan, Bahrain, Bangladesh, Bhutan, Cambodia, China, India, Iran, Iraq, Irian Jaya, Israel, Japan, Jordan, Kalimantan, Kazakhstan, North and South Korea, Kyrgyzstan, Laos, Lebanon, Malaysia, Myanmar, Mongolia, Nepal, Oman, Pakistan, the Philippines, Russia, Sarawak, Saudi Arabia, Sri Lanka, Sumatra, Syria, Tajikistan, Thailand, Asian Turkey, Turkmenistan, United Arab Emirates, Uzbekistan, Vietnam, and Yemen. Climates include virtually all types on earth, from arctic to tropical, desert to rainforest. Asia has the highest (Mount Everest) and lowest (Dead Sea) surface points in the world. Nearly 60 percent of the world's people live in Asia.

Atlantic Ocean. Second-largest body of water in the world, covering more than 25 percent of Earth's surface. Bordered by North and South America on the west, and Europe and East Africa on the east. The widest part (5,500 miles/8,800 km.) lies between West Africa and Mexico, along 20 degrees latitude. Scientists disagree on the north-south boundaries of the Atlantic; if one includes the Arctic Ocean and the Southern Ocean, the Atlantic Ocean extends about 13,300 miles (21,400 km.). The deepest spot (28,374 feet/8,648 meters) is found in the Puerto Rico Trench. The Atlantic Ocean has been a major route for trade and communications, especially between North America and Europe, for hundreds of years. This is because of its relatively narrow size and favorable currents, such as the Gulf Stream.

Australasia. Loosely defined term for the region, which, at the least, includes Australia and New Zealand; at the most, it also includes other South Pacific Islands in the region.

Australia. Smallest continent, sometimes called the "island continent." Located between the Indian and Pacific Oceans. It is the only continent occupied by a single nation, the Commonwealth of Australia. Australia is the flattest and driest continent; two-thirds is either desert or semiarid. Geologically, it is the oldest and most isolated continent. Unlike any other place on Earth, large mammals never evolved in Australia. Marsupials (pouched, warm-blooded animals) and unusual birds developed in their place.

Azores. Archipelago (group of islands) in the eastern Atlantic Ocean lying about 994 miles (1,600 km.) west of Portugal. The islands are of volcanic origin and have been known, fought over, and used by the Europeans since before the fourteenth century. Spanish explorer Christopher Columbus stopped in the Azores to wait for favorable winds before his first trip across the Atlantic Ocean.

Barents Sea. Partially enclosed section of the Arctic Ocean, bounded by Russia and Norway on the south and the Russian island of Navaya Zemlaya on the east. The Barents Sea was important in World War II because Allied convoys had to cross it, through storms and submarine patrols, to deliver war supplies to Murmansk, the only ice-free port in western Russia. It was named for the Dutch explorer Willem Barents.

Bays. See under individual names.

Beaufort Sea. Area of the Arctic Ocean located off the northern coast of Alaska and western Canada. It is usually frozen over and has no islands. Named for British admiral Sir Francis Beaufort, who devised the Beaufort Wind Scale as a means of classifying wind force at sea.

Bengal, Bay of. Northeast arm of the Indian Ocean, bounded by India on the west and Myanmar on the east. The Ganges River empties into the Bay of Bengal. The great ports of Calcutta and Madras in India, and Rangoon in Myanmar lie in the bay, making it a busy and important area for shipping for centuries.

Benguela Current. Northward-flowing current along the western coast of Southern AFRICA. Normally, the Benguela Current carries cold, rich water that wells up from the ocean depths and supports a large fishing industry. A change in winds can reduce the oxygen supply and kill huge numbers of fish, similar to what may happen off the coast of Peru during El Niño weather conditions.

Bering Sea. Portion of the northernmost PACIFIC OCEAN that is bounded by the state of Alaska on the east, Russia and the Kamchatka Peninsula on the west, and the BERING STRAIT on the north. It is a valuable fishing ground, rich in shrimp, crabs, and fish. Whales, fur seals, sea otters, and walrus are also found there.

Bering Strait. Narrowest point of connection between the BERING SEA and the ARCTIC OCEAN, located between the easternmost point of Siberia on the west and Alaska on the east. The Bering Strait is 52 miles (84 km.) wide. During the Ice Age, when the sea level was lower, humans and animals were able to walk from the Asian continent across a land bridge—now known as Beringia—to the North American continent across the frozen strait, providing the first human access to the Americas.

Bikini Atoll. Small atoll in the Marshall Islands group in the western PACIFIC OCEAN. In the 1940s, the United States began testing nuclear bombs on Bikini and neighboring atolls. The U.S. Army removed the inhabitants of Bikini, and testing occurred from 1946 to 1958. The Bikini inhabitants were allowed to return in 1969, then removed again in 1978 when high levels of radioactivity were found to remain.

Black Sea. Large inland sea situated where southeastern EUROPE meets ASIA; connected to the MEDITERRANEAN SEA through Turkey's Bosporus strait. The sea covers an area of about 178,000 square miles (461,000 sq. km.), with a maximum depth of more than 7,250 feet (2,210 meters).

Brazil Current. Extension of part of the warm, westward-flowing South EQUATORIAL CURRENT, which turns south to the coast of Brazil. The Brazil Current has very salty water because of its long flow across the equator. It joins the WEST WIND DRIFT and moves eastward across the South ATLANTIC OCEAN as part of the SOUTH ATLANTIC GYRE.

California, Gulf of. Branch of the eastern PACIFIC OCEAN that separates Baja California from mainland Mexico. Warm, nutrient-rich water supports a variety of fish, oysters, and sponges. California gray whales migrate to the gulf to give birth and breed, January through March. Fisheries and tourism are important industries in the Gulf of California. Also known as the Sea of Cortés.

California Current. Cool water that flows southeast along the western coast of NORTH AMERICA from Washington State to Baja California. The eastern portion of the NORTH PACIFIC GYRE.

Canada Basin. Part of the ocean floor that lies north of northeastern Canada and Alaska. The BEAUFORT SEA lies above the Canada Basin.

Cape Horn. Southernmost tip of SOUTH AMERICA. It is the site of notoriously severe storms and is hazardous to shipping.

Cape Verde Plateau. ATLANTIC OCEAN plateau lying off the western bulge of the AFRICAN continent. The volcanic Cape Verde Islands lie on the plateau.

Caribbean Sea. Portion of the western ATLANTIC OCEAN bounded by CENTRAL and SOUTH AMERICA to the west and south, and the islands of the Antilles chain on the north and east. Mostly tropical in climate, the Caribbean Sea supports a large variety of plant and animal life. Its islands, including Puerto Rico, the Cayman Islands, and the Virgin Islands, are popular tourist sites.

Caspian Sea. World's largest inland sea. Located east of the Caucasus Mountains at EUROPE's southeasternmost extremity, it dominates the expanses of western Central ASIA. Its basin is 750 miles (1,200 kilometers) long, and its average width is 200 miles (320 kilometers). It covers 149,200 square miles (386,400 sq. km.).

Central America. Region generally understood to constitute the irregularly shaped neck of land linking NORTH and SOUTH AMERICA, containing Belize, Guatemala, Honduras, El Salvador, Nicaragua, Costa Rica, and Panama.

Chukchi Sea. Portion of the ARCTIC OCEAN, bounded by the BERING STRAIT on the south, Siberia on the southwest, and Alaska on the southeast. The Chukchi Sea is the area of exchange between waters and sea life of the PACIFIC and ARCTIC OCEANS, and so is an area of interest to oceanographers and fishermen.

Clarion Fracture Zone. East-west-running fracture zone that begins off the west coast of Mexico and extends approximately 2.500 miles (4,023 km.) to the southwest.

Cocos Basin. Relatively small ocean basin located off the west coast of Sumatra in the northeast INDIAN OCEAN.

Coral Sea. Area of the PACIFIC OCEAN off the northeast coast of AUSTRALIA, between Australia on the southwest, Papua New Guinea and the Solomon Islands on the northeast, and New Caledonia on the east. Site of a naval battle in 1942 that prevented the Japanese invasion of Australia.

Cortés, Sea of. See CALIFORNIA, GULF OF.

Denmark Strait. Channel that separates GREENLAND and ICELAND and connects the North Atlantic Ocean with the ARCTIC OCEAN.

Dover, Strait of. Body of water between England and the European continent, separating the NORTH SEA from the ENGLISH CHANNEL. It is 33 miles (53 km.) wide at its narrowest point. The tunnel between England and France (known as the "Chunnel") was cut into the rock under the Strait of Dover.

Drake Passage. Narrow part of the SOUTHERN OCEAN that connects the ATLANTIC and PACIFIC OCEANS between the southern tip of SOUTH AMERICA and the ANTARCTIC peninsula. Named for sixteenth-century English navigator and explorer Sir Francis Drake, who discovered the passage when his ship was blown into it during a violent storm. Also called Drake Strait.

East China Sea. Area of the western PACIFIC OCEAN bounded by China on the west, the YELLOW SEA on the north, and Japan on the northeast. Large oil deposits were found under the East China Sea floor in the 1980s.

East Pacific Rise. Broad, nearly continuous undersea mountain range that extends from the southern end of Baja California southward, then curves east near ANTARCTICA. It is formed along the southeast side of the Pacific Plate and is part of the RING OF FIRE, a nearly continuous ring of volcanic and tectonic activity around the rim of the Pacific Ocean. Also called East Pacific Ridge.

East Siberian Sea. Portion of the ARCTIC OCEAN bounded by the CHUKCHI SEA on the east, Siberia on the south, and the LAPTEV SEA on the west. Much of the East Siberian Sea is covered with ice year-round.

Eastern Hemisphere. The half of the earth containing EUROPE, ASIA, and AFRICA; generally understood to fall between longitudes 20 degrees west and 160 degrees east.

El Niño. Conditions—also known as El Niño-Southern Oscillation (ENSO) events—that occur every two to ten years and cause weather and ocean temperature changes off the coast of Ecuador and Peru. Most of the time, the PERU CURRENT causes cold, nutrient-rich water to well up off the coast of Ecuador and Peru. During ENSO years, the cold upwelling is replaced by warmer surface water that does not support plankton and fish. Fisheries decline and seabirds starve. Climatic changes of El Niño can bring floods to normally dry areas and drought to wet areas. Effects can extend across NORTH and SOUTH AMERICA, and to the western PACIFIC OCEAN. During the 1990s, the ENSO event fluctuated but did not go completely away, which caused tremendous damage to fisheries and agriculture, storms and droughts in North America, and numerous hurricanes.

Emperor Seamount Chain. Largest known example of submerged underwater volcanic ridges, located in the northern PACIFIC OCEAN and extending southward from the Kamchatka Peninsula in Russia for about 2,500 miles (4,023 km.).

Enderby Land. Section of ANTARCTICA that lies between the INDIAN OCEAN and the South Polar Plateau, east of QUEEN MAUD LAND. Enderby Land lies between approximately longitude 45 and 60 degrees east.

English Channel. Strait water separating continental France from Great Britain. Runs for roughly 350 miles (560 km.), from the ATLANTIC OCEAN in the west to the Strait of Dover in the east.

Equatorial Current. Currents just north and south of the equator that flow from east to west. Equatorial currents are found in the PACIFIC and ATLANTIC OCEANS. The equatorial currents and the trade winds, which move in the same direction, greatly aid oceangoing traffic.

Eurasia. Term for the combined landmass of EUROPE and ASIA.

Europe. Sixth-largest continent, actually a large peninsula of the Eurasian landmass. Europe is densely populated and includes the countries of Albania, Andorra, Austria, Belarus, Belgium, Bulgaria, Bosnia-Herzegovina, Croatia, the Czech Republic, Denmark, Estonia, Finland, France, Germany, Greece, Hungary, Iceland, Ireland, Italy, Latvia, Lithuania, Macedonia, Malta, Moldova, Monaco, the Netherlands, Norway, Poland, Portugal, Romania, Slovakia, Spain, Switzerland, Turkey, and the United Kingdom (England, Northern Ireland, Scotland, and Wales). Climate ranges from near arctic in the north, to temperate and Mediterranean in the south.

Florida Current. Water moving northward along the east coast of Florida to Cape Hatteras, North Carolina, where it joins the GULF STREAM.

Fundy, Bay of. Large inlet on the North American Atlantic coast, northwest of Maine, separating New Brunswick and Nova Scotia in Canada. Renowned for having the largest tidal change in the world, more than 56 feet (17 meters).

Galápagos Islands. Located directly on the equator, 600 miles (965 km.) west of Ecuador. The islands are volcanic in origin and sit directly in the cold PERU CURRENT, which cools the islands and creates unusual microclimates and fogs. The extreme isolation of the islands allowed unique species to develop. Biologist Charles Darwin visited the Galápagos in the 1830s, and the unusual organisms he observed helped him to conceive the theory of evolution.

Galápagos Rift. Divergent plate boundary extending between the GALÁPAGOS ISLANDS and SOUTH AMERICA. The first hydrothermal vent community was discovered in 1977 in the Galápagos Rift. This unusual type of biological habitat is based on energy from bacteria that use heat and chemicals to make food, instead of sunlight.

Grand Banks. Portion of the northwest ATLANTIC OCEAN southeast of Nova Scotia and Newfoundland. The Grand Banks are extremely rich fishing grounds, although in the 1980s and 1990s catches of cod, flounder, and many other fish dropped dramatically due to overfishing and pollution.

Great Barrier Reef. Largest coral reef in the world, lying in the CORAL SEA off the east coast of AUSTRALIA. The reef system and its small islands stretch for more than 1,100 miles (1,750 km.) and is difficult to navigate through. The reefs are home to an incredible variety of tropical marine life, including large numbers of sharks.

Greenland. Largest island in the world that is not rated as a continent; lies between the northernmost part of the ATLANTIC OCEAN and the ARCTIC OCEAN, northeast of the North American continent. About 90 percent of Greenland is permanently covered with an ice sheet and glaciers. Residents engage in limited agriculture, growing potatoes, turnips, and cabbages. Most people live along the southwest coast, where the climate is warmed by the NORTH ATLANTIC CURRENT.

Greenland Sea. Body of water bounded by GREENLAND on the west, ICELAND on the north, and Spitsbergen on the east. It is often ice-covered.

Guinea, Gulf of. Arm of the North ATLANTIC OCEAN below the great bulge of West AFRICA.

Gulf Stream. Current of westward-moving warm water originating along the equator in the ATLANTIC OCEAN. The mass of water moves along the east coast of Florida as the FLORIDA CURRENT, then turns in a northeasterly direction off North Carolina to become the Gulf Stream. The Gulf Stream flows northeast past Newfoundland and the western edge of the British Isles. The warmer water of the Gulf Stream moderates the climate of northwestern EUROPE, causing temperatures in winter to be several degrees warmer than in areas of NORTH AMERICA at the same latitudes. The Gulf Stream decreases the time required for ships to travel from North America to Europe. This was an important factor in trade and communication in American Colonial times and has continued to be significant.

Gulfs. See under individual names.

Hatteras Abyssal Plain. Part of the floor of the northwest ATLANTIC OCEAN Basin, east of North Carolina. It rises to form shallow sandbars around Cape Hatteras, which are a notorious navigational hazard. In the seventeenth and eighteenth centuries, so many ships were lost in the area that Cape Hatteras became known as "The Graveyard of the Atlantic."

horse latitudes. Latitude belts between 30 and 35 degrees north and south latitude, where winds are usually light and variable and the climate mostly hot and dry.

Humboldt Current. See PERU CURRENT

Iceland. Island country bounded by the GREENLAND SEA on the north, the NORWEGIAN SEA on the east, and the ATLANTIC OCEAN on the south and west. Total area of 39,768 square miles (103,000 sq. km.). The nearest land mass is GREENLAND, 200 miles (320 km.) to the northwest. Situated on top of the northern part of the Atlantic Mid-Oceanic Ridge, it is characterized by major volcanic activities, geothermal springs, and glaciers.

Idzu-Bonin Trench. Ocean trench in the western PACIFIC OCEAN, about 6,082 miles (9,810 km.) long and 2,624 feet (800 meters) deep.

Indian Ocean. Third-largest of the world's oceans, bounded by the continents of AFRICA to the west, ASIA to the north, AUSTRALIA to the east, and ANTARCTICA to the south. Most of the Indian Ocean lies below the equator. It has an approximate area of 33 million square miles (76 million sq. km.) and an average depth of about 13,120 feet (4,000 meters). Its deepest point is 24,442 feet (7,450 meters), in the JAVA TRENCH. The Indian Ocean was the first major ocean to be used as a trade route, particularly by the Egyptians. About 600 BCE, the Egyptian ruler Necho sent an expedition into the Indian Ocean, and the ship circumnavigated Africa, probably the first time this feat was accomplished. Warm winds blowing over the northern part of the Indian Ocean from May to September pick up huge amounts of moisture, which falls on India and Sri Lanka as monsoons. Fishing is important and mostly is done by small, family boats. About 40 percent of the world's offshore oil production comes from the Indian Ocean.

Indonesian Trench. See JAVA TRENCH.

Intracoastal Waterway. Series of bays, sounds, and channels, part natural and part human-made, that extends along the eastern coast of the United States from the Delaware River in New Jersey, south to the tip of Florida, then around the west coast of Florida. It extends around the Gulf Coast to the Rio Grande in Texas. It runs 2,455 miles (3,951 km.) and is an important, protected route for commercial and pleasure boat traffic.

Japan, Sea of. Marginal sea of the western Pacific Ocean that is bounded by Japan on the east and the Russian mainland on the west. Its surface area is approximately 377,600 square miles (978,000 sq. km.). It has an average depth of 5,750 feet (1,750 meters) and a maximum depth of 12,276 feet (3,742 meters).

Japan Trench. Ocean trench approximately 497 miles (800 km.) long, beginning at the eastern edge of the Japanese islands and stretching southward toward the MARIANA TRENCH. Depth is 27,560 feet (8,400 meters).

Java Sea. Portion of the western PACIFIC OCEAN between the islands of Java and Borneo. The sea has a total surface area of 167,000 square miles (433,000 sq. km.) and a comparatively shallow average depth of 151 feet (46 meters).

Java Trench. Ocean trench in the INDIAN OCEAN, 2,790 miles (4,500 km.) long and 24,443 feet (7,450 meters) deep. Also called the Indonesian Trench.

Kermadec Trench. Ocean trench approximately 930 miles (1,500 km.) long, located in the southwest PACIFIC OCEAN, beginning northeast of New Zealand. It has a depth of 32,800 feet (10,000 meters). Its northern end connects with the TONGA TRENCH.

Kurile Trench. Ocean trench approximately 1,367 miles (2,200 km.) long along the northeast rim of the PACIFIC OCEAN, beginning at the north end of the Japanese island chain and extending northeastward. Depth is 34,451 feet (10,500 meters).

Labrador Current. Cold current that begins in Baffin Bay between GREENLAND and northeastern Canada and flows southward. The Labrador Current sometimes carries icebergs into North Atlantic shipping channels; such an iceberg caused the famous sinking of the great passenger ship *Titanic* in 1912.

Laptev Sea. Marginal sea of the ARCTIC OCEAN off the coast of northern Siberia. The Taymyr Peninsula bounds it on the west and the New Siberian Islands on the east. Its area is about 276,000 square miles (714,000 sq. km.). Its average depth is 1,896 feet (578 meters), and the greatest depth is 9,774 feet (2,980 meters).

Lord Howe Rise. Elevation of the floor of the western PACIFIC OCEAN that lies between AUSTRALIA and New Guinea and under the TASMAN SEA.

Mac. Robertson Land. Land near the coast of ANTARCTICA, located between the INDIAN OCEAN and the south Polar Plateau, east of ENDERBY LAND. Mac.Robertson Land lies between approximately longitude 60 and 65 degrees east.

Macronesia. Loose grouping of islands in the ATLANTIC OCEAN that includes the Azores, Madeira, the Canary Islands and Cape Verde. The term is derived from Greek words meaning "large" and "island" and should not be confused with MICRONESIA, small islands in the central and North PACIFIC OCEAN.

Madagascar. Large island nation, officially called the Malagasy Republic, located in the INDIAN OCEAN about 200 miles from the southeast coast of AFRICA. Although geographically tied to the African continent, it has a culture more closely tied to those of France and Southeast Asia. Area is 226,657 square miles (587,042 sq. km.).

Magellan, Strait of. Waterway connecting the south ATLANTIC OCEAN with the South Pacific. Ships passing through the strait, north of Tierra del Fuego Island, avoid some of the world's roughest seas around CAPE HORN.

magnetic poles. The two points on the earth, one in the NORTHERN HEMISPHERE and one in the SOUTHERN HEMISPHERE, which are defined by the internal magnetism of the earth. Each point attracts one end of a compass needle and repels the opposite end.

Malacca, Strait of. Relatively narrow passage (200 miles/322 kilometers wide) bordered by Malaysia and Sumatra and linking the SOUTH CHINA SEA and the JAVA SEA. It is one of the most heavily traveled waterways in the world, with more than one thousand ships every week.

Mariana Trench. Lowest point on Earth's surface, with a maximum depth of 36,150 feet (11,022 meters) in the Challenger Deep. The Mariana Trench is located on the western margin of the PACIFIC OCEAN southeast of Japan, and is approximately 1,584 miles (2,550 km.) long.

Marie Byrd Land. Section of ANTARCTICA located at the base of the Antarctic peninsula and shaped like a large peninsula itself. It is bounded at its base by the ROSS ICE SHELF and the Ronne Ice Shelf.

Mediterranean Sea. Large sea that separates the continents of EUROPE, AFRICA, and ASIA. It takes its name from Latin words meaning "in the middle of land"—a reference to its nearly land-locked nature. Covers about 969,100 square miles (2.5 million sq. km.) and extends 2,200 miles (3,540 km.) from west to east and about 1,000 miles (1,600 km.) from north to south at its widest. Its greatest depth is 16,897 feet (5,150 meters).

Melanesia. One of three divisions of the Pacific Islands, along with Micronesia and Polynesia; located in the western Pacific. The name Melanesia, for "dark islands," was given to the area because of its inhabitants' dark skins

Mexico, Gulf of. Nearly enclosed arm of the western Atlantic Ocean, bounded by the states of Florida, Alabama, Mississippi, Louisiana, and Texas, and Mexico and the Yucatan Peninsula. Cuba is located in the gap between the Yucatan Peninsula and Florida. Most ocean water enters through the Yucatan passage and exits the Gulf of Mexico around the tip of Florida, becoming the Florida Current. Fisheries, tourism, and oil production are important activities.

Micronesia. One of three divisions of the Pacific Islands, along with Melanesia and Polynesia. Micronesia means "small islands." Micronesia's islands are mostly atolls and coral islands, but some are of volcanic origin. The more than 2,000 islands of Micronesia are located in the Pacific Ocean east of the Philippines, mostly north of the equator.

Mid-Atlantic Ridge. Steep-sided, underwater mountain range running down the middle of the Atlantic Ocean. Formed by the divergent boundaries, or region where tectonic plates are separating.

Mozambique Current. See Agulhas Current.

New Britain Trench. Ocean trench in the southwest Pacific Ocean, about 5,158 miles (8,320 km.) long and 2,460 feet (750 meters) deep.

New Hebrides Basin. Part of the Coral Sea, located east of Australia and west of the New Hebrides island chain. The basin contains volcanic islands, both old and recent.

New Hebrides Trench. Ocean trench in the southwest Pacific Ocean, about 5,682 miles (9,165 km.) long and 3,936 feet (1,200 meters) deep.

North America. Third-largest continent, usually considered to contain all land and nearby islands in the Western Hemisphere north of the Isthmus of Panama, which connects it to South America. The major mainland countries are Canada, the United States, Mexico, Guatemala, El Salvador, Honduras, Nicaragua, Costa Rica, and Panama. Island countries include the islands of the Caribbean Sea and Greenland. Climate ranges from arctic to tropical.

North Atlantic Current. Continuation of the Gulf Stream, originating near the Grand Banks off Newfoundland. It curves eastward and divides into a northern branch, which flows into the Norwegian Sea, a southern branch, which flows eastward, and a branch that forms the Canary Current and flows south along the coast of Europe.

North Atlantic Gyre. Large mass of water, located in the Atlantic Ocean in the Northern Hemisphere, that rotates clockwise. Warm water moves toward the pole and cold water moves toward the equator.

North Pacific Current. Eastward flow of water in the Pacific Ocean in the Northern Hemisphere. It originates as the Kuroshio Current and moves from Japan toward North America.

North Pacific Gyre. Large mass of water, located in the Pacific Ocean in the Northern Hemisphere, that rotates clockwise. Warm water moves toward the pole and cold water moves toward the equator.

North Pole. Northern end of the earth's geographic axis, located at 90 degrees north latitude and longitude zero degrees. The North Pole itself is located on the Polar Abyssal Plain, about 14,000 feet (4,000 meters) deep in the Arctic Ocean. U.S. explorer Robert Edwin is credited with being the first person to reach the North Pole, in 1909, although there is historical dispute over the claim. The North Pole is different from the North magnetic pole.

North Sea. Arm of the northeastern Atlantic Ocean, bounded by Great Britain on the west and Norway, Denmark, and Germany on the east and south. The North Sea is one of the great fishing areas of the world and an important source of oil.

Northern Hemisphere. The half of the earth above the equator.

Norwegian Sea. Section of the North Atlantic Ocean. Norway borders it on the east and Iceland on the west. A submarine ridge linking

449

GREENLAND, ICELAND, the Faroe Islands, and northern Scotland separates the Norwegian Sea from the open ATLANTIC OCEAN. Cut by the ARCTIC CIRCLE, the sea is often associated with the ARCTIC OCEAN to the north. Reaches a maximum depth of about 13,020 feet (3,970 meters).

Oceania. Loosely applied term for the large island groups of the central and South Pacific; sometimes used to include AUSTRALIA and New Zealand.

Okhotsk, Sea of. Nearly enclosed area of the northwestern PACIFIC OCEAN bounded by Russia's Kamchatka Peninsula on the east and Siberia on the west. It is open to the Pacific Ocean on the south side only through Japan and the Kuril Islands, a string of islands belonging to Russia.

Pacific Ocean. Largest body of water in the world, covering more than one-third of Earth's surface—an area of about 70 million square miles (181 million sq. km.), more than the entire land area of the world. At its widest point, between Panama in CENTRAL AMERICA and the Philippines, it stretches 10,700 miles (17,200 km.). It runs 9,600 miles (15,450 km.) from the BERING STRAIT in the north to ANTARCTICA in the south. Bordered by NORTH and SOUTH AMERICA in the east, and ASIA and AUSTRALIA in the west. The average depth is about 12,900 feet (3,900 meters). It contains the deepest point on Earth (36,150 feet/11,022 meters), in the Challenger Deep of the MARIANA TRENCH, southwest of Japan. The Pacific Ocean bottom is more geologically varied than the INDIAN or ATLANTIC OCEANS; it has more volcanoes, ridges, trenches, seamounts, and islands. The vast size of the Pacific Ocean was a formidable barrier to travel, communications, and trade well into the nineteenth century. However, evidence shows that people crossed the Pacific Ocean in rafts or canoes as early as 3,000 BCE.

Pacific Rim. Modern term for the nations of ASIA and NORTH and SOUTH AMERICA that border, or are in, the PACIFIC OCEAN. Used mostly in discussions of economic growth.

Palau Trench. Ocean trench in the western PACIFIC OCEAN, about 250 miles (400 km.) long and 26,425 feet (8,054 meters) deep.

Palmer Land. Section of ANTARCTICA that occupies the base of the Antarctic peninsula.

Panama, Isthmus of. Narrow neck of land that joins CENTRAL and SOUTH AMERICA. In 1914 the Panama Canal was opened through the isthmus, creating a direct sea link between the PACIFIC OCEAN and the CARIBBEAN SEA. The canal stretches about 50 miles (80 km.) from Panama City on the Pacific to Colón on the Caribbean. More than 12,000 ships pass through the canal annually.

Persian Gulf. Large extension of the ARABIAN SEA that separates Iran from the Arabian Peninsula in the Middle East. It covers about 88,000 square miles (226,000 sq. km.) and is about 620 miles (1,000 km.) long and 125–185 miles (200–300 km.) wide.

Peru-Chile Trench. Ocean trench that runs along the eastern boundary of the PACIFIC OCEAN, off the western edge of SOUTH AMERICA. It is 3,666 miles (5,900 km.) long and 26,576 feet (8,100 meters) deep.

Peru Current. Cold, broad current that originates in the southernmost part of the SOUTH PACIFIC GYRE and flows up the west coast of SOUTH AMERICA. Off the coast of Peru, prevailing winds usually push the warmer surface water to the west. This causes the nutrient-rich, colder water of the Peru Current to well up to the surface, which provides excellent feeding for fish. At times, the upwelling ceases and biological, economic, and climatic catastrophe can result in EL NIÑO weather conditions. Also known as the Humboldt Current.

Philippine Trench. Ocean trench located on the western rim of the PACIFIC OCEAN, at the eastern margin of the PHILIPPINE ISLANDS. It is about 870 miles (1,400 km.) long and 34,451 feet (10,500 meters) deep.

Polynesia. One of three main divisions of the Pacific Islands, along with MELANESIA and MICRONESIA. The islands are spread through the central and South Pacific. Polynesia means "many islands." Mostly small, the islands are predominantly coral atolls, but some are of volcanic origin.

Puerto Rico Trench. Ocean trench in the western ATLANTIC OCEAN, about 27,500 feet (8,385 meters) deep and 963 miles (1,550 km.) long.

Queen Maud Land. Section of ANTARCTICA that lies between the ATLANTIC OCEAN and the south Polar Plateau, between approximately longitude 15 and 45 degrees east.

Red Sea. Narrow arm of water separating AFRICA from the ARABIAN PENINSULA. One of the saltiest bodies of ocean water on Earth, as a result of high evaporation and little freshwater input. It was used as a trade route for Mediterranean, Indian, and Chinese peoples for centuries before the Europeans discovered it in the fifteenth century. The Suez Canal was opened in 1869 between the MEDITERRANEAN SEA and the Red Sea, cutting the distance from the northern INDIAN OCEAN to northern EUROPE by about 5,590 miles (9,000 km.). This greatly increased the economic and military importance of the Red Sea.

Ring of Fire. Nearly continuous ring of volcanic and tectonic activity around the margins of the PACIFIC OCEAN.

Ross Ice Shelf. Thick layer of ice in the Ross SEA off the coast of ANTARCTICA. The relatively flat ice is attached to and nourished by a continental glacier.

Ross Sea. Bay in the SOUTHERN OCEAN off the coast of ANTARCTICA, located south of New Zealand. Named for English explorer James Clark Ross, the first person to break through the Antarctic ice pack in a ship, in 1841.

St. Peter and St. Paul. Cluster of rocks showing above the surface of the ATLANTIC OCEAN between Brazil and West AFRICA. Important landmarks in the days of slave ships.

Sargasso Sea. Warm, salty area of water located in the ATLANTIC OCEAN south and east of Bermuda, formed from water that circulates around the center of the NORTH ATLANTIC GYRE. Named for the seaweed, *Sargassum*, that floats on the surface in large amounts.

Scotia Sea. Area of the southernmost ATLANTIC OCEAN between the southern tip of SOUTH AMERICA and the ANTARCTIC peninsula. The area is known for severe storms.

Seas. See under individual names.

Siam, Gulf of. See THAILAND, GULF OF.

South America. Fourth-largest continent, usually considered to contain all land and nearby islands in the Western Hemisphere south of the Isthmus of Panama, which connects it to NORTH AMERICA. Countries are Argentina, Bolivia, Brazil, Chile, Colombia, Ecuador, French Guiana, Guyana, Paraguay, Peru, Suriname, Uruguay, and Venezuela. Climate ranges from tropical to cold, nearly sub-Antarctic.

South Atlantic Gyre. Large mass of water, located in the ATLANTIC OCEAN in the SOUTHERN HEMISPHERE, that rotates counterclockwise. Warm water moves toward the pole and cold water moves toward the equator.

South China Sea. Portion of the western PACIFIC OCEAN that lies along the east coast of China, Vietnam, and the southeastern part of the Gulf of Thailand. The eastern and southern edges are defined by the Philippine and Indonesian Islands.

South Equatorial Current. Part of the SOUTH ATLANTIC GYRE that is split in two by the eastern prominence of Brazil. One part moves along the northeastern coast of SOUTH AMERICA toward the CARIBBEAN SEA and the North ATLANTIC OCEAN; the other turns southward and forms the BRAZIL CURRENT.

South Pacific Gyre. Large mass of water, located in the PACIFIC OCEAN in the SOUTHERN HEMISPHERE, that rotates counterclockwise. Warm water moves toward the pole and cold water moves toward the equator.

South Pole. Southern end of the earth's geographic axis, located at 90 degrees south latitude and longitude zero degrees. The first person to reach the South Pole was Norwegian explorer Roald Amundsen, in 1911. The South Pole is different from the South MAGNETIC POLE.

Southeastern Pacific Plateau. Portion of the PACIFIC OCEAN floor closest to SOUTH AMERICA.

Southern Hemisphere. The half of the earth below the equator.

Southern Ocean. Not officially recognized as one of the major oceans, but a commonly used term for water surrounding ANTARCTICA and extending northward to 50 degrees south latitude. Also known as the Antarctic Ocean.

Straits. See under individual names.

Sunda Shelf. One of the largest continental shelves in the world, nearly 772,000 square miles (2 million sq. km.). Located in the JAVA SEA, SOUTH CHINA SEA, and Gulf of THAILAND. The area was above water in the Quaternary period, enabling large animals such as elephants and rhinoceros to migrate to Sumatra, Java, and Borneo.

Surtsey Island. Island formed by a volcanic explosion off the coast of ICELAND in 1963. It is valuable to scientists studying how island flora and fauna develop and is a popular tourist site.

Tashima Current. See TSUSHIMA CURRENT.

Tasman Sea. Area of the PACIFIC OCEAN off the southeast coast of AUSTRALIA, between Australia and Tasmania on the west and New Zealand on the east. First crossed by the Moriois people sometime before 1300 CE Also called the Tasmanian Sea.

Tasmanian Sea. See TASMAN SEA.

Thailand, Gulf of. Also known as the Gulf of Siam, inlet of the South China Sea, located between the Malay Archipelago and the Southeast Asian mainland. Bounded by Thailand, Cambodia, and Vietnam.

Tonga Trench. Ocean trench in the PACIFIC OCEAN, northeast of New Zealand. It stretches for 870 miles (1,400 km.), beginning at the northern end of the KERMADEC TRENCH. Depth is 32,810 feet (10,000 meters).

Tsushima Current. Warm current in the western PACIFIC OCEAN that flows out of the YELLOW SEA into the Sea of JAPAN in the spring and summer. Also called Tashima Current.

Walvis Ridge (Walfisch Ridge). Long, narrow undersea elevation near the southwestern coast of AFRICA, which extends about 1,900 miles (3,000 km.) in a southwesterly direction under the ATLANTIC OCEAN.

Weddell Sea. Bay in the SOUTHERN OCEAN bounded by the ANTARCTIC peninsula on the west and a northward bulge of ANTARCTICA on the east, stretching from approximately longitude 60 to 10 degrees west. One of the harshest environments on Earth; surface water temperatures stay near 32 degrees Fahrenheit (0 degrees Celsius) all year. The Weddell Sea was the site of much whaling and seal hunting in the nineteenth and twentieth centuries.

West Caroline Trench. See YAP TRENCH.

West Wind Drift. Surface portion of the ANTARCTIC CIRCUMPOLAR CURRENT, driven by westerly winds. Often extremely rough; seas as high as 98 feet (30 meters) have been reported.

Western Hemisphere. The half of the earth containing NORTH and SOUTH AMERICA; generally understood to fall between longitudes 160 degrees east and 20 degrees west.

Wilkes Land. Broad section near the coast of ANTARCTICA, which lies south of AUSTRALIA and east of the AMERICAN HIGHLANDS. Wilkes Land is the nearest landmass to the South MAGNETIC POLE.

Yap Trench. Ocean trench in the western PACIFIC OCEAN, about 435 miles (700 km.) long and 27,900 feet (8,527 meters) deep. Also called the West Caroline Trench.

Yellow Sea. Area of the PACIFIC OCEAN bounded by China on the north and west and Korea on the east. Named for the large amounts of yellow dust carried into it from central China by winds and by the Yangtze, Yalu, and Yellow Rivers. Parts of the sea often show a yellow color from the dust.

Kelly Howard

WORLD'S OCEANS AND SEAS

Name	Approximate Area		Average Depth	
	Sq. Miles	Sq. Km.	Feet	Meters
Pacific Ocean	64,000,000	165,760,000	13,215	4,028
Atlantic Ocean	31,815,000	82,400,000	12,880	3,926
Indian Ocean	25,300,000	65,526,700	13,002	3,963
Arctic Ocean	5,440,200	14,090,000	3,953	1,205
Mediterranean & Black Seas	1,145,100	2,965,800	4,688	1,429
Caribbean Sea	1,049,500	2,718,200	8,685	2,647
South China Sea	895,400	2,319,000	5,419	1,652
Bering Sea	884,900	2,291,900	5,075	1,547
Gulf of Mexico	615,000	1,592,800	4,874	1,486
Okhotsk Sea	613,800	1,589,700	2,749	838
East China Sea	482,300	1,249,200	617	188
Hudson Bay	475,800	1,232,300	420	128
Japan Sea	389,100	1,007,800	4,429	1,350
Andaman Sea	308,100	797,700	2,854	870
North Sea	222,100	575,200	308	94
Red Sea	169,100	438,000	1,611	491
Baltic Sea	163,000	422,200	180	55

MAJOR LAND AREAS OF THE WORLD

Area	Approximate Land Area		Percent of World Total
	Sq. Mi.	Sq. Km.	
World	57,308,738	148,429,000	100.0
Asia (including Middle East)	17,212,041	44,579,000	30.0
Africa	11,608,156	30,065,000	20.3
North America	9,365,290	24,256,000	16.3
Central America, South America, & Caribbean	6,879,952	17,819,000	8.9
Antarctica	5,100,021	13,209,000	8.9
Europe	3,837,082	9,938,000	6.7
Oceania, including Australia	2,967,966	7,687,000	5.2

MAJOR ISLANDS OF THE WORLD

| | | Area | |
		Sq. Mi.	Sq. Km.
Island	Location		
Greenland	North Atlantic Ocean	839,999	2,175,597
New Guinea	Western Pacific Ocean	316,615	820,033
Borneo	Western Pacific Ocean	286,914	743,107
Madagascar	Western Indian Ocean	226,657	587,042
Baffin	Canada, North Atlantic Ocean	183,810	476,068
Sumatra	Indonesia, northeast Indian Ocean	182,859	473,605
Hōnshū	Japan, western Pacific Ocean	88,925	230,316
Great Britain	North Atlantic Ocean	88,758	229,883
Ellesmere	Canada, Arctic Ocean	82,119	212,688
Victoria	Canada, Arctic Ocean	81,930	212,199
Sulawesi (Celebes)	Indonesia, western Pacific Ocean	72,986	189,034
South Island	New Zealand, South Pacific Ocean	58,093	150,461
Java	Indonesia, Indian Ocean	48,990	126,884
North Island	New Zealand, South Pacific Ocean	44,281	114,688
Cuba	Caribbean Sea	44,218	114,525
Newfoundland	Canada, North Atlantic Ocean	42,734	110,681
Luzon	Philippines, western Pacific Ocean	40,420	104,688
Iceland	North Atlantic Ocean	39,768	102,999
Mindanao	Philippines, western Pacific Ocean	36,537	94,631
Ireland	North Atlantic Ocean	32,597	84,426
Hōkkaidō	Japan, western Pacific Ocean	30,372	78,663
Hispaniola	Caribbean Sea	29,355	76,029
Tasmania	Australia, South Pacific Ocean	26,215	67,897
Sri Lanka	Indian Ocean	25,332	65,610
Sakhalin (Karafuto)	Russia, western Pacific Ocean	24,560	63,610
Banks	Canada, Arctic Ocean	23,230	60,166
Devon	Canada, Arctic Ocean	20,861	54,030
Tierra del Fuego	Southern tip of South America	18,605	48,187
Kyūshū	Japan, western Pacific Ocean	16,223	42,018
Melville	Canada, Arctic Ocean	16,141	41,805
Axel Heiberg	Canada, Arctic Ocean	15,779	40,868
Southampton	Hudson Bay, Canada	15,700	40,663

COUNTRIES OF THE WORLD

Country	Region	Population	Area Square Miles	Area Square Kilometers	Population Density Persons/ Sq. Mi.	Population Density Persons/ Sq. Km.
Afghanistan	Asia	31,575,018	249,347	645,807	127	49
Albania	Europe	2,862,427	11,082	28,703	259	100
Algeria	Africa	42,545,964	919,595	2,381,741	47	18
Andorra	Europe	76,177	179	464	425	164
Angola	Africa	29,250,009	481,354	1,246,700	60	23
Antigua and Barbuda	Caribbean	104,084	171	442	609	235
Argentina	South America	44,938,712	1,073,518	2,780,400	41	16
Armenia	Europe	2,962,100	11,484	29,743	259	100
Australia	Australia	25,576,880	2,969,907	7,692,024	9	3
Austria	Europe	8,877,036	32,386	83,879	275	106
Azerbaijan	Asia	10,027,874	33,436	86,600	300	116
Bahamas	Caribbean	386,870	5,382	13,940	73	28
Bahrain	Asia	1,543,300	300	778	5,136	1,983
Bangladesh	Asia	167,888,084	55,598	143,998	3,020	1,166
Barbados	Caribbean	287,025	166	430	1,730	668
Belarus	Europe	9,465,300	80,155	207,600	119	46
Belgium	Europe	11,515,793	11,787	30,528	976	377
Belize	Central America	398,050	8,867	22,965	44	17
Benin	Africa	11,733,059	43,484	112,622	269	104
Bhutan	Asia	821,592	14,824	38,394	55	21
Bolivia	South America	11,307,314	424,164	1,098,581	26	10
Bosnia and Herzegovina	Europe	3,511,372	19,772	51,209	179	69
Botswana	Africa	2,302,878	224,607	581,730	10.4	4
Brazil	South America	210,951,255	3,287,956	8,515,767	64	25
Brunei	Asia	421,300	2,226	5,765	189	73
Bulgaria	Europe	7,000,039	42,858	111,002	163	63
Burkina Faso	Africa	20,244,080	104,543	270,764	194	75
Burundi	Africa	10,953,317	10,740	27,816	1,020	394
Cambodia	Asia	16,289,270	69,898	181,035	233	90
Cameroon	Africa	24,348,251	179,943	466,050	135	52
Canada	North America	37,878,499	3,855,103	9,984,670	10	4
Cape Verde	Africa	550,483	1,557	4,033	352	136
Central African Republic	Africa	4,737,423	240,324	622,436	21	8
Chad	Africa	15,353,184	495,755	1,284,000	31	12
Chile	South America	17,373,831	291,930	756,096	60	23
China, People's Republic of	Asia	1,400,781,440	3,722,342	9,640,821	376	145
Colombia	South America	46,103,400	440,831	1,141,748	105	40
Comoros	Africa	873,724	719	1,861	1,215	469

Country	Region	Population	Area Square Miles	Area Square Kilometers	Population Density Persons/ Sq. Mi.	Population Density Persons/ Sq. Km.
Costa Rica	Central America	5,058,007	19,730	51,100	256	99
Côte d'Ivoire (Ivory Coast)	Africa	25,823,071	124,680	322,921	207	80
Croatia	Europe	4,087,843	21,831	56,542	186	72
Cuba	Caribbean	11,209,628	42,426	109,884	264	102
Cyprus	Europe	864,200	2,276	5,896	381	147
Czech Republic	Europe	10,681,161	30,451	78,867	350	135
Dem. Republic of the Congo	Africa	86,790,567	905,446	2,345,095	96	37
Denmark	Europe	5,814,461	16,640	43,098	350	135
Djibouti	Africa	1,078,373	8,880	23,000	122	47
Dominica	Caribbean	71,808	285	739	251	97
Dominican Republic	Caribbean	10,358,320	18,485	47,875	559	216
Ecuador	South America	17,398,588	106,889	276,841	163	63
Egypt	Africa	99,873,587	387,048	1,002,450	258	100
El Salvador	Central America	6,704,864	8,124	21,040	826	319
Equatorial Guinea	Africa	1,358,276	10,831	28,051	124	48
Eritrea	Africa	3,497,117	46,757	121,100	75	29
Estonia	Europe	1,324,820	17,505	45,339	75	29
Eswatini (Swaziland)	Africa	1,159,250	6,704	17,364	174	67
Ethiopia	Africa	107,534,882	410,678	1,063,652	262	101
Fed. States of Micronesia	Pacific Islands	105,300	271	701	388	150
Fiji	Pacific Islands	884,887	7,078	18,333	124	48
Finland	Europe	5,527,405	130,666	338,424	41	16
France	Europe	67,022,000	210,026	543,965	319	123
Gabon	Africa	2,067,561	103,347	267,667	21	8
Gambia	Africa	2,228,075	4,127	10,690	539	208
Georgia	Europe	3,729,600	26,911	69,700	140	54
Germany	Europe	83,073,100	137,903	357,168	603	233
Ghana	Africa	30,280,811	92,098	238,533	329	127
Greece	Europe	10,724,599	50,949	131,957	210	81
Grenada	Caribbean	108,825	133	344	818	316
Guatemala	Central America	17,679,735	42,042	108,889	420	162
Guinea	Africa	12,218,357	94,926	245,857	129	50
Guinea-Bissau	Africa	1,604,528	13,948	36,125	114	44
Guyana	South America	782,225	83,012	214,999	9.3	3.6
Haiti	Caribbean	11,263,077	10,450	27,065	1,077	416
Honduras	Central America	9,158,345	43,433	112,492	210	81
Hungary	Europe	9,764,000	35,919	93,029	272	105
Iceland	Europe	360,390	39,682	102,775	9.1	3.5
India	Asia	1,357,041,500	1,269,211	3,287,240	1,069	413
Indonesia	Asia	268,074,600	735,358	1,904,569	365	141
Iran	Asia	83,096,438	636,372	1,648,195	131	50

Country	Region	Population	Area		Population Density	
			Square Miles	Square Kilometers	Persons/ Sq. Mi.	Persons/ Sq. Km.
Iraq	Asia	39,309,783	169,235	438,317	233	90
Ireland	Europe	4,921,500	27,133	70,273	181	70
Israel	Asia	9,141,680	8,522	22,072	1,073	414
Italy	Europe	60,252,824	116,336	301,308	518	200
Jamaica	Caribbean	2,726,667	4,244	10,991	642	248
Japan	Asia	126,140,000	145,937	377,975	865	334
Jordan	Asia	10,587,132	34,495	89,342	307	119
Kazakhstan	Asia	18,592,700	1,052,090	2,724,900	18	7
Kenya	Africa	47,564,296	224,647	581,834	212	82
Kiribati	Pacific Islands	120,100	313	811	383	148
Kuwait	Asia	4,420,110	6,880	17,818	642	248
Kyrgyzstan	Asia	6,309,300	77,199	199,945	83	32
Laos	Asia	6,492,400	91,429	236,800	70	27
Latvia	Europe	1,910,400	24,928	64,562	78	30
Lebanon	Asia	6,855,713	4,036	10,452	1,740	672
Lesotho	Africa	2,263,010	11,720	30,355	194	75
Liberia	Africa	4,475,353	37,466	97,036	119	46
Libya	Africa	6,470,956	683,424	1,770,060	9.6	3.7
Liechtenstein	Europe	38,380	62	160	622	240
Lithuania	Europe	2,793,466	25,212	65,300	111	43
Luxembourg	Europe	613,894	998	2,586	614	237
Madagascar	Africa	25,680,342	226,658	587,041	114	44
Malawi	Africa	17,563,749	45,747	118,484	383	148
Malaysia	Asia	32,715,210	127,724	330,803	256	99
Maldives	Asia	378,114	115	298	3,287	1,269
Mali	Africa	19,107,706	482,077	1,248,574	39	15
Malta	Europe	493,559	122	315	3,911	1,510
Marshall Islands	Pacific Islands	55,500	70	181	795	307
Mauritania	Africa	3,984,233	397,955	1,030,700	10.4	4
Mauritius	Africa	1,265,577	788	2,040	1,606	620
Mexico	North America	126,577,691	759,516	1,967,138	166	64
Moldova	Europe	2,681,735	13,067	33,843	205	79
Monaco	Europe	38,300	0.78	2.02	49,106	18,960
Mongolia	Asia	3,000,000	603,902	1,564,100	4.9	1.9
Montenegro	Europe	622,182	5,333	13,812	117	45
Morocco	Africa	35,773,773	172,414	446,550	207	80
Mozambique	Africa	28,571,310	308,642	799,380	93	36
Myanmar (Burma)	Asia	54,339,766	261,228	676,577	207	80
Namibia	Africa	2,413,643	318,580	825,118	7.5	2.9
Nauru	Pacific Islands	11,000	8	21	1,357	524
Nepal	Asia	29,609,623	56,827	147,181	521	201

Country	Region	Population	Area Square Miles	Area Square Kilometers	Population Density Persons/ Sq. Mi.	Population Density Persons/ Sq. Km.
Netherlands	Europe	17,370,348	16,033	41,526	1,083	418
New Zealand	Pacific Islands	4,952,186	104,428	270,467	47	18
Nicaragua	Central America	6,393,824	46,884	121,428	137	53
Niger	Africa	21,466,863	458,075	1,186,408	47	18
Nigeria	Africa	200,962,000	356,669	923,768	565	218
North Korea	Asia	25,450,000	46,541	120,540	546	211
North Macedonia	Europe	2,077,132	9,928	25,713	210	81
Norway	Europe	5,328,212	125,013	323,782	41	16
Oman	Asia	4,183,841	119,499	309,500	36	14
Pakistan	Asia	218,198,000	310,403	803,940	703	271
Palau	Pacific Islands	17,900	171	444	104	40
Panama	Central America	4,158,783	28,640	74,177	145	56
Papua New Guinea	Pacific Islands	8,558,800	178,704	462,840	47	18
Paraguay	South America	7,052,983	157,048	406,752	44	17
Peru	South America	32,162,184	496,225	1,285,216	65	25
Philippines	Asia	108,785,760	115,831	300,000	939	363
Poland	Europe	38,386,000	120,728	312,685	319	123
Portugal	Europe	10,276,617	35,556	92,090	290	112
Qatar	Asia	2,740,479	4,468	11,571	614	237
Republic of the Congo	Africa	5,399,895	132,047	342,000	41	16
Romania	Europe	19,405,156	92,043	238,391	210	81
Russia[1]	Europe/Asia	146,877,088	6,612,093	17,125,242	23	9
Rwanda	Africa	12,374,397	10,169	26,338	1,217	470
St Kitts and Nevis	Caribbean	56,345	104	270	541	209
St Lucia	Caribbean	180,454	238	617	756	292
St Vincent and Grenadines	Caribbean	110,520	150	389	736	284
Samoa	Pacific Islands	199,052	1,093	2,831	181	70
San Marino	Europe	34,641	24	61	1,471	568
Sahrawi Arab Dem. Rep.[2]	Africa	567,421	97,344	252,120	6	2.3
São Tomé and Príncipe	Africa	201,784	386	1,001	523	202
Saudi Arabia	Asia	34,218,169	830,000	2,149,690	41	16
Senegal	Africa	16,209,125	75,955	196,722	212	82
Serbia	Europe	6,901,188	29,913	77,474	231	89
Seychelles	Africa	96,762	176	455	552	213
Sierra Leone	Africa	7,901,454	27,699	71,740	285	110
Singapore	Asia	5,638,700	279	722.5	20,212	7,804
Slovakia	Europe	5,450,421	18,933	49,036	287	111
Slovenia	Europe	2,084,301	7,827	20,273	267	103
Solomon Islands	Pacific Islands	682,500	10,954	28,370	62	24
Somalia	Africa	15,181,925	246,201	637,657	62	24
South Africa	Africa	58,775,022	471,359	1,220,813	124	48

Country	Region	Population	Area Square Miles	Area Square Kilometers	Population Density Persons/ Sq. Mi.	Population Density Persons/ Sq. Km.
South Korea	Asia	51,811,167	38,691	100,210	1,339	517
South Sudan	Africa	12,575,714	248,777	644,329	52	20
Spain	Europe	46,934,632	195,364	505,990	241	93
Sri Lanka	Asia	21,803,000	25,332	65,610	860	332
Sudan	Africa	40,782,742	710,251	1,839,542	57	22
Suriname	South America	568,301	63,251	163,820	9.1	3.5
Sweden	Europe	10,344,405	173,860	450,295	59	23
Switzerland	Europe	8,586,550	15,940	41,285	539	208
Syria	Asia	17,070,135	71,498	185,180	238	92
Taiwan	Asia	23,596,266	13,976	36,197	1,689	652
Tajikistan	Asia	9,127,000	55,251	143,100	166	64
Tanzania	Africa	55,890,747	364,900	945,087	153	59
Thailand	Asia	66,455,280	198,117	513,120	335	130
Timor-Leste	Asia	1,167,242	5,760	14,919	202	78
Togo	Africa	7,538,000	21,853	56,600	344	133
Tonga	Pacific Islands	100,651	278	720	362	140
Trinidad and Tobago	Caribbean	1,359,193	1,990	5,155	683	264
Tunisia	Africa	11,551,448	63,170	163,610	183	71
Turkey	Europe/Asia	82,003,882	302,535	783,562	271	105
Turkmenistan	Asia	5,851,466	189,657	491,210	31	12
Tuvalu	Pacific Islands	10,200	10	26	1,020	392
Uganda	Africa	40,006,700	93,263	241,551	429	166
Ukraine[3]	Europe	41,990,278	232,820	603,000	180	70
United Arab Emirates	Asia	9,770,529	32,278	83,600	303	117
United Kingdom	Europe	66,435,600	93,788	242,910	708	273
United States	North America	330,546,475	3,796,742	9,833,517	87	34
Uruguay	South America	3,518,553	68,037	176,215	52	20
Uzbekistan	Asia	32,653,900	172,742	447,400	189	73
Vanuatu	Pacific Islands	304,500	4,742	12,281	65	25
Vatican City	Europe	1,000	0.17	0.44	5,887	2,273
Venezuela	South America	32,219,521	353,841	916,445	91	35
Vietnam	Asia	96,208,984	127,882	331,212	751	290
Yemen	Asia	28,915,284	175,676	455,000	166	64
Zambia	Africa	16,405,229	290,585	752,612	57	22
Zimbabwe	Africa	15,159,624	150,872	390,757	101	39

Notes: (1) Including the population and area of Autonomous Republic of Crimea and City of Sevastopol, Ukraine's administrative areas on the Crimean Peninsula, which are claimed by Russia; (2) Administration is split between Morocco and the Sahrawi Arab Democratic Republic (Western Sahara), both of which claim the entire territory; (3) Excludes Crimea.
Source: U.S. Census Bureau, International Data Base

PAST AND PROJECTED WORLD POPULATION GROWTH, 1950-2050

Year	Approximate World Population	Ten-Year Growth Rate (%)
1950	2,556,000,053	18.9
1960	3,039,451,023	22.0
1970	3,706,618,163	20.2
1980	4,453,831,714	18.5
1990	5,278,639,789	15.2
2000	6,082,966,429	12.6
2010	6,848,932,929	10.7
2020	7,584,821,144	8.7
2030	8,246,619,341	7.3
2040	8,850,045,889	5.6
2050	9,346,399,468	—

Note: The listed years are the baselines for the estimated ten-year growth rate figures; for example, the rate for 1950-1960 was 18.9%.
Source: U.S. Census Bureau, International Data Base

WORLD'S LARGEST COUNTRIES BY AREA

			Area	
Rank	Country	Region	Sq. Miles	Sq. Km.
1	Russia	Europe/Asia	6,612,093	17,125,242
2	Canada	North America	3,855,103	9,984,670
3	United States	North America	3,796,742	9,833,517
4	China	Asia	3,722,342	9,640,821
5	Brazil	South America	3,287,956	8,515,767
6	Australia	Australia	2,969,907	7,692,024
7	India	Asia	1,269,211	3,287,240
8	Argentina	South America	1,073,518	2,780,400
9	Kazakhstan	Asia	1,052,090	2,724,900
10	Algeria	Africa	919,595	2,381,741
11	Democratic Rep. of the Congo	Africa	905,446	2,345,095
12	Saudi Arabia	Asia	830,000	2,149,690
13	Mexico	North America	759,516	1,967,138
14	Indonesia	Asia	735,358	1,904,569
15	Sudan	Africa	710,251	1,839,542
16	Libya	Africa	683,424	1,770,060
17	Iran	Asia	636,372	1,648,195
18	Mongolia	Asia	603,902	1,564,100
19	Peru	South America	496,225	1,285,216
20	Chad	Africa	495,755	1,284,000
21	Mali	Africa	482,077	1,248,574
22	Angola	Africa	481,354	1,246,700
23	South Africa	Africa	471,359	1,220,813
24	Niger	Africa	458,075	1,186,408
25	Colombia	South America	440,831	1,141,748
26	Bolivia	South America	424,164	1,098,581
27	Ethiopia	Africa	410,678	1,063,652
28	Mauritania	Africa	397,955	1,030,700
29	Egypt	Africa	387,048	1,002,450
30	Tanzania	Africa	364,900	945,087

Source: U.S. Census Bureau, International Data Base

WORLD'S SMALLEST COUNTRIES BY AREA

Rank	Country	Region	Area Sq. Miles	Area Sq. Km.
1	Vatican City*	Europe	0.17	0.44
2	Monaco*	Europe	0.78	2.02
3	Nauru	Pacific Islands	8	21
4	Tuvalu	Pacific Islands	10	26
5	San Marino*	Europe	24	61
6	Liechtenstein*	Europe	62	160
7	Marshall Islands	Pacific Islands	70	181
8	Saint Kitts and Nevis	Central America	104	270
9	Maldives	Asia	115	298
10	Malta	Europe	122	315
11	Grenada	Caribbean	133	344
12	Saint Vincent and the Grenadines	Central America	150	389
13	Barbados	Caribbean	166	430
14	Antigua and Barbuda	Caribbean	171	442
15	Palau	Pacific Islands	171	444
16	Seychelles	Africa	176	455
17	Andorra*	Europe	179	464
18	Saint Lucia	Central America	238	617
19	Federated States of Micronesia	Pacific Islands	271	701
20	Tonga	Pacific Islands	278	720

Note: Asterisks (*) denote countries on continents; all other countries are islands or island groups.
Source: U.S. Census Bureau, International Data Base

WORLD'S LARGEST COUNTRIES BY POPULATION

Rank	Country	Region	Population
1	China	Asia	1,401,028,280
2	India	Asia	1,357,746,150
3	United States	North America	329,229,067
4	Indonesia	Asia	268,074,600
5	Pakistan	Asia	218,385,000
6	Brazil	South America	211,032,216
7	Nigeria	Africa	200,962,000
8	Bangladesh	Asia	167,983,726
9	Russia	Europe/Asia	146,877,088
10	Mexico	North America	126,577,691
11	Japan	Asia	126,020,000
12	Philippines	Asia	108,210,625
13	Ethiopia	Africa	107,534,882
14	Egypt	Africa	99,930,038
15	Vietnam	Asia	96,208,984
16	Democratic Rep. of the Congo	Africa	86,790,567
17	Germany	Europe	83,149,300
18	Iran	Asia	83,142,818
19	Turkey	Europe/Asia	82,003,882
20	France	Europe	67,060,000
21	Thailand	Asia	66,461,867
22	United Kingdom	Europe	66,435,600
23	Italy	Europe	60,252,824
24	South Africa	Africa	58,775,022
25	Tanzania	Africa	55,890,747
26	Myanmar	Asia	54,339,766
27	South Korea	Asia	51,811,167
28	Kenya	Africa	47,564,296
29	Spain	Europe	46,934,632
30	Colombia	South America	46,127,200

Source: U.S. Census Bureau, International Data Base

World's Smallest Countries by Population

Rank	Country	Region	Population
1	Vatican City	Europe	1,000
2	Tuvalu	Pacific Islands	10,200
3	Nauru	Pacific Islands	11,000
4	Palau	Pacific Islands	17,900
5	San Marino	Europe	34,641
6	Monaco	Europe	38,300
7	Liechtenstein	Europe	38,380
8	Marshall Islands	Pacific Islands	55,500
9	Saint Kitts and Nevis	Central America	56,345
10	Dominica	Caribbean	71,808
11	Andorra	Europe	76,177
12	Seychelles	Africa	96,762
13	Tonga	Pacific Islands	100,651
14	Antigua and Barbuda	Caribbean	104,084
15	Federated States of Micronesia	Pacific Islands	105,300
16	Grenada	Caribbean	108,825
17	Saint Vincent and the Grenadines	Central America	110,520
18	Kiribati	Pacific Islands	120,100
19	Saint Lucia	Central America	180,454
20	Samoa	Pacific Islands	199,052
21	São Tomé and Príncipe	Africa	201,784
22	Barbados	Caribbean	287,025
23	Vanuatu	Pacific Islands	304,500
24	Iceland	Europe	360,390
25	Maldives	Asia	378,114
26	Bahamas	Caribbean	386,870
27	Belize	Central America	398,050
28	Brunei	Asia	421,300
29	Malta	Europe	493,559
30	Cape Verde	Africa	550,483

Source: U.S. Census Bureau, International Data Base.

WORLD'S MOST DENSELY POPULATED COUNTRIES

				Area		Persons Per Square	
Rank	Country	Region	Population	Sq. Miles	Sq. Km.	Mile	Km.
1	Monaco	Europe	38,300	0.78	2.02	49,106	18,960
2	Singapore	Asia	5,638,700	279	722.5	20,212	7,804
3	Vatican City	Europe	1,000	0.17	0.44	5,887	2,273
4	Bahrain	Asia	1,543,300	300	778	5,136	1,983
5	Malta	Europe	493,559	122	315	3,911	1,510
6	Maldives	Asia	378,114	115	298	3,287	1,269
7	Bangladesh	Asia	167,983,726	55,598	143,998	3,021	1,167
8	Lebanon	Asia	6,855,713	4,036	10,452	1,740	672
9	Barbados	Caribbean	287,025	166	430	1,730	668
10	Taiwan	Asia	23,596,266	13,976	36,197	1,689	652
11	Mauritius	Africa	1,265,577	788	2,040	1,606	620
12	San Marino	Europe	34,641	24	61	1,471	568
13	Nauru	Pacific Islands	11,000	8	21	1,344	519
14	South Korea	Asia	51,811,167	38,691	100,210	1,339	517
15	Rwanda	Africa	12,374,397	10,169	26,338	1,217	470
16	Comoros	Africa	873,724	719	1,861	1,215	469
17	Netherlands	Europe	17,426,881	16,033	41,526	1,087	420
18	Haiti	Caribbean	11,263,077	10,450	27,065	1,077	416
19	Israel	Asia	9,149,500	8,522	22,072	1,074	415
20	India	Asia	1,357,746,150	1,269,211	3,287,240	1,070	413
21	Burundi	Africa	10,953,317	10,740	27,816	1,020	394
22	Tuvalu	Pacific Islands	10,200	10	26	1,015	392
23	Belgium	Europe	11,515,793	11,849	30,689	974	376
24	Philippines	Asia	108,210,625	115,831	300,000	934	361
25	Japan	Asia	126,020,000	145,937	377,975	862	333
26	Sri Lanka	Asia	21,803,000	25,332	65,610	860	332
27	El Salvador	Central America	6,704,864	8,124	21,040	826	319
28	Grenada	Caribbean	108,825	133	344	818	316
29	Marshall Islands	Pacific Islands	55,500	70	181	795	307
30	Saint Lucia	Central America	180,454	238	617	756	292

Source: U.S. Census Bureau, International Data Base

WORLD'S LEAST DENSELY POPULATED COUNTRIES

Rank	Country	Region	Population	Area Sq. Miles	Area Sq. Km.	Persons Per Square Mile	Persons Per Square Km.
1	Mongolia	Asia	3,000,000	603,902	1,564,100	4.9	1.9
2	Western Sahara	Africa	567,421	97,344	252,120	6	2.3
3	Namibia	Africa	2,413,643	318,580	825,118	7.5	2.9
4	Australia	Australia	25,594,366	2,969,907	7,692,024	9	3
5	Suriname	South America	568,301	63,251	163,820	9.1	3.5
6	Iceland	Europe	360,390	39,682	102,775	9.1	3.5
7	Guyana	South America	782,225	83,012	214,999	9.3	3.6
8	Libya	Africa	6,470,956	683,424	1,770,060	9.6	3.7
9	Canada	North America	37,898,384	3,855,103	9,984,670	10	4
10	Mauritania	Africa	3,984,233	397,955	1,030,700	10.4	4
11	Botswana	Africa	2,302,878	224,607	581,730	10.4	4
12	Kazakhstan	Asia	18,592,700	1,052,090	2,724,900	18	7
13	Central African Republic	Africa	4,737,423	240,324	622,436	21	8
14	Gabon	Africa	2,067,561	103,347	267,667	21	8
15	Russia	Europe/Asia	146,877,088	6,612,093	17,125,242	23	9
16	Bolivia	South America	11,307,314	424,164	1,098,581	26	10
17	Chad	Africa	15,353,184	495,755	1,284,000	31	12
18	Turkmenistan	Asia	5,851,466	189,657	491,210	31	12
19	Oman	Asia	4,183,841	119,499	309,500	36	14
20	Mali	Africa	19,107,706	482,077	1,248,574	39	15
21	Argentina	South America	44,938,712	1,073,518	2,780,400	41	16
22	Saudi Arabia	Asia	34,218,169	830,000	2,149,690	41	16
23	Republic of the Congo	Africa	5,399,895	132,047	342,000	41	16
24	Finland	Europe	5,527,405	130,666	338,424	41	16
25	Norway	Europe	5,328,212	125,013	323,782	41	16
26	Paraguay	South America	7,052,983	157,048	406,752	44	17
27	Belize	Central America	398,050	8,867	22,965	44	17
28	Algeria	Africa	42,545,964	919,595	2,381,741	47	18
29	Niger	Africa	21,466,863	458,075	1,186,408	47	18
30	Papua New Guinea	Pacific Islands	8,558,800	178,704	462,840	47	18

Source: U.S. Census Bureau, International Data Base

WORLD'S MOST POPULOUS CITIES

Rank	City	Country	Region	Population
1	Chongqing	China	Asia	30,484,300
2	Shanghai	China	Asia	24,256,800
3	Beijing	China	Asia	21,516,000
4	Chengdu	China	Asia	16,044,700
5	Karachi	Pakistan	Asia	14,910,352
6	Guangzhou	China	Asia	14,043,500
7	Istanbul	Turkey	Europe	14,025,000
8	Tokyo	Japan	Asia	13,839,910
9	Tianjin	China	Asia	12,784,000
10	Mumbai	India	Asia	12,478,447
11	São Paulo	Brazil	South America	12,252,023
12	Moscow	Russia	Europe/Asia	12,197,596
13	Kinshasa	Dem. Rep. of Congo	Africa	11,855,000
14	Baoding	China	Asia	11,194,372
15	Lahore	Pakistan	Asia	11,126,285
16	Wuhan	China	Asia	11,081,000
17	Delhi	India	Asia	11,034,555
18	Harbin	China	Asia	10,635,971
19	Suzhou	China	Asia	10,459,890
20	Cairo	Egypt	Africa	10,230,350
21	Seoul	South Korea	Asia	10,197,604
22	Jakarta	Indonesia	Asia	10,075,310
23	Lima	Peru	South America	9,174,855
24	Mexico City	Mexico	North America	9,041,395
25	Ho Chi Minh City	Vietnam	Asia	8,993,082
26	Dhaka	Bangladesh	Africa	8,906,039
27	London	United Kingdom	Europe	8,825,001
28	Bangkok	Thailand	Asia	8,750,600
29	Xi'an	China	Asia	8,705,600
30	New York	United States	North America	8,622,698
31	Bangalore	India	Asia	8,425,970
32	Shenzhen	China	Asia	8,378,900
33	Nanjing	China	Asia	8,230,000
34	Tehran	Iran	Asia	8,154,051
35	Rio de Janeiro	Brazil	South America	6,718,903
36	Shantou	China	Asia	5,391,028
37	Kolkata	India	Asia	4,486,679
38	Shijiazhuang	China	Asia	4,303,700
39	Los Angeles	United States	North America	3,884,307
40	Buenos Aires	Argentina	South America	3,054,300

MAJOR LAKES OF THE WORLD

Lake	Location	Surface Area		Maximum Depth	
		Sq. Mi.	Sq. Km.	Feet	Meters
Caspian Sea	Central Asia	152,239	394,299	3,104	946
Superior	North America	31,820	82,414	1,333	406
Victoria	East Africa	26,828	69,485	270	82
Huron	North America	23,010	59,596	750	229
Michigan	North America	22,400	58,016	923	281
Aral	Central Asia	13,000	33,800	223	68
Tanganyika	East Africa	12,700	32,893	4,708	1,435
Baikal	Russia	12,162	31,500	5,712	1,741
Great Bear	North America	12,000	31,080	270	82
Nyasa	East Africa	11,600	30,044	2,316	706
Great Slave	North America	11,170	28,930	2,015	614
Chad	West Africa	9,946	25,760	23	7
Erie	North America	9,930	25,719	210	64
Winnipeg	North America	9,094	23,553	204	62
Ontario	North America	7,520	19,477	778	237
Balkhash	Central Asia	7,115	18,428	87	27
Ladoga	Russia	7,000	18,130	738	225
Onega	Russia	3,819	9,891	361	110
Titicaca	South America	3,141	8,135	1,214	370
Nicaragua	Central America	3,089	8,001	230	70
Athabasca	North America	3,058	7,920	407	124
Rudolf	Kenya, East Africa	2,473	6,405	—	—
Reindeer	North America	2,444	6,330	—	—
Eyre	South Australia	2,400	6,216	varies	varies
Issyk-Kul	Central Asia	2,394	6,200	2,297	700
Urmia	Southwest Asia	2,317	6,001	49	15
Torrens	Australia	2,200	5,698	—	—
Vänern	Sweden	2,141	5,545	322	98
Winnipegosis	North America	2,086	5,403	59	18
Mobutu Sese Seko	East Africa	2,046	5,299	180	55
Nettilling	North America	1,950	5,051	—	—

Note: The sizes of some lakes vary with the seasons.

Major Rivers of the World

River	Region	Source	Outflow	Approximate Length	
				Miles	Km.
Nile	N. Africa	Tributaries of Lake Victoria	Mediterranean Sea	4,180	6,690
Mississippi-Missouri-Red Rock	N. America	Montana	Gulf of Mexico	3,710	5,970
Yangtze Kiang	East Asia	Tibetan Plateau	China Sea	3,602	5,797
Ob	Russia	Altai Mountains	Gulf of Ob	3,459	5,567
Yellow (Huang He)	East Asia	Kunlun Mountains, west China	Gulf of Chihli	2,900	4,667
Yenisei	Russia	Tannu-Ola Mountains, western Tuva, Russia	Arctic Ocean	2,800	4,506
Paraná	S. America	Confluence of Paranaiba and Grande Rivers	Río de la Plata	2,795	4,498
Irtysh	Russia	Altai Mountains, Russia	Ob River	2,758	4,438
Congo	Africa	Confluence of Lualaba and Luapula Rivers, Congo	Atlantic Ocean	2,716	4,371
Heilong (Amur)	East Asia	Confluence of Shilka and Argun Rivers	Tatar Strait	2,704	4,352
Lena	Russia	Baikal Mountains, Russia	Arctic Ocean	2,652	4,268
Mackenzie	N. America	Head of Finlay River, British Columbia, Canada	Beaufort Sea	2,635	4,241
Niger	West Africa	Guinea	Gulf of Guinea	2,600	4,184
Mekong	Asia	Tibetan Plateau	South China Sea	2,500	4,023
Mississippi	N. America	Lake Itasca, Minnesota	Gulf of Mexico	2,348	3,779
Missouri	N. America	Confluence of Jefferson, Gallatin, and Madison Rivers, Montana	Mississippi River	2,315	3,726
Volga	Russia	Valdai Plateau, Russia	Caspian Sea	2,291	3,687
Madeira	S. America	Confluence of Beni and Maumoré Rivers, Bolivia-Brazil boundary	Amazon River	2,012	3,238
Purus	S. America	Peruvian Andes	Amazon River	1,993	3,207
São Francisco	S. America	S.W. Minas Gerais, Brazil	Atlantic Ocean	1,987	3,198

| River | Region | Source | Outflow | Approximate Length | |
				Miles	Km.
Yukon	N. America	Junction of Lewes and Pelly Rivers, Yukon Terr., Canada	Bering Sea	1,979	3,185
St. Lawrence	N. America	Lake Ontario	Gulf of St. Lawrence	1,900	3,058
Rio Grande	N. America	San Juan Mountains, Colorado	Gulf of Mexico	1,885	3,034
Brahmaputra	Asia	Himalayas	Ganges River	1,800	2,897
Indus	Asia	Himalayas	Arabian Sea	1,800	2,897
Danube	Europe	Black Forest, Germany	Black Sea	1,766	2,842
Euphrates	Asia	Confluence of Murat Nehri and Kara Su Rivers, Turkey	Shatt-al-Arab	1,739	2,799
Darling	Australia	Eastern Highlands, Australia	Murray River	1,702	2,739
Zambezi	Africa	Western Zambia	Mozambique Channel	1,700	2,736
Tocantins	S. America	Goiás, Brazil	Pará River	1,677	2,699
Murray	Australia	Australian Alps, New S. Wales	Indian Ocean	1,609	2,589
Nelson	N. America	Head of Bow River, western Alberta, Canada	Hudson Bay	1,600	2,575
Paraguay	S. America	Mato Grosso, Brazil	Paraná River	1,584	2,549
Ural	Russia	Southern Ural Mountains, Russia	Caspian Sea	1,574	2,533
Ganges	Asia	Himalayas	Bay of Bengal	1,557	2,506
Amu Darya (Oxus)	Asia	Nicholas Range, Pamir Mountains, Turkmenistan	Aral Sea	1,500	2,414
Japurá	S. America	Andes, Colombia	Amazon River	1,500	2,414
Salween	Asia	Tibet, south of Kunlun Mountains	Gulf of Martaban	1,500	2,414
Arkansas	N. America	Central Colorado	Mississippi River	1,459	2,348
Colorado	N. America	Grand County, Colorado	Gulf of California	1,450	2,333
Dnieper	Russia	Valdai Hills, Russia	Black Sea	1,419	2,284
Ohio-Allegheny	N. America	Potter County, Pennsylvania	Mississippi River	1,306	2,102
Irrawaddy	Asia	Confluence of Nmai and Mali rivers, northeast Burma	Bay of Bengal	1,300	2,092
Orange	Africa	Lesotho	Atlantic Ocean	1,300	2,092

River	Region	Source	Outflow	Approximate Length	
				Miles	Km.
Orinoco	S. America	Serra Parima Mountains, Venezuela	Atlantic Ocean	1,281	2,062
Pilcomayo	S. America	Andes Mountains, Bolivia	Paraguay River	1,242	1,999
Xi Jiang	East Asia	Eastern Yunnan Province, China	China Sea	1,236	1,989
Columbia	N. America	Columbia Lake, British Columbia, Canada	Pacific Ocean	1,232	1,983
Don	Russia	Tula, Russia	Sea of Azov	1,223	1,968
Sungari	East Asia	China-North Korea boundary	Amur River	1,215	1,955
Saskatchewan	N. America	Canadian Rocky Mountains	Lake Winnipeg	1,205	1,939
Peace	N. America	Stikine Mountains, British Columbia, Canada	Great Slave River	1,195	1,923
Tigris	Asia	Taurus Mountains, Turkey	Shatt-al-Arab	1,180	1,899

HIGHEST PEAKS IN EACH CONTINENT

Continent	Mountain	Location	Height Feet	Height Meters
Asia	Everest	Tibet & Nepal	29,028	8,848
South America	Aconcagua	Argentina	22,834	6,960
North America	McKinley	Alaska	20,320	6,194
Africa	Kilimanjaro	Tanzania	19,340	5,895
Europe	Elbrus	Russia & Georgia	18,510	5,642
Antarctica	Vinson Massif	Ellsworth Mountains	16,066	4,897
Australia	Kosciusko	New South Wales	7,316	2,228

Note: The world's highest sixty-six mountains are all in Asia.

MAJOR DESERTS OF THE WORLD

Desert	Location	Approximate Area		Type
		Sq. Miles	Sq. Km.	
Antarctic	Antarctica	5,400,000	14,002,200	polar
Sahara	North Africa	3,500,000	9,075,500	subtropical
Arabian	Southwest Asia	1,000,000	2,593,000	subtropical
Great Western (Gibson, Great Sandy, and Great Victoria)	Australia	520,000	1,348,360	subtropical
Gobi	East Asia	500,000	1,296,500	cold winter
Patagonian	Argentina, South America	260,000	674,180	cold winter
Kalahari	Southern Africa	220,000	570,460	subtropical
Great Basin	Western United States	190,000	492,670	cold winter
Thar	South Asia	175,000	453,775	subtropical
Chihuahuan	Mexico	175,000	453,775	subtropical
Karakum	Central Asia	135,000	350,055	cold winter
Colorado Plateau	Southwestern United States	130,000	337,090	cold winter
Sonoran	United States and Mexico	120,000	311,160	subtropical
Kyzylkum	Central Asia	115,000	298,195	cold winter
Taklimakan	China	105,000	272,265	cold winter
Iranian	Iran	100,000	259,300	cold winter
Simpson	Eastern Australia	56,000	145,208	subtropical
Mojave	Western United States	54,000	140,022	subtropical
Atacama	Chile, South America	54,000	140,022	cold coastal
Namib	Southern Africa	13,000	33,709	cold coastal
Arctic	Arctic Circle			polar

HIGHEST WATERFALLS OF THE WORLD

Waterfall	Location	Source	Height Feet	Meters
Angel	Canaima National Park, Venezuela	Rio Caroni	3,212	979
Tugela	Natal National Park, South Africa	Tugela River	3,110	948
Utigord	Norway	glacier	2,625	800
Monge	Marstein, Norway	Mongebeck	2,540	774
Mutarazi	Nyanga National Park, Zimbabwe	Mutarazi River	2,499	762
Yosemite	Yosemite National Park, California, U.S.	Yosemite Creek	2,425	739
Espelands	Hardanger Fjord, Norway	Opo River	2,307	703
Lower Mar Valley	Eikesdal, Norway	Mardals Stream	2,151	655
Tyssestrengene	Odda, Norway	Tyssa River	2,123	647
Cuquenan	Kukenan Tepuy, Venezuela	Cuquenan River	2,000	610
Sutherland	Milford Sound, New Zealand	Arthur River	1,904	580
Kjell	Gudvanger, Norway	Gudvangen Glacier	1,841	561
Takkakaw	Yoho Natl Park, British Columbia, Canada	Takkakaw Creek	1,650	503
Ribbon	Yosemite National Park, California, U.S.	Ribbon Stream	1,612	491
Upper Mar Valley	near Eikesdal, Norway	Mardals Stream	1,536	468
Gavarnie	near Lourdes, France	Gave de Pau	1,388	423
Vettis	Jotunheimen, Norway	Utla River	1,215	370
Hunlen	British Columbia, Canada	Hunlen River	1,198	365
Tin Mine	Kosciusko National Park, Australia	Tin Mine Creek	1,182	360
Silver Strand	Yosemite National Park, California, U.S.	Silver Strand Creek	1,170	357
Basaseachic	Baranca del Cobre, Mexico	Piedra Volada Creek	1,120	311
Spray Stream	Lauterburnnental, Switzerland	Staubbach Brook	985	300
Fachoda	Tahiti, French Polynesia	Fautaua River	985	300
King Edward VIII	Guyana	Courantyne River	850	259
Wallaman	near Ingham, Australia	Wallaman Creek	844	257
Gersoppa	Western Ghats, India	Sharavati River	828	253
Kaieteur	Guyana	Rio Potaro	822	251
Montezuma	near Rosebery, Tasmania	Montezuma River	800	240
Wollomombi	near Armidale, Australia	Wollomombi River	722	2203

Source: Fifth Continent Australia Pty Limited

Glossary

Places whose names are printed in SMALL CAPS *are subjects of their own entries in this glossary.*

Ablation. Loss of ice volume or mass by a GLACIER. Ablation includes melting of ice, SUBLIMATION, DEFLATION (removal by WIND), EVAPORATION, and CALVING. Ablation occurs in the lower portions of glaciers.

Abrasion. Wearing away of ROCKS in STREAMS by grinding, especially when rocks and SEDIMENT are carried along by stream water. The STREAMBED and VALLEY are carved out and eroded, and the rocks become rounded and smoothed by abrasion.

Absolute location. Position of any PLACE on the earth's surface. The absolute location can be given precisely in terms of DEGREES, MINUTES, and SECONDS of LATITUDE (0 to 90 degrees north or south) and of LONGITUDE (0 to 180 degrees east or west). The EQUATOR is 0 degrees latitude; the PRIME MERIDIAN, which runs through Greenwich in England, is 0 degrees longitude.

Abyss. Deepest part of the OCEAN. Modern TECHNOLOGY—especially sonar—has enabled accurate mapping of the ocean floors, showing that there are MOUNTAIN CHAINS, or RIDGES, in all the oceans, as well as deep CANYONS or TRENCHES closer to the edges of the oceans.

Acid rain. PRECIPITATION containing high levels of nitric or sulfuric acid; a major environmental problem in parts of North America, Europe, and Asia. Natural precipitation is slightly acidic (about 5.6 on the pH SCALE), because CARBON DIOXIDE—which occurs naturally in the ATMOSPHERE—is dissolved to form a weak carbonic acid.

Adiabatic. Change of TEMPERATURE within the ATMOSPHERE that is caused by compression or expansion without addition or loss of heat.

Advection. Horizontal movement of AIR from one PLACE to another in the ATMOSPHERE, associated with WINDS.

Advection fog. FOG that forms when a moist AIR mass moves over a colder surface. Commonly, warm moist air moves over a cool OCEAN CURRENT, so the air cools to SATURATION POINT and fog forms. This phenomenon, known as sea fog, occurs along subtropical west COASTS.

Aerosol. Substances held in SUSPENSION in the ATMOSPHERE, as solid particles or liquid droplets.

Aftershock. EARTHQUAKE that follows a larger earthquake and originates at or near the focus of the latter; many aftershocks may follow a major earthquake, decreasing in frequency and magnitude with time.

Agglomerate. Type of ROCK composed of volcanic fragments, usually of different sizes and rough or angular.

Aggradation. Accumulation of SEDIMENT in a STREAMBED. Aggradation often results from reduced flow in the channel during dry periods. It also occurs when the STREAM's load (BEDLOAD and SUSPENDED LOAD) is greater than the stream capacity. A BRAIDED STREAM pattern often results.

Air current. Air currents are caused by differential heating of the earth's surface, which causes heated air to rise. This causes WINDS at the surface as well as higher in the earth's ATMOSPHERE.

Air mass. Large body of air with distinctive homogeneous characteristics of TEMPERATURE, HUMIDITY, and stability. It forms when air remains stationary over a source REGION for a period of time, taking on the conditions of that region. An air mass can extend over a million square miles with a depth of more than a mile. Air masses are classified according to moisture content (*m* for maritime or *c* for continental) and temperature

(*A* for ARCTIC, *P* for polar, *T* for tropical, or *E* for equatorial). The air masses affecting North America are mP, cP, and mT. The interaction of AIR masses produces WEATHER. The line along which air masses meet is a FRONT.

Albedo. Measure of the reflective properties of a surface; the ratio of reflected ENERGY (INSOLATION) to the total incoming energy, expressed as a percentage. The albedo of Earth is 33 percent.

Alienation (land). Land alienation is the appropriation of land from its original owners by a more powerful force. In preindustrial societies, the ownership of agricultural land is of prime importance to subsistence farmers.

Alkali flat. Dry LAKEBED in an arid REGION, covered with a layer of SALTS. A well-known example is the Alkali Flat area of White Sands National Monument in New Mexico; it is the bed of a large lake that formed when the GLACIERS were melting. It is covered with a form of gypsum crystals called selenite. This material is blown off the surface into large SAND DUNES. Also called a salina. See also BITTER LAKE.

Allogenic sediment. SEDIMENT that originates outside the PLACE where it is finally deposited; SAND, SILT, and CLAY carried by a STREAM into a LAKE are examples.

Alluvial fan. Common LANDFORM at the mouth of a CANYON in arid REGIONS. Water flowing in a narrow canyon immediately slows as it leaves the canyon for the wider VALLEY floor, depositing the SEDIMENTS it was transporting. These spread out into a fan shape, usually with a BRAIDED STREAM pattern on its surface. When several alluvial fans grow side by side, they can merge into one continuous sloping surface between the HILLS and the valley. This is known by the Spanish word *bajada*, which means "slope."

Alluvial system. Any of various depositional systems, excluding DELTAS, that form from the activity of RIVERS and STREAMS. Much alluvial SEDIMENT is deposited when rivers top their BANKS and FLOOD the surrounding countryside. Buried alluvial sediments may be important water-bearing RESERVOIRS or may contain PETROLEUM.

Alluvium. Material deposited by running water. This includes not only fertile SOILS, but also CLAY, SILT, or SAND deposits resulting from FLUVIAL processes. FLOODPLAINS are covered in a thick layer of alluvium.

Altimeter. Instrument for measuring ALTITUDE, or height above the earth's surface, commonly used in airplanes. An altimeter is a type of ANEROID BAROMETER.

Altitudinal zonation. Existence of different ECOSYSTEMS at various ELEVATIONS above SEA LEVEL, due to TEMPERATURE and moisture differences. This is especially pronounced in Central America and South America. The hot and humid COASTAL PLAINS, where bananas and sugarcane thrive, is the *tierra caliente*. From about 2,500 to 6,000 feet (750–1,800 meters) is the *tierra templada*; crops grown here include coffee, wheat, and corn, and major cities are situated in this zone. From about 6,000 to 12,000 feet (1,800–3,600 meters) is the *tierra fria*; here only hardy crops such as potatoes and barley are grown, and large numbers of animals are kept. From about 12,000 to 15,000 feet (3,600 to 4,500 meters) lies the *tierra helada*, where hardy animals such as sheep and alpaca graze. Above 15,000 feet (4,500 meters) is the frozen *tierra nevada*; no permanent life is possible in the permanent SNOW and ICE FIELDS there.

Angle of repose. Maximum angle of steepness that a pile of loose materials such as SAND or ROCK can assume and remain stable; the angle varies with the size, shape, moisture, and angularity of the material.

Antecedent river. STREAM that was flowing before the land was uplifted and was able to erode at the pace of UPLIFT, thus creating a deep CANYON. Most deep canyons are attributed to antecedent rivers. In the Davisian CYCLE OF EROSION, this process was called REJUVENATION.

Anthropogeography. Branch of GEOGRAPHY founded in the late nineteenth century by German geographer Friedrich Ratzel. The field is closely related to human ECOLOGY—the study of humans, their DISTRIBUTION over the earth, and

478

their interaction with their physical ENVIRONMENT.

Anticline. Area where land has been UPFOLDED symmetrically. Its center contains stratigraphically older ROCKS. See also SYNCLINE.

Anticyclone. High-pressure system of rotating WINDS, descending and diverging, shown on a WEATHER chart by a series of closed ISOBARS, with a high in the center. In the NORTHERN HEMISPHERE, the rotation is CLOCKWISE; in the SOUTHERN HEMISPHERE, the rotation is COUNTERCLOCKWISE. An anticyclone brings warm weather.

Antidune. Undulatory upstream-moving bed form produced in free-surface flow of water over a SAND bed in a certain RANGE of high flow speeds and shallow flow depths.

Antipodes. TEMPERATE ZONE of the SOUTHERN HEMISPHERE. The term is now usually applied to the countries of Australia and New Zealand. The ancient Greeks had believed that if humans existed there, they must walk upside down. This idea was supported by the Christian Church in the Middle Ages.

Antitrade winds. WINDS in the upper ATMOSPHERE, or GEOSTROPHIC winds, that blow in the opposite direction to the TRADE WINDS. Antitrade winds blow toward the northeast in the NORTHERN HEMISPHERE and toward the southeast in the SOUTHERN HEMISPHERE.

Aperiodic. Irregularly occurring interval, such as found in most WEATHER CYCLES, rendering them virtually unpredictable.

Aphelion. Point in the earth's 365-DAY REVOLUTION when it is at its greatest distance from the SUN. This is caused by Earth's elliptical ORBIT around the Sun. The distance at aphelion is 94,555,000 miles (152,171,500 km.) and usually falls on July 4. The opposite of PERIHELION.

Aposelene. Earth's farthest point from the MOON.

Aquifer. Underground body of POROUS ROCK that contains water and allows water PERCOLATION through it. The largest aquifer in the United States is the Ogallala Aquifer, which extends south from South Dakota to Texas.

Arête. Serrated or saw-toothed ridge, produced in glaciated MOUNTAIN areas by CIRQUES eroding on either side of a RIDGE or mountain RANGE. From the French word for knife-edge.

Arroyo. Spanish word for a dry STREAMBED in an arid area. Called a WADI in Arabic and a WASH in English.

Artesian well. WELL from which GROUNDWATER flows without mechanical pumping, because the water comes from a CONFINED AQUIFER, and is therefore under pressure. The Great Artesian Basin of Australia has hundreds of artesian wells, called BORES, that provide drinking water for sheep and cattle. The name comes from the Artois REGION of France, where the phenomenon is common. A subartesian well is sunk into an UNCONFINED AQUIFER and requires a pump to raise water to the surface.

Asteroid belt. REGION between the ORBITS of Mars and Jupiter containing the majority of ASTEROIDS.

Asthenosphere. Part of the earth's UPPER MANTLE, beneath the LITHOSPHERE, in which PLATE movement takes place. Also known as the low-velocity zone.

Astrobleme. Remnant of a large IMPACT CRATER on Earth.

Astronomical unit (AU). Unit of measure used by astronomers that is equivalent to the average distance from the SUN to Earth (93 million miles/150 million km.).

Atmospheric pressure. Weight of the earth's ATMOSPHERE, equally distributed over earth's surface and pressing down as a result of GRAVITY. On average, the atmosphere has a force of 14.7 pounds per square inch (1 kilogram per centimeter) squared at SEA LEVEL, also expressed as 1013.2 millibars. Variations in atmospheric pressure, high or low, cause WINDS and WEATHER changes that affect CLIMATE. Pressure decreases rapidly with ALTITUDE or distance from the surface: Half of the total atmosphere is found below 18,000 feet (5,500 meters); more than 99 percent of the atmosphere is below 30 miles (50 km.) of the surface. Atmospheric pressure is measured with a BAROMETER.

Atoll. Ring-shaped growth of CORAL REEF, with a LAGOON in the middle. Charles Darwin, who observed many Pacific atolls during his voyage on the *Beagle* in the nineteenth century, suggested that they were created from FRINGING REEFS around volcanic ISLANDS. As such islands sank beneath the water (or as SEA LEVELS rose), the coral continued growing upward. SAND resting atop an atoll enables plants to grow, and small human societies have arisen on some atolls. The world's largest atoll, Kwajalein in the Marshall Islands, measures about 40 by 18 miles (65 by 30 km.), but perhaps the most famous atoll is Bikini Atoll—the SITE of nuclear-bomb testing during the 1950s.

Aurora. Glowing and shimmering displays of colored lights in the upper ATMOSPHERE, caused by interaction of the SOLAR WIND and the charged particles of the IONOSPHERE. Auroras occur at high LATITUDES. Near the North Pole they are called aurora borealis or northern lights; near the South Pole, aurora australis or southern lights.

Austral. Referring to an object or occurrence that is located in the SOUTHERN HEMISPHERE or related to Australia.

Australopithecines. Erect-walking early human ancestors with a cranial capacity and body size within the RANGE of modern apes rather than of humans.

Avalanche. Mass of SNOW and ice falling suddenly down a MOUNTAIN slope, often taking with it earth, ROCKS, and trees.

Bank. Elevated area of land beneath the surface of the OCEAN. The term is also used for elevated ground lining a body of water.

Bar (climate). Measure of ATMOSPHERIC PRESSURE per unit surface area of 1 million dynes per square centimeter. Millibars (thousandths of a bar) are the MEASUREMENT used in the United States. Other countries use kilopascals (kPa); one kilopascal is ten millibars.

Bar (land). RIDGE or long deposit of SAND or gravel formed by DEPOSITION in a RIVER or at the COAST. Offshore bars and baymouth bars are common coastal features.

Barometer. Instrument used for measuring ATMOSPHERIC PRESSURE. In the seventeenth century, Evangelista Torricelli devised the first barometer—a glass tube sealed at one end, filled with mercury, and upended into a bowl of mercury. He noticed how the height of the mercury column changed and realized this was a result of the pressure of air on the mercury in the bowl. Early MEASUREMENTS of atmospheric pressure were, therefore, expressed as centimeters of mercury, with average pressure at SEA LEVEL being 29.92 inches (760 millimeters). This cumbersome barometer was replaced with the ANEROID BAROMETER—a sealed and partially evacuated box connected to a needle and dial, which shows changes in atmospheric pressure. See also ALTIMETER.

Barrier island. Long chain of SAND islands that forms offshore, close to the COAST. LAGOONS or shallower MARSHES separate the barrier islands from the mainland. Such LOCATIONS are hazardous for SETTLEMENTS because they are easily swept away in STORMS and HURRICANES.

Basalt. IGNEOUS EXTRUSIVE ROCK formed when LAVA cools; often black in color. Sometimes basalt occurs in tall hexagonal columns, such as the Giant's Causeway in Ireland, or the Devils Postpile at Mammoth, California.

Basement. Crystalline, usually PRECAMBRIAN, IGNEOUS and METAMORPHIC ROCKS that occur beneath the SEDIMENTARY ROCK on the CONTINENTS.

Basin order. Approximate measure of the size of a STREAM BASIN, based on a numbering scheme applied to RIVER channels as they join together in their progress downstream.

Batholith. Large LANDFORM produced by IGNEOUS INTRUSION, composed of CRYSTALLINE ROCK, such as GRANITE; a large PLUTON with a surface area greater than 40 square miles (100 sq. km.). Most mountain RANGES have a batholith underneath.

Bathymetric contour. Line on a MAP of the OCEAN floor that connects points of equal depth.

Beaufort scale. SCALE that measures WIND force, expressed in numbers from 0 to 12. The original Beaufort scale was based on descriptions of the state of the SEA. It was adapted to land conditions, using descriptions of chimney smoke, leaves of trees, and similar factors. The scale was devised in the early nineteenth century by Sir Francis Beaufort, a British naval officer.

Belt. Geographical REGION that is distinctive in some way.

Bergeron process. PRECIPITATION formation in COLD CLOUDS whereby ice crystals grow at the expense of supercooled water droplets.

Bight. Wide or open BAY formed by a curve in the COASTLINE, such as the Great Australian Bight.

Biogenic sediment. SEDIMENT particles formed from skeletons or shells of microscopic plants and animals living in seawater.

Biostratigraphy. Identification and organization of STRATA based on their FOSSIL content and the use of fossils in stratigraphic correlation.

Bitter lake. Saline or BRACKISH LAKE in an arid area, which may dry up in the summer or in periods of DROUGHT. The water is not suitable for drinking. Another name for this feature is "salina." See also ALKALI FLAT.

Block lava. LAVA flows whose surfaces are composed of large, angular blocks; these blocks are generally larger than those of AA flows and have smooth, not jagged, faces.

Block mountain. MOUNTAIN or mountain RANGE with one side having a gentle slope to the crest, while the other slope, which is the exposed FAULT SCARP, is quite steep. It is formed when a large block of the earth's CRUST is thrust upward on one side only, while the opposite side remains in place. The Sierra Nevada in California are a good example of block mountains. Also known as fault-block mountain.

Blowhole. SEA CAVE or tunnel formed on some rocky, rugged COASTLINES. The pressure of the seawater rushing into the opening can force a jet of seawater to rise or spout through an opening in the roof of the cave. Blowholes are found in Scotland, Tasmania, and Mexico, and on the Hawaiian ISLANDS of Kauai and Maui.

Bluff. Steep slope that marks the farthest edge of a FLOODPLAIN.

Bog. Damp, spongy ground surface covered with decayed or decaying VEGETATION. Bogs usually are formed in cool CLIMATES through the in-filling, or silting up, of a LAKE. Moss and other plants grow outward toward the edge of the lake, which gradually becomes shallower, until the surface is completely covered. Bogs also can form on cold, damp MOUNTAIN surfaces. Many bogs are filled with PEAT.

Bore. Standing WAVE, or wall, of water created in a narrow ESTUARY when the strong incoming, or FLOOD, TIDE meets the RIVER water flowing outward; it moves upstream with the advancing tide, and downstream with the EBB TIDE. South America's Amazon River and Asia's Mekong River have large bores. In North America, the bore in the Bay of Fundy is visited by many tourists each year. Its St. Andrew's wharf is designed to handle changes in water level of as much as 53 feet (15 meters) in one DAY.

Boreal. Alluding to an item or event that is in the NORTHERN HEMISPHERE.

Bottom current. Deep-sea current that flows parallel to BATHYMETRIC CONTOURS.

Brackish water. Water with SALT content between that of SALT WATER and FRESH WATER; it is common in arid areas on the surface, in coastal MARSHES, and in salt-contaminated GROUNDWATER.

Braided stream. STREAM having a CHANNEL consisting of a maze of interconnected small channels within a broader STREAMBED. Braiding occurs when the stream's load exceeds its capacity, usually because of reduced flow.

Breaker. WAVE that becomes oversteepened as it approaches the SHORE, reaching a point at which it cannot maintain its vertical shape. It then breaks, and the water washes toward the shore.

Breakwater. Large structure, usually of ROCK, built offshore and parallel to the COAST, to absorb WAVE ENERGY and thus protect the SHORE. Between the breakwater and the shore is an area of calm water, often used as a boat anchorage or

HARBOR. A similar but smaller structure is a seawall.

Breeze. Gentle WIND with a speed of 4 to 31 miles (6 to 50 km.) per hour. On the BEAUFORT SCALE, the numbers 2 through 6 represent breezes of increasing strength.

Butte. Flat-topped HILL, smaller than a MESA, found in arid REGIONS.

Caldera. Large circular depression with steep sides, formed when a VOLCANO explodes, blowing away its top. The ERUPTION of Mount St. Helens produced a caldera. Crater Lake in Oregon is a caldera that has filled with water. From the Spanish word for kettle.

Calms of Cancer. Subtropical BELT of high pressure and light WINDS, located over the OCEAN near 25 DEGREES north LATITUDE. Also known as the HORSE LATITUDES.

Calms of Capricorn. Subtropical BELT of high pressure and light WINDS, located over the OCEAN near 25 DEGREES south LATITUDE.

Calving. Loss of glacial mass when GLACIERS reach the SEA and large blocks of ice break off, forming ICEBERGS.

Cancer, tropic of. PARALLEL of LATITUDE at 23.5 DEGREES north; this line is the latitude farthest north on the earth where the noon SUN is ever directly overhead. The REGION between it and the tropic of CAPRICORN is known as the TROPICS.

Capricorn, tropic of. Line of LATITUDE at 23.5 DEGREES south; this line is the latitude farthest south on the earth where the noon SUN is ever directly overhead. The REGION between it and the tropic of CANCER is known as the TROPICS.

Carbon dating. Method employed by physicists to determine the age of organic matter—such as a piece of wood or animal tissue—to determine the age of an archaeological or paleontological SITE. The method works on the principle that the amount of radioactive carbon in living matter diminishes at a steady, and measurable, rate after the matter dies. Technique is also known as carbon-14 dating, after the radioactive carbon-14 isotope it uses. Also known as radiocarbon dating.

Carrying capacity. Number of animals that a given area of land can support, without additional feed being necessary. Lush GRASSLAND may have a carrying capacity of twenty sheep per acre, while more arid, SEMIDESERT land may support only two sheep per acre. The term sometimes is used to refer to the number of humans who can be supported in a given area.

Catastrophism. Theory, popular in the eighteenth and nineteenth centuries, that explained the shape of LANDFORMS and CONTINENTS and the EXTINCTION of species as the results of intense or catastrophic events. The biblical FLOOD of Noah was one such event, which supposedly explained many extinctions. Catastrophism is linked closely to the belief that the earth is only about 6,000 years old, and therefore tremendous forces must have acted swiftly to create present LANDSCAPES. An alternative or contrasting theory is UNIFORMITARIANISM.

Catchment basin. Area of land receiving the PRECIPITATION that flows into a STREAM. Also called catchment or catchment area.

Central place theory. Theory that explains why some SETTLEMENTS remain small while others grow to be middle-sized TOWNS, and a few become large cities or METROPOLISES. The explanation is based on the provision of goods and services and how far people will travel to acquire these. The German geographer Walter Christaller developed this theory in the 1930s.

Centrality. Measure of the number of functions, or services, offered by any CITY in a hierarchy of cities within a COUNTRY or a REGION. See also CENTRAL PLACE THEORY.

Chain, mountain. Another term for mountain RANGE.

Chemical farming. Application of artificial FERTILIZERS to the SOIL and the use of chemical products such as insecticides, fungicides, and herbicides to ensure crop success. Chemical farming is practiced mainly in high-income countries, because the cost of the chemical products is high. Farmers in low-income economies rely more on natural organic fertilizers such as animal waste.

Chemical weathering. Chemical decomposition of solid ROCK by processes involving water that change its original materials into new chemical combinations.

Chlorofluorocarbons (CFCs). Manufactured compounds, not occurring in nature, consisting of chlorine, fluorine, and carbon. CFCs are stable and have heat-absorbing properties, so they have been used extensively for cooling in refrigeration and air-conditioning units. Previously, they were used as propellants for aerosol products. CFCs rise into the STRATOSPHERE where ULTRAVIOLET RADIATION causes them to react with OZONE, changing it to oxygen and exposing the earth to higher levels of ultraviolet (UV) radiation. Therefore, the manufacture and use of CFCs was banned in many countries. The commercial name for CFCs is Freon.

Chorology. Description or mapping of a REGION. Also known as chorography.

Chronometer. Highly accurate CLOCK or timekeeping device. The first accurate and effective chronometers were constructed in the mid-eighteenth century by John Harrison, who realized that accurate timekeeping was the secret to NAVIGATION at SEA.

Cinder cone. Small conical HILL produced by PYROCLASTIC materials from a VOLCANO. The material of the cone is loose SCORIA.

Circle of illumination. Line separating the sunlit part of the earth from the part in darkness. The circle of illumination moves around the earth once in every approximately 24 hours. At the VERNAL and autumnal EQUINOXES, the circle of illumination passes through the POLES.

Cirque. Circular BASIN at the head of an ALPINE GLACIER, shaped like an armchair. Many cirques can be seen in MOUNTAIN areas where glaciers have completely melted since the last ICE AGE.

City Beautiful movement. Planning and architectural movement that was at its height from around 1890 to the 1920s in the United States. It was believed that classical architecture, wide and carefully laid-out streets, parks, and urban mon-uments would reflect the higher values of the society and be a civilizing, even uplifting, experience for the citizens of such cities. Civic pride was fostered through remodeling or modernizing older URBAN AREAS. Chicago, Illinois, and Pasadena, California, are cities where the planners of the City Beautiful movement left their imprint.

Clastic. ROCK or sedimentary matter formed from fragments of older rocks.

Climatology. Study of Earth CLIMATES by analysis of long-term WEATHER patterns over a minimum of thirty years of statistical records. Climatologists—scientists who study climate—seek similarities to enable grouping into climatic REGIONS. Climate patterns are closely related to natural VEGETATION. Computer TECHNOLOGY has enabled investigation of phenomena such as the EL NIÑO effect and global climate change. The KÖPPEN CLIMATE CLASSIFICATION system is the most commonly used scheme for climate classification.

Climograph. Graph that plots TEMPERATURE and PRECIPITATION for a selected LOCATION. The most commonly used climographs plot monthly temperatures and monthly precipitation, as used in the KÖPPEN CLIMATE CLASSIFICATION. Also spelled "climagraph." The term climagram is rarely used.

Clinometer. Instrument used by surveyors to measure the ELEVATION of land or the inclination (slope) of the land surface.

Cloud seeding. Injection of CLOUD-nucleating particles into likely clouds to enhance PRECIPITATION.

Cloudburst. Heavy rain that falls suddenly.

Coal. One of the FOSSIL FUELS. Coal was formed from fossilized plant material, which was originally FOREST. It was then buried and compacted, which led to chemical changes. Most coal was formed during the CARBONIFEROUS PERIOD (286 million to 360 million years ago) when the earth's CLIMATE was wetter and warmer than at present.

Coastal plain. Large area of flat land near the OCEAN. Coastal plains can form in various ways, but FLUVIAL DEPOSITION is an important process. In the United States, the coastal plain extends from Texas to North Carolina.

Coastal wetlands. Shallow, wet, or flooded shelves that extend back from the freshwater-saltwater interface and may consist of MARSHES, BAYS, LAGOONS, tidal flats, or MANGROVE SWAMPS.

Cognitive map. Mental image that each person has of the world, which includes LOCATIONS and connections. These maps expand as children mature, from plans of their rooms, to their houses, to their neighborhoods. Adults know certain parts of the CITY and the streets connecting them.

Coke. Type of fuel produced by heating COAL.

Col. Lower section of a RIDGE, usually formed by the headward EROSION of two CIRQUE GLACIERS at an ARÊTE. Sometimes called a saddle.

Colonialism. Control of one COUNTRY over another STATE and its people. Many European countries have created colonial empires, including Great Britain, France, Spain, Portugal, the Netherlands, and Russia.

Columbian exchange. Interaction that occurred between the Americas and Europe after the voyages of Christopher Columbus. Food crops from the New World transformed the diet of many European countries.

Comet. Small body in the SOLAR SYSTEM, consisting of a solid head with a long gaseous tail. The elliptical ORBIT of a comet causes it to range from very close to the SUN to very far away. In ancient times, the appearance of a comet in the sky was thought to be an omen of great events or changes, such as war or the death of a king.

Comfort index. Number that expresses the combined effects of TEMPERATURE and HUMIDITY on human bodily comfort. The index number is obtained by measuring ambient conditions and comparing these to a chart.

Commodity chain. Network linking labor, production, delivery, and sale for any product. The chain begins with the production of the raw material, such as the extraction of MINERALS by miners, and extends to the acquisition of the finished product by a consumer.

Complex crater. IMPACT CRATER of large diameter and low depth-to-diameter ratio caused by the presence of a central UPLIFT or ring structure.

Composite cone. Cone or VOLCANO formed by volcanic explosions in which the LAVA is of different composition, sometimes fluid, sometimes PYROCLASTS such as cinders. The alternation of layers allows a concave shape for the cone. These are generally regarded as the world's most beautiful volcanoes. Composite volcanoes are sometimes called STRATOVOLCANOES.

Condensation nuclei. Microscopic particles that may have originated as DUST, soot, ASH from fires or VOLCANOES, or even SEA SALT; an essential part of CLOUD formation. When AIR rises and cools to the DEW POINT (saturation), the moisture droplets condense around the nuclei, leading to the creation of raindrops or snowflakes. A typical air mass might contain 10 billion condensation nuclei in a single cubic yard (1 cubic meter) of air.

Cone of depression. Cone-shaped depression produced in the WATER TABLE by pumping from a WELL.

Confined aquifer. AQUIFER that is completely filled with water and whose upper BOUNDARY is a CONFINING BED; it is also called an artesian aquifer.

Confining bed. Impermeable layer in the earth that inhibits vertical water movement.

Confluence. PLACE where two STREAMS or RIVERS flow together and join. The smaller of the two streams is called a TRIBUTARY.

Conglomerate. Type of SEDIMENTARY ROCK consisting of smaller rounded fragments naturally cemented together by another MINERAL. If the cemented fragments are jagged or angular, the rock is called breccia.

Conical projection. MAP PROJECTION that can be imagined as a cone of paper resting like a witch's hat on a globe with a light source at its center; the images of the CONTINENTS would be projected onto the paper. In reality, maps are constructed mathematically. A conic projection can show only part of one HEMISPHERE. This projection is suitable for constructing a MAP of the United States, as a good EQUAL-AREA representation can be achieved. Also called conic projection.

Consequent river. RIVER that flows across a LANDSCAPE because of GRAVITY. Its direction is determined by the original slope of the land. TRIBUTARY streams, which develop later as EROSION proceeds, are called subsequent streams.

Continental climate. CLIMATE experienced over the central REGIONS of large LANDMASSES; drier and subject to greater seasonal extremes of TEMPERATURE than at the CONTINENTAL MARGINS.

Continental rift zones. Continental rift zones are PLACES where the CONTINENTAL CRUST is stretched and thinned. Distinctive features include active VOLCANOES and long, straight VALLEY systems formed by normal FAULTS. Continental rifting in some cases has evolved into the breaking apart of a CONTINENT by SEAFLOOR SPREADING to form a new OCEAN.

Continental shelf. Shallow, gently sloping part of the seafloor adjacent to the mainland. The continental shelf is geologically part of the CONTINENT and is made of CONTINENTAL CRUST, whereas the OCEAN floor is OCEANIC CRUST. Although continental shelves vary greatly in width, on average they are about 45 miles (75 km.) wide and have slopes of 7 minutes (about one-tenth of a degree). The average depth of a continental shelf is about 200 feet (60 meters). The outer edge of the continental shelf is marked by a sharp change in angle where the CONTINENTAL SLOPE begins. Most continental shelves were exposed above current SEA LEVEL during the PLEISTOCENE EPOCH and have been submerged by rising sea levels over the past 18,000 years.

Continental shield. Area of a CONTINENT that contains the oldest ROCKS on Earth, called CRATONS. These are areas of granitic rocks, part of the CONTINENTAL CRUST, where there are ancient MOUNTAINS. The Canadian Shield in North America is an example.

Convectional rain. Type of PRECIPITATION caused when AIR over a warm surface is warmed and rises, leading to ADIABATIC cooling, CONDENSATION, and, if the air is moist enough, rain.

Convective overturn. Renewal of the bottom waters caused by the sinking of SURFACE WATERS that have become denser, usually because of decreased TEMPERATURE.

Convergence (climate). AIR flowing in toward a central point.

Convergence (physiography). Process that occurs during the second half of a SUPERCONTINENT CYCLE, whereby crustal PLATES collide and intervening OCEANS disappear as a result of plate SUBDUCTION.

Convergent plate boundary. Compressional PLATE BOUNDARY at which an oceanic PLATE is subducted or two continental plates collide.

Convergent plate margin. Area where the earth's LITHOSPHERE is returned to the MANTLE at a SUBDUCTION ZONE, forming volcanic "ISLAND ARCS" and associated HYDROTHERMAL activity.

Conveyor belt current. Large CYCLE of water movement that carries warm water from the north Pacific westward across the Indian Ocean, around Southern Africa, and into the Atlantic, where it warms the ATMOSPHERE, then returns at a deeper OCEAN level to rise and begin the process again.

Coordinated universal time (UTC). International basis of time, introduced to the world in 1964. The basis for UTC is a small number of ATOMIC CLOCKS. Leap seconds are occasionally added to UTC to keep it synchronized with universal time.

Core-mantle boundary. SEISMIC discontinuity 1,790 miles (2,890 km.) below the earth's surface that separates the MANTLE from the OUTER CORE.

Core region. Area, generally around a COUNTRY's CAPITAL CITY, that has a large, dense POPULATION and is the center of TRADE, financial services, and production. The rest of the country is referred to as the PERIPHERY. On a larger scale, the CONTINENT of Europe has a core region, which includes London, Paris, and Berlin; Iceland, Portugal, and Greece are peripheral LOCATIONS.

Coriolis effect. Apparent deflection of moving objects above the earth because of the earth's ROTAtion. The deflection is to the right in the NORTH-ERN HEMISPHERE and to the left in the SOUTH-ERN HEMISPHERE. The deflection is inversely proportional to the speed of the earth's rotation, being negligible at the EQUATOR but at its maximum near the POLES. The Coriolis effect is a major influence on the direction of surface WINDS. Sometimes called Coriolis force.

Corrasion. EROSION and lowering of a STREAMBED by FLUVIAL action, especially by ABRASION of the bedload (material transported by the STREAM) but also including SOLUTION by the water.

Cosmogony. Study of the origin and nature of the SOLAR SYSTEM.

Cotton Belt. Part of the United States extending from South Carolina through Georgia, Alabama, Mississippi, Tennessee, Louisiana, Arkansas, Texas, and Oklahoma, where cotton was grown on PLANTATIONS using slave labor before the Civil War. After that war, the South stagnated for almost a century. Racial SEGREGATION contributed to cultural isolation from the rest of the United States. Cotton is still produced in this REGION, but California has overtaken the Southern STATES as a cotton producer, and other agricultural products, such as soybeans and poultry, have become dominant crops in the old Cotton Belt. In-migration, due to the SUN BELT attraction, has led to rapid urban growth.

Counterurbanization. Out-migration of people from URBAN AREAS to smaller TOWNS or RURAL areas. As large modern cities are perceived to be overcrowded, stressful, polluted, and dangerous, many of their residents move to areas they regard as more favorable. Such moves are often related to individuals' retirements; however, younger workers and families are also part of counterurbanization.

Crater morphology. Structure or form of CRATERS and the related processes that developed them.

Craton. Large, geologically old, relatively stable CORE of a continental LITHOSPHERIC PLATE, sometimes termed a CONTINENTAL SHIELD.

Creep. Slow, gradual downslope movement of SOIL materials under gravitational stress. Creep tests are experiments conducted to assess the effects of time on ROCK properties, in which environmental conditions (surrounding pressure, TEMPERATURE) and the deforming stress are held constant.

Crestal plane. Plane or surface that goes through the highest points of all beds in a fold; it is coincident with the axial plane when the axial plane is vertical.

Cross-bedding. Layers of ROCK or SAND that lie at an angle to horizontal bedding or to the ground.

Crown land. Land belonging to a NATION'S MONARCHY.

Crude oil. Unrefined OIL, as it occurs naturally. Also called PETROLEUM.

Crustal movements. PLATE TECTONICS theorizes that Earth's CRUST is not a single rigid shell, but comprises a number of large pieces that are in motion, separating or colliding. There are two types of crust—the older continental and the much younger OCEANIC CRUST. When PLATES diverge, at SEAFLOOR SPREADING zones, new (oceanic) crust is created from the MAGMA that flows out at the MID-OCEAN RIDGES. When plates converge and collide, denser oceanic crust is SUBDUCTED under the lighter CONTINENTAL CRUST. The boundaries at the areas where plates slide laterally, neither diverging nor converging, are called TRANSFORM FAULTS. The San Andreas Fault represents the world's best-known transform BOUNDARY. As a result of crustal movements, the earth can be deformed in several ways. Where PLATE BOUNDARIES converge, compression can occur, leading to FOLDING and the creation of SYNCLINES and ANTICLINES. Other stresses of the crust can lead to fracture, or faulting, and accompanying EARTHQUAKES. LANDFORMS created in this way include HORSTS, GRABEN, and BLOCK MOUNTAINS.

Culture hearth. LOCATION in which a CULTURE has developed; a CORE REGION from which the culture later spread or diffused outward through a larger REGION. Mesopotamia, the Nile Valley, and the Peruvian ALTIPLANO are examples of culture hearths.

Curie point. TEMPERATURE at which a magnetic MINERAL locks in its magnetization. Also known as Curie temperature.

Cycle of erosion. Influential MODEL of LANDSCAPE change proposed by William Morris Davis near the end of the nineteenth century. The UPLIFT of a relatively flat surface, or PLAIN, in an area of moderate RAINFALL and TEMPERATURE, led to gradual EROSION of the initial surface in a sequence Davis categorized as Youth, Maturity, and Old Age. The final landscape was called PENEPLAIN. Davis also recognized the stage of REJUVENATION, when a new uplift could give new ENERGY to the cycle, leading to further downcutting and erosion. The model also was used to explain the sequence of LANDFORMS developed in REGIONS of ALPINE GLACIERS. The model has been criticized as misleading, since CRUSTAL MOVEMENT is continuous and more frequent than Davis perhaps envisaged, but remained useful as a description of TOPOGRAPHY. Also known as the Davisian cycle or geomorphic cycle.

Cyclonic rain. In the NORTHERN HEMISPHERE winter, two low-pressure systems or CYCLONES—the Aleutian Low and the Icelandic Low—develop over the OCEAN near 60 DEGREES north LATITUDE. The polar FRONT forms where the cold and relatively dry ARCTIC AIR meets the warmer, moist air carried by westerly WINDS. The warm air is forced upward, cools, and condenses. These cyclonic STORMS often move south, bringing winter PRECIPITATION to North America, especially to the STATES of Washington and Oregon.

Cylindrical projection. MAP PROJECTION that represents the earth's surface as a rectangle. It can be imagined as a cylinder of paper wrapped around a globe with a light source at its center; the images of the CONTINENTS would be projected onto the paper. In reality, MAPS are constructed mathematically. It is impossible to show the North Pole or South Pole on a cylindrical projection. Although the map is conformal, distortion of area is extreme beyond 50 DEGREES north and south LATITUDES. The Mercator projection, developed in the sixteenth century by the Flemish cartographer Gerardus Mercator, is the best-known cylindrical projection. It has been popular with seamen because the shortest route between two PORTS (the GREAT CIRCLE route) can be plotted as straight lines that show the COMPASS direction that should be followed. Use of this projection for other purposes, however, can lead to misunderstandings about size; for example, compare Greenland on a globe and on a Mercator map.

Datum level. Baseline or level from which other heights are measured, above or below. MEAN SEA LEVEL is the datum commonly used in surveying and in the construction of TOPOGRAPHIC MAPS.

Daylight saving time. System of seasonal adjustments in CLOCK settings designed to increase hours of evening sunlight during summer months. In the spring, clocks are set ahead one hour; in the fall, they are put back to standard time. In North America, these changes are made on the first Sunday in April and the last Sunday in October. The U.S. Congress standardized daylight saving time in 1966; however, parts of Arizona, Indiana, and Hawaii do not follow the system.

Débâcle. In a scientific context, this French word means the sudden breaking up of ice in a RIVER in the spring, which can lead to serious, sudden flooding.

Debris avalanche. Large mass of SOIL and ROCK that falls and then slides on a cushion of AIR downhill rapidly as a unit.

Debris flow. Flowing mass consisting of water and a high concentration of SEDIMENT with a wide RANGE of size, from fine muds to coarse gravels.

Declination, magnetic. Measure of the difference, in DEGREES, between the earth's NORTH MAGNETIC POLE and the North Pole on a MAP; this difference changes slightly each year. The needle of a magnetic COMPASS points to the earth's geomagnetic pole, which is not exactly the same as the North Pole of the geographic GRID or the set of lines of LATITUDE and LONGITUDE. The geomagnetic poles, north and south, mark the ends of the AXIS of the earth's MAGNETIC FIELD, but this field is not stationary. In fact, the geomagnetic poles have completely reversed hundreds of times throughout earth history. Lines of equal magnetic declination are called ISOGONIC LINES.

Declination of the Sun. LATITUDE of the SUBSOLAR POINT, the PLACE on the earth's surface where the SUN is directly overhead. In the course of a year, the declination of the Sun migrates from 23.5 DEGREES north LATITUDE, at the (northern) summer SOLSTICE, to 23.5 degrees south latitude, at the (northern) WINTER SOLSTICE. Hawaii is the only part of the United States that experiences the Sun directly overhead twice a year.

Deep-focus earthquakes. EARTHQUAKES occurring at depths ranging from 40 to 400 miles (70–700 km.) below the earth's surface. This RANGE of depths represents the zone from the base of the earth's CRUST to approximately one-quarter of the distance into Earth's MANTLE. Deep-focus earthquakes provide scientists information about the PLANET's interior structure, its composition, and SEISMICITY. Observation of deep-focus earthquakes has played a fundamental role in the discovery and understanding of PLATE TECTONICS.

Deep-ocean currents. Deep-ocean currents involve significant vertical and horizontal movements of seawater. They distribute oxygen- and nutrient-rich waters throughout the world's OCEANS, thereby enhancing biological productivity.

Defile. Narrow MOUNTAIN PASS or GORGE through which troops could march only in single file.

Deflation. EROSION by WIND, resulting in the removal of fine particles. The LANDFORM that typically results is a deflation hollow.

Deforestation. Removal or destruction of FORESTS. In the late twentieth century, there was widespread concern about tropical deforestation—destruction of the tropical RAINFOREST—especially that of Brazil. Forest clearing in the TROPICS is uneconomic because of low SOIL fertility. Deforestation causes severe EROSION and environmental damage; it also destroys habitat, which leads to the EXTINCTION of both plant and animal species.

Degradation. Process of CRATER EROSION from all processes, including WIND and other meteorological mechanisms.

Degree (geography). Unit of LATITUDE or LONGITUDE in the geographic GRID, used to determine ABSOLUTE LOCATION. One degree of latitude is about 69 miles (111 km.) on the earth's surface. It is not exactly the same everywhere, because the earth is not a perfect sphere. One degree of longitude varies greatly in length, because the MERIDIANS converge at the POLES. At the EQUATOR, it is 69 miles (111 km.), but at the North or South Pole it is zero.

Degree (temperature). Unit of MEASUREMENT of TEMPERATURE, based on the CELSIUS SCALE, except in the United States, which uses the FAHRENHEIT SCALE. On the Celsius scale, one degree is one-hundredth of the difference between the freezing point of water and the boiling point of water.

Demographic measure. Statistical data relating to POPULATION.

Demographic transition. MODEL of POPULATION change that fits the experience of many European countries, showing changes in birth and death rates. In the first stage, in preindustrial countries, population size was stable because both BIRTH RATES and DEATH RATES were high. Agricultural reforms, together with the INDUSTRIAL REVOLUTION and subsequent medical advances, led to a rapid fall in the death rate, so that the second and third stages of the model were periods of rapid population growth, often called the POPULATION EXPLOSION. In the fourth stage of the model, birth rates fall markedly, leading again to stable population size.

Dendritic drainage. Most common pattern of STREAMS and their TRIBUTARIES, occurring in areas of uniform ROCK type and regular slope. A MAP, or aerial photograph, shows a pattern like the veins on a leaf—smaller streams join the main stream at an acute angle.

Denudation. General word for all LANDFORM processes that lead to a lowering of the LANDSCAPE, including WEATHERING, mass movement, EROSION, and transport.

Deposition. Laying down of SEDIMENTS that have been transported by water, WIND, or ice.

Deranged drainage. LANDSCAPE whose integrated drainage network has been destroyed by irregular glacial DEPOSITION, yielding numerous shallow LAKE BASINS.

Derivative maps. MAPS that are prepared or derived by combining information from several other maps.

Desalinization. Process of removing SALT and MINERALS from seawater or from saline water occurring in AQUIFERS beneath the land surface to render it fit for AGRICULTURE or other human use.

Desert climate. Low PRECIPITATION, low HUMIDITY, high daytime TEMPERATURES, and abundant sunlight are characteristics of desert climates. The hot DESERTS of the world generally are located on the western sides of CONTINENTS, at LATITUDES from fifteen to thirty DEGREES north or south of the EQUATOR. One definition, based on precipitation, defines deserts as areas that receive between 0 and 9 inches (0 to 250 millimeters) of precipitation per year. REGIONS receiving more precipitation are considered to have a SEMIDESERT climate, in which some AGRICULTURE is possible.

Desert pavement. Surface covered with smoothed PEBBLES and gravels, found in arid areas where DEFLATION (WIND EROSION) has removed the smaller particles. Called a "gibber plain" in Australia.

Desertification. Increase in DESERT areas worldwide, largely as a result of overgrazing or poor agricultural practices in semiarid and marginal CLIMATES. DEFORESTATION, DROUGHT, and POPULATION increase also contribute to desertification. The REGION of Africa just south of the Sahara Desert, known as the SAHEL, is the largest and most dramatic demonstration of desertification.

Detrital rock. SEDIMENTARY ROCK composed mainly of grains of silicate MINERALS as opposed to grains of calcite or CLAYS.

Devolution. Breaking up of a large COUNTRY into smaller independent political units is the final and most extreme form of devolution. The Soviet Union devolved from one single country into fifteen separate countries in 1991. At an intermediate level, devolution refers to the granting of political autonomy or self-government to a REGION, without a complete split. The reopening of the Scottish Parliament in 1999 and the Northern Ireland parliament in 2000 are examples of devolution; the Parliament of the United Kingdom had previously met only in London and made laws there for all parts of the country. Canada experienced devolution with the creation of the new territory of Nunavut, whose residents elect the members of their own legislative assembly.

Dew point. TEMPERATURE at which an AIR mass becomes saturated and can hold no more moisture. Further cooling leads to CONDENSATION. At ground level, this produces DEW.

Diagenesis. Conversion of unconsolidated SEDIMENT into consolidated ROCK after burial by the processes of compaction, cementation, recrystallization, and replacement.

Diaspora. Dispersion of a group of people from one CULTURE to a variety of other REGIONS or to other lands. A Greek word, used originally to refer to the Jewish diaspora. Jewish people now live in many countries, although they have Israel as a HOMELAND. Similar to this are the diasporas of the Irish and the overseas Chinese.

Diastrophism. Deformation of the earth's CRUST by faulting or FOLDING.

Diatom ooze. Deposit of soft mud on the OCEAN floor consisting of the shells of diatoms, which are microscopic single-celled creatures with SILICA-rich shells. Diatom ooze deposits are located in the southern Pacific around Antarctica and in the northern Pacific. Other PELAGIC, or deep-ocean, SEDIMENTS include CLAYS and calcareous ooze.

Dike (geology). LANDFORM created by IGNEOUS intrusion when MAGMA or molten material within the earth forces its way in a narrow band through overlying ROCK. The dike can be exposed at the surface through EROSION.

Dike (water). Earth wall or DAM built to prevent flooding; an EMBANKMENT or artificial LEVEE. Sometimes specifically associated with structures built in the Netherlands to prevent the entry of seawater. The land behind the dikes was reclaimed for AGRICULTURE; these new fields are called POLDERS.

Distance-decay function. Rate at which an activity diminishes with increasing distance. The effect that distance has as a deterrent on human activity is sometimes described as the FRICTION OF DISTANCE. It occurs because of the time and cost of overcoming distances between people and their desired activity. An example of the distance-decay function is the rate of visitors to a football stadium. The farther people have to travel, the less likely they are to make this journey.

Distributary. STREAM that takes waters away from the main CHANNEL of a RIVER. A DELTA usually comprises many distributaries. Also called distributary channel.

Diurnal range. Difference between the highest and lowest TEMPERATURES registered in one twenty-four-hour period.

Diurnal tide. Having only one high tide and one low tide each lunar DAY; TIDES on some parts of the Gulf of Mexico are diurnal.

Divergent boundary. BOUNDARY that results where two PLATES are moving apart from each other, as is the case along MID-OCEANIC RIDGES.

Divergent margin. Area where the earth's CRUST and LITHOSPHERE form by SEAFLOOR SPREADING.

Doline. Large SINKHOLE or circular depression formed in LIMESTONE areas through the CHEMICAL WEATHERING process of carbonation.

Dolomite. MINERAL consisting of calcium and magnesium carbonate compounds that often forms from PRECIPITATION from seawater; it is abundant in ancient ROCKS.

Downwelling. Sinking of OCEAN water.

Drainage basin. Area of the earth's surface that is drained by a STREAM. Drainage basins vary greatly in size, but each is separated from the next by RIDGES, or drainage DIVIDES. The CATCHMENT of the drainage basin is the WATERSHED.

Drift ice. ARCTIC or ANTARCTIC ice floating in the open SEA.

Drumlin. Low HILL, shaped like half an egg, formed by DEPOSITION by CONTINENTAL GLACIERS. A drumlin is composed of TILL, or mixed-size materials. The wider end faces upstream of the glacier's movement; the tapered end points in the direction of the ice movement. Drumlins usually occur in groups or swarms.

Dust devil. Whirling cloud of DUST and small debris, formed when a small patch of the earth's surface becomes heated, causing hot AIR to rise; cooler air then flows in and begins to spin. The resulting dust devil can grow to heights of 150 feet (50 meters) and reach speeds of 35 miles (60 km.) per hour.

Dust dome. Dome of AIR POLLUTION, composed of industrial gases and particles, covering every large CITY in the world. The pollution sometimes is carried downwind to outlying areas.

Earth pillar. Formation produced when a boulder or caprock prevents EROSION of the material directly beneath it, usually CLAY. The clay is easily eroded away by water during RAINFALL, except where the overlying ROCK protects it. The result is a tall, slender column, as high as 20 feet (6.5 meters) in exceptional cases.

Earth radiation. Portion of the electromagnetic spectrum, from about 4 to 80 microns, in which the earth emits about 99 percent of its RADIATION.

Earth tide. Slight deformation of Earth resulting from the same forces that cause OCEAN TIDES, those that are exerted by the MOON and the SUN.

Earthflow. Term applied to both the process and the LANDFORM characterized by fluid downslope movement of SOIL and ROCK over a discrete plane of failure; the landform has a HUMMOCKY surface and usually terminates in discrete lobes.

Earth's heat budget. Balance between the incoming SOLAR RADIATION and the outgoing terrestrial reradiation.

Eclipse, lunar. Obscuring of all or part of the light of the MOON by the shadow of the earth. A lunar eclipse occurs at the full moon up to three times a year. The surface of the Moon changes from gray to a reddish color, then back to gray. The sequence may last several hours.

Eclipse, solar. At least twice a year, the SUN, MOON, and Earth are aligned in one straight line. At that time, the Moon obscures all the light of the Sun along a narrow band of the earth's surface, causing a total eclipse; in REGIONS of Earth adjoining that area, there is a partial eclipse. A corona (halo of light) can be seen around the Sun at the total eclipse. Viewing a solar eclipse with naked eyes is extremely dangerous and can cause blindness.

Ecliptic, plane of. Imaginary plane that would touch all points in the earth's ORBIT as it moves around the SUN. The angle between the plane of the ecliptic and the earth's AXIS is 66.5 DEGREES.

Edge cities. Forms of suburban downtown in which there are nodal concentrations of office space and shopping facilities. Edge cities are located close to major freeways or highway intersections, on the outer edges of METROPOLITAN AREAS.

Effective temperature. TEMPERATURE of a PLANET based solely on the amount of SOLAR RADIATION that the planet's surface receives; the effective temperature of a planet does not include the GREENHOUSE temperature enhancement effect.

Ejecta. Material ejected from the CRATER made by a meteoric impact.

Ekman layer. REGION of the SEA, from the surface to about 100 meters down, in which the WIND directly affects water movement.

Eluviation. Removal of materials from the upper layers of a SOIL by water. Fine material may be removed by SUSPENSION in the water; other material is removed by SOLUTION. The removal by solution is called LEACHING. Eluviation from an upper layer leads to illuviation in a lower layer.

Enclave. Piece of territory completely surrounded by another COUNTRY. Two examples are Lesotho, which is surrounded by the Republic of South Africa, and the Nagorno-Karabakh REGION, populated by Armenians but surrounded by Azerbaijan. The term is also used for smaller regions, such as ethnic neighborhoods within larger cities. See also EXCLAVE.

Endemic species. Species confined to a restricted area in a restricted ENVIRONMENT.

Endogenic sediment. SEDIMENT produced within the water column of the body in which it is deposited; for example, calcite precipitated in a LAKE in summer.

Environmental degradation. Situation that occurs in slum areas and SQUATTER SETTLEMENTS because of poverty and inadequate INFRASTRUCTURE. Too-rapid human POPULATION growth can lead to the accumulation of human waste and garbage, the POLLUTION of GROUNDWATER, and DENUDATION of nearby FORESTS. As a result, LIFE EXPECTANCY in such degraded areas is lower than in the RURAL communities from which many of the settlers came. INFANT MORTALITY is particularly high. When people leave an area because of such environmental degradation, that is referred to as ecomigration.

Environmental determinism. Theory that the major influence on human behavior is the physical ENVIRONMENT. Some evidence suggests that TEMPERATURE, PRECIPITATION, sunlight, and TOPOGRAPHY influence human activities. Originally espoused by early German geographers, this theory has led to some extreme stances, however, by authors who have sought to explain the dominance of Europeans as a result of a cool temperate CLIMATE.

Eolian (aeolian). Relating to, or caused by, WIND. In Greek mythology, Aeolus was the ruler of the winds. EROSION, TRANSPORT, and DEPOSITION are common eolian processes that produce LANDFORMS in DESERT REGIONS.

Eolian deposits. Material transported by the WIND.

Eolian erosion. Mechanism of EROSION or CRATER DEGRADATION caused by WIND.

Eon. Largest subdivision of geologic time; the two main eons are the PRECAMBRIAN (c. 4.6 billion years ago to 544 million years ago) and the PHANEROZOIC (c. 544 million years ago to the present).

Ephemeral stream. Watercourse that has water for only a DAY or so.

Epicontinental sea. Shallow SEAS that are located on the CONTINENTAL SHELF, such as the North Sea or Hudson Bay. Also called an EPEIRIC SEA.

Epifauna. Organisms that live on the seafloor.

Epilimnion. Warmer surface layer of water that occurs in a LAKE during summer stratification; during spring, warmer water rises from great depths, and it heats up through the summer SEASON.

Equal-area projection. MAP PROJECTION that maintains the correct area of surfaces on7 a MAP, although shape distortion occurs. The property of such a map is called equivalence.

Erg. Sandy DESERT, sometimes called a SEA of SAND. Erg deserts account for less than 30 percent of the world's deserts. "Erg" is an Arabic word.

Eruption, volcanic. Emergence of MAGMA (molten material) at the earth's surface as LAVA. There are various types of volcanic eruptions, depending on the chemistry of the magma and its viscosity. Scientists refer to effusive and explosive eruptions. Low-viscosity magma generally produces effusive eruptions, where the lava emerges gently, as in Hawaii and Iceland, although explosive events can occur at those SITES as well. Gently sloping SHIELD VOLCANOES are formed by effusive eruptions; FLOODS, such as the Columbia Plateau, can also result. Explosive eruptions are generally associated with SUBDUCTION. Much gas, including steam, is associated with magma formed from OCEANIC CRUST, and the compressed gas helps propel the explosion. COMPOSITE CONES, such as Mount Saint Helens, are created by explosive eruptions.

Escarpment. Steep slope, often almost vertical, formed by faulting. Sometimes called a FAULT SCARP.

Esker. Deposit of coarse gravels that has a sinuous, winding shape. An esker is formed by a STREAM of MELTWATER that flowed through a tunnel it formed under a CONTINENTAL GLACIER. Now that the continental glaciers have melted, eskers can be found exposed at the surface in many PLACES in North America.

Estuarine zone. Area near the COASTLINE that consists of estuaries and coastal saltwater WETLANDS.

Etesian winds. WINDS that blow from the north over the Mediterranean during July and August.

Ethnocentrism. Belief that one's own ETHNIC GROUP and its CULTURE are superior to any other group.

Ethnography. Study of different CULTURES and human societies.

Eustacy. Any change in global SEA LEVEL resulting from a change in the absolute volume of available sea water. Also known as eustatic sea-level change.

Eustatic movement. Changes in SEA LEVEL.

Exclave. Territory that is part of one COUNTRY but separated from the main part of that country by another country. Alaska is an exclave of the United States; Kaliningrad is an exclave of Russia. See also ENCLAVE.

Exfoliation. When GRANITE rocks cooled and solidified, removal of the overlyingrock that was present reduced the pressure on the granite mass, allowing it to expand and causing sheets or layers of rock to break off. An exfoliation DOME, such as Half Dome in Yosemite National Park, is the resultant LANDFORM.

Exotic stream. RIVER that has its source in an area of high RAINFALL and then flows through an arid REGION or DESERT. The Nile River is the most famous exotic STREAM. In the United States, the Colorado River is a good example of an exotic stream.

Expansion-contraction cycles. Processes of wetting-drying, heating-cooling, or freezing-thawing, which affect SOIL particles differently according to their size.

Extrusive rock. Fine-grained, or glassy, ROCK which was formed from a MAGMA that cooled on the surface of the earth.

Fall line. Edge of an area of uplifted land, marked by WATERFALLS where STREAMS flow over the edge.

Fata morgana. Large mirage. Originally, the name given to a multiple mirage phenomenon often observed over the Straits of Messina and supposed to be the work of the fairy ("fata") Morgana. Another famous fata morgana is located in Antarctica.

Fathometer. Instrument that uses sound waves or sonar to determine the depth of water or the depth of an object below the water.

Fault drag. Bending of ROCKS adjacent to a FAULT.

Fault line. Line of breakage on the earth's surface. FAULTS may be quite short, but many are extremely long, even hundreds of miles. The origin of the faulting may lie at a considerable depth below the surface. Movement along the fault line generates EARTHQUAKES.

Fault plane. Angle of a FAULT. When fault blocks move on either side of a fault or fracture, the movement can be vertical, steeply inclined, or sometimes horizontal. In a NORMAL FAULT, the fault plane is steep to almost vertical. In a REVERSE FAULT, one block rides over the other, forming an overhanging FAULT SCARP. The angle of inclination of the fault plane from the horizontal is called the dip. The inclination of a fault plane is generally constant throughout the length of the fault, but there can be local variations in slope. In a STRIKE-SLIP FAULT the movement is horizontal, so no fault scarp is produced, although the FAULT LINE may be seen on the surface.

Fault scarp. FAULTS are produced through breaking or fracture of the surface ROCKS of the earth's CRUST as a result of stresses arising from tectonic movement. A NORMAL FAULT, one in which the earth movement is predominantly vertical, produces a steep fault scarp. A STRIKE-SLIP FAULT does not produce a fault scarp.

Feldspar. Family name for a group of common MINERALS found in such ROCKS as GRANITE and composed of silicates of aluminum together with potassium, sodium, and calcium. Feldspars are the most abundant group of minerals within the earth's CRUST. There are many varieties of feldspar, distinguished by variations in chemistry and crystal structure. Although feldspars have some economic uses, their principal importance lies in their role as rock-forming minerals.

Felsic rocks. IGNEOUS ROCKS rich in potassium, sodium, aluminum, and SILICA, including GRANITES and related rocks.

Fertility rate. DEMOGRAPHIC MEASURE of the average number of children per adult female in any given POPULATION. Religious beliefs, education, and other cultural considerations influence fertility rates.

Fetch. Distance along a large water surface over which a WIND of almost uniform direction and speed blows.

Feudalism. Social and economic system that prevailed in Europe before the INDUSTRIAL REVOLUTION. The land was owned and controlled by a minority comprising noblemen or lords; all other people were peasants or serfs, who worked as agricultural laborers on the lords' land. The peasants were not free to leave, or to do anything without their lord's permission. Other REGIONS such as China and Japan also had a feudal system in the past.

Firn. Intermediate stage between SNOW and glacial ice. Firn has a granular TEXTURE, due to compaction. Also called NÉVÉ.

Fission, nuclear. Splitting of an atomic nucleus into two lighter nuclei, resulting in the release of neutrons and some of the binding ENERGY that held the nucleus together.

Fissure. Fracture or crack in ROCK along which there is a distinct separation.

Flash flood. Sudden rush of water down a STREAM CHANNEL, usually in the DESERT after a short but intense STORM. Other causes, such as a DAM failure, could lead to a flash flood.

Flood control. Attempts by humans to prevent flooding of STREAMS. Humans have consistently settled on FLOODPLAINS and DELTAS because of the fertile SOIL for AGRICULTURE, and attempts at flood control date back thousands of years. In strictly agricultural societies such as ancient Egypt, people built VILLAGES above the FLOOD levels, but transport and industry made riverside LOCATIONS desirabl and engineers devised technological means to try to prevent flood damage. Artificial LEVEES, RESERVOIRS, and DAMS of ever-increasing size were built on

RIVERS, as well as bypass CHANNELS leading to artificial floodplains. In many modern dam construction projects, the production of HYDRO-ELECTRIC POWER was more important than flood control. Despite modern TECHNOLOGY, floods cause the largest loss of human life of all natural disasters, especially in low-income countries such as Bangladesh.

Flood tide. Rising or incoming tide. Most parts of the world experience two flood TIDES in each 24-hour period.

Floodplain. Flat, low-lying land on either side of a STREAM, created by the DEPOSITION of ALLUVIUM from floods. Also called ALLUVIAL PLAIN.

Fluvial. Pertaining to running water; for example, fluvial processes are those in which running water is the dominant agent.

Fog deserts. Coastal DESERTS where FOG is an important source of moisture for plants, animals, and humans. The fog forms because of a cold OCEAN CURRENT close to the SHORE. The Namib Desert of southwestern Africa, the west COAST of California, and the Atacama Desert of Peru are coastal deserts.

Föhn wind. WIND warmed and dried by descent, usually on the LEE side of a MOUNTAIN. In North America, these winds are called the CHINOOK.

Fold mountains. ROCKS in the earth's CRUST can be bent by compression, producing folds. The Swiss Alps are an example of complex FOLDING, accompanied by faulting. Simple upward folds are ANTICLINES, downward folds are SYNCLINES; but subsequent EROSION can produce LANDSCAPES with synclinal MOUNTAINS.

Folding. Bending of ROCKS in the earth's CRUST, caused by compression. The rocks are deformed, sometimes pushed up to form mountain RANGES.

Foliation. TEXTURE or structure in which MINERAL grains are arranged in parallel planes.

Food web. Complex network of FOOD CHAINS. Food chains are interconnected, because many organisms feed on a variety of others, and in turn may be eaten by any of a number of predators.

Forced migration. MIGRATION that occurs when people are moved against their will. The Atlantic slave trade is an example of forced migration. People were shipped from Africa to countries in Europe, Asia, and the New World as forced immigrants. Within the United States, some NATIVE AMERICANS were forced by the federal government to migrate to new reservations.

Ford. Short shallow section of a RIVER, where a person can cross easily, usually by walking or riding a horse. To cross a STREAM in such a manner.

Formal region. Cultural REGION in which one trait, or group of traits, is uniform. LANGUAGE might be the basis of delineation of a formal cultural region. For example, the Francophone region of Canada constitutes a formal region based on one single trait. One might also identify a formal Mormon region centered on the STATE of Utah, combining RELIGION and LANDSCAPE as defining traits. Cultural geographers generally identify formal regions using a combination of traits.

Fossil fuel. Deposit rich in hydrocarbons, formed from organic materials compressed in ROCK layers—COAL, OIL, and NATURAL GAS.

Fossil record. Fossil record provides evidence that addresses fundamental questions about the origin and history of life on the earth: When life evolved; how new groups of organisms originated; how major groups of organisms are related. This record is neither complete nor without biases, but as scientists' understanding of the limits and potential of the fossil record grows, the interpretations drawn from it are strengthened.

Fossilization. Processes by which the remains of an organism become preserved in the ROCK record.

Fracture zones. Large, linear zones of the seafloor characterized by steep CLIFFS, irregular TOPOGRAPHY, and FAULTS; such zones commonly cross and displace oceanic RIDGES by faulting.

Free association. Relationship between sovereign NATIONS in which one nation—invariably the larger—has responsibility for the other nation's defense. The Cook Islands in the South Pacific have such a relationship with New Zealand.

Friction of distance. Distance is of prime importance in social, political, economic, and other relationships. Large distance has a negative effect on human activity. The time and cost of overcoming distance can be a deterrent to various activities. This has been called the friction of distance.

Frigid zone. Coldest of the three CLIMATE zones proposed by the ancient Greeks on the basis of their theories about the earth. There were two frigid zones, one around each POLE. The Greeks believed that human life was possible only in the TEMPERATE ZONE.

Fringing reef. Type of CORAL REEF formed at the SHORELINE, extending out from the land in shallow water. The top of the coral may be exposed at low TIDE.

Frontier Thesis. Thesis first advanced by the U.S. historian Frederick Jackson Turner, who declared that U.S. history and the U.S. character were shaped by the existence of empty, FRONTIER lands that led to exploration and westward expansion and DEVELOPMENT. The closing of the frontier occurred when transcontinental railroads linked the East and West Coasts and SETTLEMENTS spread across the United States. This thesis was used by later historians to explain the history of South Africa, Canada, and Australia. Critics of the Frontier Thesis point out that minorities and women were excluded from this view of history.

Frost wedging. Powerful form of PHYSICAL WEATHERING of ROCK, in which the expansion of water as it freezes in JOINTS or cracks shatters the rock into smaller pieces. Also known as frost shattering.

Fumarole. Crack in the earth's surface from which steam and other gases emerge. Fumaroles are found in volcanic areas and areas of GEOTHERMAL activity, such as Yellowstone National Park.

Fusion energy. Heat derived from the natural or human-induced union of atomic nuclei; in effect, the opposite of FISSION energy.

Gall's projection. MAP PROJECTION constructed by projecting the earth onto a cylinder that intersects the sphere at 45 DEGREES north and 45 degrees south LATITUDE. The resulting map has less distortion of area than the more familiar CYLINDRICAL PROJECTION of Mercator.

Gangue. Apparently worthless ROCK or earth in which valuable gems or MINERALS are found.

Garigue. VEGETATION cover of small shrubs found in Mediterranean areas. Similar to the larger *maquis.*

Genus (plural, genera). Group of closely related species; for example, *Homo* is the genus of humans, and it includes the species *Homo sapiens* (modern humans) and *Homo erectus* (Peking Man, Java Man).

Geochronology. Study of the time SCALE of the earth; it attempts to develop methods that allow the scientist to reconstruct the past by dating events such as the formation of ROCKS.

Geodesy. Branch of applied mathematics that determines the exact positions of points on the earth's surface, the size and shape of the earth, and the variations of terrestrial GRAVITY and MAGNETISM.

Geoid. Figure of the earth considered as a MEAN SEA LEVEL surface extended continuously through the CONTINENTS.

Geologic terrane. Crustal block with a distinct group of ROCKS and structures resulting from a particular geologic history; assemblages of TERRANES form the CONTINENTS.

Geological column. Order of ROCK layers formed during the course of the earth's history.

Geomagnetic elements. MEASUREMENTS that describe the direction and intensity of the earth's MAGNETIC FIELD.

Geomagnetism. External MAGNETIC FIELD generated by forces within the earth; this force attracts materials having similar properties, inducing them to line up (point) along field lines of force.

Geostationary orbit. ORBIT in which a SATELLITE appears to hover over one spot on the PLANET's EQUATOR; this procedure requires that the orbit be high enough that its period matches the planet's rotational period, and have no inclination relative to the equator; for Earth, the ALTITUDE is 22,260 miles (35,903 km.).

Geostrophic. Force that causes directional change because of the earth's ROTATION.

Geotherm. Curve on a TEMPERATURE-depth graph that describes how temperature changes in the subsurface.

Geothermal power. Power having its source in the earth's internal heat.

Glacial erratic. ROCK that has been moved from its original position and transported by becoming incorporated in the ice of a GLACIER. Deposited in a new LOCATION, the rock is noteworthy because its geology is completely different from that of the surrounding rocks. Glacial erratics provide information about the direction of glacial movement and strength of the flow. They can be as small as PEBBLES, but the most interesting erratics are large boulders. Erratics become smoothed and rounded by the transport and EROSION.

Glaciation. This term is used in two senses: first, in reference to the cyclic widespread growth and advance of ICE SHEETS over the polar and high- to mid-LATITUDE REGIONS of the CONTINENTS; second, in reference to the effect of a GLACIER on the TERRAIN it transverses as it advances and recedes.

Global Positioning System (GPS). Group of SATELLITES that ORBIT Earth every twenty-four hours, sending out signals that can be used to locate PLACES on Earth and in near-Earth orbits.

Global warming. Trend of Earth CLIMATES to grow increasingly warm as a result of the GREENHOUSE EFFECT. One of the most dramatic effects of global warming is the melting of the POLAR ICE CAPS and a consequent rise the level of the world's OCEANS.

Gondwanaland. Hypothesized ancient CONTINENT in the SOUTHERN HEMISPHERE that geologists theorize broke into at least two large segments; one segment became India and pushed northward to collide with the Eurasian LANDMASS, while the other, Africa, moved westward.

Graben. Roughly symmetrical crustal depression formed by the lowering of a crustal block between two NORMAL FAULTS that slope toward each other.

Granules. Small grains or pellets.

Gravimeter. Device that measures the attraction of GRAVITY.

Gravitational differentiation. Separation of MINERALS, elements, or both as a result of the influence of a gravitational field wherein heavy phases sink or light phases rise through a melt.

Great circle. Largest circle that goes around a sphere. On the earth, all lines of LONGITUDE are parts of great circles; however, the EQUATOR is the only line of LATITUDE that is a great circle.

Green mud. SOILS that develop under conditions of excess water, or waterlogged soils, can display colors of gray to blue to green, largely because of chemical reactions involving iron. Fine CLAY soils and muds in areas such as BOGS or ESTUARIES can be called green mud. This soil-forming process is called gleization.

Greenhouse effect. Trapping of the SUN's rays within the earth's ATMOSPHERE, with a consequence rise in TEMPERATURES that leads to GLOBAL WARMING.

Greenhouse gas. Atmospheric gas capable of absorbing electromagnetic radiation in the infrared part of the spectrum.

Greenwich mean time. Also known as universal time, the solar mean time on the MERIDIAN running through Greenwich, England—which is used as the basis for calculating time throughout most of the world.

Grid. Pattern of horizontal and vertical lines forming squares of uniform size.

Groundwater movement. Flow of water through the subsurface, known as groundwater movement, obeys set principles that allow hydrologists to predict flow directions and rates.

Groundwater recharge. Water that infiltrates from the surface of the earth downward through SOIL and ROCK pores to the WATER TABLE, causing its level to rise.

Growth pole. LOCATION where high-growth economic activity is deliberately encouraged and promoted. Governments often establish growth poles by creating industrial parks, open cities, special economic zones, new TOWNS, and other incentives. The plan is that the new industries will further stimulate economic growth in a cumulative trend. Automobile plants are a traditional form of growth industry but have been overtaken by high-tech industries and BIOTECHNOLOGY. In France, the term "technopole" is used for a high-tech growth pole. A related concept is SPREAD EFFECTS.

Guyot. Drowned volcanic ISLAND with a flat top caused by WAVE EROSION or coral growth. A type of SEAMOUNT.

Gyre. Large semiclosed circulation patterns of OCEAN CURRENTS in each of the major OCEAN BASINS that move in opposite directions in the Northern and Southern hemispheres.

Haff. Term used for various WETLANDS or LAGOONS located around the southern end of the Baltic Sea, from Latvia to Germany. Offshore BARS of SAND and shingle separate the haffs from the open SEA. One of the largest is the Stettiner Haff, which covers the BORDER REGION between Germany and Poland and is separated from the Baltic by the low-lying ISLAND of Usedom. The Kurisches Haff (in English, the Courtland Lagoon) is located on the Lithuanian border.

Harmonic tremor. Type of EARTHQUAKE activity in which the ground undergoes continuous shaking in response to subsurface movement of MAGMA.

Headland. Elevated land projecting into a body of water.

Headwaters. Source of a RIVER. Also called headstream.

Heat sink. Term applied to Antarctica, whose cold CLIMATE causes warm AIR masses flowing over it to chill quickly and lose ALTITUDE, affecting the entire world's WEATHER.

Heterosphere. Major realm of the ATMOSPHERE in which the gases hydrogen and helium become predominant.

High-frequency seismic waves. EARTHQUAKE WAVES that shake the ROCK through which they travel most rapidly.

Histogram. Bar graph in which vertical bars represent frequency and the horizontal axis represents categories. A POPULATION PYRAMID, or age-sex pyramid, is a histogram, as is a CLIMOGRAPH.

Historical inertia. Term used by economic geographers when heavy industries, such as steelmaking and large manufacture, that require huge capital investments in land and plant continue in operation for long periods, even after they become out of date, uncompetitive, or obsolete.

Hoar frost. Similar to DEW, except that moisture is deposited as ice crystals, not liquid dew, on surfaces such as grass or plant leaves. When moist AIR cools to saturation level at TEMPERATURES below the freezing point, CONDENSATION occurs directly as ice. Technically, hoar frost is not the same as frozen dew, but it is difficult to distinguish between the two.

Hogback. Steeply sloping homoclinal RIDGE, with a slope of 45 DEGREES or more. The angle of the slope is the same as the dip of the ROCK STRATA. These LANDFORMS develop in REGIONS where the underlying rocks, usually SEDIMENTARY, have been folded into anticlinal ridges and synclinal VALLEYS. Differential EROSION causes softer rock layers to wear away more rapidly than the harder layers of rock that form the hogback ridge. A similar feature with a gentler slope is called a CUESTA.

Homosphere. Lower part of the earth's ATMOSPHERE. In this area, 60 miles (100 km.) thick, the component gases are uniformly mixed together, largely through WINDS and turbulent AIR CURRENTS. Above the homosphere is the REGION of the atmosphere called the HETEROSPHERE. There, the individual gases separate out into layers on the basis of their molecular weight. The lighter gases, hydrogen and helium, are at the top of the heterosphere.

Hook. A long, narrow deposit of SAND and SILT that grows outward into the OCEAN from the land is called a SPIT or sandspit. A hook forms when currents or WAVES cause the deposited material to curve back toward the land. Cape Cod is the most famous spit and hook in the United States.

Horse latitudes. Parts of the OCEANS from about 30 to 35 DEGREES north or south of the EQUATOR. In these latitudes, AIR movement is usually light WINDS, or even complete calm, because there are semipermanent high-pressure cells called ANTI-CYCLONES, which are marked by dry subsiding air and fine clear WEATHER. The atmospheric circulation of an anticyclone is divergent and CLOCKWISE in the NORTHERN HEMISPHERE, so to the north of the horse latitudes are the westerly winds and to the south are the northeast TRADE WINDS. In the SOUTHERN HEMISPHERE, the circulation is reversed, producing the easterly winds and the southeast trade winds. It is believed that the name originated because when ships bringing immigrants to the Americas were becalmed for any length of time, horses were thrown overboard because they required too much FRESH WATER. Also called the CALMS OF CANCER.

Horst. FAULT block or piece of land that stands above the surrounding land. A horst usually has been uplifted by tectonic forces, but also could have originated by downward movement or lowering of the adjacent lands. Movement occurs along the parallel faults on either side of a horst. If the land is downthrown instead of uplifted, a VALLEY known as a GRABEN is formed. "Horst" comes from the German word for horse, because the flat-topped feature resembles a vaulting horse used in gymnastics.

Hot spot. PLACE on the earth's surface where heat and MAGMA rise from deep in the interior, perhaps from the lower MANTLE. Erupting VOLCA-NOES may be present, as in the formation of the Hawaiian Islands. More commonly, the heat from the rising magma causes GROUNDWATER to form HOT SPRINGS, GEYSERS, and other thermal and HYDROTHERMAL features. Yellowstone National Park is located on a hot spot. Also known as a MANTLE PLUME.

Hot spring. SPRING where hot water emerges at the earth's surface. The usual cause is that the GROUNDWATER is heated by MAGMA. A GEYSER is a special type of hot spring at which the water heats under pressure and that periodically spouts hot water and steam. Old Faithful is the best known of many geysers in Yellowstone National Park. In some countries, GEOTHERMAL EN-ERGY from hot springs is used to generate electricity. Also called thermal spring.

Humus. Uppermost layer of a SOIL, containing decaying and decomposing organic matter such as leaves. This produces nutrients, leading to a fertile soil. Tropical soils are low in humus, because the rate of decay is so rapid. Soils of GRASSLANDS and DECIDUOUS FOREST develop thick layers of humus. In a SOIL PROFILE, the layer containing humus is the O Horizon.

Hydroelectric power. Electricity generated when falling water turns the blades of a turbine that converts the water's potential ENERGY to mechanical energy. Natural WATERFALLS can be used, but most hydroelectric power is generated by water from DAMS, because the flow of water from a dam can be controlled. Hydroelectric generation is a RENEWABLE, clean, cheap way to produce power, but dam construction inundates land, often displacing people, who lose their homes, VILLAGES, and farmland. Aquatic life is altered and disrupted also; for example, Pacific salmon cannot return upstream on the Columbia River to their spawning REGION. In a few coastal PLACES, TIDAL ENERGY is used to generate hydroelectricity; La Rance in France is the oldest successful tidal power plant.

Hydrography. Surveying of underwater features or those parts of the earth that are covered by water, especially OCEAN depths and OCEAN CURRENTS. Hydrographers make MAPS and CHARTS of the ocean floor and COASTLINES, which are used by mariners for NAVIGATION. For centuries, mariners used a leadline, a long rope with a lead weight at the bottom. The line was thrown overboard and the depth of water measured. The unit of MEASUREMENT was FATHOMS (6 feet/1.8 meters), which is one-thousandth of a NAUTICAL MILE. The invention of sonar (underwater echo sounding) has enabled mapping of large areas, and hydrographers currently use both television cameras and SATELLITE data.

Hydrologic cycle. Continuous circulation of the earth's HYDROSPHERE, or waters, through EVAP-ORATION, CONDENSATION, and PRECIPITATION.

Other parts of the hydrologic cycle include RUN-OFF, INFILTRATION, and TRANSPIRATION.

Hydrostatic pressure. Pressure imposed by the weight of an overlying column of water.

Hydrothermal vents. Areas on the OCEAN floor, typically along FAULT LINES or in the vicinity of undersea VOLCANOES, where water that has percolated into the ROCK reemerges much hotter than the surrounding water; such heated water carries various dissolved MINERALS, including metals and sulfides.

Hyetograph. Chart showing the DISTRIBUTION of RAINFALL over time. Typically, a hyetograph is constructed for a single STORM, showing the amount of total PRECIPITATION accumulating throughout the period. A hyetograph shows how rainfall intensity varies throughout the duration of a storm.

Hygrometer. Instrument for measuring the RELATIVE HUMIDITY of AIR, or the amount of water vapor in the ATMOSPHERE at any time.

Hypsometer. Instrument used for measuring ALTITUDE (height above SEA LEVEL), using boiling water that circulates around a THERMOMETER. Since ATMOSPHERIC PRESSURE falls with increased altitude, the boiling point of water is lower. The hypsometer relies on this difference in boiling point to calculate ELEVATION. A more common instrument for measuring altitude is the ALTIMETER.

Ice blink. Bright, usually yellowish-white glare or reflection on the underside of a CLOUD layer, produced by light reflected from an ice-covered surface such as pack ice. A similar phenomenon of reflection from a snow-covered surface is called snow blink.

Ice-cap climate. Earth's most severe CLIMATE, where the mean monthly TEMPERATURE is never above 32 DEGREES Fahrenheit (0 degrees Celsius). This climate is found in Greenland and Antarctica, which are high PLATEAUS, where KATABATIC WINDS blow strongly and frequently. At these high LATITUDES, INSOLATION (SOLAR ENERGY) is received for a short period in the summer months, but the high reflectivity of the ice and SNOW means that much is reflected back instead of being absorbed by the surface. No VEGETATION can grow, because the LANDSCAPE is permanently covered in ice and snow. Because AIR temperatures are so cold, PRECIPITATION is usually less than 5 inches (13 centimeters) annually. The POLES are REGIONS of stable, high-pressure air, where dry conditions prevail, but strong winds that blow the snow around are common. In the KÖPPEN CLIMATE CLASSIFICATION, the ice-cap climate is signified by the letters *EF.*

Ice sheet. Huge CONTINENTAL GLACIER. The only ice sheets remaining cover most of Antarctica and Greenland. At the peak of the last ICE AGE, around 18,000 years ago, ice covered as much as one-third of the earth's land surfaces. In the NORTHERN HEMISPHERE, there were two great ice sheets—the Laurentide ice sheet, covering North America, and the Scandinavian ice sheet, covering northwestern Europe and Scandinavia.

Ice shelf. Portion of an ICE SHEET extending into the OCEAN.

Ice storm. STORM characterized by a fall of freezing rain, with the formation of glaze on Earth objects.

Icefoot. Long, tapering extension of a GLACIER floating above the seawater where it enters the OCEAN. Eventually, it breaks away and forms an ICEBERG.

Igneous rock. ROCKS formed when molten material or MAGMA cools and crystallizes into solid rock. The type of rock varies with the composition of the magma and, more important, with the rate of cooling. Rocks that cool slowly, far beneath the earth's surface, are igneous INTRUSIVE ROCKS. These have large crystals and coarse grains. GRANITE is the most typical igneous intrusive rock. When cooling is more rapid, usually closer to or at the surface, finer-grained igneous EXTRUSIVE ROCKS such as rhyolite are formed. If the magma flows out to the surface as LAVA, it may cool quickly, forming a glassy rock called obsidian. If there is gas in the lava, rocks full of holes from bubbles of escaping gases form; PUMICE and BASALT are common igneous extrusive rocks.

Impact crater. Generally circular depression formed on the surface of a PLANET by the impact of a high-velocity projectile such as a METEORITE, ASTEROID, or COMET.

Impact volcanism. Process in which major impact events produce huge CRATERS along with MAGMA RESERVOIRS that subsequently produce volcanic activity. Such cratering is clearly visible on the MOON, Mars, Mercury, and probably Venus. It is assumed that Earth had similar craters, but EROSION has erased most of the evidence.

Import substitution. Economic process in which domestic producers manufacture or supply goods or services that were previously imported or purchased from overseas and foreign producers.

Index fossil. Remains of an ancient organism that are useful in establishing the age of ROCKS; index fossils are abundant and have a wide geographic DISTRIBUTION, a narrow stratigraphic RANGE, and a distinctive form.

Indian summer. Short period, usually not more than a week, of unusually warm WEATHER in late October or early November in the NORTHERN HEMISPHERE. Before the Indian summer, TEMPERATURES are cooler and there can be occurrences of FROST. Indian summer DAYS are marked by clear to hazy skies and calm to light WINDS, but nights are cool. The weather pattern is a high-pressure cell or ridge located for a few days over the East Coast of North America. The name originated in New England, referring to the practice of NATIVE AMERICANS gathering foods for winter storage over this brief spell. Similar weather in England is called an Old Wives' summer.

Infant mortality. DEMOGRAPHIC MEASURE calculated as the number of deaths in a year of infants, or children under one year of age, compared with the total number of live births in a COUNTRY for the same year. Low-income countries have high infant mortality rates, more than 100 infant deaths per thousand.

Infauna. Organisms that live in the seafloor.

Infiltration. Movement of water into and through the SOIL.

Initial advantage. In terms of economic DEVELOPMENT, not all LOCATIONS are suited for profitable investment. Some locations offer initial advantages, including an existing skilled labor pool, existing consumer markets, existing plants, and situational advantages. These advantages can also lead to clustering of a number of industries at a particular location and to further economic growth, which will provide the preconditions of initial advantage for further economic development.

Inlier. REGION of old ROCKS that is completely surrounded by younger rocks. These are often PLACES where ORES or MINERALS are found in commercial quantities.

Inner core. The innermost layer of the earth; the inner core is a solid ball with a radius of about 900 miles.

Inselberg. Exposed rocky HILL in a DESERT area, made of resistant ROCKS, rising steeply from the flat surrounding countryside. There are many inselbergs in Africa, but Uluru (Ayers Rock) in Australia is possibly the most famous inselberg. The word is German for "island mountain." A special type of inselberg is a bornhardt.

Insolation. ENERGY received by the earth from the SUN, which heats the earth's surface. The average insolation received at the top of the earth's ATMOSPHERE at an average distance from the Sun is called the SOLAR CONSTANT. Insolation is predominantly shortwave radiation, with wavelengths in the RANGE of 0.39 to 0.76 micrometers, which corresponds to the visible spectrum. Less than half of the incoming SOLAR ENERGY reaches the earth's surface-insolation is reflected back into space by CLOUDS; smaller amounts are reflected back by surfaces, absorbed, or scattered by the atmosphere. Insolation is not distributed evenly over the earth, because of Earth's curved surface. Where the rays are perpendicular, at the SUBSOLAR POINT, insolation is at the maximum. The word is a shortened form of incoming (or intercepted) SOLAR RADIATION.

Insular climate. Island climates are influenced by the fact that no PLACE is far from the SEA. Therefore, both the DIURNAL (daily) TEMPERATURE RANGE and the annual temperature range are small.

Insurgent state. STATE that arises when an uprising or guerrilla movement gains control of part of the territory of a COUNTRY, then establishes its own form of control or government. In effect, the insurgents create a state within a state. In Colombia, for example, the government and armed forces have been unable to control several REGIONS where insurgents have created their own domains. This is generally related to coca growing and the production of cocaine. Civilian farmers are unable to resist the drug-financed "armies."

Interfluve. Higher area between two STREAMS; the surface over which water flows into the stream. These surfaces are subject to RUNOFF and EROSION by RILL action and GULLYING. Over time, interfluves are lowered.

Interlocking spur. STREAM in a hilly or mountainous REGION that winds its way in a sinuous VALLEY between the different RIDGES, slowly eroding the ends of the spurs and straightening its course. The view of interlocking spurs looking upstream is a favorite of artists, as colors change with the receding distance of each interlocking spur.

Intermediate rock. IGNEOUS ROCK that is transitional between a basic and a silicic ROCK, having a SILICA content between 54 and 64 percent.

Internal migration. Movement of people within a COUNTRY, from one REGION to another. Internal MIGRATION in high-income economies is often urban-to-RURAL, such as the migration to the SUN BELT in the United States. In low-income economies, rural-to-URBAN migration is more common.

Intertillage. Mixed planting of different seeds and seedling crops within the same SWIDDEN or cleared patch of agricultural land. Potatoes, yams, corn, rice, and bananas might all be planted. The planting times are staggered throughout the year to increase the variety of crops or nutritional balance available to the subsistence farmer and his or her family.

Intrusive rock. IGNEOUS ROCK which was formed from a MAGMA that cooled below the surface of the earth; it is commonly coarse-grained.

Irredentism. Expansion of one COUNTRY into the territory of a nearby country, based on the residence of nationals in the neighboring country. Hitler used irredentist claims to invade Czechoslovakia, because small groups of German-speakers lived there in the Sudetenland. The term comes from Italian, referring to Italy's claims before World War I that all Italian-speaking territory should become part of Italy.

Isallobar. Imaginary line on a MAP or meteorological chart joining PLACES with an equal change in ATMOSPHERIC PRESSURE over a certain time, often three hours. Isallobars indicate a pressure tendency and are used in WEATHER FORECASTING.

Island arc. Chain of VOLCANOES next to an oceanic TRENCH in the OCEAN BASINS; an oceanic PLATE descends, or subducts, below another oceanic plate at ISLAND arcs.

Isobar. Imaginary line joining PLACES of equal ATMOSPHERIC PRESSURE. WEATHER MAPS show isobars encircling areas of high or low pressure. The spacing between isobars is related to the pressure gradient.

Isobath. Line on a MAP or CHART joining all PLACES where the water depth is the same; a kind of underwater CONTOUR LINE. This kind of map is a BATHYMETRIC CONTOUR.

Isoclinal folding. When the earth's CRUST is folded, the size and shape of the folds vary according to the force of compression and nature of the ROCKS. When the surface is compressed evenly so that the two sides of the fold are parallel, isoclinal folding results. When the sides or slopes of the fold are unequal or dissimilar in shape and angle, this can be an asymmetrical or overturned fold. See also ANTICLINE; SYNCLINE.

Isotherm. Line joining PLACES of equal TEMPERATURE. A world MAP with isotherms of average monthly temperature shows that over the OCEANS, temperature decreases uniformly from the EQUATOR to the POLES, and higher temperatures occur over the CONTINENTS in summer and lower temperatures in winter because of the unequal heating properties of land and water.

Isotropic surface. Hypothetical flat surface or PLAIN, with no variation in any physical attribute. An isotropic surface has uniform ELEVATION, SOIL type, CLIMATE, and VEGETATION. Economic geographic models study behavior on an isotropic surface before applying the results to the real world. For example, in an isotropic model, land value is highest at the CITY center and falls regularly with increasing distance from there. In the real world, land values are affected by elevation, water features, URBAN regulations, and other factors. The von Thuenen model of the Isolated State is based on a uniform plain or isotropic surface.

Isthmian links. Chains of ISLANDS between substantial LANDMASSES.

Isthmus. Narrow strip of land connecting two larger bodies of land. The Isthmus of Panama connects North and South America; the Isthmus of Suez connects Africa and Asia. Both of these have been cut by CANALS to shorten shipping routes.

Jet stream. WINDS that move from west to east in the upper ATMOSPHERE, 23,000 to 33,000 feet (7,000–10,000 meters) above the earth, at about 200 miles (300 km.) per hour. They are narrow bands, elliptical in cross section, traveling in irregular paths. Four jet streams of interest to earth scientists and meteorologists are the polar jet stream and the subtropical jet stream in the Northern and SOUTHERN HEMISPHERES. The polar jet stream is located at the TROPOPAUSE, the BOUNDARY between the TROPOSPHERE and the STRATOSPHERE, along the polar FRONT. There is a complex interaction between surface winds and jet streams. In winter the NORTHERN HEMISPHERE polar front can move as far south as Texas, bringing BLIZZARDS and extreme WEATHER conditions. In summer, the polar jet stream is located over Canada. The subtropical jet stream is located at the tropopause around 30 DEGREES north or south LATITUDE, but it also migrates north or south, depending on the SEASON.

At times, the polar and subtropical jet streams merge for a few DAYS. Aircraft take advantage of the jet stream, or avoid it, depending on the direction of their flight. Upper atmosphere winds are also known as GEOSTROPHIC winds.

Joint. Naturally occurring fine crack in a ROCK, formed by cooling or by other stresses. SEDIMENTARY ROCKS can split along bedding planes; other joints form at right angles to the STRATA, running vertically through the rocks. In IGNEOUS ROCKS such as GRANITE, the stresses of cooling and contraction cause three sets of joints, two vertical and one parallel to the surface, which leads to the formation of distinctive LANDFORMS such as TORS. BASALT often demonstrates columnar jointing, producing tall columns that are mostly hexagonal in section. The presence of joints in BEDROCK hastens WEATHERING, because water can penetrate into the joints. This is particularly obvious in LIMESTONE, where joints are rapidly enlarged by SOLUTION. FROST WEDGING is a type of PHYSICAL WEATHERING that can split large boulders through the expansion when water in a joint freezes to form ice. Compare with FAULTS, which occur through tectonic activity.

Jurassic. Second of the three PERIODS that make up

Kame. Small HILL of gravel or mixed-size deposits, SAND, and gravel. Kames are found in areas previously covered by CONTINENTAL GLACIERS or ICE SHEETS, near what was the outer edge of the ice. They may have formed by materials dropping out of the melting ice, or in a deltalike deposit by a STREAM of MELTWATER. These deposits of which kames are made are called drift. Small LAKES called KETTLES are often found nearby. A closely spaced group of kames is called a kame field.

Karst. LANDSCAPE of SINKHOLES, underground STREAMS and caverns, and associated features created by CHEMICAL WEATHERING, especially SOLUTION, in REGIONS where the BEDROCK is LIMESTONE. The name comes from a region in the southwest of what is now Slovenia, the Krs (Kras) Plateau, but the karst region extends south through the Dinaric Alps bordering the

Adriatic Sea, into Bosnia-Herzegovina and Montenegro. Where limestone is well jointed, RAINFALL penetrates the JOINTS and enters the GROUNDWATER, carrying the MINERALS, especially calcium, away in solution. Most of the famous CAVES and caverns of the world are found in karst areas. The Carlsbad Caverns in New Mexico are a good example. Kentucky, Tennessee, and Florida also have well-known areas of karst. In some tropical countries, a form called tower karst is found. Tall conical or steep-sided HILLS of limestone rise above the flat surrounding landscape. Around 15 percent of the earth's land surface is karst TOPOGRAPHY.

Katabatic wind. GRAVITY DRAINAGE WINDS similar to MOUNTAIN BREEZES but stronger in force and over a larger area than a single VALLEY. Cold AIR collects over an elevated REGION, and the dense cold air flows strongly downslope. The ICE-SHEETS of Antarctica and Greenland produce fierce katabatic winds, but they can occur in smaller regions. The BORA is a strong, cold, squally downslope wind on the Dalmatian COAST of Yugoslavia in winter.

Kettle. Small depression, often a small LAKE, produced as a result of continental GLACIATION. It is formed by an isolated block of ice remaining in the ground MORAINE after a GLACIER has retreated. Deposited material accumulates around the ice, and when it finally melts, a steep hole remains, which often fills with water. Walden Pond, made famous by writer Henry David Thoreau, is a glacial kettle.

Khamsin. Hot, dry, DUST-laden WIND that blows in the eastern Sahara, in Egypt, and in Saudi Arabia, bringing high TEMPERATURES for three or four DAYS. Winds can reach GALE force in intensity. The word Khamsin is Arabic for "fifty" and refers to the period between March and June when the khamsin can occur.

Knickpoint. Abrupt change in gradient of the bed of a RIVER or STREAM. It is marked by a WATER-FALL, which over time is eroded by FLUVIAL action, restoring the smooth profile of the riverbed. The knickpoint acts as a TEMPORARY BASE LEVEL for the upper part of the stream.

Knickpoints can occur where a hard layer of ROCK is slower to erode than the rocks downstream, for example at Niagara Falls. Other knickpoints and waterfalls can develop as a result of tectonic forces. UPLIFT leads to new EROSION by a stream, creating a knickpoint that gradually moves upstream. The bed of a tributary GLACIER is often considerably higher than the VALLEY of the main glacier, so that after the glaciers have melted, a waterfall emerges over this knickpoint from the smaller hanging valley to join the main stream. Yosemite National Park has several such waterfalls.

Köppen climate classification. Commonly used scheme of CLIMATE classification that uses statistics of average monthly TEMPERATURE, average monthly PRECIPITATION, and total annual precipitation. The system was devised by Wladimir Köppen early in the twentieth century.

La Niña. WEATHER phenomenon that is the opposite part of EL NIÑO. When the SURFACE WATER in the eastern Pacific Ocean is cooler than average, the southeast TRADE WINDS blow strongly, bringing heavy rains to countries of the western Pacific. Scientists refer to the whole RANGE of TEMPERATURE, pressure, WIND, and SEA LEVEL changes as the SOUTHERN OSCILLATION (ENSO). The term "El Niño" gained wide currency in the U.S. media after a strong ENSO warm event in 1997–1998. A weak ENSO cold event, or La Niña, followed it in 1998. Means "the little girl" in Spanish. Alternative terms are "El Viejo" and "anti-El Niño."

Laccolith. LANDFORM of INTRUSIVE volcanism formed when viscous MAGMA is forced between overlying sedimentary STRATA, causing the surface to bulge upward in a domelike shape.

Lahar. Type of mass movement in which a MUD-FLOW occurs because of a volcanic explosion or ERUPTION. The usual cause is that the heat from the LAVA or other pyroclastic material melts ice and SNOW at the VOLCANO'S SUMMIT, causing a hot mudflow that can move downslope with great speed. The eruption of Mount Saint Helens in 1985 was accompanied by a lahar.

Lake basin. Enclosed depression on the surface of the land in which SURFACE WATERS collect; BASINS are created primarily by glacial activity and tectonic movement.

Lakebed. Floor of a LAKE.

Land bridge. Piece of land connecting two CONTINENTS, which permits the MIGRATION of humans, animals, or plants from one area to another. Many former land bridges are now under water, because of the rise in SEA LEVEL after the last ICE AGE. The Bering Strait connecting Asia and North America was an important land bridge for the latter continent.

Land hemisphere. Because the DISTRIBUTION of land and water surfaces on Earth is quite asymmetrical on either side of the EQUATOR, the NORTHERN HEMISPHERE might well be called the land hemisphere. For many centuries, Europeans refused to believe that there was not an equal area of land in the SOUTHERN HEMISPHERE. Explorers such as James Cook were dispatched to seek such a "Great South Land."

Landmass. Large area of land—an ISLAND or a CONTINENT.

Landsat. Space-exploration project begun in 1972 to MAP the earth continuously with SATELLITE imaging. The satellites have collected data about the earth: its AGRICULTURE, FORESTS, flat lands, MINERALS, waters, and ENVIRONMENT. These were the first satellites to aid in Earth sciences, helping to produce the best maps available and assisting farmers around the world to improve their crop yields.

Language family. Group of related LANGUAGES believed to have originated from a common prehistoric language. English belongs in the Indo-European language family, which includes the languages spoken by half of the world's peoples.

Lapilli. Small ROCK fragments that are ejected during volcanic ERUPTIONS. A lapillus ranges from about the size of a pea to not larger than a walnut. Some lapilli form by accretion of VOLCANIC ASH around moisture droplets, in a manner similar to hailstone formation. Lapilli sometimes form into a textured rock called lapillistone.

Laterite. Bright red CLAY SOIL, rich in iron oxide, that forms in tropical CLIMATES, where both TEMPERATURE and PRECIPITATION are high year-round, as ROCKS weather. It can be used in brick making and is a source of iron. When the soil is rich in aluminum, it is called BAUXITE. When laterite or bauxite forms a hard layer at the surface, it is called duricrust. Australia and sub-Saharan Africa have large areas of duricrust, some of which is thought to have formed under previous conditions during the TRIASSIC period.

Laurasia. Hypothetical SUPERCONTINENT made up of approximately the present CONTINENTS of the NORTHERN HEMISPHERE.

Lava tube. Cavern structure formed by the draining out of liquid LAVA in a pahoehoe flow.

Layered plains. Smooth, flat REGIONS believed to be composed of materials other than sulfur compounds.

Leaching. Removal of nutrients from the upper horizon or layer of a SOIL, especially in the humid TROPICS, because of heavy RAINFALL. The remaining soil is often bright red in color because iron is left behind. Despite their bright color, tropical soils are infertile.

Leeward. Rear or protected side of a MOUNTAIN or RANGE is the leeward side. Compare to WINDWARD.

Legend. Explanation of the different colors and symbols used on a MAP. For example, a map of the world might use different colors for high-income, middle-income, and low-income economies. A historical map might use different colors for countries that were once colonies of Britain, France, or Spain.

Light year. Distance traveled by light in one year; widely used for measuring stellar distances, it is equal to roughly 6 trillion miles (9.5 million km.).

Lignite. Low-grade COAL, often called brown coal. It is mined and used extensively in eastern Germany, Slovakia, and the Moscow Basin.

Liquefaction. Loss in cohesiveness of water-saturated SOIL as a result of ground shaking caused by an EARTHQUAKE.

Lithification. Process whereby loose material is transformed into solid ROCK by compaction or cementation.

Lithology. Description of ROCKS, such as rock type, MINERAL makeup, and fluid in rock pores.

Lithosphere. Solid outermost layer of the earth. It varies in thickness from a few miles to more than 120 miles (200 km.). It is broken into pieces known as TECTONIC PLATES, some of which are extremely large, while others are quite small. The upper layer of the lithosphere is the CRUST, which may be CONTINENTAL CRUST or OCEANIC CRUST. Below the crust is a layer called the ASTHENOSPHERE, which is weaker and plastic, enabling the motion of tectonic plates.

Littoral. Adjacent to or related to a SEA.

Llanos. Grassy REGION in the Orinoco Basin of Venezuela and part of Colombia. SAVANNA VEGETATION gradually gives way to scrub at the outer edges of the *llanos.* The area is relatively undeveloped.

Loam. SOIL TEXTURE classification, indicating a soil that is approximately equal parts of SAND, SILT, and CLAY. Farmers generally consider a sandy loam to be the best soil texture because of its water-retaining qualities and the ease with which it can be cultivated.

Local sea-level change. Change in SEA LEVEL only in one area of the world, usually by land rising or sinking in that specific area.

Lode deposit. Primary deposit, generally a VEIN, formed by the filling of a FISSURE with MINERALS precipitated from a HYDROTHERMAL solution.

Loess. EOLIAN, or wind-blown, deposit of fine, silt-sized, light-colored material. Loess covers about 10 percent of the earth's land surface. The loess PLATEAU of China is good agricultural land, although susceptible to EROSION. Loess has the property of being able to form vertical CLIFFS or BLUFFS, and many people have built dwellings in the steep cliffs above the Huang He (Yellow) River. In the United States, loess deposits are found in the VALLEYS of the Platte, Missouri, Mississippi, and Ohio Rivers, and on the Columbia Plateau. A German word, meaning loose or unconsolidated, which comes from loess deposits along the Rhine River.

Longitudinal bar. Midchannel accumulation of SAND and gravel with its long end oriented roughly parallel to the RIVER flow.

Longshore current. Current in the OCEAN close to the SHORE, in the surf zone, produced by WAVES approaching the COAST at an angle. Also called a LITTORAL current. The longshore current combined with wave action can move large amounts of SAND and other BEACH materials down the coast, a process called LONGSHORE DRIFT.

Longshore drift. The movement of SEDIMENT parallel to the BEACH by a LONGSHORE CURRENT.

Maar. Explosion vent at the earth's surface where a volcanic cone has not formed. A small ring of pyroclastic materials surrounds the maar. Often a LAKE occupies the small CRATER of a maar. A larger form is called a TUFF RING.

Macroburst. Updrafts and downdrafts within a CUMULONIMBUS CLOUD or THUNDERSTORM can cause severe TURBULENCE. A DOWNBURST within a thunderstorm when windspeeds are greater than 130 miles (210 km.) per hour and over areas of 2.5 square miles (5 sq. km.) or more is called a macroburst. See also MICROBURST.

Magnetic poles. Locations on the earth's surface where the earth's MAGNETIC FIELD is perpendicular to the surface. The magnetic poles do not correspond exactly to the geographic North Pole and South Pole, or earth's AXIS; the difference is called magnetic variation or DECLINATION.

Magnetic reversal. Change in the earth's MAGNETIC FIELD from the North Pole to the South MAGNETIC POLE.

Magnetic storm. Rapid changes in the earth's MAGNETIC FIELD as a result of the bombardment of the earth by electrically charged particles from the SUN.

Magnetosphere. REGION surrounding a PLANET where the planet's own MAGNETIC FIELD predominates over magnetic influences from the SUN or other planets.

Mantle convection. Thermally driven flow in the earth's MANTLE thought to be the driving force of PLATE TECTONICS.

Mantle plume. Rising jet of hot MANTLE material that produces tremendous volumes of basaltic LAVA. See also HOT SPOT.

Map projection. Mathematical formula used to transform the curved surface of the earth onto a flat plane or sheet of paper. Projections are divided into three classes: CYLINDRICAL, CONICAL, and AZIMUTHAL.

Marchland. FRONTIER area where boundaries are poorly defined or absent. The marches themselves were a type of BOUNDARY REGION. Marchlands have changed hands frequently throughout history. The name is related to the fact that armies marched across them.

Mass balance. Summation of the net gain and loss of ice and SNOW mass on a GLACIER in a year.

Mass extinction. Die-off of a large percentage of species in a short time.

Mass wasting. Downslope movement of Earth materials under the direct influence of GRAVITY.

Massif. French term used in geology to describe very large, usually IGNEOUS INTRUSIVE bodies.

Meandering river. RIVER confined essentially to a single CHANNEL that transports much of its SEDIMENT load as fine-grained material in SUSPENSION.

Mechanical weathering. Another name for PHYSICAL WEATHERING, or the breaking down of ROCK into smaller pieces.

Mechanization. Replacement of human labor with machines. Mechanization occurred in AGRICULTURE as tractors, reapers, picking machinery, and similar technological inventions took the place of human farm labor. Mechanization in industry was part of the INDUSTRIAL REVOLUTION, as spinning and weaving machines were introduced into the textile industry.

Medical geography. Branch of geography specializing in the study of health and disease, with a particular emphasis on the areal spread or DIFFUSION of disease. The spatial perspective of geography can lead to new medical insights. Geographers working with medical researchers in Africa have made great contributions to understanding the role of disease on that CONTINENT. John Snow's studies of the origin and spread of cholera in London in 1854 mark the beginnings of medical geography.

Megalopolis. Conurbation formed when large cities coalesce physically into one huge built-up area. Originally coined by the French geographer Jean Gottman in the early 1960s for the northeastern part of the United States, from Boston to Washington, D.C.

Mesa. Flat-topped HILL with steep sides. EROSION removes the surrounding materials, while the mesa is protected by a cap of harder, more resistant ROCK. Usually found in arid REGIONS. A larger LANDFORM of this type is a PLATEAU; a smaller feature is a BUTTE. The Colorado Plateau and Grand Canyon in particular are rich in these landforms. From the Spanish word for table.

Mesosphere. Atmospheric layer above the STRATOSPHERE where TEMPERATURE drops rapidly.

Mestizo. Person of mixed European and Amerindian ancestry, especially in countries of LATIN AMERICA.

Metamorphic rock. Any ROCK whose mineralogy, MINERAL chemistry, or TEXTURE has been altered by heat, pressure, or changes in composition; metamorphic rocks may have IGNEOUS, SEDIMENTARY, or other, older metamorphic rocks as their precursors.

Metamorphic zone. Areas of ROCK affected by the same limited RANGE of TEMPERATURE and pressure conditions, commonly identified by the presence of a key individual MINERAL or group of minerals.

Meteor. METEOROID that enters the ATMOSPHERE of a PLANET and is destroyed through frictional heating as it comes in contact with the various gases present in the atmosphere.

Meteorite. Fragment of an ASTEROID that survives passage through the ATMOSPHERE and strikes the surface of the earth.

Meteoroid. Small planetary body that enters Earth's ATMOSPHERE because its path intersects the earth's ORBIT. Friction caused by the earth's atmosphere on the meteoroid creates a glowing METEOR, or "shooting star." This is a common phenomenon, and most meteors burn away completely. Those that are large enough to reach the ground are called METEORITES.

Microburst. Brief but intense downward WIND, lasting not more than fifteen minutes over an area of 0.6 to 0.9 square mile (1.5–8 sq. km.). Usually associated with THUNDERSTORMS, but are quite unpredictable. The sudden change in wind direction associated with a microburst can create wind shear that causes airplanes to crash, especially if it occurs during takeoff or landing. See also MACROBURST.

Microclimate. CLIMATE of a small area, at or within a few yards of the earth's surface. In this REGION, variations of TEMPERATURE, PRECIPITATION, and moisture can have a pronounced effect on the bioclimate, influencing the growth or well-being of plants and animals, including humans. DEW or FROST, RAIN SHADOW effects, wind-tunneling between tall buildings, and similar phenomena are studied by microclimatologists. Horticulturists know the variations in aspect that affect INSOLATION and temperature, so that certain plants grow best on south-facing walls, for example. The growing of grapes for wine production is a major industry where microclimatology is essential. The study of microclimatology was pioneered by the German meteorologist Rudolf Geiger.

Microcontinent. Independent LITHOSPHERIC PLATE that is smaller than a CONTINENT but possesses continental-type CRUST. Examples include Cuba and Japan.

Microstates. Tiny countries. In 2000, seventeen independent countries each had an area of less than 200 square miles (520 sq. km.). The smallest microstate is Vatican City, with an area of 0.2 square miles (0.5 sq. km.). Most of the world's microstates are island NATIONS, including Nauru, Tuvalu, Marshall Islands, Saint Kitts and Nevis, Seychelles, Maldives, Malta, Grenada, Saint Vincent and the Grenadines, Barbados, Antigua and Barbuda, and Palau.

Mineral species. Mineralogic division in which all the varieties in any one species have the same basic physical and chemical properties.

Monadnock. Isolated HILL far from a STREAM, composed of resistant BEDROCK. Monadnocks are found in humid temperate REGIONS. A similar LANDFORM in an arid region is an INSELBERG.

Monogenetic. Pertaining to a volcanic ERUPTION in which a single vent is used only once.

Moraine. Materials transported by a GLACIER, and often later deposited as a RIDGE of unsorted ROCKS and smaller material. Lateral moraine is found at the side of the glacier; medial moraine occurs when two glaciers join. Other types of moraine include ABLATION moraine, ground moraine, and push, RECESSIONAL, and TERMINAL MORAINE.

Mountain belts. Products of PLATE TECTONICS, produced by the CONVERGENCE of crustal PLATES. Topographic MOUNTAINS are only the surficial expression of processes that profoundly deform and modify the CRUST. Long after the mountains themselves have been worn away, their former existence is recognizable from the structures that mountain building forms within the ROCKS of the crust.

Nappe. Huge sheet of ROCK that was the upper part of an overthrust fold, and which has broken and traveled far from its original position due to the tremendous forces. The Swiss Alps have nappes in many LOCATIONS.

Narrows. STRAIT joining two bodies of water.

Nation-state. Political entity comprising a COUNTRY whose people are a national group occupying the area. The concept originated in eighteenth century France; in practice, such cultural homogeneity is rare today, even in France.

Natural increase, rate of. DEMOGRAPHIC MEASURE of POPULATION growth: the difference between births and deaths per year, expressed as a percentage of the POPULATION. The rate of natural increase for the United States in 2000 was 0.6 percent. In countries where the population is decreasing, the DEATH RATE is greater than the BIRTH RATE.

Natural selection. Main process of biological evolution; the production of the largest number of offspring by individuals with traits that are best adapted to their ENVIRONMENTS.

Nautical mile. Standard MEASUREMENT at SEA, equalling 6,076.12 feet (1.85 km.). The mile used for land measurements is called a statute mile and measures 5,280 feet (1.6 km.).

Neap tide. TIDE with the minimum RANGE, or when the level of the high tide is at its lowest.

Near-polar orbit. Earth ORBIT that lies in a plane that passes close to both the north and south POLES.

Nekton. PELAGIC organisms that can swim freely, without having to rely on OCEAN CURRENTS or WINDS. Nekton includes shrimp; crabs; oysters; MARINE reptiles such as turtles, crocodiles, and snakes; and even sharks; porpoises; and whales.

Net migration. Net balance of a COUNTRY or REGION's IMMIGRATION and EMIGRATION.

Nomadism. Lifestyle in which pastoral people move with grazing animals along a defined route, ensuring adequate pasturage and water for their flocks or herds. This lifestyle has decreased greatly as countries discourage INTERNATIONAL MIGRATION. A more restricted form of nomadism is TRANSHUMANCE.

North geographic pole. Northernmost REGION of the earth, located at the northern point of the PLANET's AXIS of ROTATION.

North magnetic pole. Small, nonstationary area in the Arctic Circle toward which a COMPASS needle points from any LOCATION on the earth.

Notch. Erosional feature found at the base of a SEA CLIFF as a result of undercutting by WAVE EROSION, bioabrasion from MARINE organisms, and dissolution of ROCK by GROUNDWATER seepage. Also known as a nip.

Nuclear energy. ENERGY produced from a naturally occurring isotope of uranium. In the process of nuclear FISSION, the unstable uranium isotope absorbs a neutron and splits to form tin and molybdenum. This releases more neurons, so a chain reaction proceeds, releasing vast amounts of heat energy. Nuclear energy was seen in the 1950s as the energy of the future, but safety fears and the problem of disposal of radioactive nuclear waste have led to public condemnation of nuclear power plants.

Nuée ardente. Hot cloud of ROCK fragments, ASH, and gases that suddenly and explosively erupt from some VOLCANOES and flow rapidly down their slopes.

Nunatak. Isolated MOUNTAIN PEAK or RIDGE that projects through a continental ICE SHEET. Found in Greenland and Antarctica.

Obduction. Tectonic collisional process, opposite in effect to SUBDUCTION, in which heavier OCEANIC CRUST is thrust up over lighter CONTINENTAL CRUST.

Oblate sphere. Flattened shape of the earth that is the result of ROTATION.

Occultation. ECLIPSE of any astronomical object other than the SUN or the MOON caused by the Moon or any PLANET, SATELLITE, or ASTEROID.

Ocean basins. Large worldwide depressions that form the ultimate RESERVOIR for the earth's water supply.

Ocean circulation. Worldwide movement of water in the SEA.

Ocean current. Predictable circulation of water in the OCEAN, caused by a combination of WIND friction, Earth's ROTATION, and differences in TEMPERATURE and density of the waters. The five great oceanic circulations, known as GYRES, are in the North Pacific, North Atlantic, South Pacific, South Atlantic, and Indian Oceans. Because of the CORIOLIS EFFECT, the direction of circulation is CLOCKWISE in the NORTHERN HEMISPHERE and COUNTERCLOCKWISE in the SOUTHERN HEMISPHERE, except in the Indian Ocean, where the direction changes annually with the pattern of winds associated with the Asian MONSOON. Currents flowing toward the EQUATOR are cold currents; those flowing away from the equator are warm currents. An important current is the warm Gulf Stream, which flows north from the Gulf of Mexico along the East Coast of the United States; it crosses the North Atlantic, where it is called the North Atlantic Drift, and brings warmer conditions to the western parts of Europe. The West Coast of the United States is affected by the cool, south-flowing California Current. The cool Humboldt, or Peru, Current, which flows north along the South American coast, is an important indicator of whether there will be an EL NIÑO event. Deep currents, below 300 feet (100 meters), are extremely complicated and difficult to study.

Oceanic crust. Portion of the earth's CRUST under its OCEAN BASINS.

Oceanic island. ISLANDS arising from seafloor volcanic ERUPTIONS, rather than from continental shelves. The Hawaiian Islands are the best-known examples of oceanic islands.

Off-planet. Pertaining to REGIONS off the earth in orbital or planetary space.

Ore deposit. Natural accumulation of MINERAL matter from which the owner expects to extract a metal at a profit.

Orogeny. MOUNTAIN-building episode, or event, that extends over a period usually measured in tens of millions of years; also termed a revolution.

Orographic precipitation. Phenomenon caused when an AIR mass meets a topographic barrier, such as a mountain RANGE, and is forced to rise; the air cools to saturation, and orographic precipitation falls on the WINDWARD side as rain or SNOW. The lee side is a RAIN SHADOW. This effect is noticeable on the West Coast of the United States, which has RAINFOREST on the windward side of the MOUNTAINS and DESERTS on the lee.

Orography. Study of MOUNTAINS that incorporates assessment of how they influence and are affected by WEATHER and other variables.

Oscillatory flow. Flow of fluid with a regular back-and-forth pattern of motion.

Overland flow. Flow of water over the land surface caused by direct PRECIPITATION.

Oxbow lake. LAKE created when floodwaters make a new, shorter CHANNEL and abandon the loop of a MEANDER. Over time, water in the oxbow lake evaporates, leaving a dry, curving, low-lying area known as a meander scar. Oxbow lakes are common on FLOODPLAINS. Another name for this feature is a cut-off.

Ozone hole. Decrease in the abundance of ANTARCTIC OZONE as sunlight returns to the POLE in early springtime

Ozone layer. Narrow band of the STRATOSPHERE situated near 18 miles (30 km.) above the earth's surface, where molecules of OZONE are concentrated. The average concentration is only one in 4 million, but this thin layer protects the earth by absorbing much of the ultraviolet light from the SUN and reradiating it as longer-wavelength radiation. Scientists were disturbed to discover that the ozonosphere was being destroyed by photochemical reaction with CHLOROFLUORO-CARBONS (CFCs). The OZONE HOLES over the South and North Poles negatively affect several animal species, as well as humans; skin cancer risk is increasing rapidly as a consequence of depletion of the ozone layer. Stratospheric ozone should not be confused with ozone at lower levels, which is a result of PHOTOCHEMICAL SMOG. Also called the ozonosphere.

P wave. Fastest elastic wave generated by an EARTHQUAKE or artificial ENERGY source; basically an acoustic or shock wave that compresses and stretches solid material in its path.

Pangaea. Name used by Alfred Wegener for the SUPERCONTINENT that broke apart to create the present CONTINENTS.

Parasitic cone. Small volcanic cone that appears on the flank of a larger VOLCANO, or perhaps inside a CALDERA.

Particulate matter. Mixture of small particles that adversely affect human health. The particles may come from smoke and DUST and are in their highest concentrations in large URBAN AREAS, where they contribute to the "DUST DOME." Increased occurrences of illnesses such as asthma and bronchitis, especially in children, are related to high concentrations of particulate matter.

Pastoralism. Type of AGRICULTURE involving the raising of grazing animals, such as cattle, goats, and sheep. Pastoral nomads migrate with their domesticated animals in order to ensure sufficient grass and water for the animals.

Paternoster lakes. Small circular LAKES joined by a STREAM. These lakes are the result of glacial EROSION. The name comes from the resemblance to rosary beads and the accompanying prayer (the Our Father).

Pedestal crater. A CRATER that has assumed the shape of a pedestal as a result of unique shaping processes caused by WIND.

Pedology. Scientific study of SOILS.

Pelagic. Relating to life-forms that live on or in open SEAS, rather than waters close to land.

Peneplain. In the geomorphic CYCLE, or cycle of LANDFORM development, described by W. M. Davis, the final stage of EROSION led to the creation of an extensive land surface with low RELIEF. Davis named this a peneplain, meaning "almost a plain." It is now known that tectonic forces are so frequent that there would be insufficient time for such a cycle to complete all stages required to complete this landform.

Percolation. Downward movement of part of the water that falls on the surface of the earth, through the upper layers of PERMEABLE SOIL and ROCKS under the influence of GRAVITY. Eventually, it accumulates in the zone of SATURATION as GROUNDWATER.

Perforated state. STATE whose territory completely surrounds another state. The classic example of a perforated state is South Africa, within which lies the COUNTRY of Lesotho. Technically, Italy is perforated by the MICROSTATES of San Marino and Vatican City.

Perihelion. Point in Earth's REVOLUTION when it is closest to the SUN (usually on January 3). At perihelion, the distance between the earth and the Sun is 91,500,000 miles (147,255,000 km.). The opposite of APHELION.

Periodicity. The recurrence of related phenomena at regular intervals.

Permafrost. Permanently frozen SUBSOIL. The condition occurs in perennially cold areas such as the ARCTIC. No trees can grow because their roots cannot penetrate the permafrost. The upper portion of the frozen SOIL can thaw briefly in the summer, allowing many smaller plants to thrive in the long daylight. Permafrost occurs in about 25 percent of the earth's land surface, and the condition even hampers construction in REGIONS such as Siberia and ARCTIC Canada.

Perturb. To change the path of an orbiting body by a gravitational force.

Petrochemical. Chemical substance obtained from NATURAL GAS or PETROLEUM.

Petrography. Description and systematic classification of ROCKS.

Photochemical smog. Mixture of gases produced by the interaction of sunlight on the gases emanating from automobile exhausts. The gases include OZONE, nitrogen dioxide, carbon monoxide, and peroxyacetyl nitrates. Many large cities suffer from poor AIR quality because of photochemical smog. Severe health problems arise from continued exposure to photochemical smog.

Photometry. Technique of measuring the brightness of astronomical objects, usually with a photoelectric cell.

Phylogeny. Study of the evolutionary relationships among organisms.

Phylum. Major grouping of organisms, distinguished on the basis of basic body plan, grade of anatomical complexity, and pattern of growth or development.

Physiography. The PHYSICAL GEOGRAPHY of a PLACE—the LANDFORMS, water features, CLIMATE, SOILS, and VEGETATION.

Piedmont glacier. GLACIER formed when several ALPINE GLACIERS join together into a spreading glacier at the base of a MOUNTAIN or RANGE. The Malaspina glacier in Alaska is a good example of a piedmont glacier.

Place. In geographic terms, space that is endowed with physical and human meaning. Geographers study the relationship between people, places, and ENVIRONMENTS. The five themes that geographers use to examine the world are LOCATION, place, human/environment interaction, movement, and REGIONS.

Placer. Accumulation of valuable MINERALS formed when grains of the minerals are physically deposited along with other, nonvaluable mineral grains.

Planetary wind system. Global atmospheric circulation pattern, as in the BELT of prevailing westerly WINDS.

Plantation. Form of AGRICULTURE in which a large area of agricultural land is devoted to the production of a single cash crop, for export. Many plantation crops are tropical, such as bananas, sugarcane, and rubber. Coffee and tea plantations require cooler CLIMATES. Formerly, slave labor was used on most plantations, and the owners were Europeans.

Plate boundary. REGION in which the earth's crustal PLATES meet, as a converging (SUBDUCTION ZONE), diverging (MID-OCEAN RIDGE), TRANSFORM FAULT, or collisional interaction.

Plate tectonics. Theory proposed by German scientist Alfred Wegener in 1910. Based on extensive study of ancient geology, STRATIGRAPHY, and CLIMATE, Wegener concluded that the CONTINENTS were formerly one single enormous LANDMASS, which he named PANGAEA. Over the past 250 million years, Pangaea broke apart, first into LAURASIA and GONDWANALAND, and subsequently into the present continents. Earth scientists now believe that the earth's CRUST is composed of a series of thin, rigid PLATES that are in motion, sometimes diverging, sometimes colliding.

Plinian eruption. Rapid ejection of large volumes of VOLCANIC ASH that is often accompanied by the collapse of the upper part of the VOLCANO. Named either for Pliny the Elder, a Roman naturalist who died while observing the ERUPTION of Mount Vesuvius in 79 CE, or for Pliny the Younger, his nephew, who chronicled the eruption.

Plucking. Term used to describe the way glacial ice can erode large pieces of ROCK as it makes its way downslope. The ice penetrates JOINTS, other openings on the floor, or perhaps the side wall, and freezes around the block of stone, tearing it away and carrying it along, as part of the glacial MORAINE. The rocks contribute greatly to glacial ABRASION, causing deep grooves or STRIATIONS in some places. The jagged torn surface left behind is subject to further plucking. ALPINE GLA-CIERS can erode steep VALLEYS called glacial TROUGHS.

Plutonic. IGNEOUS ROCKS made of MINERAL grains visible to the naked eye. These igneous rocks have cooled relatively slowly. GRANITE is a good example of a plutonic rock.

Pluvial period. Episode of time during which rains were abundant, especially during the last ICE AGE, from a few million to about 10,000 years ago.

Polar stratospheric clouds. CLOUDS of ice crystals formed at extremely low TEMPERATURES in the polar STRATOSPHERE.

Polder. Lands reclaimed from the SEA by constructing DIKES to hold back the sea and then pumping out the water retained between the dikes and the land. Before AGRICULTURE is possible, the SOIL must be specially treated to remove the SALT. Some polders are used for recreational land; cities also have been built on polders. The largest polders are in the Netherlands, where the northern part, known as the Low Netherlands, covers almost half of the total area of this COUNTRY.

Polygenetic. Pertaining to volcanism from several physically distinct vents or repeated ERUPTIONS from a single vent punctuated by long periods of quiescence.

Polygonal ground. Distinctive geological formation caused by the repetitive freezing and thawing of PERMAFROST.

Possibilism. Concept that arose among French geographers who rejected the concept of ENVIRONMENTAL DETERMINISM, instead asserting that the relationship between human beings and the ENVIRONMENT is interactive.

Potable water. FRESH WATER that is being used for domestic consumption.

Potholes. Circular depressions formed in the bed of a RIVER when the STREAM flows over BEDROCK. The scouring of PEBBLES as a result of water TURBULENCE wears away the sides of the depression, deepening it vertically and producing a smooth, rounded pothole. (In modern parlance, the term is also applied to holes in public roads.)

Primary minerals. MINERALS formed when MAGMA crystallizes.

Primary wave. Compressional type of EARTHQUAKE wave, which can travel in any medium and is the fastest wave.

Primate city. CITY that is at least twice as large as the next-largest city in that COUNTRY. The "law of the primate city" was developed by U.S. geographer Mark Jefferson, to analyze the phenomenon of countries where one huge city dominates the political, economic, and cultural life of that country. The size and dominance of a primate city is a PULL FACTOR and ensures its continuing dominance.

Principal parallels. The most important lines of LATITUDE. PARALLELS are imaginary lines, parallel to the EQUATOR. The principal parallels are the equator at zero DEGREES, the tropic of CANCER at 23.5 degrees North, the tropic of CAPRICORN at 23.5 degrees south, the Arctic Circle at 66.5 degrees north, and the Antarctic Circle at 66.5 degrees south.

Protectorate. COUNTRY that is a political DEPENDENCY of another NATION; similar to a COLONY, but usually having a less restrictive relationship with its overseeing power.

Proterozoic eon. Interval between 2.5 billion and 544 million years ago. During this PERIOD in the GEOLOGIC RECORD, processes presently active on Earth first appeared, notably the first clear evidence for PLATE TECTONICS. ROCKS of the Proterozoic eon also document changes in conditions on Earth, particularly an apparent increase in atmospheric oxygen.

Pull factors. Forces that attract immigrants to a new COUNTRY or LOCATION as permanent settlers. They include economic opportunities, educational facilities, land ownership, gold rushes, CLIMATE conditions, democracy, and similar factors of attraction.

Push factors. Forces that encourage people to migrate permanently from their HOMELANDS to settle in a new destination. They include war, persecution for religious or political reasons, hunger, and similar negative factors.

Pyroclasts. Materials that are ejected from a VOLCANO into the AIR. Pyroclastic materials return to Earth at greater or lesser distances, depending on their size and the height to which they are thrown by the explosion of the volcano. The largest pyroclasts are volcanic bombs. Smaller pieces are volcanic blocks and scoria. These generally fall back onto the volcano and roll down the sides. Even smaller pyroclasts are LAPILLI, cinders, and VOLCANIC ASH. The finest pyroclastic materials may be carried by WINDS for great distances, even completely around the earth, as was the case with DUST from the Krakatoa explosion in 1883 and the early 1990s explosions of Mount Pinatubo in the Philippines.

Qanat. Method used in arid REGIONS to bring GROUNDWATER from mountainous regions to lower and flatter agricultural land. A qanat is a long tunnel or series of tunnels, perhaps more than a mile long. The word *qanat* is Arabic, but the first qanats are thought to have been constructed in Farsi-speaking Persia more than 2,000 years ago. Qanats are still used there, as well as in Afghanistan and Morocco.

Quaternary sector. Economic activity that involves the collection and processing of information. The rapid spread of computers and the Internet caused a major increase in the importance of employment in the quaternary sector.

Radar imaging. Technique of transmitting radar toward an object and then receiving the reflected radiation so that time-of-flight MEASUREMENTS provide information about surface TOPOGRAPHY of the object under study.

Radial drainage. The pattern of STREAM courses often reveals the underlying geology or structure of a REGION. In a radial drainage pattern, streams radiate outward from a center, like spokes on a wheel, because they flow down the slopes of a VOLCANO.

Radioactive minerals. MINERALS combining uranium, thorium, and radium with other elements. Useful for nuclear TECHNOLOGY, these minerals furnish the basic isotopes necessary not only for nuclear reactors but also for advanced medical treatments, metallurgical analysis, and chemicophysical research.

Rain gauge. Instrument for measuring RAINFALL, usually consisting of a cylindrical container open to the sky.

Rain shadow. Area of low PRECIPITATION located on the LEEWARD side of a topographic barrier such as a mountain RANGE. Moisture-laden WINDS are forced to rise, so they cool ADIABATICALLY, leading to CONDENSATION and precipitation on the WINDWARD side of the barrier. When the AIR descends on the other side of the MOUNTAIN, it is dry and relatively warm. The area to the east of the Rocky Mountains is in a rain shadow.

Range, mountain. Linear series of MOUNTAINS close together, formed in an OROGENY, or mountain-building episode. Tall mountain ranges such as the Rocky Mountains are geologically much younger than older mountain ranges such as the Appalachians.

Rapids. Stretches of RIVERS where the water flow is swift and turbulent because of a steep and rocky CHANNEL. The turbulent conditions are called WHITE WATER. If the change in ELEVATION is greater, as for small WATERFALLS, they are called CATARACTS.

Recessional moraine. Type of TERMINAL MORAINE that marks a position of shrinkage or wasting or a GLACIER. Continued forward flow of ice is maintained so that the debris that forms the moraine continues to accumulate. Recessional moraines occur behind the terminal moraine.

Recumbent fold. Overturned fold in which the upper part of the fold is almost horizontal, lying on top of the nearest adjacent surface.

Reef (geology). VEIN of ORE, for example, a reef of gold.

Reef (marine). Underwater ridge made up of sand, rocks, or coral that rises near to the water's surface.

Refraction of waves. Bending of waves, which can occur in all kinds of waves. When OCEAN WAVES approach a COAST, they start to break as they approach the SHORE because the depth decreases.

The wave speed is retarded and the WAVE CREST seems to bend as the wavelength decreases. If waves are approaching a coast at an oblique angle, the crest line bends near the shore until it is almost parallel. If waves are approaching a BAY, the crests are refracted to fit the curve of the bay.

Regression. Retreat of the SEA from the land; it allows land EROSION to occur on material formerly below the sea surface.

Relative humidity. Measure of the HUMIDITY, or amount of moisture, in the ATMOSPHERE at any time and place compared with the total amount of moisture that same AIR could theoretically hold at that TEMPERATURE. Relative humidity is a ratio that is expressed as a percentage. When the air is saturated, the relative humidity reaches 100 percent and rain occurs. When there is little moisture in the air, the relative humidity is low, perhaps 20 percent. Relative humidity varies inversely with temperature, because warm air can hold more moisture than cooler air. Therefore, when temperatures fall overnight, the air often becomes saturated and DEW appears on grass and other surfaces. The human COMFORT INDEX is related to the relative humidity. Hot temperatures are more bearable when relative humidity is low. Media announcers frequently use the term "humidity" when they mean relative humidity.

Replacement rate. The rate at which females must reproduce to maintain the size of the POPULATION. It corresponds to a FERTILITY RATE of 2.1.

Reservoir rock. Geologic ROCK layer in which OIL and gas often accumulate; often SANDSTONE or LIMESTONE.

Retrograde orbit. ORBIT of a SATELLITE around a PLANET that is in the opposite sense (direction) in which the planet rotates.

Retrograde rotation. ROTATION of a PLANET in a direction opposite to that of its REVOLUTION.

Reverse fault. Feature produced by compression of the earth's CRUST, leading to crustal shortening. The UPTHROWN BLOCK overhangs the downthrown block, producing a FAULT SCARP where the overhang is prone to LANDSLIDES. When the movement is mostly horizontal, along a low angle FAULT, an overthrust fault is formed. This is commonly associated with extreme FOLDING.

Reverse polarity. Orientation of the earth's MAGNETIC FIELD so that a COMPASS needle points to the SOUTHERN HEMISPHERE.

Ria coast. Ria is a long narrow ESTUARY or RIVER MOUTH. COASTS where there are many rias show the effects of SUBMERGENCE of the land, with the SEA now occupying former RIVER VALLEYS. Generally, there are MOUNTAINS running at an angle to the coast, with river valleys between each RANGE, so that the ria coast is a succession of estuaries and promontories. The submergence can result from a rising SEA LEVEL, which is common since the melting of the PLEISTOCENE GLACIERS, or it can be the result of SUBSIDENCE of the land. There is often a great TIDAL RANGE in rias, and in some, a tidal BORE occurs with each TIDE. The eastern coast of the United States, from New York to South Carolina, is a ria coast. The southwest coast of Ireland is another. The name comes from Spain, where rias occur in the south.

Richter scale. SCALE used to measure the magnitude of EARTHQUAKES; named after U.S. physicist Charles Richter, who, together with Beno Gutenberg, developed the scale in 1935. The scale is a quantitative measure that replaced the older MERCALLI SCALE, which was a descriptive scale. Numbers range from zero to nine, although there is no upper limit. Each whole number increase represents an order of magnitude, or an increase by a factor of ten. The actual MEASUREMENT was logarithm to base 10 of the maximum SEISMIC WAVE amplitude (in thousandths of a millimeter) recorded on a standard SEISMOGRAPH at a distance of 60 miles (100 km.) from the earthquake EPICENTER.

Rift valley. Long, low REGION of the earth's surface; a VALLEY or TROUGH with FAULTS on either side. Unlike valleys produced by EROSION, rift valleys are produced by tectonic forces that have caused the faults or fractures to develop in the ROCKS of Earth's CRUST. TENSION can lead to the block of land between two faults dropping in ELEVATION compared to the surrounding blocks, thus forming the rift valley. A small LANDFORM produced

in this way is called a GRABEN. A rift valley is a much larger feature. In Africa, the Great Rift Valley is partially occupied by Lake Malawi and Lake Tanganyika, as well as by the Red Sea.

Ring dike. Volcanic LANDFORM created when MAGMA is intruded into a series of concentric FAULTS. Later EROSION of the surrounding material may reveal the ring dike as a vertical feature of thick BASALT rising above the surroundings.

Ring of Fire. Zone of volcanic activity and associated EARTHQUAKES that marks the edges of various TECTONIC PLATES around the Pacific Ocean, especially those where SUBDUCTION is occurring.

Riparian. Term meaning related to the BANKS of a STREAM or RIVER. Riparian VEGETATION is generally trees, because of the availability of moisture. RIPARIAN RIGHTS allow owners of land adjacent to a river to use water from the river.

River terraces. LANDFORMS created when a RIVER first produces a FLOODPLAIN, by DEPOSITION of ALLUVIUM over a wide area, and then begins downcutting into that alluvium toward a lower BASE LEVEL. The renewed EROSION is generally because of a fall in SEA LEVEL, but can result from tectonic UPLIFT or a change in CLIMATE pattern due to increased PRECIPITATION. On either side of the river, there is a step up from the new VALLEY to the former alluvium-covered floodplain surface, which is now one of a pair of river terraces. This process may occur more than once, creating as many as three sets of terraces. These are called depositional terraces, because the terrace is cut into river deposits. Erosional terraces, in contrast, are formed by lateral migration of a river, from one part of the valley to another, as the river creates a floodplain. These terraces are cut into BEDROCK, with only a thin layer of alluvium from the point BAR deposits, and they do not occur in matching pairs.

River valleys. VALLEYS in which STREAMS flow are produced by those streams through long-term EROSION and DEPOSITION. The LANDFORMS produced by FLUVIAL action are quite diverse, ranging from spectacular CANYONS to wide, gently sloping valleys. The patterns formed by stream networks are complex and generally reflect the BEDROCK geology and TERRAIN characteristics.

Rock avalanche. Extreme case of a rockfall. It occurs when a large mass of ROCK moves rapidly down a steeply sloping surface, taking everything that lies in its path. It can be started by an EARTHQUAKE, rock-blasting operations, or vibrations from thunder or artillery fire.

Rock cycle. Cycle by which ROCKS are formed and reformed, changing from one type to another over long PERIODS of geologic time. IGNEOUS ROCKS are formed by cooling from molten MAGMA. Once exposed at the surface, they are subject to WEATHERING and EROSION. The products of erosion are compacted and cemented to form SEDIMENTARY ROCKS. The heat and pressure accompanying a volcanic intrusion causes adjacent rocks to be altered into METAMORPHIC ROCKS.

Rock slide. Event that occurs when water lubricates an unconsolidated mass of weathered ROCK on a steep slope, causing rapid downslope movement. In a RIVER VALLEY where there are steep SCREE slopes being constantly carried away by a swiftly flowing STREAM, the undercutting at the base can lead to constant rockslides of the surface layer of rock. A large rockslide is a ROCK AVALANCHE.

S waves. Type of SEISMIC disturbance of the earth when an EARTHQUAKE occurs. In an S wave, particles move about at right angles to the direction in which the wave is traveling. S waves cannot pass through the earth's CORE, which is why scientists believe the INNER CORE is liquid. Also called transverse wave, shear wave, or secondary wave.

Sahel. Southern edge of the Sahara Desert; a great stretch of semiarid land extending from the Atlantic Ocean in Senegal and Mauritania through Mali, Burkina Faso, Nigeria, Niger, Chad, and Sudan. Northern Ethiopia, Eritrea, Djibouti, and Somalia usually are included also. This transition zone between the hot DESERT and the tropical SAVANNA has low summer RAINFALL of less than 8 inches (200 millimeters) and a natural VEGETATION of low grasses with some small shrubs. The REGION traditionally has been used for PASTORALISM, raising goats, camels, and occasionally sheep. Since a prolonged DROUGHT in the 1970s, DESERTIFICATION, SOIL EROSION, and FAMINE have plagued the Sahel. The narrow band between the northern Sahara and the Mediterranean North African COAST is also called Sahel. "Sahel" is the Arabic word for edge.

Saline lake. LAKE with elevated levels of dissolved solids, primarily resulting from evaporative concentration of SALTS; saline lakes lack an outlet to the SEA. Well-known examples include Utah's Great Salt Lake, California's Mono Lake and Salton Sea, and the Dead Sea in the Middle East.

Salinization. Accumulation of SALT in SOIL. When IRRIGATION is used to grow crops in semiarid to arid REGIONS, salinization is frequently a problem. Because EVAPORATION is high, water is drawn upward through the soil, depositing dissolved salts at or near the surface. Over years, salinization can build up until the soil is no longer suitable for AGRICULTURE. The solution is to maintain a plentiful flow of water while ensuring that the water flows through the soil and is drained away.

Salt domes. Formations created when deeply buried salt layers are forced upwards. SALT under pressure is a plastic material, one that can flow or move slowly upward, because it is lighter than surrounding SEDIMENTARY ROCKS. The salt forms into a plug more than a half mile (1 km.) wide and as much as 5 miles (8 km.) deep, which passes through overlying sedimentary rock layers, pushing them up into a dome shape as it passes. Some salt domes emerge at the earth's surface; others are close to the surface and are easy to mine for ROCK SALT. OIL and NATURAL GAS often accumulate against the walls of a salt dome. Salt domes are numerous around the COAST of the Gulf of Mexico, in the North Sea REGION, and in Iran and Iraq, all of which are major oil-producing regions.

Sand dunes. Accumulations of SAND in the shape of mounds or RIDGES. They occur on some COASTS and in arid REGIONS. Coastal dunes are formed when the prevailing WINDS blow strongly on-shore, piling up sand into dunes, which may become stabilized when grasses grow on them. DESERT sand dunes are a product of DEFLATION, or wind EROSION removing fine materials to leave a DESERT PAVEMENT in one region and sand deposits in another. Sand dunes are classified by their shape into barchans, or crescent-shaped dunes; seifs or LONGITUDINAL DUNES; TRANSVERSE DUNES; star dunes; and sand drifts or sand sheets.

Sapping. Natural process of EROSION at the bases of HILL slopes or CLIFFS whereby support is removed by undercutting, thereby allowing overlying layers to collapse; SPRING SAPPING is the facilitation of this process by concentrated GROUNDWATER flow, generally at the heads of VALLEYS.

Saturation, zone of. Underground REGION below the zone of AERATION, where all pore space is filled with water. This water is called GROUNDWATER; the upper surface of the zone of saturation is the WATER TABLE.

Scale. Relationship between a distance on a MAP or diagram and the same distance on the earth. Scale can be represented in three ways. A linear, or graphic, scale uses a straight line, marked off in equally spaced intervals, to show how much of the map represents a mile or a kilometer. A representative fraction (RF) gives this scale as a ratio. A verbal scale uses words to explain the relationship between map size and actual size. For example, the RF 1:63,360 is the same as saying "one inch to the mile."

Scarp. Short version of the word "ESCARPMENT," a short steep slope, as at the edge of a PLATEAU. EARTHQUAKES lead to the formation of FAULT SCARPS.

Schist. METAMORPHIC ROCK that can be split easily into layers. Schist is commonly produced from the action of heat and pressure on SHALE or SLATE. The rock looks flaky in appearance. Mica-schists are shiny because of the development of visible mica. Other schists include talc-schist, which contains a large amount of talc,

and hornblende-schist, which develops from basaltic rocks.

Scree. Broken, loose ROCK material at the base of a slope or CLIFF. It is often the result of FROST WEDGING of BEDROCK cliffs, causing rockfall. Another name for scree is TALUS.

Sedimentary rocks. ROCKS formed from SEDIMENTS that are compressed and cemented together in a process called LITHIFICATION. Sedimentary rocks cover two-thirds of the earth's land surface but are only a small proportion of the earth's CRUST. SANDSTONE is a common sedimentary rock. Sedimentary rocks form STRATA, or layers, and sometimes contain FOSSILS.

Seif dunes. Long, narrow RIDGES of SAND, built up by WINDS blowing at different times of year from two different directions. Seif dunes occur in parallel lines of sand over large areas, running for hundreds of miles in the Sahara, Iran, and central Australia. Another name for seif dunes is LONGITUDINAL DUNES. The Arabic word means sword.

Seismic activity. Movements within the earth's CRUST that often cause various other geological phenomena to occur; the activity is measured by SEISMOGRAPHS.

Seismology. The scientific study of EARTHQUAKES. It is a branch of GEOPHYSICS. The study of SEISMIC WAVES has provided a great deal of knowledge about the composition of the earth's interior.

Shadow zone. When an EARTHQUAKE occurs at one LOCATION, its waves travel through the earth and are detected by SEISMOGRAPHS around the world. Every earthquake has a shadow zone, a band where neither P nor S WAVES from the earthquake will be detected. This shadow zone leads scientists to draw conclusions about the size, density, and composition of the earth's CORE.

Shale oil. SEDIMENTARY ROCK containing sufficient amounts of hydrocarbons that can be extracted by slow distillation to yield OIL.

Shallow-focus earthquakes. EARTHQUAKES having a focus less than 35 miles (60 km.) below the surface.

Shantytown. URBAN SQUATTER SETTLEMENT, usually housing poor newcomers.

Shield. Large part of the earth's CONTINENTAL CRUST, comprising very old ROCKS that have been eroded to REGIONS of low RELIEF. Each CONTINENT has a shield area. In North America, the Canadian Shield extends from north of the Great Lakes to the Arctic Ocean. Sometimes known as a CONTINENTAL SHIELD.

Shield volcano. VOLCANO created when the LAVA is quite viscous or fluid and highly basaltic. Such lava spreads out in a thin sheet of great radius but comparatively low height. As flows continue to build up the volcano, a low DOME shape is created. The greatest shield volcanoes on Earth are the ISLANDS of Hawaii, which rise to a height of almost 30,000 feet (10,000 meters) above SEA LEVEL.

Shock city. CITY that typifies disturbing changes in social and cultural conditions or in economic conditions. In the nineteenth century, the shock city of the United States was Chicago.

Sierra. Spanish word for a mountain RANGE with a serrated crest. In California, the Sierra Nevada is an important range, containing Mount Whitney, the highest PEAK in the continental United States.

Sill. Feature formed by INTRUSIVE volcanic activity. When LAVA is forced between two layers of ROCK, it can form a narrow horizontal layer of BASALT, parallel with the adjacent beds. Although it resembles a windowsill in its flatness, a sill may be hundreds of miles long and can range in thickness from a few centimeters to considerable thickness.

Siltation. Build-up of SILT and SAND in creeks and waterways as a result of SOIL EROSION, clogging water courses and creating DELTAS at RIVER MOUTHS. Siltation often results from DEFORESTATION or removal of tree cover. Such ENVIRONMENTAL DEGRADATION causes loss of agricultural productivity, worsening of water supply, and other problems.

Sima. Abbreviation for SILICA and *ma*gnesium. These are the two principal constituents of heavy ROCKS such as BASALT, which forms much of the OCEAN floor. Lighter, more abundant rock is SIAL.

Sinkhole. Circular depression in the ground surface, caused by WEATHERING of LIMESTONE, mainly through the effects of SOLUTION on JOINTS in the ROCK. If a STREAM flows above ground and then disappears down a sinkhole, the feature is called a swallow hole. In everyday language, many events that cause the surface to collapse are called sinkholes, even though they are rarely in limestone and rarely caused by weathering.

Sinking stream. STREAM or RIVER that loses part or all of its water to pathways dissolved underground in the BEDROCK.

Situation. Relationship between a PLACE, such as a TOWN or CITY, and its RELATIVE LOCATION within a REGION. A situation on the COAST is desirable in terms of overseas TRADE.

Slip-face. LEEWARD side of a SAND DUNE. As the WIND piles up sand on the WINDWARD side, it then slips down the rear or slip-face. The angle of the slip-face is gentler than the angle of the windward slope.

Slump. Type of LANDSLIDE in which the material moves downslope with a rotational motion, along a curved slip surface.

Snout. Terminal end of a GLACIER.

Snow line. The height or ELEVATION at which snow remains throughout the year, without melting away. Near the EQUATOR, the snow line is more than 15,000 feet (almost 5,000 meters); at higher LATITUDES, the snow line is correspondingly lower, reaching SEA LEVEL at the POLES. The actual snow line varies with the time of year, retreating in summer and coming lower in winter.

Soil horizon. SOIL consists of a series of layers called horizons. The uppermost layer, the O horizon, contains organic materials such as decayed leaves that have been changed into HUMUS. Beneath this is the A horizon, the TOPSOIL, where farmers plow and plant seeds. The B HORIZON often contains MINERALS that have been washed downwards from the A horizon, such as calcium, iron, and aluminum. The A and B horizons together comprise a solum, or true soil. The C horizon is weathered BEDROCK, which contains pieces of the original ROCK from which the soil formed. Another name for the C horizon is REGOLITH. Beneath this is the R horizon, or bedrock.

Soil moisture. Water contained in the unsaturated zone above the WATER TABLE.

Soil profile. Vertical section of a SOIL, extending through its horizon into the unweathered parent material.

Soil stabilization. Engineering measures designed to minimize the opportunity and/or ability of EXPANSIVE SOILS to shrink and swell.

Solar energy. One of the forms of ALTERNATIVE or RENEWABLE ENERGY. In the late 1990s, the world's largest solar power generating plant was located at Kramer Junction, California. There, solar energy heats huge OIL-filled containers with a parabolic shape, which produces steam to drive generating turbines. An alternative is the production of energy through photovoltaic cells, a TECHNOLOGY that was first developed for space exploration. Many individual homes, especially in isolated areas, use this technology.

Solar system. SUN and all the bodies that ORBIT it, including the PLANETS and their SATELLITES, plus numerous COMETS, ASTEROIDS, and METEOROIDS.

Solar wind. Gases from the SUN's ATMOSPHERE, expanding at high speeds as streams of charged particles.

Solifluction. Word meaning flowing SOIL. In some REGIONS of PERMAFROST, where the ground is permanently frozen, the uppermost layer thaws during the summer, creating a saturated layer of soil and REGOLITH above the hard layer of frozen ground. On slopes, the material can flow slowly downhill, creating a wavy appearance along the hillslope.

Solution. Form of CHEMICAL WEATHERING in which MINERALS in a ROCK are dissolved in water. Most substances are soluble, but the combination of water with CARBON DIOXIDE from the ATMOSPHERE means that RAINFALL is slightly acidic, so

that the chemical reaction is often a combination of solution and carbonation.

Sound. Long expanse of the SEA, close to the COAST, such as a large ESTUARY. It can also be the expanse of sea between the mainland and an ISLAND.

Source rock. ROCK unit or bed that contains sufficient organic carbon and has the proper thermal history to generate OIL or gas.

Spatial diffusion. Notion that things spread through space and over time. An understanding of geographic change depends on this concept. Spatial diffusion can occur in various ways. Geographers distinguish between expansion diffusion, relocation diffusion, and hierarchical diffusion.

Spheroidal weathering. Form of ROCK WEATHERING in which layers of rock break off parallel to the surface, producing a rounded shape. It results from a combination of physical and CHEMICAL WEATHERING. Spheroidal weathering is especially common in GRANITE, leading to the creation of TORS and similar rounded features. Onion-skin weathering is a term sometimes used, especially when this is seen on small rocks.

Spring tide. TIDE of maximum RANGE, occurring when lunar and solar tides reinforce each other, a few DAYS after the full and new MOONS.

Squall line. Line of vigorous THUNDERSTORMS created by a cold downdraft that spreads out ahead of a fast-moving COLD FRONT.

Stacks. Pieces of ROCK surrounded by SEA water, which were once part of the mainland. WAVE EROSION has caused them to be isolated. Also called sea stacks.

Stalactite. Long, tapering piece of calcium carbonate hanging from the roof of a LIMESTONE CAVE or cavern. Stalactites are formed as water containing the MINERAL in solution drips downward. The water evaporates, depositing the dissolved minerals.

Stalagmite. Column of calcium carbonate growing upward from the floor of a LIMESTONE CAVE or cavern.

Steppe. Huge REGION of GRASSLANDS in the midlatitudes of Eurasia, extending from central

Europe to northeast China. The region is not uniform in ELEVATION; most of it is rolling PLAINS, but some mountain RANGES also occur. These have not been a barrier to the migratory lifestyle of the herders who have occupied the steppe for many centuries. The Asian steppe is colder than the European steppe, because of greater elevation and greater continentality. The best-known rulers from the steppe were the Mongols, whose empire flourished in the thirteenth and fourteenth centuries. Geographers speak of a steppe CLIMATE, a semiarid climate where the EVAPORATION rate is double that of PRECIPITATION. South of the steppe are great DESERTS; to the north are midlatitude mixed FORESTS. In terms of climate and VEGETATION, the steppe is like the short-grass PRAIRIE vegetation west of the Mississippi River. Also called steppes.

Storm surge. General rise above normal water level, resulting from a HURRICANE or other severe coastal STORM.

Strait. Relatively narrow body of water, part of an OCEAN or SEA, separating two pieces of land. The world's busiest SEAWAY is the Johore Strait between the Malay Peninsula and the island of Sumatra.

Strata. Layers of SEDIMENT deposited at different times, and therefore of different composition and TEXTURE. When the sediments are laid down, strata are horizontal, but subsequent tectonic processes can lead to tilting, FOLDING, or faulting. Not all SEDIMENTARY ROCKS are stratified. Singular form of the word is stratum.

Stratified drift. Material deposited by glacial MELTWATERS; the water separates the material according to size, creating layers.

Stratigraphy. Study of sedimentary STRATA, which includes the concept of time, possible correlation of the ROCK units, and characteristics of the rocks themselves.

Stratovolcano. Type of VOLCANO in which the ERUPTIONS are of different types and produce different LAVAS. Sometimes an eruption ejects cinder and ASH; at other times, viscous lava flows down the sides. The materials flow, settle, and fall to produce a beautiful symmetrical LANDFORM with a broad circular base and concave slopes tapering upward to a small circular CRATER. Mount Rainier, Mount Saint Helens, and Mount Fuji are stratovolcanoes. Also known as a COMPOSITE CONE.

Streambed. Channel through which a STREAM flows. Dry streambeds are variously known as ARROYOS, DONGAS, WASHES, and WADIS.

Strike. Term used when earth scientists study tilted or inclined beds of SEDIMENTARY ROCK. The strike of the inclined bed is the direction of a horizontal line along a bedding plane. The strike is at right angles to the dip of the rocks.

Strike-slip fault. In a strike-slip fault, the surface on either side of the fault moves in a horizontal plane. There is no vertical displacement to form a FAULT SCARP, as there is with other types of faults. The San Andreas Fault is a strike-slip fault. Also called a transcurrent fault.

Subduction zone. CONVERGENT PLATE BOUNDARY where an oceanic PLATE is being thrust below another plate.

Sublimation. Process by which water changes directly from solid (ice) to vapor, or vapor to solid, without passing through a liquid stage.

Subsolar point. Point on the earth's surface where the SUN is directly overhead, making the Sun's rays perpendicular to the surface. The subsolar point receives maximum INSOLATION, compared with other PLACES, where the Sun's rays are oblique.

Sunspots. REGIONS of intense magnetic disturbances that appear as dark spots on the solar surface; they occur approximately every eleven years.

Supercontinent. Vast LANDMASS of the remote geologic past formed by the collision and amalgamation of crustal PLATES. Hypothesized supercontinents include PANGAEA, GONDWANALAND, and LAURASIA.

Supersaturation. State in which the AIR's RELATIVE HUMIDITY exceeds 100 percent, the condition necessary for vapor to begin transformation to a liquid state.

Supratidal. Referring to the SHORE area marginal to shallow OCEANS that are just above high-tide level.

Swamp. WETLAND where trees grow in wet to waterlogged conditions. Swamps are common close to the RIVER on FLOODPLAINS, as well as in some coastal areas.

Swidden. Area of land that has been cleared for SUBSISTENCE AGRICULTURE by a farmer using the technique of slash-and-burn. A variety of crops is planted, partly to reduce the risk of crop failure. Yields are low from a swidden because SOIL fertility is low and only human labor is used for CLEARING, planting, and harvesting. See also INTERTILLAGE.

Symbolic landscapes. LANDSCAPES centered on buildings or structures that are so visually emblematic that they represent an entire CITY.

Syncline. Downfold or TROUGH shape that is formed through compression of ROCKS. An upfold is an ANTICLINE.

Tableland. Large area of land with a mostly flat surface, surrounded by steeply sloping sides, or ESCARPMENTS. A small PLATEAU.

Taiga. Russian name for the vast BOREAL FORESTS that cover Siberia. The marshy ground supports a tree VEGETATION in which the trees are CONIFEROUS, comprising mostly pine, fir, and larch.

Talus. Broken and jagged pieces of ROCK, produced by WEATHERING of steep slopes, that fall to the base of the slope and accumulate as a talus cone. In high MOUNTAINS, a ROCK GLACIER may form in the talus. See also SCREE.

Tarn. Small circular LAKE, formed in a CIRQUE, which was previously occupied by a GLACIER.

Tectonism. The formation of MOUNTAINS because of the deformation of the CRUST of the earth on a large scale.

Temporary base level. STREAMS or RIVERS erode their beds down toward a BASE LEVEL—in most cases, SEA LEVEL. A section of hard ROCK may slow EROSION and act as a temporary, or local, base level. Erosion slows upstream of the temporary base level. A DAM is an artificially constructed temporary base level.

Tension. Type of stress that produces a stretching and thinning or pulling apart of the earth's CRUST. If the surface breaks, a NORMAL FAULT is created, with one side of the surface higher than the other.

Tephra. General term for volcanic materials that are ejected from a vent during an ERUPTION and transported through the AIR, including ASH (volcanic), BLOCKS (volcanic), cinders, LAPILLI, scoria, and PUMICE.

Terminal moraine. RIDGE of unsorted debris deposited by a GLACIER. When a glacier erodes it moves downslope, carrying ROCK debris and creating a ground MORAINE of material of various sizes, ranging from big angular blocks or boulders down to fine CLAY. At the terminus of the glacier, where the ice is melting, the ground moraine is deposited, building the ridge of unsorted debris called a terminal moraine.

Terrain. Physical features of a REGION, as in a description of rugged terrain. It should not be confused with TERRANE.

Terrane. Piece of CONTINENTAL CRUST that has broken off from one PLATE and subsequently been joined to a different plate. The terrane has quite different composition and structure from the adjacent continental materials. Alaska is composed mostly of terranes that have accreted, or joined, the North American plate.

Terrestrial planet. Any of the solid, rocky-surfaced bodies of the inner SOLAR SYSTEM, including the PLANETS Mercury, Venus, Earth, and Mars and Earth's SATELLITE, the MOON.

Terrigenous. Originating from the WEATHERING and EROSION of MOUNTAINS and other land formations.

Texture. One of the properties of SOILS. The three textures are SAND, SILT, and CLAY. Texture is measured by shaking the dried soil through a series of sieves with mesh of reducing diameters. A mixture of sand, silt, and clay gives a LOAM soil.

Thermal equator. Imaginary line connecting all PLACES on Earth with the highest mean daily TEMPERATURE. The thermal equator moves south of the EQUATOR in the SOUTHERN HEMISPHERE summer, especially over the CONTINENTS of South America, Africa, and Australia. In the northern summer, the thermal equator moves far into Asia, northern Africa, and North America.

Thermal pollution. Disruption of the ECOSYSTEM caused when hot water is discharged, usually as a thermal PLUME, into a relatively cooler body of water. The TEMPERATURE change affects the aquatic ecosystem, even if the water is chemically pure. Nuclear power-generating plants use large volumes of water in the process and are important sources of thermal pollution.

Thermocline. Depth interval at which the TEMPERATURE of OCEAN water changes abruptly, separating warm SURFACE WATER from cold, deep water.

Thermodynamics. Area of science that deals with the transformation of ENERGY and the laws that govern these changes; equilibrium thermodynamics is especially concerned with the reversible conversion of heat into other forms of energy.

Thermopause. Outer limit of the earth's ATMOSPHERE.

Thermosphere. Atmospheric zone beyond the MESOSPHERE in which TEMPERATURE rises rapidly with increasing distance from the earth's surface.

Thrust belt. Linear BELT of ROCKS that have been deformed by THRUST FAULTS.

Thrust fault. FAULT formed when extreme compression of the earth's CRUST pushes the surface into folds so closely spaced that they overturn and the ROCK then fractures along a fault.

Tidal force. Gravitational force whose strength and direction vary over a body and thus act to deform the body.

Tidal range. Difference in height between high TIDE and low tide at a given point.

Tidal wave. Incorrect name for a TSUNAMI.

Till. Mass of unsorted and unstratified SEDIMENTS deposited by a GLACIER. Boulders and smaller rounded ROCKS are mixed with CLAY-sized materials.

Timberline. Another term for tree line, the BOUNDARY of tree growth on MOUNTAIN slopes. Above the timberline, TEMPERATURES are too cold for tree growth.

Tombolo. Strip of SAND or other SEDIMENT that connects an ISLAND or SEA stack to the mainland. Mont-Saint-Michel is linked to the French mainland by a tombolo.

Topography. Description of the natural LANDSCAPE, including LANDFORMS, RIVERS and other waters, and VEGETATION cover.

Topological space. Space defined in terms of the connectivity between LOCATIONS in that space. The nature and frequency of the connections are measured, while distance between locations is not considered an important factor. An example of topological space is a transport network diagram, such as a bus route or a MAP of an underground rail system. Networks are most concerned with flows, and therefore with connectivity.

Toponyms. PLACE names. Sometimes, names of features and SETTLEMENTS reveal a good deal about the history of a REGION. For example, the many names starting with "San" or "Santa" in the Southwest of the United States recall the fact that Spain once controlled that area. The scientific study of place names is toponymics.

Tor. Rocky outcrop of blocks of ROCK, or corestones, exposed and rounded by WEATHERING. Tors frequently form in GRANITE, where three series of JOINTS often developed as the rock originally cooled when it was formed.

Transform faults. FAULTS that occur along DIVERGENT PLATE boundaries, or SEAFLOOR SPREADING zones. The faults run perpendicular to the spreading center, sometimes for hundreds of miles, some for more than five hundred miles. The motion along a transform fault is lateral or STRIKE-SLIP.

Transgression. Flooding of a large land area by the SEA, either by a regional downwarping of continental surface or by a global rise in SEA LEVEL.

Transmigration. Policy of the government of Indonesia to encourage people to move from the densely overcrowded ISLAND of Java to the sparsely populated other islands.

Transverse bar. Flat-topped body of SAND or gravel oriented transverse to the RIVER flow.

Trophic level. Different types of food relations that are found within an ECOSYSTEM. Organisms that derive food and ENERGY through PHOTOSYNTHESIS are called autotrophs (self-feeders) or producers. Organisms that rely on producers as their source of energy are called heterotrophs (feeders on others) or consumers. A third trophic level is represented by the organisms known as decomposers, which recycle organic waste.

Tropical cyclone. STORM that forms over tropical OCEANS and is characterized by extreme amounts of rain, a central area of calm AIR, and spinning WINDS that attain speeds of up to 180 miles (300 km.) per hour.

Tropical depression. STORM with WIND speeds up to 38 miles (64 km.) per hour.

Tropopause. BOUNDARY layer between the TROPOSPHERE and the STRATOSPHERE.

Troposphere. Lowest and densest of Earth's atmospheric layers, marked by considerable TURBULENCE and a decrease in TEMPERATURE with increasing ALTITUDE.

Tsunami. SEISMIC SEA WAVE caused by a disturbance of the OCEAN floor, usually an EARTHQUAKE, although undersea LANDSLIDES or volcanic ERUPTIONS can also trigger tsunami.

Tufa. LIMESTONE or calcium carbonate deposit formed by PRECIPITATION from an alkaline LAKE. Mono Lake is famous for the dramatic tufa towers exposed by the lowering of the level of lake water. Also known as TRAVERTINE.

Tumescence. Local swelling of the ground that commonly occurs when MAGMA rises toward the surface.

Tunnel vent. Central tube in a volcanic structure through which material from the earth's interior travels.

U-shaped valley. Steep-sided VALLEY carved out by a GLACIER. Also called a glacial TROUGH.

Ubac slope. Shady side of a MOUNTAIN, where local or microclimatic conditions permit lower TIMBERLINES and lower SNOW LINES than occur on a sunny side.

Ultimate base level. Level to which a STREAM can erode its bed. For most RIVERS, this is SEA LEVEL. For streams that flow into a LAKE, the ultimate base level is the level of the lakebed.

Unconfined aquifer. AQUIFER whose upper BOUNDARY is the WATER TABLE; it is also called a water table aquifer.

Underfit stream. STREAM that appears to be too small to have eroded the VALLEY in which it flows. A RIVER flowing in a glaciated valley is a good example of underfit.

Uniformitarianism. Theory introduced in the early nineteenth century to explain geologic processes. It used to be believed that the earth was only a few thousand years old, so the creation of LANDFORMS would have been rapid, even catastrophic. This theory, called CATASTROPHISM, explained most landforms as the result of the Great Flood of the Bible, when Noah, his family, and animals survived the deluge. Uniformitarian- ism, in contrast, stated that the processes in operation today are slow, so the earth must be immensely older than a mere few thousand years.

Universal time (UT). See GREENWICH MEAN TIME.

Universal Transverse Mercator. Projection in which the earth is divided into sixty zones, each six DEGREES of LONGITUDE wide. In a traditional Mercator projection, the earth is seen as a sphere with a cylinder wrapped around the EQUATOR. UTM can be visualized as a series of six-degree side strips running transverse, or north-south.

Unstable air. Condition that occurs when the AIR above rising air is unusually cool so that the rising air is warmer and accelerates upward.

Upthrown block. When EARTHQUAKE motion produces a FAULT, the block of land on one side is displaced vertically relative to the other. The higher is the upthrown block; the lower is the downthrown block.

Upwelling. OCEAN phenomenon in which warm SURFACE WATERS are pushed away from the

COAST and are replaced by cold waters that carry more nutrients up from depth.

Urban heat island. Cities experience a different MICROCLIMATE from surrounding REGIONS. The CITY TEMPERATURE is typically higher by a few DEGREES, both DAY and night, because of factors such as surfaces with higher heat absorption, decreased WIND strength, human heat-producing activities such as power generation, and the layer of AIR POLLUTION (DUST DOME).

Vadose zone. The part of the SOIL also known as the zone of AERATION, located above the WATER TABLE, where space between particles contains AIR.

Valley train. Fan-shaped deposit of glacial MORAINE that has been moved down-valley and redeposited by MELTWATER from the GLACIER.

Van Allen radiation belts. Bands of highly energetic, charged particles trapped in Earth's MAGNETIC FIELD. The particles that make up the inner BELT are energetic protons, while the outer belt consists mainly of electrons and is subject to DAY-night variations.

Varnish, desert. Shiny black coating often found over the surface of ROCKS in arid REGIONS. This is a form of OXIDATION or CHEMICAL WEATHERING, in which a coating of manganese oxides has formed over the exposed surface of the rock.

Varve. Pair of contrasting layers of SEDIMENT deposited over one year's time; the summer layer is light, and the winter layer is dark.

Ventifacts. PEBBLES on which one or more sides have been smoothed and faceted by ABRASION as the WIND has blown SAND particles.

Volcanic island arc. Curving or linear group of volcanic ISLANDS associated with a SUBDUCTION ZONE.

Volcanic rock. Type of IGNEOUS ROCK that is erupted at the surface of the earth; volcanic rocks are usually composed of larger crystals inside a fine-grained matrix of very small crystals and glass.

Volcanic tremor. Continuous vibration of long duration, detected only at active VOLCANOES.

Volcanology. Scientific study of VOLCANOES.

Voluntary migration. Movement of people who decide freely to move their place of permanent residence. It results from PULL FACTORS at the chosen destination, together with PUSH FACTORS in the home situation.

Warm temperate glacier. GLACIER that is at the melting TEMPERATURE throughout.

Water power. Generally means the generation of electricity using the ENERGY of falling water. Usually a DAM is constructed on a RIVER to provide the necessary height difference. The potential energy of the falling water is converted by a water turbine into mechanical energy. This is used to power a generator, which produces electricity. Also called HYDROELECTRIC POWER. Another form of water power is tidal power, which uses the force of the incoming and outgoing TIDE as its source of energy.

Water table. The depth below the surface where the zone of AERATION meets the zone of SATURATION. Above the water table, there may be some SOIL MOISTURE, but most of the pore space is filled with air. Below the water table, pore space of the ROCKS is occupied by water that has percolated down through the overlying earth material. This water is called GROUNDWATER. In practice, the water table is rarely as flat as a table, but curved, being far below the surface in some PLACES and even intersecting the surface in others. When GROUNDWATER emerges at the surface, because it intersects the water table, this is called a SPRING. The depth of the water table varies from SEASON to season, and with pumping of water from an AQUIFER.

Watershed. The whole surface area of land from which RAINFALL flows downslope into a STREAM. The watershed comprises the STREAMBED or CHANNEL, together with the VALLEY sides, extending up to the crest or INTERFLUVE, which separates that watershed from its neighbor. Each watershed is separated from the next by the drainage DIVIDE. Also called a DRAINAGE BASIN.

Waterspout. TORNADO that forms over water, or a tornado formed over land which then moves over water. The typical FUNNEL CLOUD, which reaches down from a CUMULONIMBUS CLOUD, is a narrow rotating STORM, with WIND speeds reaching hundreds of miles per hour.

Wave crest. Top of a WAVE.

Wave-cut platform. As SEA CLIFFS are eroded and worn back by WAVE attack, a wave-cut platform is created at the base of the cliffs. ABRASION by ROCK debris from the cliffs scours the platform further, as waves wash to and fro and TIDES ebb and flow. The upper part of the wave-cut platform is exposed at high tide. These areas contain rockpools, which are rich in interesting MARINE life-forms. Offshore beyond the platform, a wave-built TERRACE is formed by DEPOSITION.

Wave height. Vertical distance between one WAVE CREST and the adjacent WAVE TROUGH.

Wave length. Distance between two successive WAVE CRESTS or two successive WAVE TROUGHS.

Wave trough. The low part of a WAVE, between two WAVE CRESTS.

Weather analogue. Approach to WEATHER FORECASTING that uses the WEATHER behavior of the past to predict what a current weather pattern will do in the future.

Weather forecasting. Attempt to predict WEATHER patterns by analysis of current and past data.

Wilson cycle. Creation and destruction of an OCEAN BASIN through the process of SEAFLOOR SPREADING and SUBDUCTION of existing ocean basins.

Wind gap. Abandoned WATER GAP. The Appalachian Mountains contain both wind gaps and water gaps.

Windbreak. Barrier constructed at right angles to the prevailing WIND direction to prevent damage to crops or to shelter buildings. Generally, a row of trees or shrubs is planted to form a windbreak. The feature is also called a shelter belt.

Windchill. MEASUREMENT of apparent TEMPERATURE that quantifies the effects of ambient WIND and temperature on the rate of cooling of the human body.

World Aeronautical Chart. International project undertaken to map the entire world, begun during World War II.

World city. CITY in which an extremely large part of the world's economic, political, and cultural activity occurs. In the year 2018, the top ten world cities were London, New York City, Tokyo, Paris, Singapore, Amsterdam, Seoul, Berlin, Hong Kong, and Sydney.

Xenolith. Smaller piece of ROCK that has become embedded in an IGNEOUS ROCK during its formation. It is a piece of older rock that was incorporated into the fluid MAGMA.

Xeric. Description of SOILS in REGIONS with a MEDITERRANEAN CLIMATE, with moist cool winters and long, warm, dry summers. Since summer is the time when most plants grow, the lack of SOIL MOISTURE is a limiting factor on plant growth in a xeric ENVIRONMENT.

Xerophytic plants. Plants adapted to arid conditions with low PRECIPITATION. Adaptations include storage of moisture in tissue, as with cactus plants; long taproots reaching down to the WATER TABLE, as with DESERT shrubs; or tiny leaves that restrict TRANSPIRATION.

Yardangs. Small LANDFORMS produced by WIND EROSION. They are a series of sharp RIDGES, aligned in the direction of the wind.

Yazoo stream. TRIBUTARY that flows parallel to the main STREAM across the FLOODPLAIN for a considerable distance before joining that stream. This occurs because the main stream has built up NATURAL LEVEES through flooding, and because RELIEF is low on the floodplain. The yazoo stream flows in a low-lying wet area called backswamps. Named after the Yazoo River, a tributary of the Mississippi.

Zero population growth. Phenomenon that occurs when the number of deaths plus EMIGRATION is matched by the number of births plus IMMIGRATION. Some European countries have reached zero population growth.

BIBLIOGRAPHY

THE NATURE OF GEOGRAPHY

Adams, Simon, Anita Ganeri, and Ann Kay. *Geography of the World*. London: DK, 2010. Print.

Harley, J. B., and David Woodward, eds. *The History of Cartography: Cartography in the Traditional Islamic and South Asian Societies*. Vol. 2, book 1. Chicago: University of Chicago Press, 1992. Offers a critical look at maps, mapping, and mapmakers in the Islamic world and South Asia.

_____, eds. *The History of Cartography: Cartography in the Traditional East and Southeast Asian Societies*. Vol. 2, book 2. Chicago: University of Chicago Press, 1994. Similar in thrust and breadth to volume 2, book 1.

Marshall, Tim, and John Scarlett. *Prisoners of Geography: Ten Maps That Tell You Everything You Need to Know About Global Politics*. London : Elliott and Thompson Limited, 2016. Print

Nijman, Jan. *Geography: Realms, Regions, and Concepts*. Hoboken, NJ : Wiley, 2020. Print.

Snow, Peter, Simon Mumford, and Peter Frances. *History of the World Map by Map*. New York: DK Smithsonian, 2018.

Woodward, David, et al., eds. *The History of Cartography: Cartography in the Traditional African, American, Arctic, Australian, and Pacific Societies*. Vol. 2, book 3. Chicago: University of Chicago Press, 1998. Investigates the roles that maps have played in the wayfinding, politics, and religions of diverse societies such as those in the Andes, the Trobriand Islanders of Papua-New Guinea, the Luba of central Africa, and the Mixtecs of Central America.

PHYSICAL GEOGRAPHY

Christopherson, Robert W, and Ginger H. Birkeland. *Elemental Geosystems*. Hoboken, NJ : Wiley, 2016. Print.

Lutgens, Frederick K., and Edward J. Tarbuck. *Foundations of Earth Science*. Upper Saddle River, N.J.: Prentice-Hall, 2017. Undergraduate text for an introductory course in earth science, consisting of seven units covering basic principles in geology, oceanography, meteorology, and astronomy, for those with little background in science.

McKnight, Tom. *Physical Geography: A Landscape Appreciation*. 12th ed. New York: Prentice Hall, 2017. Now-classic college textbook that has become popular because of its illustrations, clarity, and wit. Comes with a CD-ROM that takes readers on virtual-reality field trips.

Robinson, Andrew. *Earth Shock: Climate Complexity and the Force of Nature*. New York: W. W. Norton, 1993. Describes, illustrates, and analyzes the forces of nature responsible for earthquakes, volcanoes, hurricanes, floods, glaciers, deserts, and drought. Also recounts how humans have perceived their relationship with these phenomena throughout history.

Weigel, Marlene. *UxL Encyclopedia of Biomes*. Farmington Hills, Mich.: Gale Group, 1999. This three-volume set should meet the needs of seventh grade classes for research. Covers all biomes such as the forest, grasslands, and desert. Each biome includes sections on development of that particular biome, type, and climate, geography, and plant and animal life.

Woodward, Susan L. *Biomes of Earth*. Westport, CT: Greenwood Press, 2003. Print.

HUMAN GEOGRAPHY

Blum, Richard C, and Thomas C. Hayes. *An Accident of Geography: Compassion, Innovation, and the Fight against Poverty*. Austin, TX: Greenleaf Book Group, 2016. Print.

Dartnell, Lewis. *Origins: How the Earth Made Us*. New York: Hachette Book Group, 2019. Print.

Glantz, Michael H. *Currents of Change: El Niño's Impact on Climate and Society*. New York: Cambridge University Press, 1996. Aids readers in understanding the complexities of the earth's weather pattern, how it relates to El Niño, and the impact upon people around the globe.

Morland, Paul. *The Human Tide: How Population Shaped the Modern World*. New York: PublicAffairs, 2019. Print.

Novaresio, Paolo. *The Explorers: From the Ancient World to the Present*. New York: Stewart, Tabori and Chang, 1996. Describes amazing journeys and exhilarating discoveries from the earliest days of seafaring to the first landing on the moon and beyond.

Rosin, Christopher J, Paul Stock, and Hugh Campbell. *Food Systems Failure: The Global Food Crisis and the Future of Agriculture*. New York: Routledge, 2014. Print.

ECONOMIC GEOGRAPHY

Diamond, Jared M. *Guns, Germs, and Steel: The Fates of Human Societies*. New York: Norton, 2011. Print.

Esping-Andersen, Gosta. *Social Foundations of Postindustrial Economies*. New York: Cambridge University Press, 1999. Examines such topics as social risks and welfare states, the structural bases of postindustrial employment, and recasting welfare regimes for a postindustrial era.

Michaelides, Efstathios E. S. *Alternative Energy Sources*. Berlin: Springer Berlin, 2014. Print. This book offers a clear view of the role each form of alternative energy may play in supplying energy needs in the near future. It details the most common renewable energy sources as well as examines nuclear energy by fission and fusion energy.

Robertson, Noel, and Kenneth Blaxter. *From Dearth to Plenty: The Modern Revolution in Food Production*. New York: Cambridge University Press, 1995. Tells a story of scientific discovery and its exploitation for technological advance in agriculture. It encapsulates the history of an important period, 1936-86, when government policy sought to aid the competitiveness of the agricultural industry through fiscal measures and by encouraging scientific and technical innovation.

REGIONAL GEOGRAPHY

Biger, Gideon, ed. *The Encyclopedia of International Boundaries*. New York: Facts on File, 1995. Entries for approximately 200 countries are arranged alphabetically, each beginning with introductory information describing demographics, political structure, and political and cultural history. The boundaries of each state are then described with details of the geographical setting, historical background, and present political situation, including unresolved claims and disputes.

Leinen, Jo, Andreas Bummel, and Ray Cunningham. *A World Parliament: Governance and Democracy in the 21st Century*. Berlin Democracy Without Borders, 2018. Print.

Pitts, Jennifer. *Boundaries of the International: Law and Empire*. Cambridge, Mass: Harvard University Press, 2018. Print.

Index